Proceedings in Life Sciences

Plant Growth Substances 1979

Proceedings of the 10th International Conference on Plant Growth Substances, Madison, Wisconsin, July 22-26, 1979

Edited by F. Skoog

With 209 Figures

Springer-Verlag
Berlin Heidelberg New York 1980

Professor Dr. FOLKE SKOOG
Department of Botany
University of Wisconsin
Madison, Wisconsin 53706/USA

Cover motive: A vine tendril coiled round a branch to support the budding flower shoot. (After C. Darwin)

ISBN-13: 978-3-642-67722-9 e-ISBN-13: 978-3-642-67720-5
DOI: 10.1007/978-3-642-67720-5

Library of Congress Cataloging in Publication Data. International Conference on Plant Growth Substances, 10th, Madison, Wis., 1979. Plant growth substances, 1979. (Proceedings in life sciences) Includes bibliographies and index. 1. Plant regulators. I. Skoog, Folke Karl, 1908–. II. Title. QK745.I55 1979 581.3'1 80-24252.

Softcover reprint of the hardcover 1st edition 1980

2131/3130-543210

Preface

The Tenth International Conference on Plant Growth Substances was held July 22–26, 1979 at the Wisconsin Center of the University of Wisconsin-Madison under the joint sponsorship of The International Plant Growth Substances Association (IPGSA) and the Graduate School of the University. More than 500 persons, including 423 registered participants, attended the Conference. Financial support was generously provided by the organizations listed under Acknowledgments.

The Conference was planned and hosted by a Local Committee in response to a request from Professor Dennis Carr, Secretary of IPGSA, in 1976, that the Tenth Conference be held on this campus in 1979. To achieve comprehensive, systematic coverage of the subject and yet provide maximum opportunity for individual contributions by participants, reports were presented under ten topics, each with sessions of oral reports and poster demonstrations. Chairmen appointed by the Local Committee organized the material to be presented and arranged for a series of integrated, invited reports on each topic. They presided and led discussions at the sessions, and they also greatly assisted in the editing of the invited reports which are presented in full in these Proceedings. Unfortunately it was economically impractical to publish all reports, but the 244 submitted abstracts have been printed and distributed to participants.

Novel features of the Tenth IPGS Conference, recognizing rapidly expanding practical utilizaton of plant growth substances research, were sessions on hormonal regulation presented under the three topics Morphogenesis, Reproductive Development and Applications in Agriculture. A special closing event was a symposium on plant movements suggested, organized, and chaired by Professor Arthur W. Galston, commemorating the publication in 1880 of *The Power of Movement in Plants* by Charles Darwin, assisted by Francis Darwin. Also in recognition of this historic event as a beginning of plant hormone research, in the Plenary Session opening the Conference, Professor J. Heslop-Harrison analyzed Darwin's contributions to research on plant growth, and Professor K.V. Thimann described subsequent development of plant hormone research. President Masuda in his opening remarks briefly reviewed the 30-year history of IPGSA itself.

The local committee is most grateful to all contributors to the Conference and wishes to express special thanks to the following persons for valuable help and services: Professor Paul E. Pilet, organizer of the Ninth IPGS Conference for useful advice; Dr. Ruth Y. Schmitz for collaboration on planning and preparation of the program and compilation of abstracts; Dr. Barbara J. Taller, for supervising the poster demonstrations and for the index and other editorial work on the Proceedings; Ms. Lucy Taylor, for illustration and other art work; Professor A.C. Leopold, Secretary of IPGSA, for the printing and mailing of circulars; Mr. Robert Lee, Director, Ms. Pat Gaitan, Program Director, and the staff of the Wisconsin Center and Mr. George Gurda and staff of the Resident Halls, for the use of facilities and effective services; Mrs. Mary Evert, Mary Ellen Gerloff for guided tours and services for associate members, and all others who contributed to the successful operation of the Conference.

Madison, September 1980

FOLKE SKOOG

Acknowledgements

The Organizing Committee gratefully acknowledges generous financial support of the Tenth IPGS Conference by the following organizations: Campbell Inst. for Agricultural Research, Camden, N.J., Ciba-Geigy Corporation, Greensboro, N.C., E.I. du Pont de Nemours, Wilmington, Delaware, Eli Lilly & Co., Indianapolis, In., FMC Corporation, Philadelphia, Pa., ICI Americas, Inc. Goldsboro, N.C., S.C. Johnson & Son, Inc., Racine, Wisconsin, Merck & Company, Inc., Rahway, New Jersey, Mitsubishi Corporation, Tokyo, Japan, Mobay Chemical Corporation, Kansas City, Mo., Monsanto Company, St. Louis, Mo., Rhodia, Inc., New York, New York, Union Carbide Agricultural Products Co., Jacksonville, Fl., Uniroyal Chemical, Bethany, Ct., USDA Division of Education Administration, Washington, D.C., Velsicol Chemical Corporation, Chicago, Ill.

IPGSA Council 1976–1979

President Y. Masuda (Japan); Vice President F. Skoog (USA); Secretary A.C. Leopold (USA); Members: D.J. Carr (Australia), G. Deleuze (Venezuela), M. Johri (India), V. Kefeli (USSR), S. Lavee (Israel), E. Libbert (E. Germany), J. MacMillan (United Kingdom), L.G. Paleg (Australia), B.O. Phinney (USA), P.E. Pilet (Switzerland), N. Takahashi (Japan), and F.W. Wightman (Canada).

Local Organizing Committee

Drs. W.M. Becker, R.H. Burris, G.C. Gerloff, J.P. Helgeson, K. Keegstra, E.H. Newcomb, R.Y. Schmitz, L. Sequeira, and F. Skoog.

Contents

Ethylene
Chairman: H. KENDE

Abscisic Acid
Chairman: F.T. ADDICOTT

New Growth Factors
Chairman: C.A. WEST

Hormonal Regulation in Plant Reproductive Development
Chairman: A. LANG

Hormonal Regulation of Morphogenesis
Chairman: D.E. FOSKET

Agricultural Uses of Plant Growth Regulators
Chairman: P.W. MORGAN

Contributors

You will find the addresses at the beginning of the respective contribution.

Origin and Development of Plant Growth Substance Research

Chairman: R. H. BURRIS

Darwin and the Movement of Plants: A Retrospect [1]

J. HESLOP-HARRISON [2]

One hundred years ago this week, Charles Darwin was approaching the completion of what was to become perhaps his most influential botanical work, *The Power of Movement in Plants* (1). On July 17th, 1879, he wrote to Professor Carus concerning a possible translation, but not without misgivings: "Together with my son Francis, I am preparing a rather large volume on the general movements of Plants, and I think that we have made out a good many new points and views.

"I fear that our views will meet with a good deal of opposition in Germany; but we have been working very hard for some years at the subject.

"I shall be much pleased if you think the book worth translating, and proof-sheets shall be sent you, whenever they are ready." (2)

In the following May he informed his good friend A. de Candolle that the book had gone to press. It was published on November 6th, 1880; the first printing was quickly sold, and the book went immediately to a second printing. The work provoked considerable public interest. Intriguingly, it was even discussed in a leader in *The Times* – a fate that would be impossible today, not only because at this point in history there is no *Times,* but also because it is scarcely conceivable that any topic in the plant sciences could attract sufficient press interest to stimulate a modern leader writer.

It is now widely accepted that Darwin's work marked an important milestone in the study of growth and movement in higher plants, a milestone on the path that was to lead in the due course of time to the discovery of auxins. If so, it is appropriate that we should recall it this week in Madison, on the occasion of an international meeting devoted to plant growth substances. In this talk I propose to look back over the hundred years that separates us from the period of *The Power of Movement* and try to gain some perspective on Darwin's work.

I shall deal with four aspects: his motivation; his experiments and their interpretation; the intellectual environment of the plant physiological world of the time, and the fate of his ideas in the closing years of last century and the opening ones of this.

As Darwin wrote in his letter to Carus, the publication of *The Power of Movement* followed a long period of experimental work. But his interest in the general matter of plant movement had extended over a still longer time. It had two rather different origins: his curiosity about the behaviour of climbing plants, and his fascination with insectivorous plants. Darwin's devotion to climbing plants extended back to the time

---

1 Text of a lecture given for the Centenary of the publication of Charles Darwin's *The Power of Movement in Plants,* 1880

2 Welsh Plant Breeding Station, University College of Wales, Aberystwyth, SY23 3EB, United Kingdom

when he was working on the *Origin of Species* (3), and was in fact stimulated by a paper by Asa Gray on the coiling of tendrils published in the *Proceedings of the American Academy of Arts and Sciences* in 1858. The great supporting work for the *Origin* was *The Variation of Animals and Plants under Domestication* (4). The compilation and writing of this he found a bore; and he had no hesitation in saying so to his closer confidants. Adjuring J.D. Hooker at Kew not to send him any more plant material, we find him saying, "... it is mere virtue that makes me not wish to examine any more ... for I like it far better than writing about varieties of cocks and hens and ducks." Certainly the experimental work with plants seems to have provided Darwin with just the kind of relaxation he needed when depressed by the tedium of the work on variation. But that with climbing plants, beginning as no more than a dilettante interest, burgeoned into a respectable bit of research, and was given to the Linnean Society as a paper in 1865 (5). Asa Gray received a copy in the same year and responded enthusiastically. Developed and extended, the work appeared in book form in 1875 under the title, *The Movements and Habits of Climbing Plants* (6).

The work on climbing plants greatly affected Darwin's thought. Particularly was he impressed by the effects of contact stimuli. He records that a piece of platinum wire weighing 1.2 mg caused a response in the tendril of *Passiflora gracilis,* a response seen within 25 s after contact. Darwin was delighted with this, for it confirmed what Asa Gray had recorded for a cucurbit of the genus *Sicyos,* a tendril of which produced a visible movement within 30 s of a touch. Darwin was able to locate sensitive areas on tendrils and stems, and to show that stimuli could be caused to interact, equal contact on two opposite surfaces leading to an abolition of the curvature response.

The observations on climbing plants led to an important event: the first disagreement between Darwin and the pre-eminent German plant physiologist, Julius von Sachs. Since Darwin's interactions with Sachs are vital in any attempt to understand the events in plant growth physiology between 1865 and Darwin's death, I must refer to some of the background. At the time when Darwin began his work on plants, botany in the United Kingdom was in a sadly unbalanced state. The early impetus given to physiological studies by Nehemiah Grew and Stephen Hales had been exhausted. The centre of gravity of the science had moved towards morphology and systematics, and the active men of the day were almost wholly non-experimental in outlook. The same was true in the Nordic countries, where the influence of Linnaeus was still felt; and also of the U.S. and France, although to a lesser extent for the latter. In remarkable contrast, physiological botany – in its widest sense, including developmental anatomy – was in a state of vigorous and healthy growth in Germany and Austria; and the greatest eminence of the time was Sachs. Sachs, Professor of Botany in the Würzburg Botanical Institute, was a professional scientist in essentially the modern style. He had his own institute and his own journal; he ran a well-equipped laboratory, and had his own court of students, co-workers, and visitors. He was also a dominant figure: intolerant of criticism and jealous of competition. He viewed his own work with great respect, and revealed a lively sense of priority in publication.

In contrast with this splendid figure, Darwin was the veriest amateur. He had no institute; his laboratory was his study at Down House, and his planting space, the small greenhouse and tiny garden adjoining. He had no workshop; no technicians, and his only experimental assistant was his son, Francis. The difference in the circumstances of the two was striking, and to Sachs, Darwin must have looked the merest upstart. Yet the situation was not at all a David and Goliath one. Darwin actually had great strengths transcending even those of Sachs. He had behind him the prestige and fame of the evolution theory. He had a vigorous correspondence with all of the great botanists of the time – and indeed with all those who mattered in his areas of the natural sciences. He had access to the libraries of London, and through his friendship with Hooker at Kew, to the largest existing collection of living plants.

But Darwin knew well enough he was an intruder in the field of botany, and admitted as much in his letters. Writing to the head of the propagating department at the Edinburgh Royal Botanic Garden in 1862, he said, "I know only odds and ends of botany . . . you know far more." To him, Sachs was "a distinguished authority", to be treated always with courtesy and deference. Yet he was prepared to differ.

The occasion first arose in connection with the climbing plants, and concerned the mechanism of tendril curvature. The idea that the curvature of plant parts was due to differential growth had been expressed earlier in the century, and was indeed contained in A.P. de Candolle's partial etiolation theory of phototropism, in his *Physiologie Végétale* of 1832 (7). Sachs had accepted the essence of this theory, but had given it a totally new precision in his research communications and in his *Lehrbuch der Botanik*, the first edition of which was published in 1868 (8). Tendrils, he said, curved and coiled because the cells on the convex side increased in length more than those on the concave. Daringly, Darwin contested this. He wrote in *Climbing Plants*, "Sachs attributes all the movements of tendrils to rapid growth on the side opposite to that which becomes concave. These movements consist of revolving nutation, the bending to and from the light, and in opposition to gravity, those caused by a touch, and spiral contraction. It is rash to differ from so great an authority, but I cannot believe that at least one of these movements – curvature from a touch – is thus caused . . . One of my chief reasons for doubting whether curvature from a touch is the result of growth, is the extraordinary rapidity of the movement . . . (It appears to me) that the curvature of the tendril from a touch depends on the contraction of the cells along the concave side."

Darwin's evidence for this view was undoubtedly inadequate, and his argument was feeble enough. Yet the idea of rapid movement of plant parts through the *contraction* of cells was not a novelty for him. In the course of his work on insectivorous plants he had examined the movement of the "tentacles" – stalked glands – of *Drosera*. He found that tiny stimuli applied to the gland head induced curvature at the base, and this he seems to have accepted was due to the contraction of cells on the inner side. It was not, then, unnatural to suppose that the same might be true for tendril curvature.

Although Sachs did indeed admit the possibility of curvature through cell contraction, his resentment against Darwin for doubting his interpretations was very great. In Lecture No. XXXVIII of his *Vorlesungen über Pflanzenphysiologie*, published some years later in 1882 (9), he treated Darwin's work with contempt, giving *The Movements and Habits of Climbing Plants* no more than a single-line reference in a footnote.

Darwin's work on climbing plants led on to the study of plant movements in general, but in pursuit of a peculiarly Darwinian hypothesis. The early observations on stem climbers brought him to a detailed study of nutation – the roving movement of the apex Sachs had called revolving nutation, and which Darwin, contributing another little pinprick, renamed circumnutation. This in turn caused him to offer a generalisation which illustrates his individualistic mode of thought. The generalisation was that *all* movements of higher plants are derivatives of the movement of circumnutation. According to this view, the climbing plants owe their special kinds of behaviour to an amplification of circumnutation, and the curvatures that occur in response to light, gravity and humidity are "polarised" forms of circumnutation. In some respects this interpretation offers no explanation at all; but it clearly pleased Darwin himself, since it arose so naturally from his ideas about species transformation and the concept of progressive adaptive change under the pressure of natural selection. It must have seemed strange – even absurd – to the German school, whose inclination it was to seek first for physical and chemical explanations of biological phenomen. Sachs dismissed the thesis, accusing Darwin of extending the idea of nutation to excessive importance – but not failing to note at the same time that it was he, Sachs, who had made the first communication on nutational phenomena, in 1865.

Darwin's serious work on curvatures began in the early 1870s, and in it he was assisted by his son, Francis, who later became a plant physiologist in his own right. His attention was quickly caught by the sensitivity of the stem and root apices to stimuli, and it was in consequence of this and the experiments he and Francis carried out on stem and root curvatures that he became convinced of the physical separateness of "perceptive" and "motor" tissues and came to accept the idea of a linking, mobile, influence. But I must first recall the historical antecedents.

Two of Darwin's early observations, not directly connected with tropisms, indicated to him that the parts of plants must be in communication with each other. The first was made in 1860, and was reported in letters to his friends, including J.D. Hooker and Asa Gray in the U.S. It concerned the movements of *Drosera* tentacles I have already mentioned: stimulated at the *head,* they curve at the *base.* How else could this be explained except by the passage of a message between the two parts? The response, Darwin suggested, was analogous to that of the animal nervous system; but he made clear that he did not ascribe such to the tentacle. The second observation dates from 1861, and was mentioned in his book, *On the Various Contrivances by which Orchids are Fertilised by Insects* (10). The flowers of the orchid genus *Catasetum* have "sensory" horns, and when these are stimulated, the pollen masses are discharged. Such a response, Darwin realised, must indicate the passage of some influence through the intervening tissues.

Darwin was certainly not alone in making observations like these, and one can interpret passages in various earlier writings as suggesting that the growth responses of plants depend on the transmission of mobile stimuli. But the first unequivocal statement in connection with tropisms came from Theophil Ciesielski. Ciesielski was a Pole who worked in Berlin and Wroclaw. He gained his Ph.D. for a thesis entitled *Studies on the downward curvature of roots* in 1871. The results were published in 1872 in the new botanical journal, *Beiträge zur Biologie der Pflanzen* (11). Ciesielski's work is of substantial importance, and it is unfortunate that only rarely does he nowadays receive credit for his discoveries. His essential findings were (a) that the curvature of roots under the influence of gravity took place in the region of extension growth (b) that it resulted from a greater extension of the cells on the upper side than on the lower (c) that decapitated roots of peas, beans and lentils lose their sensitivity to gravity (d) that the sensitivity is regained when a new tip is regenerated and (e) that if roots are exposed unilaterally to gravity and then decapitated before any curvature is developed, the stumps still curve, whatever posture they may be in. These results scarcely allow any other conclusion than that the control of growth curvature in the root depends on a stimulus transmitted backwards from the tip.

Sachs must have known of Ciesielski's work from the date of its publication, and it seems that he must immediately have set about repeating the experiments. He met with failure, and reported the fact in the journal of the Würzburg Botanical Institute in 1873. In consequence, Sachs gave the interpretation no credence, and Ciesielski's work receives scant treatment in his *Textbook* of 1874, while his name gains no mention at all in Lecture No. XXXIX, dealing with tropisms, in the 1882 German edition of the *Vorlesungen über Pflanzenphysiologie.*

Darwin probably did not become aware of Ciesielski's work until some time after its publication, but when he did he found the conclusions very acceptable. It is now easy to see why. The experiment on *Drosera* and tendrils convinced him that plants were sensitive to touch, and at some point, probably before 1870, he seems to have concluded that roots were similarly sensitive. He was well enough aware of the nature of growth movements, and fully accepted Sach's interpretation that the curvature of roots occurred only in the regions of growth. What he did not see was how curvatures in this region could be so controlled in the soil as to permit the root to find its way among the soil particles without the additional ability to sense contact. Some of the experiments of the 1870's were designed to test this. He and Francis proceeded by attaching pieces of card to the root tips of beans, maize and other species, or by wounding the root tips chemically or mechanically. By these means they induced remarkable curvatures, executed in the zone of greatest extension growth behind the tip, and this left Darwin assured that the apex did have special sensitivity. The curvatures were such as to take the root away from the stimulus, indicating a more rapid growth rate on the same side. The sensitivity in the broad bean extended back for up to 1.5 mm. Wounding behind this zone, in the growth region itself, led to a curvature *towards* the source of irritation. In the light of these experiments, Darwin had no hesitation in proposing in *The Power of Movement* that some "influence" must pass from the tip to the growth zone. The sensitivity of the tip to contact, he believed, could provide the guidance system during growth through the soil.

Darwin was thus very receptive of Ciesielski's findings, and he and Francis set about repeating and extending his experiments. They worked with species of Leguminosae, Malvaceae, Cucurbitaceae and Gramineae, and obtained results that fully bore out those of Ciesielski. When the tip was removed from the root, the sensitivity to gravity was lost; similarly, when the tip was damaged by chemical treatment, no geotropic response could be elicited. Again the conclusion seemed inescapable: in the response to gravity, "it is the tip alone that is acted upon, and . . . this part transmits some influence to the adjoining parts, causing them to curve downwards."

But Sachs had failed in attempting to repeat Ciesielski's experiments: how could this be? Darwin did not hesitate to suggest why: ". . . it seems probable that Sachs unintentionally amputated the radicles on which he experimented not in a strictly transverse direction." It is interesting to speculate upon how Sachs – proud, we may suppose, of his reputation as a meticulous experimenter – reacted to this.

At some period during the 1870's, the Darwins had turned their attention to phototropism, or heliotropism as the response was then named. With the first recognition that light was responsible for certain plant movements, various theories were offered to explain the effect. De Candolle proposed that curvatures of stems towards the light took place because the shaded side became partly etiolated (7). Etiolated stems elongate more than those in the light; therefore the shaded side would elongate more than the illuminated, and curvature towards the light would then follow. This idea had at first been accepted by Sachs, but later he began to develop another view, and I shall shortly return to this.

Darwin seems simply to have transferred to phototropism his ideas about the sensitivity of stem and root tips. There is no indication in *The Power of Movement* or in the correspondence of the period as to why he selected grass coleoptiles for the main part of the experimental work, but it is evident that he and Francis found in the coleoptile a simple structure that could be used to follow light effects just as effectively as the bean root had been used by Ciesielski to examine the influence of gravity. The first experiments reported were on the coleoptiles of *Phalaris canariensis*; later those of *Avena sativa* were used, and in this way the organ that has played so important a part in the study of growth was introduced into plant physiology. The experiments are perhaps too well known to require detailed review, but I must recapitulate the critical findings. Firstly, a considerable series of experiments was run in which the tips – upper 0.15–0.2 inch – were protected from light by foil caps. This prevented curvature, even though the bases were unilaterally illuminated. Then the corresponding effect was demonstrated when the tips were removed, a demonstration paralleling that of Ciesielski with roots. In further experiments it was shown that shielding the base or tips did not prevent the development of curvature following unilateral illumination. Clearly the sensitivity lay in the coleptile tip, where "some matter . . . is acted upon by light, and this transmits its effect to the lower part." Darwin was himself astonished by the extent of this sensitivity. *Phalaris* coleptiles were exposed in a darkened room to the light from a very small lamp at a distance of 12 feet, where the "light was so obscure that we could not see the seedlings themselves, nor read the large Roman figures on the face of a watch." But within eight hours, curvature was evident. In such experiments Darwin came close to discovering the principle that the curvature elicited is related in general to light intensity × time; but this seems nowhere to have been stated, and it was left to Blaauw to make the connection some thirty years later.

I must now return to Sachs. He had moved away from the partial etiolation theory of phototropism, mostly because of the emerging idea of a unity of mechanisms among the growth movements of plants, expressed especially by A.B. Frank in his *Beiträge zur Pflanzenphysiologie* of 1868 (13). Instead Sachs had been gripped by the fact that curvature responses were apparently quantitatively related to the *angles* of exposure to light and gravity, and this led him to a general theory, in the statement of which he found himself "in the agreeable position of being able to depend step by step on my own detailed observations." The heart of the idea was that for light, as for gravity, it was the directionality of the stimulus, not the difference of intensity on the two sides of a unilaterally illuminated organ, that determined the response. So, it was the cells in the responding region that directly perceived the stimulus. Sachs hesitated to propose a mechanism for the effects, but noted that the relative turgescence of the cells on the two sides of an organ must be altered to bring about the observed growth; this he regarded as "obvious." This had been his conviction for some time, and the idea had been supported by the work of his student, Hugo de Vries (14), later himself to gain great fame in other connections.

Sachs was not likely, then, to be receptive to the idea that the apex controlled the tropic responses of root and coleoptile, and that the growing region merely responded to transmitted stimuli. In itself, the conception that mobile factors within the plant could control growth and morphogenesis was not at all foreign to him: after all, it was he who had introduced the idea of organ-forming substances, moving under the influence of gravity and light and responsible for the induction of roots and shoots. It was the suggestion that the apex could sense stimuli and control growth at which he baulked. Ciesielski's experiments had failed with him; and he found it intolerable that the Darwins, amateurs in his eyes, had seemingly got them to work. "In such experiments with roots" he wrote in Lecture No. XXXIX, "not only is great precaution necessary, but also the experience of years and extensive knowledge of vegetable physiology, to avoid falling into errors, as did Charles Darwin and his son Francis, who, on the basis of experiments which were unskilfully made and improperly explained, came to the conclusion, as wonderful as it was sensational, that the growing point of the root, like the brain of an animal, dominates the various movements in the root." So Darwin's prediction in his letter to Carus that his views would meet with a good deal of opposition in Germany was proved correct.

Other reactions in Europe were less hostile, and the response of Julius Wiesner, Professor of Botany in Vienna at the time, was positive and helpful. Unlike Sachs, Wiesner took the work seriously, and paid Darwin the considerable compliment of publishing a book of comment and criticism, *Das Bewegungsvermögen der Pflanzen* (15). Although much of what Wiesner had to say was in the way of refutation, his attitude charmed Darwin, who wrote an appreciative letter to him, in which he said, ". . . let me thank you cordially for the manner in which you have everywhere treated me. You have shown how a man may differ from another in the most decided manner, and yet express his difference with the most perfect courtesy. Not a few English and German naturalists might learn a useful lesson from your example; for the coarse language often used by scientific men towards each other does no good, and only degrades science."

Wiesner did not accept the evidence for the transmission of an "influence" in tropic responses, carefully although he must have studied it; evidently, like Sachs, he supposed that the tissues that perceived the stimulus and those that responded were one and the same. In retrospect it seems curious that it took so long to establish the general principle suggested by the work of Ciesielski and the Darwins; yet it is true that a satisfactory resolution was not reached until more than thirty years after the publication of Ciesielski's Ph.D. dissertation. The German plant physiologist who most fully appreciated the implications of a separation of regions of perception and response in the plant was Wilhelm Pfeffer. Himself a student of Sachs, he nevertheless readily accepted the findings reported in *The Power of Movement* for incorporation in his own *Pflanzenphysiologie* (16). Nevertheless, it was some years before he – or for that matter anyone else – set out seriously to repeat the work under critical conditions. Darwin's experiments on decapitated roots met the obvious objection that so severe a treatment might well sufficiently injure the root to suspend its growth, and were this true, then necessarily any capacity for curvature would be lost. In 1895, F. Czapek, working in Pfeffer's laboratory in Leipzig, carried out the experiment in a new manner, using the flexibile roots of lupin (17). The tip was inserted in a close-fitting glass "boot", and then either turned into the horizontal position with the upper part of the root remaining vertical, or held vertical with the remainder horizontal. The predictable results were obtained; curvatures developed, or failed to develop, according to the posture of the tip and not that of the growing zone. In 1889, Francis Darwin did the corresponding experiment with the coleoptile of the grass, *Setaria viridis,* and showed that if the tip were constrained in the horizontal position, the remainder of the coleoptile curled into a spiral due to continued differential growth in the sub-apical zone (18). Later, Piccard was to carry out a still more convincing experiment using Knight's Wheel, a device allowing growing plants to be exposed to centrifugal force (19). Piccard arranged growing bean roots so that they crossed the axis of rotation of the wheel with the tip subject to centrifugal force in one sense and the remainder of the root in the opposite. The results proved that the main sensitivity did indeed reside in the tip, but indicated that the growing zone also had some capacity for perception.

Another student of Pfeffer's, W. Rothert, working about the same time as Czapek, repeated the Darwins' work on phototropism under strict laboratory conditions (20). He confirmed all of the major results, but added new facts. Ciesielski had found that geotropic sensitivity was ultimately recovered in decapitated bean roots, and Rothert showed that the same was true for phototropic sensitivity in decapitated coleoptiles. Perhaps more importantly, he was able to demonstrate that, although the greatest sensitivity lay in the tip, the lower parts of the coleoptile also possess some capacity for bending in response to unilateral light. So, as with the geotropic response in the root, it had been proved by the end of the 19th Century that light sensitivity in the coleoptile is not confined to the tip, although greatest there.

The conception of Sachs and Wiesner that cells might respond directly to light was by no means killed by these experiments. In the early years of the new century, the direct effect of light upon growth was studied in detail in F.A.F.C. Went's laboratory in Utrecht, mainly by A.H. Blaauw (21). This work was characterised by unprecedented precision in measurement, both of the light and of the plant's reaction to it. Blaauw showed that the linear growth rate of stems and stem-like organs was strongly influenced by light, the overall effect of illumination usually being a retardation of growth. This "light-growth reaction" clearly provided a basis for explaining phototropism; curvature towards the light would take place if the side of the stem towards the source of illumination were to be retarded more than the shaded one. So, strangely, through a lineage of thought leading back to Sachs, a version of de Candolle's original theory of heliotropism was reinstated – a theory Sachs himself had categorically rejected a quarter of a century earlier.

The hypothesis of apical sensitivity in the phototropic reaction was unnecessary for Blaauw's theory; yet the experiments of Pfeffer and Rothert could not be set aside. The reconciliation was not to come until the 1920's, with the work of C. van Dillewijn (22). It was van Dillewijn who showed that *both* the tip and the sub-apical parts of the coleoptile are involved in the light-growth reaction. The first recordable response, he found, was attributable to light reaching the growth zone directly; this was followed by a slower, but much stronger, reaction, due to the illumination of the tip. Because the tip response was the dominant one, it was that likely to be concerned in the ultimate response to unilateral light, namely the positive phototropic curvature.

Yet another important trail led from Rothert's work on the transmission of the stimulus from the tip of the coleoptile. Two familiar names enter the scene at this time, that of H. Fitting, formerly a student at Leipzig, and that of Fitting's student P. Boysen-Jensen, who worked in Copenhagen. Both experimented on the transmission of the stimulus. Fitting found that foil strips inserted in slits cut below the apex of the coleoptile interfered with the phototropic response, but he did not at first draw the appropriate conclusion (23). This was left to Boysen-Jensen, who showed that the stimulus from the tip, whatever form it might take, could pass across a water gap, but not through an impermeable septum (24). He interpreted his results as indicating that the stimulus was one promoting growth, and that it passed down the shaded side.

The field of plant growth physiology was now beginning to fill up with workers, almost all Continental. With so many thinking more or less along the same lines, the question of who first specified that the "influence" postulated by Darwin and acknowledged by Rothert and others was in fact a chemical stimulant controlling cell extension cannot, I think, be answered simply. Certainly the idea was current before 1911, and Boysen-Jensen's claim is very good (25).

But it is an intriguing fact that the episode of a clinching nature arose not in the work on coleoptiles and tropisms at all, but from Fitting's studies on orchids, published in 1909 and 1910 (26). Darwin, half a century earlier, had been intrigued by the trigger mechanism of the flowers of the orchid *Catasetum*, which showed so dramatically that plant tissues were capable of rapidly transmitting stimuli. Fitting's attention was attracted by another characteristic of orchid flowers: the way that the development of the central column can be initiated by pollination, a considerably slower response than that which Darwin had studied, but one which involved growth over a period of time. Fitting showed that the response could be induced, albeit in limited degree, by the application of dead pollen to the stigma, and even by water extracts of pollen. And so, for the first time, a plant growth substance was collected outside of living tissue. Fitting made comparisons with the chemical messengers of the animal body, to which E.H. Starling had applied the term "hormone" in 1905 (27). He likened his chemical messenger to an animal hormone, and this was the first use of the term in the plant sciences.

What, then, of tropic curvatures? The general idea that these might depend on the movement of growth factors had currency before the beginning of the 1914–1918 War, and during the war work continued at Leipzig, where A. Paál conducted experiments in the direct lineage of those of Rothert, Fitting, and Boysen-Jensen (28). Paál worked with the massive seedling of the grass, *Coix lachrymajobi*, rather more easily handled than that of *Avena* because of its size. He was responsible for connecting aspects of the earlier work on the control of straight growth with that on tropic curvatures. In *Coix*, the hypocotyl is more extended, and is involved in the growth movements. Paál showed that if the coleoptile is removed and placed eccentrically on the hypocotyl, curvatures occurred away from the side on which it was applied, both in the light and in darkness. Furthermore, he proved that decapitation itself retarded growth, but that normal extension growth was resumed when the detached apex was rested back in place.

These experiments led to the now familiar idea that growth is controlled by a continuous flow of some chemical growth factor from the apex. Paál himself was convinced that it had to be a growth promoter; but even at this point there were set-backs; Brauner, for example, produced a reasoned case for supposing that the substances leaving the apex were inhibitors (29). But the work of H. Söding, reported in 1925 (30), confirmed Paál's interpretation. With *Avena* as test material, Söding carefully measured the growth of intact coleoptiles, and compared this with the growth of decapitated coleoptiles, and those that had been decapitated but had received their tips back, either directly, or separated by a film of gelatine. The results left little doubt. Intact coleoptiles, and decapitated coleoptiles with their tips replaced directly or with intervening gelatine, grew more rapidly than those without their tips; therefore the factor diffusing from the tip must be a growth promoter.

The next definitive steps were taken at Utrecht, where F.A.F.C. Went had, in fact, shown himself to be less than sympathetic to the concept of plant hormones. However, it was to be his son, F.W. Went who, inspired by the experiments of Paál and Söding, carried out the classical work that led to the extraction of the coleoptile growth hormone and developed the methods for assaying it (31). But that trail is one for Professor Thimann to pursue in the following article.

So let us revert to Darwin. How important were his contributions for growth physiology in the last analysis? What Francis Darwin referred to as the "central thesis" of *The Power of Movement* was the linking together of tropic movements with circumnutation. Charles Darwin saw circumnutation as a universal property of growing stems and roots; of it he wrote, ". . . apparently every growing part of every plant is circumnutating, though often on a small scale . . . In this universally present movement we have the basis or groundwork for the acquirement, according to the requirements of the plant, of the most diversified movements." Circumnutation, Darwin knew, resulted from the movement of the region of maximum growth around the stem or other organ, and with the understanding of the factors concerned in cell enlargement given by the work of Wiesner and de Vries, he was able to write in strikingly modern terms, ". . . increased growth, first on one side and then on another, is a secondary effect . . . the increased turgescence of cells, together with the (increased) extensibility of their walls, is the primary cause of circumnutation." Then, ". . . we know that there is always movement in progress, and its amplitude, or direction, or both, have only to be modified for the good of the plant in relation to internal or external stimuli." But here lies the difficulty: for the last statement comes dangerously close to assuming as an axiom the very phenomenon an expression of which it purports to explain. For us, as for Sachs, it is not a *useful* hypothesis, and perhaps not therefore worthy, in itself, of a place in experimental science.

Yet Darwin's own experimental work on growth movements was not truly linked to the unifying hypothesis at all. The "hypothesis" is an assertion, and as stated is scarcely capable of test. Darwin did not seek a test in any real sense. What he and Francis did was to explore in greater detail than any predecessors the characteristic of tropic and other growth movements, and to prove the function of the root and coleoptile apices in the control of curvatures. In the work on geotropism, Darwin was preceded by Ciesielski, to whom he gives full credit for priority, The researches on phototropism were wholly original, and broke new ground both in conception and execution.

The work was truly seminal, in leading directly to the concept of separate perception and motor regions and the transmitted signal that must link the two. Such was acknowledged by Pfeffer, Jost (32), and other leading plant physiologists of the second half of the 19th Century. It is not at all difficult to follow the trail from the Darwins, through Pfeffer and his gifted students and the Danish and Dutch schools to the modern period of plant growth substance research.

Ciesielski did not die until 1916, and if he kept in contact with the field at all, could have had the satisfaction of seeing his work validated and accepted. Not so Darwin. He had little cause for gratification in the immediate reception of his work on the Continent, and particularly not in Germany. Sachs's contempt he must have found wounding; and his letter to Wiesner shows how grateful he was for a treatment that, if critical, was at least polite. In the same letter he acknowledged certain errors, and indicated a readiness to be converted on other points; but his closing words are peculiarly sad: "Finally, I wish that I had enough strength and spirit to commence a fresh set of experiments, and publish the results, with a full recantation of my errors when convinced of them; but I am too old for such an undertaking, nor do I suppose that I shall be able to do much, or any more, original work."

In less than six months after writing these words he died, on April 19th, 1882.

References

1. Darwin, C.: The Power of Movement in Plants. London, 1st Edn. 1880. German edition, transl. V. Carus, Bewegungsvermögen der Pflanzen. Stuttgart, 1881
2. The quotations from letters are taken from The Life and Letters of Charles Darwin. Francis Darwin (ed.). London, 3 Vols, 1887, and from letters of Charles Darwin held in the Archives of the Royal Botanic Gardens, Kew
3. Darwin, C.: On the Origin of Species. London, 6th Edn. 1872 (1st Edn. 1859)
4. Darwin, C.: The Variation of Animals and Plants Under Domestication. London, 2 Vols. 1868
5. Darwin, C.: J. Linn. Soc. *9*, 1 (1867)
6. Darwin, C.: The Movements and Habits of Climbing Plants. London, 1875
7. Candolle, A.P. de: Physiologie Végétale. Paris, 3 Vols. 1832
8. Sachs, J. von: Lehrbuch der Botanik. Leipzig, 4th Edn. 1874. English edition, transl. A.W. Bennett and W.T. Thiselton-Dyer, Textbook of Botany. Oxford 1875
9. Sachs, J. von: Vorlesungen über Pflanzenphysiologie. Leipzig, 1882. English edition, transl. H. Marshall Ward, Lectures on the Physiology of Plants. Oxford, 1862
10. Darwin, C.: On the various Contrivances by which Orchids are Fertilised by Insects. London, 1862
11. Ciesielski, T.: Beitr. Biol. Pflanz. *1*, 1 (1872)
12. Sachs, J. von: Arb. Bot. Inst. Würzburg *3*, 432 (1873)
13. Frank, A.B.: Beiträge zur Pflanzenphysiologie. Leipzig, 1868
14. Vries, H. de: Untersuchungen über die mechanischen Ursachen der Zellstreckung. Leipzig, 1877. Also Bot. Z. *37*, 830 (1879)
15. Wiesner, J.: Das Bewegungsvermögen der Pflanzen. Vienna, 1881
16. Pfeffer, W.: Pflanzenphysiologie. Leipzig, 2 Vols. 1897–1904
17. Czapek, F.: Pringsheims Jahrb. *28*, 243 (1895)
18. Darwin, F.: Ann. Bot. *13*, 567 (1899)
19. Piccard, A.: Pringsheims Jahrb. *40*, 94 (1904)
20. Rothert, W.: Beitr. Biol. Pflanz. *7*, 184 (1894)
21. Blaauw, A.H.: Rec. Trav. Bot. Neerl. *5*, 257 (1908)
22. Dillewijn, C. van: Rec. Trav. Bot. Neerl. *24*, 323 (1927)

23. Fitting, H.: Jahrb. Wiss. Bot. *44*, 177 (1907)
24. Boysen-Jensen, P.: K. Dan. Vidensk. Selsk. *3*, 1 (1911)
25. Boysen-Jensen, P.: Growth Hormones in Plants. Transl. G.S. Avery et al. New York, London, 1936
26. Fitting, H.: Z. Bot. *1*, 1 (1909); Z. Bot. *2*, 225 (1910)
27. Starling, E.H.: Lancet *1905 II*, 339
28. Paál, A.: Jahrb. Wiss. Bot. *58*, 406 (1917)
29. Brauner, L.: Z. Bot. *14*, 496 (1922)
30. Söding, H.: Jahrb. Wiss. Bot. *64*, 587 (1925)
31. Went, F.W.: Rec. Trav. Bot. Neerl. *25*, 1 (1928)
32. Jost, L.: Vorlesungen über Pflanzenphysiologie. Jena, 1913

The Development of Plant Hormone Research in the Last 60 Years

K. V. THIMANN [1]

The development of ideas traced by Dr. Heslop-Harrison turned out to be the forerunner of a real revolution in our concepts of how the higher plant is integrated and organized. It will not be possible to enter on such a detailed enquiry as the previous speaker, because the field has become so much broader. The concept of a growth hormone arose directly out of the work of Arpad Paál and, like the work of Darwin and Boysen-Jensen before him, was founded firmly on the nature of tropisms. (The 1909 work of Fitting on the post-floration phenomena in orchids, caused by a postulated "pollenhormon," was quite separate and is an exception to the clear line that can be traced through tropisms, especially phototropism.) Unlike the early developments in animal hormone studies, the plant hormone concept depended from the first on the idea of a quantitative response, i.e., some proportional relationship between the quantity of hormone and the intensity of the response. By contrast, gastric secretion, thyroid action, male and female sexual development (e.g., comb formation in cockerels) were observed rather as all-or-none phenomena, at least at first; later, of course, bioassays were developed whose quantitative nature allowed the beginnings of chemical isolation.

The quantitative relationship was always implicit in the explanation of tropistic curvature, and was crystallized in N.G. Cholodny's bold generalization of the nature of tropisms, namely that all such responses result from quantitative differences between the amounts of growth hormone, and therefore the rates of growth, on the two sides of a curving organ. Cholodny's 1927 paper, *Growth Hormones and Tropisms in Plants* (1) came to the following conclusions:

1. We can consider it certain that different tropisms in plants can be traced to growth regulation by so-called growth hormones.
2. It is most probable that unequal (asymmetric) growth, which we can observe in various plant organs under the influence of definite stimuli (photoinduction, geoinduction, etc.) and that results in tropistic curvatures, is caused by unequal distribution of growth hormones, which are produced by definite cells in the organ continuously, and independently of the above-mentioned stimuli. (Translation mine.)

The contents of this mainly theoretical paper rested on about ten years' study of tropisms both to light and to gravity, and with both roots and shoots. Among the matters with which Cholodny was much concerned was the idea, stemming from the concept of "stimulus" (espoused particularly by Blaauw, and widely current at that time), to the effect that phototropism resulted from a direct effect of light on growth.

1 Thimann Laboratories, University of California, Santa Cruz, California 95064, USA

He showed in elegant experiments that if seedlings were exposed to *gradual* increase of light intensity there was absolutely no change in growth rate although a nearly normal phototropic curvature developed. Thus as he stated (above) *production* of the hormone is held to be independent of the stimulus.

This evolving hormone concept did not develop in a vacuum, of course. In 1922 Peter Stark had tried to extract a growth hormone and to test for it by applying the extracts to one side of a decapitated oat coleoptile, but without success. No doubt the extracts were too dilute, and oxidative destruction (later studied by Bonner and Galston) may have contributed to the negative result. Miss Seubert (2) did succeed, however, in showing that there was a growth-promoting property in solutions of amylase and other enzymes. For this she used Stark's procedure, thus demonstrating that it was in fact capable of producing large curvatures.

It was then that F.W. Went (3) made his classical contribution. Looking back, we can see that the background was well prepared for it. In Went's father's laboratory at Utrecht, van Asperen de Boer had been studying the growth of *Phycomyces,* and van

Fig. 1. A view of the Botanisch Laboratorium at Utrecht. The auxin work was conducted mainly in the basement

Dillewijn had just completed a thesis on phototropism which included a measurement of the difference in light intensities on the two sides of a unilaterally illuminated *Avena* coleoptile. Although Prof. Went senior was basically a mycologist, his interests must have been extremely broad, embracing growth phenomena in both lower and higher plants. It was natural therefore that the Utrecht laboratory (Fig. 1) should have been the center of much of the basic work on growth substances.

Last October 13th (1978) the biological section of the Royal Dutch Academy of Sciences, together with the Royal Netherlands Botanical Society, held a special meeting to celebrate "Fifty years after Went's dissertation," surely a most unusual recognition (Fig. 2). But Went's discoveries were of exceptional importance. Basically he set out to bring the growth hormone down from a biological concept to an experimental reality. While extracts of ground-up tissue had not shown any growth-substance activity in Stark's hands, Went (3) succeeded in demonstrating the reality of the growth

Fig. 2. Frits Went at the special meeting of the Royal Dutch Academy and Botanical Society in 1978. The quotation at right is discussed in text. Photo courtesy of Bob Biersma

substance by showing that if intact tips of *Avena* coleoptiles were allowed to stand for an hour or two on small blocks of agar, then growth hormone activity could readily be detected afterwards in the agar. For this he adapted and improved Stark's procedure of applying the test material to one side of a decapitated coleoptile, in darkness. (There was actually a weak red light, which had been found not to be phototropically active.) The beauty of his method was the fact that the resulting curvature could be easily measured, and thus the quantitative concept was demonstrated experimentally. Went showed (Fig. 3) that the curvature was proportional to the time a given number of coleoptile tips had stood on the agar, and further that if the hormone content was diluted by allowing the agar block to equilibrate with a block of plain agar then the curvature was halved and indeed could be halved again. The success of these experiments had consequences of two fundamentally different types.

Anzahl Spitzen.	60 Min.	30 Min.	23 Min.	60 Min.1 Mal verd.	60 Min.2 Mal verd.
6 Spitzen	11.2 ± 0.5	6.1 ± 0.4	4.6 ± 0.4	5.5 ± 0.4	2.8 ± 0.8
12 Spitzen	17.1 ± 0.8			11.2 ± 0.4	5.8 ± 0.4

Fig. 3. The table, from Went's thesis, which laid the foundation for quantitative work with auxin. From (3). (N.B. 1 Mal verd. = once diluted)

In the first place, it enabled the Cholodny theory of tropisms to be directly verified. When coleoptile tips on agar were illuminated from one side, the agar in contact with the dark side could produce a larger curvature on test plants than that in contact with the lighted side. Herman Dolk (4), who was making a parallel study of the geotropic response, was very soon afterwards able to show the corresponding behavior in geotropism, namely that the agar in contact with the lower side of a horizontal coleoptile could subsequently cause a larger curvature on test plants than that in contact with the upper side. Dolk added two other critical points. In parallel with Cholodny's work on phototropism he showed that if the plants were carefully turned into the horizontal position there was no overall change in growth rate, i.e., there was no need for a "geo-growth reaction." Also, the asymmetrical distribution of growth substance could be brought about by gravity even in subapical segments when these were supplied with an external source of auxin. In this respect geotropism differed from phototropism. Dolk also clearly demonstrated the migration of geotropic curvature down the plant. Three years later van Overbeek (5) brought a dicotyledonous plant into the picture for the first time, showing that when a *Raphanus* seedling curves toward light the growth substance, tested on *Avena,* undergoes accumulation on the shaded side just as in *Avena.*

The second consequence of Went's successful procedure was that the curvatures produced by *extracts* were proportional to their growth substance content, and hence the test could be directly put to use in monitoring studies of the purification and isolation of the substance. Went's proportionality curve, an arithmetic relationship between concentration and curvature up to a "maximum angle" (Fig. 4), has been confirmed in numerous laboratories, the absolute values varying somewhat with the experimental conditions. Such studies were very soon put in hand both at the Utrecht lab and in

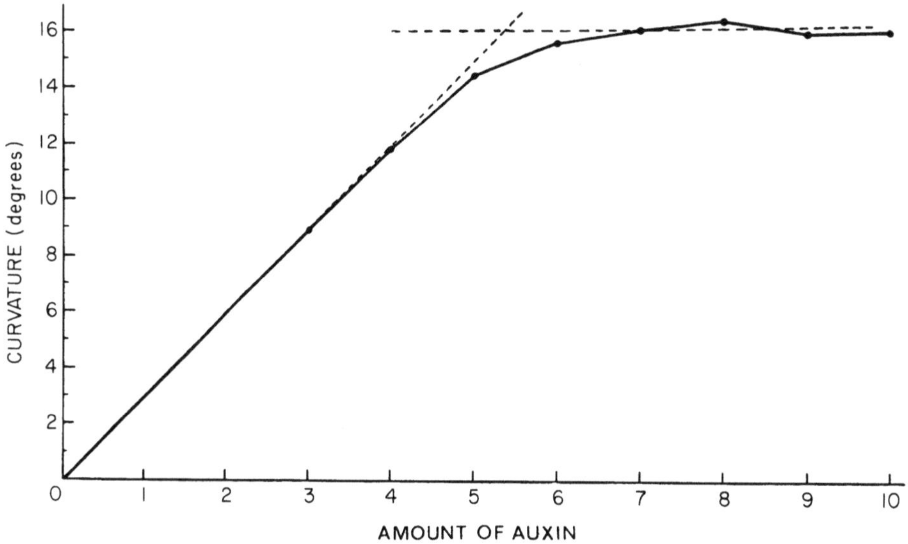

Fig. 4. The relationship between auxin concentration and the curvature of decapitated *Avena* coleoptiles under standard conditions. From (3)

Pasadena at the California Institute of Technology, where T.H. Morgan (the chairman of Biology) had offered Dolk (Fig. 5) an appointment. My own participation in this field dates from that time, since I joined Dolk as chemist in early 1931 to make a joint assault on the nature of the growth substance. The Institute built a special laboratory for Dolk, with the until-then unknown facility of humidity-controlled dark rooms in the basement. Because this building was on the side of the road reserved for residences, the laboratory was made to look like a small house (Figs. 6 and 7). Dolk

Fig. 5. Herman and Fransje Dolk in 1932. Photo courtesy J. Bonner

Fig. 6

Fig. 7

lost no time in establishing the bioassay, but the rapid progress made at first was later slowed down by his untimely death in an auto accident.

At Utrecht preliminary studies by Kögl and Haagen-Smit had shown human urine to be a rich source of the growth substance, while the observation that the fungus *Rhizopus suinus* yielded good amounts into its culture medium was utilized at Pasadena. At first there was a curious contretemps, which remains unexplained to this day, in that the growth activity was ascribed to two related compounds, believed to contain a cyclopentene ring with a five-carbon sidechain. This development at Utrecht could not be confirmed at Pasadena, and a little later both laboratories announced that the plant growth hormone was a compound already known to chemistry, indole-3-acetic acid (6, 7). At Utrecht, this was thought to be additional to the two cyclopentene derivatives, and hence the name "hetero-auxin" was applied to indole-acetic acid, but gradually, as it was identified with increasing frequency in extracts from different plant materials, it became clear that this was indeed the naturally occurring hormone. (The general name *auxin* was coined by Kögl as indicating *increase,* and has the same root as our word *auction.*)

The team at Pasadena was soon strengthened by the arrival of Went, who replaced Dolk, and by the graduate students James Bonner and Folke Skoog, and by Jan van Overbeek, who arrived from Holland following the completion of his work on *Raphanus* just mentioned (Figs. 8, 9, 10). The isolation and identification were completed in late 1934, and the physiological aspects of the field began to broaden rapidly. In Utrecht, van der Weij proved that the transport of the auxin was basipetally polar, and

Fig. 8. James F. Bonner in 1933. From California Tech. archives

Figs. 6 and 7. Exterior and interior of the small "growth substance" laboratory at California Institute of Technology in the 1930's. The figure at the bench is Dr. Charles Schneider, the first technician. Photo courtesy Jan van Overbeek

Fig. 9. Folke Skoog and Anton Lang. Recent photo

Fig. 10. Jan van Overbeek, a recent photo. From Annu. Rev. Plant Physiol *27*, 1 (1976)

worked out many details of the process, nearly all of which have stood the test of time (8). In Pasadena, we showed that auxin and the postulated root-forming hormone were identical (7). But more surprisingly, the Utrecht group reported that when the roots of seedlings dip into an auxin solution, their growth is inhibited (9). In the same year, Skoog and I showed that the inhibiting influence of the apex of a dicotyledonous plant on the development of lateral buds in the axils of its leaves could be completely replaced by auxin (10). These two findings established that auxin is not only a promoter of growth but also an inhibitor. The second discovery also provided a mechanism for the earlier observation of Dostal (11) in Bohemia, to the effect that leaves somewhat inhibit the development of buds in their axils. The finding that the secretion of auxin by leaves is at its maximum when they are very young and decreases steadily with age completed that aspect, as well as providing a basis for the much later work on the hormonal control of abscission.

But a further surprise was in store. At the 1935 Botanical Congress in Amsterdam, Robin Snow (12) from Oxford read a short paper which showed conclusively that the formation of cambium in dicot seedlings was promoted by auxin; thus auxin was a controlling factor not only in elongation but also in cell division (Fig. 11). [About 20 years later auxin was shown to induce xylem (13), thus also active in differentiation (Fig. 12).] Thus within a space of two years the role of auxin was expanded from that of a hormone specific for elongation to one involved in a varied group of apparently unrelated processes. As you know, still more processes, especially at the biochemical level, were added later.

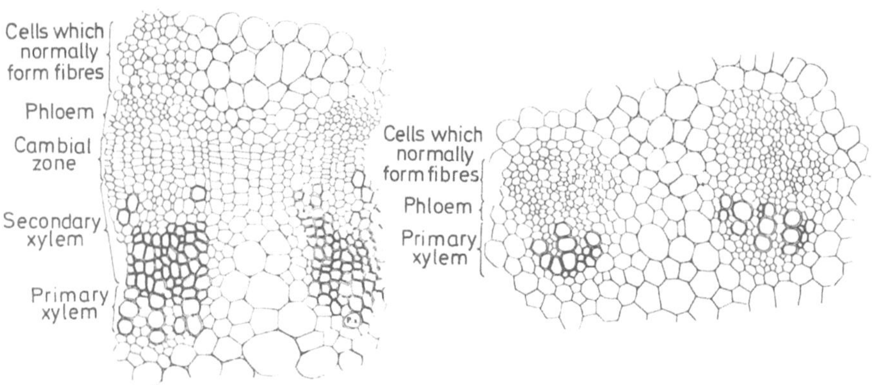

Fig. 11. Robin Snow's demonstration of the activation of cambium in *Helianthus* seedlings by auxin. *Right,* control; *left,* 1 p.p.m. IAA applied. From (12)

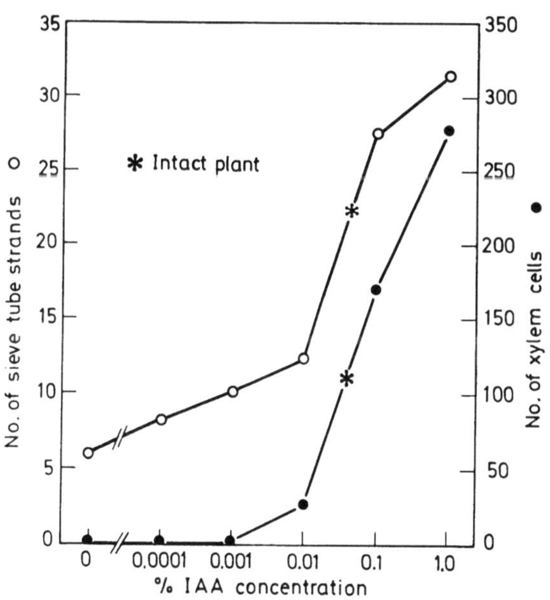

Fig. 12. Induction of the formation of xylem and phloem in decapitated *Coleus* plants by serial concentrations of IAA. The *asterisk* shows the amounts normally present in intact plants. From (13)

Simultaneously with the growth of knowledge about auxin in the western world, there was the beginnings of knowledge about an entirely different type of hormone in the Orient. For many years this was quite unknown to us in the West. Kurosawa's observations on the Bakanae disease of rice date back to 1926, when after many trials he was able to reproduce symptoms of the disease with the culture medium in which *Gibberella fujikuroi* (the perfect form of a variant of *Fusarium moniliforme*) had been grown. There was no fungal infection involved, yet the treated plants showed the typical excessive elongation, pale green color and inhibition of root growth characteristic of Bakanae. Work on the nature of the active material was taken up by a number of workers in Japan, especially Shimada at Hokkaido in 1932–34 and Takahashi at Mie from 1929 to 1936. Then, within a few years of the isolation of indole-acetic acid from yeast and *Rhizopus*, the first isolation of gibberellic acid from the *Gibberella* culture medium was announced by Yabuta and Sumiki and their colleagues at Tokyo University (1938) (Fig. 13). One difficulty was the presence in the culture medium of

Fig. 13. Teijiro Yabuta, left, and Yusuke Sumiki, the first to isolate a gibberellin. From (14)

a strong growth inhibitor, fusaric acid, which obscured the gibberellin action; another was the need for developing a medium which would give high yields, and finally, the genuine chemical difficulties engendered by the complex formula. [See (14) for a review of this early work.] Indeed, it was 20 years more before the chemistry of the main product was cleared up. In 1955 Stodola's group at Illinois, USA, Sumiki's group in Tokyo, and a group at Imperial Chemical Industries in Britain all agreed on a formula for gibberellic acid, soon thereafter called GA_3. Its structure was presented by Cross et al. (15) the following year, but even so the precise location of the lactone group had to be changed later.

The first that most workers in the western world knew of these fundamental discoveries was the publication in 1954 by the British group of a short paper in *Chem-*

istry and Industry (16) with this striking picture (Fig. 14). Thus was ushered in a period of physiological studies, some of which were based on the bioassay that had been used in the isolation work. However, the discovery that certain dwarf cultivars of corn responded more sensitively than rice seedlings accelerated this work. The parallel with auxin, in that GA was isolated from a fungal growth medium as was IAA from *Rhizopus* and from yeast, was strikingly continued in the study of the different responses.

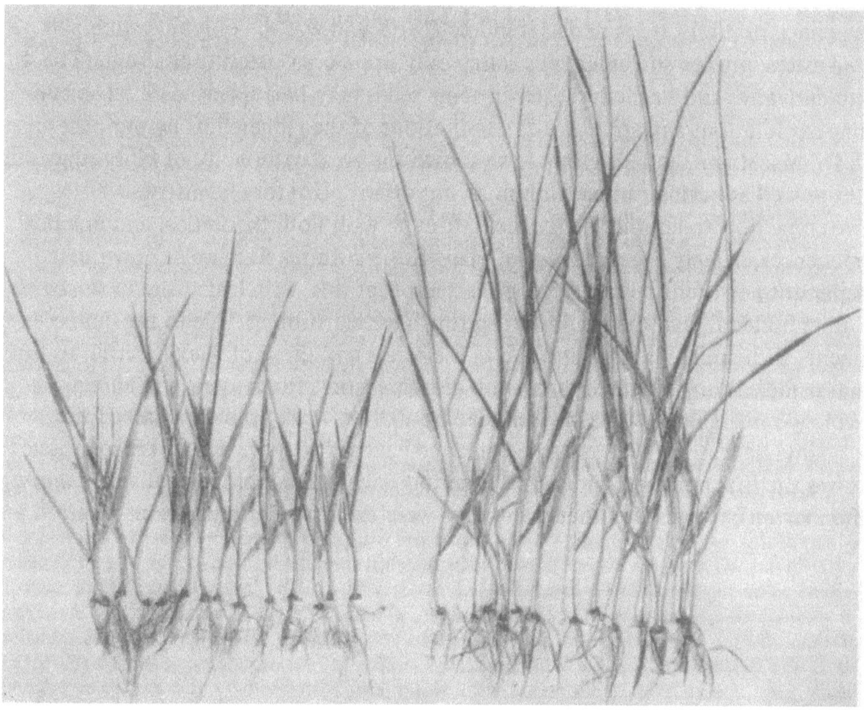

Fig. 14. The picture that brought the gibberellins to the attention of most western plant physiologists. Seedlings of Victor wheat. *Left* controls, *right* GA_3 added to the nutrient solution. From (16)

For as soon as appreciable quantities of GA became available, it developed that its activity was by no means limited to the elongation of graminaceous plants. First the elongation of dicots, as shown by Brian's group with dwarf peas and by Lona with rosette plants, reemphasized the quantitative nature of the response. Then in quick succession came the promotion of flowering of long-day plants in short day conditions, the identification of the amylase-hydrolyzing factor in barley seeds with gibberellin, and the curiously complex interactions with red light in which gibberellin on the one hand removed the growth inhibition caused by red light, yet acted like red light in allowing the germination of light-requiring seeds. The attractive generalization that dwarf plants are simply unable to synthesize gibberellin had to be dropped when some dwarfs were found not to respond, and the similar generalization that GA acts like light had to be dropped when it was found that it also promoted the germination of light-inhibited seeds. The effect of auxin on cell division, first seen in cambial activation, was paralleled by the finding of Lang (Fig. 9) that vigorous cell division in the apex precedes the initiation of flowering by gibberellins.

In one major respect the gibberellin story differs from that of auxin; large numbers, 50 or so, of naturally occurring gibberellins with varying biological potencies exist, but only a very few auxins have been found. Indeed, until Wightman's (17) convincing demonstration of the natural occurrence of phenylacetic acid in biologically active concentrations, it seemed as if indole-acetic acid was the only real auxin, all related esters, nitriles and aldehydes acting only upon conversion to the acid. (The 4-chloro derivative and its methyl ester in some seeds may be a special case.) The continuing explorations and structure determinations of the gibberellins by our past president, Dr. MacMillan, and his group have shown the great catholicity of biosynthesis in plants, as well as having earned for him an important prize for chemistry.

We come now to another major breakthrough with both theoretical and practical consequences, namely the discovery of plant tissue cultures. It is well known that Haberlandt's persistent but unsuccessful attempts at this, with leaf tissue of *Bocconia* and other material, had the result of inspiring Ross Harrison at Yale to try similar experiments with animal tissues, and led, in his hands and those of Alexis Carrel, to true animal tissue cultures. Indeed, those who dispense funds for medical and animal-physiology researches need to be reminded how often work on plants has led the way – amylase was the first extracted enzyme nearly 150 years ago, tobacco mosaic virus was the first extracted virus (and note too Beijerinck's bold idea of a *contagium fluidum vivum*), it and one other plant virus were the first viruses to be crystallized –

Fig. 15. Roger J. Gautheret at the time of his first continuous culture of plant tissues. Photo courtesy of Dr. Gautheret

and now tissue cultures. Although the continuous culture of roots had been achieved by Robbins and by White with the addition of yeast extract to the medium (later shown to owe its action mainly to thiamine), the continuous culture of plant *tissues*, properly speaking, was not possible until auxin was added to the otherwise well-formulated culture medium. This was done independently and almost simultaneously by Gautheret (Fig. 15) and Nobécourt (18, 19) both in France (1937), and since then the French have maintained great vigor and productivity in the field. The premature loss of both Morel and Nitsch has been a severe one to this Association and to plant science in general.

Success with tissue cultures has led to two major ramifications. First, the discovery by Blakeslee, Conklin and van Overbeek of the marked stimulation exerted by coconut milk led Steward (Fig. 16) and coworkers to produce vast numbers of plantlets from small clumps of "free cells" (20). Indeed this could be done with single cells [Fig. 17; (21)].

Fig. 16. Frederick C. Steward *(left)* and a colleague with an orchid grown from a culture of free cells. Photo courtesy of Dr. Steward

The second major ramification from the success in achieving continuous culture of plant tissues was the discovery of the cytokinins. Once again, as with auxin and the gibberellins, a preparation from fungi was crucial to the solution of a mystery specifically bearing on the green plants. The first indications of a new growth substance, following Skoog's work on growth promotion of tissue cultures by adenine, came from the isolation of a highly active preparation, with the properties of a purine, from yeast extract. DNA preparations similarly were remarkably active in stimulating growth and differentiation of tissue cultures, but only after they had been aged or had been autoclaved. Hence activity was not ascribed to DNA itself. Thus Skoog and Miller with the aid of Wisconsin biochemists Strong et al., working in buildings only a few hundred meters from this one, set about isolating the active material called kinetin from herring sperm DNA and identified it as 6-furfurylaminopurine (22). At this point, in marked contrast to auxin work, where many months intervened between the identification of a natural substance (IAA) and the proof of activity of a purely synthetic analogue (indene-3-acetic acid), and where it was several years before the large group of synthetic compounds culminating in 2,4-D appeared, in the case of cytokinins it was only *3 days* before F.M. Strong and his group synthesized 6-benzylaminopurine, the first and still widely used synthetic cytokinin.

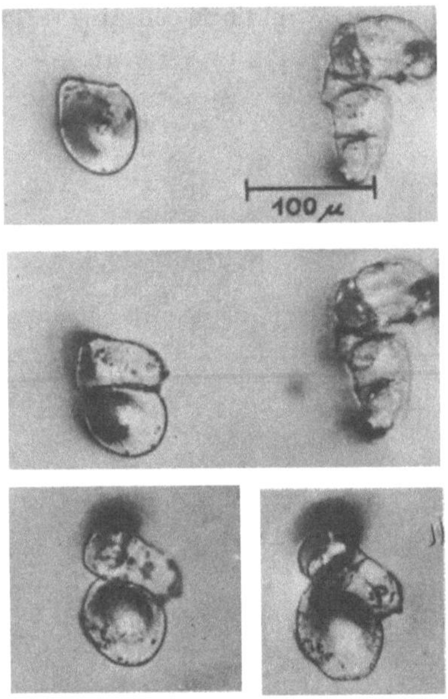

Fig. 17. Jakob Reinert's demonstration that a single cell of *Haplopappus* can develop into a group of cells. Read *down on left, up on right.* From (21)

It is of interest that the factor being searched for was one which would release lateral buds from apical dominance, so that it was not at first anticipated that this same substance would initiate cell division in old vacuolated cells of tobacco pith, as seen in Fig. 18, and would promote bud formation. However, these results led to the very practical finding that whether cultures would give rise to roots, shoots, or just callus tissue depended on the ratio between the concentrations of auxin and cytokinin (23) (Fig. 19). Although the name cytokinin indicated the compounds' ability to promote cytokinesis, the pattern of discoveries which subsequently followed was similar to that with auxin and gibberellin. In succession, cytokinins were found to promote DNA synthesis (the basis of its action on cell division), to relieve the apical inhibition of lateral buds (another case of balance with auxin), to permit or promote germination of many seeds, especially those whose germination is affected by light, to promote expansion of leaves, to inhibit the development of lateral roots, to cause certain outgrowths in intact plants, to suppress the formation or counteract effects of ethyl-

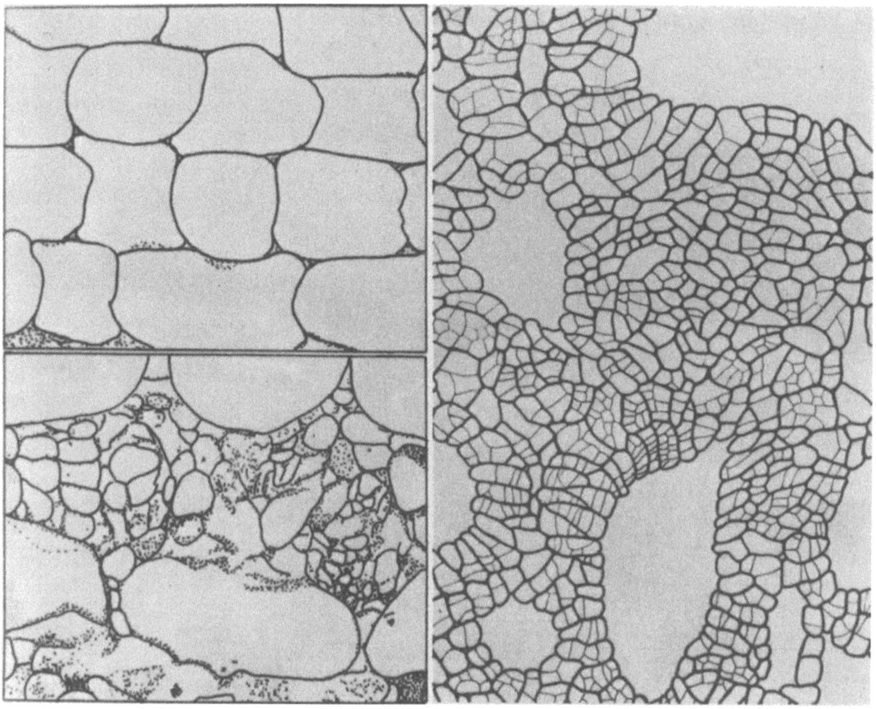

Fig. 18. Multiple subdivision of old cells. *Left* Skoog and Miller's tobacco pith culture; *above* on basal medium; *below* with auxin and kinetin added. *Right* Gautheret's *Amorphophallus* tuber, after treatment with auxin and coconut milk; (*thick lines* are the old cell walls)

ene, and to maintain chlorophyll synthesis, thus opposing the process of senescence in leaves. Extensive studies on the relation between structure and activity in a large group of synthetic purines have been somewhat parallel to those on synthetic auxins. Letham's identification of zeatin as a natural product of immature corn kernels (24) serves to explain the earlier observation of Mitchell and Skaggs on the biological activity found in extracts of unripe seeds; in retrospect this was doubtless due to the presence of cytokinins. More recently the potency of coconut milk (which can stimulate free cells to grow into whole plants) has been proven by Letham, 1967 and by Miller and others to be due to the cytokinin. Thus the cytokinins have been well established as natural hormones, and now their ubiquitous presence in many RNAs points to a fundamental mode of action, the exact nature of which, however, is still proving elusive.

The damage done to plants by coal-gas led to a discovery of quite a different sort. As first described by Fahnestock, the damage consisted mainly of the premature abscission of leaves in greenhouse plants, and this was later observed in shade trees. But it was observations of several types of effects on plants (which had been put in an enclosed space with apples) that led to the recognition that fruits evolve a gas having growth effects similar to those of coal gas. Neljubov identified the active constituent of coal-gas as ethylene, and Gane's later (1934) identification of ethylene in the volatile pro-

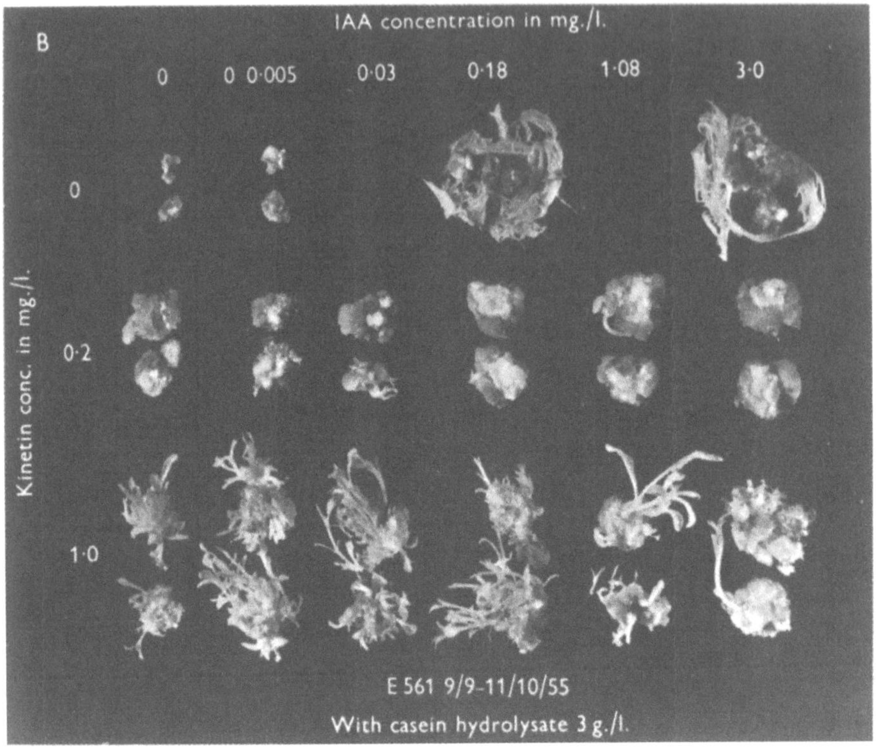

Fig. 19. Cultures of tobacco pith on agar, showing the development of buds when the concentration of kinetin is relatively high and that of auxin is low. *Abscissa* concentration of IAA (increasing to right); *ordinates* concentration of kinetin (increasing from top row downwards). From (23); also in (30)

ducts from apples was a turning point in this area (25). In many of these early experiments the downward curvature ("epinasty") of petioles was the significant response, and this is indeed a sensitive and characteristic reaction, as in Denny and Miller's experiments of 1935 (26), which demonstrated effects of this type from the emanations of numerous leaves, flowers and fruits (Fig. 20).

The dramatic case of the ripening of citrus fruits in railroad cars warmed with kerosene stoves, which did not occur when the stoves were replaced by steam heat, gave a quite independent approach, for the combustion of the kerosene was found to give rise, inter alia, to ethylene (see reviews by Burg, 1962 and Abeles, 1972).

For about twenty years, physiologists were reluctant to accept the concept of a *gaseous* hormone, and as yet there is no parallel among animal hormones, though the volatile attraction substances of insects come close. In any event, the pattern of gradual broadening of our conception of the functions of ethylene has followed curiously closely that established with other hormones. For ethylene not only causes the "epinasty" of petioles and ripening of fruits (though perhaps not the initial stages), but it is also responsible for the inhibition of growth of seedlings, the rotation of the direction

Fig. 20. The use of epinasty of petioles of potato to detect the evolution of ethylene by plant parts. *Left to right,* controls, and plants enclosed with increasing numbers of dandelion flowers. From (26)

of auxin transport through 90°, the senescence of many flowers and the post-floration changes in orchids, to name but a few. The latter recalls the very early work of Fitting on the "pollen-hormon" (29), which now seems to rest on the primary liberation of auxin and its secondary effect on the production of ethylene. Indeed, auxin acts to stimulate ethylene production by promoting the formation of aminocyclopropane-carboxylic acid in many tissues, and here we encounter one of the great complexities of hormone work, namely that the hormones act not only in the *presence* of one another, but may even regulate the *formation* of one another. Thus cytokinins regulate the synthesis of both auxin and thiamine in tissue cultures, and gibberellins increase the diffusible auxin content in several plants.

The discovery of abscisic acid through the analysis of abscission and of dormancy is too recent to figure in an historical sketch. Unlike the other hormones, abscisic acid was not discovered through any interaction with fungi; like ethylene, it was approached from more than one direction, and like the auxins, so far at least, there appear to be only a few compounds of its biological type. There are also growth inhibitors and growth modifiers of other types, all of them naturally occurring and therefore undoubtedly functioning in specific cases: the phenols and the flavonoids, colchicine, heliangine et al. (30). These may not be hormones properly speaking, but all will have to be taken into account in any overall understanding of the endogenous control of the growth of green plants. Frits Went recently restated his continuing interest in the "calines" postulated by Sachs in 1880, saying that he could not make up his mind about their existence but tended to believe in them in alternate years! The quotation in Fig. 2 reads "Only in even years I believe in the existence of the substance."

What can we say about the field of plant growth hormones as a whole? I specifically omit here any consideration of growth-modifying substances other than those that occur and function in the plant itself.

First, the history of the field shows how important it has been *to follow up leads.* The patient investigation of a significant discrepancy, the continued improvement of a bioassay, the repeated attempt to clarify an idea or a result, these have been the source of much of our knowledge.

Second, the main driving force has been curiosity. Many applications have come, primarily to further laboratory experimentation, but eventually utilized in large-scale plant industry – rooting of cuttings, weed control, regulation of vegetative growth, propagation of cells, tissues and organs, regulation of reproductive development, and fruit ripening and preservation. But these applications have not been the initiating reasons for the research. As was pointed out long ago, one would not start to develop new types of herbicides by studying the tropisms of seedlings in an air-conditioned dark room. (Imagine Senator Proxmire's reaction to such a proposal!) Yet the line of development has been direct. It is the fundamental approach which leads to broadest applications.

A more specific conclusion concerns the interactions between the fungi and the higher plants. The fungi evidently can in some cases produce, in excess, compounds which in higher plants are produced only under strict control, as befits hormones of high activity. I wonder, therefore, whether the reverse is also true; perhaps we should be examining higher plants for the growth-controlling hormones of the fungi. These latter are a group of compounds that almost certainly must exist, but of which we have extremely little knowledge at present. It may be that there is a fertile field for investigation here.

In regard to the mode of action of hormones, it is interesting and suggestive that auxin induces a movement of protons out of the cytoplasm and abscisic acid induces a movement of potassium ions into the cytoplasm – in one case in elongating cells and in the other case in guard cells. Ethylene modifies the movement of what are probably indole-acetate ions, too. Should we be looking for the movement of ions across membranes in other hormonal responses? At least it is an attractive lead, and one that could well be compatible with the characteristic multiple effects that we have just surveyed.

I will conclude by quoting from a Sigma Xi lecture by our colleague John Decker, given at Arizona some 20 years ago:

"Our usual method of reporting research leaves out all the fun and adventure. It leaves in only the bare bones of orderly fact, as we can arrange them best after prolonged and sober study. This process may engender an image of the scholarly Scientist who operates according to a coldly logical Scientific Method . . . But what keeps you and me plugging away at research is not some lofty sense of scholarly achievement, it is an intensely human reward of fun and adventure."

I hope that we may all continue to enjoy the fun and adventure of growth substance research for many years to come.

References

1. Cholodny, N.G.: Biol. Zentralbl. *47*, 604–626 (1927)
2. Seubert, E.: Z. Bot. *17*, 49–88 (1925)
3. Went, F.W.: Rec. Trav. Bot. Neerl. *25*, 1–116 (1928)
4. Dolk, H.E.: Diss. Utrecht (1930). Engl. transl. In: Rec. Trav. Bot. Neerl. *33*, 509–585 (1936)
5. Overbeek, J. van: Rec. Trav. Bot. Neerl. *30*, 537–626 (1933)
6. Kögl, F., Haagen-Smit, A.J., Erxleben, H.: Z. Physiol. Chem. *228*, 90–103 (1934)

7. Thimann, K.V.: J. Biol. Chem. *109*, 279–291 (1935); Thimann, K.V., Koepfli, J.B.: Nature (Lond.) *135*, 101 (1935)
8. Weij, H.G. van der: Rec. Trav. Bot. Neerl. *29*, 379–496 (1932); *31*, 810–857 (1934)
9. Kögl, F., Haagen-Smit, A.J., Erxleben, H.: Z. Physiol. Chem. *228*, 104–112 (1934)
10. Thimann, K.V., Skoog, F.: Proc. Roy. Soc. London Ser. B *114*, 317–339 (1934); Skoog, F., Thimann, K.V.: Proc. Natl. Acad. Sci. USA *20*, 480–485 (1934)
11. Dostál, R.: Acta Soc. Sci. Nat. Moravicae *3*, 83–209 (1926)
12. Snow, R.: New Phytol. *34*, 347–360 (1935)
13. Thompson, N.P., Jacobs, W.P.: Plant Physiol. *41*, 673–682 (1966)
14. Stowe, B.B., Yamaki, T.: Annu. Rev. Plant Physiol. *8*, 181–210 (1957); Science *129* (March 27), 807–816 (1957)
15. Cross, B.E., Grove, J.F., MacMillan J., Mulholland, T.F.C., Sheppard, N.: Proc. Chem. Soc. *1958*, 221 (1958)
16. Brian, P.W., Elson, G.W., Hemming, H.G., Radley, M.: J. Sci. Food Agric. *12*, 602–612 (1954)
17. Wightman, F., Rauthan, B.S.: In: Plant Growth Substances 1973, pp 15–27. Tokyo: Hirokawa Publ. Co 1974
18. Gautheret, R.-J.: C. R. Soc. Biol. *127*, 259–262 (1938); C.R. Acad. Sci. *208*, 118–120 (1939)
19. Nobécourt, P.: Bull. Soc. Bot. Fr. *85*, 182–185 (1938); C.R. Soc. Biol. *130*, 1270–1274 (1939)
20. Steward, F.C., Shantz, E.M.: In: The Chemistry and Mode of Action of Plant Growth Substances. Wain, R.L., Wightman, F. (eds.), pp. 165–186. Ottawa: Runge Press 1956
21. Reinert, J.: In: Proc. Int. Conf. Plant Tissue Cult. White, P.R., Grove, A.R. (eds.), pp. 1–8. 1963
22. Miller, C.O., Skoog, F., Okumura, F.S., Saltza, M.H. von, Strong, F.M.: J. Am. Chem. Soc. *77*, 2662 (1955); *78*, 1375–1380 (1956)
23. Skoog, F., Miller, C.O.: Symp. Soc. Exp. Biol. *11*, 118–141 (1957)
24. Letham, D.S., Shannon, J.S., McDonald, T.R.: Proc. Chem. Soc. 230 (1964)
25. Gane, R.: Nature (Lond.) *134*, 1008 (1934)
26. Denny, F.E., Miller, L.P.: Contrib. Boyce Thompson Inst. *7*, 97–102 (1935)
27. Burg, S.P.: Annu. Rev. Plant Physiol. *13*, 265–302 (1962)
28. Abeles, F.: Annu. Rev. Plant Physiol. *23*, 259–292 (1972)
29. Fitting, H.: Z. Bot. *1*, 1–86 (1909); *2*, 225–267 (1910)
30. Thimann, K.V.: In: The Natural Plant Hormones. Plant Physiol. Vol. VI B, pp. 129–145. New York: Academic Press 1972

Auxins

Chairman: L. N. VANDERHOEF

Homeostatic Control of Concentrations of Indole-3-Acetic Acid

R.S. BANDURSKI [1]

Introduction

Improved analytical techniques permit a reexamination of the early experiments of Skoog, van Overbeek, Went, Thimann, Haagen-Smit and Bonner (cf. 45). This reexamination demonstrates the importance of the ester and amide conjugates of indole-3-acetic acid (IAA), the "bound auxins". Conjugates of IAA have been shown to have four metabolic roles: (1) IAA conjugates, and not tryptophan, are sources of IAA for the seed during germination (11); (2) an IAA conjugate is the "seed auxin precursor" (11, 31); (3) conjugation of IAA protects it against peroxidative attack (8); and (4) reversible synthesis and hydrolysis of IAA conjugates provides a hormonal homeostatic system that is responsive to environmental controls (5). These roles have been adduced from studies of seedlings of *Zea mays* and their general applicability has not been established. However, we predict, in a process so basic as hormonal metabolism, that a modified interpretation of the biblical enunciation of comparative biochemistry, "all flesh is grass" (20), will be applicable.

I will discuss: (1) the structure of the IAA conjugates, (2) the concentration and turnover of seedling indoles, (3) the identity of the "seed auxin precursor" and (4) hormonal homeostasis.

Structures of IAA Conjugates

For *Zea mays*, sweet corn, the concentration and structures of the indolylic compounds, occurring in amounts greater than a few μg per kg, are known (9, 10, 24, 43). There are more than 16 different conjugates of indole-3-acetic acid (IAA), including the isomeric esters of IAA and myo-inositol, IAA esters of myo-inositol glycosides, and IAA esters of high molecular weight glucans (Fig. 1). Oats contain an IAA ester of glycoprotein (32) and rice contains IAA-myo-inositol and other esters (17). This is all that is known of the structure of the naturally occurring IAA adducts except for a few suggestive, but not complete, characterizations (cf. 41). In addition, studies of exogenously applied IAA show, for example, formation of IAA-aspartate and IAA-glucose (1, 46), although it is not known if these are the same as the endogenous conjugates.

1 Department of Botany and Plant Pathology, Michigan State University, East Lansing, Michigan 48824, USA

COMPOUND	STRUCTURE	AMOUNT IN DRY SEED MG/KG		PERCENT OF TOTAL	
Indole-3-acetic acid	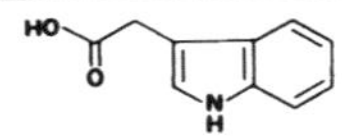		0.5		0.8%
Indoleacetylinositols	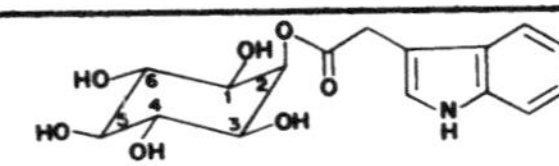		10.1		15.2%
2-*O*-(indole-3-acetyl)-*myo*-inositol		7.0		10.5%	
1-DL-(indole-3-acetyl)-*myo*-inositol		3.1		4.7%	
Indoleacetylinositol-arabinosides	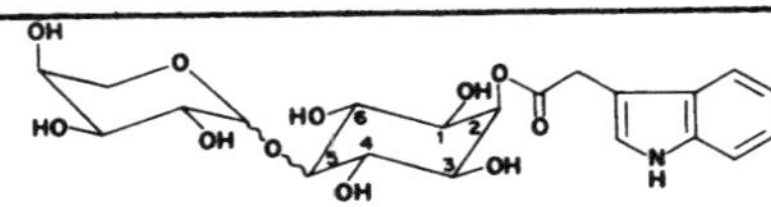		15.4		23.2%
5-*O*-β-L-arabinopyranosyl-2-*O*-(indole-3-acetyl)-*myo*-inositol		11.7		17.6%	
5-*O*-β-L-arabinopyranosyl-1-DL-(indole-3-acetyl)-*myo*-inositol		3.7		5.6%	
5-*O*-β-L-galactopyranosyl-2-*O*-(indole-3-acetyl)-*myo*inositol	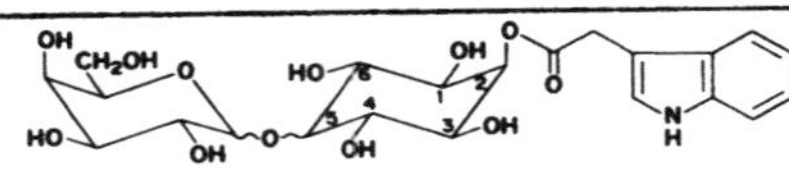		5.4		8.1%
Trace compounds	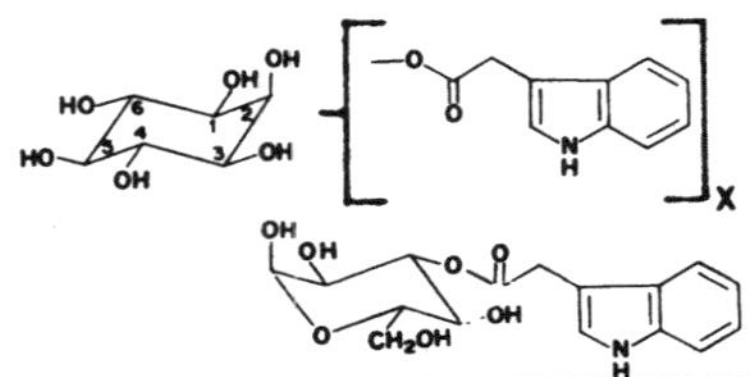		0.2		0.3%
Di-*O*-(indole-3-acetyl)-*myo*-inositol		0.08		-	
Tri-*O*-(indole-3-acetyl)-*myo*-inositol		0.03		-	
2-*O*-(indole-3-acetyl)-D-glucopyranose		0.02		-	
4-*O*-(indole-3-acetyl)-D-glucopyranose		0.02		-	
6-*O*-(indole-3-acetyl)-D-glucopyranose		0.05		-	
LOW M.W. COMPOUNDS -- TOTAL			31.2		47.6%
(indole-3-acetyl)-glucan	β 1 4 cellulosic glucan with 7 to 50 glucose units per IAA		35.0		52.5%

Fig. 1. A summary of the amounts and concentrations of the indolylic constituents of kernels of *Zea mays*

Concentration and Turnover of Indoles

Knowing the structure of the indoles of *Zea* seedlings enables determination of their concentration and turnover. Such knowledge was necessary to permit translation of the radioactivity of an applied, and transported, indole into the amount of that indole.

Kernels of corn were germinated for 4 days, then one-third of the kernel was removed and the labeled indolylic compound applied to the exposed, semi-liquid endosperm surface. With this technique, no new membrane barriers are interposed, nor need the applied indole permeate cut and damaged cells. Homogenization of the plant

Table 1. Distribution of IAA and ester- and amide-linked conjugates of IAA in selected plant species (4)

Species	Tissue	IAA content		
		Free IAA [a]	Ester IAA [b]	Amide IAA [c]
			μg/kg [d]	
Cereals				
Avena sativa	Vegetative tissue	16	5	69
Avena sativa	Seed	440	7620	n.d. [f]
Hordeum vulgare	Seed (milled)	40 [e]	329	– [g]
Oryza sativa	Seed	1703	2739	–
Panicum miliaceum	Seed	366	3198	–
Triticum aestivum	Seed	123	511	–
Zea mays	Vegetative tissue	24	328	60
Zea mays	Seed	500 to 1000	71600 to 78500	–
Legumes				
Glycine max	Seed	4	50 [e]	524
Phaseolus vulgaris	Seed	20 [e]	30 [e]	136
Pisum sativum	Vegetative tissue	35	5	43
Pisum sativum	Seed	93	n.d.	202
Others				
Cocos nucifera	Liquid endosperm	0	905	–
Fagopyrum esculentum	Seed	40	127	25
Helianthus annuus	Seed	30 [e]	110 [e]	–
Lycopersicum esculentum	Fruit	trace	trace	–
Saccharomyces cerevisea	Packed cells	290	n.d.	–

[a] No alkaline hydrolysis
[b] IAA after hydrolysis with 1 N alkali minus the free IAA
[c] IAA after hydrolysis with 7 N alkali minus the free and ester IAA
[d] Seedlings and fruits are fresh weight, seeds are air dry and yeast cells contain 30% dry matter
[e] A visual estimate of IAA on a TLC plate as colorimetry was precluded by contaminants
[f] n.d. (not detectible). Where the ester content is high, small amounts of IAA escape detection
[g] A dash (–) indicates the assay was not done

immediately after isotope application, and reisolating the isotopically labeled compound permits the amount of that compound to be determined by the isotope dilution method (33). This method has previously been utilized in studies establishing that IAA-conjugates are present in *all* plants examined [(3, 4), Table 1], including plants limited in growth rate by IAA (4).

An extension of the isotope dilution technique permits determination of the rate of turnover of the applied compound. If the plant is incubated after application of the isotope, then the rate of change of specific activity of the applied compound, as a function of time, measures the rate of turnover of that compound (38, 47).

Table 2. Rate of change as a function of time of the specific activity of ^{14}C-labeled indolylic compounds applied to the endosperm of germinating *Zea* kernels (11, 31)

Compound	Incubation time h	Specific activity dpm/μg	k h^{-1}	$t_{1/2}$ h
IAA	0	31,000		
	4	8,900	0.22	3.2
	8	5,400		
Tryptophan	0	15,200		
	8	5,060	0.14	5.0
IAA-myo-inositol	0	935		
	8	590	0.06	12.0

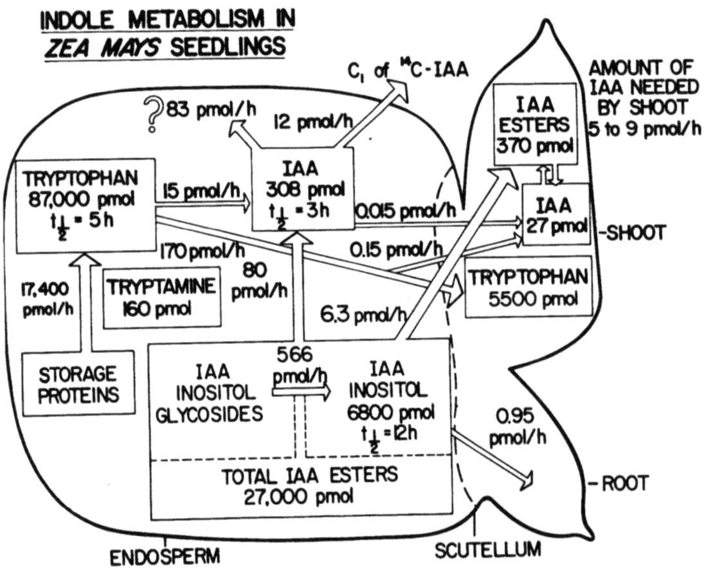

Fig. 2. A diagrammatic representation of the amounts and turnover of the indolylic constituents of kernels of *Zea mays* and of the transport of these compounds from kernel to shoot (11, 31). The amount (pool size) of each indolylic component of the seedling is shown in the boxes, and the time required for biosynthesis of an amount of the component equal to the original pool size is shown as $t_{1/2}$. The values above the arrows indicate the rate at which the compounds are being converted to other compounds or transported from endosperm to shoot or root. This figure summarizes (11, 18, 31)

Examples of such data are shown in Table 2. IAA has previously been shown to be steady state in concentration in the endosperm of germinating *Zea* kernels (11, 42). The data of Table 2 show that the specific activity of labeled IAA applied to the endosperm decreases as a function of time, and the rate of decrease obeys first-order kinetics with a rate constant of $k = 0.22$ and $t_{1/2} = 3.2$ h (11). Thus, as shown in Fig. 2, in 3.2 h the entire pool of 300 pmol · $kernel^{-1}$ of IAA was replaced by 300 pmol of new IAA. This "churning" of IAA has not previously been observed, and we suggest it has significance for seed germination. The question then arises, where does the IAA in the kernel come from? Is it biosynthesized from tryptophan or by hydrolysis of conjugates? We previously found that high molecular weight, Ehrlich-reactive-IAA conjugates of the endosperm are being hydrolyzed at the rate of 135 pmol · $kernel^{-1}$ · h^{-1} and this would be adequate to provide the "new" IAA for the endosperm (42). From the data in Fig. 2, based on techniques identical to those described above, it can be seen that the tryptamine pool is too small to serve as an important source of IAA for the seed (11) and that the overall rate of conversion of tryptophan to IAA is small. Thus, for 4-day-old *Zea* seedlings, the IAA conjugates, and not tryptophan or its immediate metabolites, are the sources of endosperm IAA. Young seedlings of *Zea* represent a "closed" system for studies of IAA and IAA conjugate metabolism with little de novo synthesis of IAA from tryptophan.

The "Seed Auxin Precursor"

Berger and Avery, Cholodny, Haagen-Smit, Skoog, and van Overbeek, among others (6, 7, 16, 36, 44) demonstrated the existence of a "seed auxin precursor". Skoog's experiments, in particular, showed that the seed auxin precursor moved from the seed to the shoot, there to be converted to an active auxin (36). Chemical characterization of the precursor was not achieved (15, 37), but it was found that tryptophan could be converted to IAA (34, 40) and could serve as a model for the native precursor which would be converted to auxin. IAA was not generally accepted at that time as an endogenous auxin in higher plants.

Knowledge of the identity of the indolylic compounds of *Zea* and their turnover (9, 10, 24, 43) and the commercial availability of labeled IAA and tryptophan permitted approaching this problem anew. Additionally, labeled IAA-myo-inositol ester was needed, since this is a major seed auxin conjugate. Dr. Nowacki and Mr. Cohen therefore, have synthesized the ^{14}C-labeled IAA-myo-inositols (30). Comparison of IAA, tryptophan, and IAA-myo-inositol as possible seed auxin precursor was now possible.

Labeled IAA, tryptophan or IAA-myo-inositol was applied to cut endosperm surfaces and after 8 h the shoot tissue harvested. Following addition of carrier IAA, the IAA was reisolated from the shoot. Knowing the pool size and rate of turnover of each of these compounds in the endosperm permitted approximation of the specific activity of each compound as it left the endosperm and entered the scutellum and then the seedling. Furthermore since the amount of carrier IAA was known, it was possible to correct for losses during the multi-step reisolation procedures (18). Assumptions involved in such calculations have been discussed elsewhere (38, 47).

When ^{3}H-IAA, ^{3}H-tryptophan and ^{14}C-IAA-myo-inositol were applied to the endosperm, the rates at which total IAA appeared in the shoot was 0.015, 0.15 and 6.3 pmol · shoot^{-1} · h^{-1} respectively. IAA production from tryptophan was probably overestimated because of the large radiological, non-enzymatic, conversion of tryptophan to IAA (11). We conclude, for germinating *Zea* kernels, that IAA-myo-inositol is the major source of IAA for the shoot and not tryptophan, tryptamine or free IAA.

Is the 6.2 pmol · h^{-1} · seedling^{-1} of IAA-myo-inositol transported from endosperm to shoot sufficient to supply the seedling's needs? Gillespie and Thimann (14) estimated, by bioassay, that 5 pmol · h^{-1} of IAA diffused into agar blocks from each *Zea* coleoptile tip. Presumably, that amount of precursor must move from endosperm to tip, since what goes down must come up. We previously estimated IAA transport from endosperm to shoot at 9 pmol · h^{-1} · shoot^{-1} based on the amount of IAA needed to maintain a steady state concentration in the tissue. The observed rate of transport of IAA-myo-inositol from endosperm to shoot appears adequate to provide the required 5 to 10 pmol · h^{-1} · shoot^{-1}, particularly since the possible transport of IAA-myo-inositol glycosides has not even been studied.

Ester Hydrolysis and Synthesis as a Mechanism for Hormonal Homeostasis

In *Zea* seedlings, the concentration of free IAA appears to be controlled by the relative rates of synthesis and hydrolysis of the IAA-conjugates and not by de novo formation of IAA. The evidence is as follows: (1) neither tryptophan nor tryptamine are major sources of seedling IAA (11); (2) the enzymes to make and hydrolyze the conjugates are present in *Zea* tissue (19, 23, 28); (3) the conjugate synthesizing and conjugate hydrolizing enzymes are active in vivo since equilibrium concentrations of IAA can be attained by providing the tissue with IAA or its conjugates (11, 31) and (4) the steady-state-equilibrium concentrations of IAA can be perturbed by an environmental stimulus, such as light, and this perturbation is accompanied by changes in growth rates of the tissue [(5) and this paper]. To our knowledge, this is the first demonstration of hormonal homeostasis involving formation and hydrolysis of a covalently bonded hormone conjugate (2, 5, 23). Suggestions that hormone conjugates are "storage" forms of the hormone have previously been made (13, 35) but lacked experimental proof that conjugation and hydrolysis is a reversible system as outlined above and described below.

In Vitro Studies of IAA-Conjugate Synthesis. In 1975, Kopcewicz et al. (23) published the first data proving in vitro enzymatic synthesis of an IAA conjugate by an enzyme system from *Zea.* Their system required ^{14}C-IAA, ATP, Mg^{2+}, myo-inositol, and CoASH (Table 3). Both myo-inositol and glucose enhanced ester formation, indicating that this unfractionated enzyme system synthesized both IAA-myo-inositol and IAA-glucose. Michalczuk (28) has subsequently found a second route to IAA-myo-inositol, and these data will be published elsewhere.

In Vitro Studies of IAA-Conjugate Hydrolysis. There are no detailed studies of the enzymatic hydrolysis of IAA-conjugates. Proof that conjugate hydrolysis does occur

Table 3. Co-factor requirements for the in vitro enzymatic synthesis of IAA-myo-inositol. From (23) and with permission of Plant Physiology

Reaction mixture	Radioactivity			
	Expt. I		Expt. II	
	cpm	% control	cpm	% control
Complete, undialyzed enzyme	3106		2620	
Complete, dialyzed enzyme	1602	100	1428	100
Without ATP	210	13	268	19
Without CoA	230	14	244	17
Without inositol	428	27	336	24
Without inositol plus glucose	1138	71	816	57
Complete, zero time	124	8		
Complete, boiled enzyme	0	0		

in vivo is provided by the data shown below and is suggested by the observation of Hamilton et al. (19) that ether-induced autolysis of *Zea* seedlings caused hydrolysis of IAA esters and by Cholodny's previous observation that water moistened endosperm blocks liberated auxins (7). Knowledge of the activity and localization of enzymes that hydrolyze the IAA conjugates is lacking. Does the IAA-myo-inositol hydrolase occur in the tip of the plant, there to hydrolyze the conjugate and provide IAA for transport to the growing sections below? That is a question for future studies.

In Vivo Equilibrium Between IAA and IAA-myo-Inositol. The very existence of IAA-myo-inositol conjugates proves that enzyme systems for their formation are present. We wished to ascertain whether the enzymes were active in vivo in a tissue limited in growth rate by IAA and whether the relative amounts of IAA and its conjugates could be environmentally controlled since this would establish these enzymes as controllers of growth rate. Growing shoots of *Zea* seemed ideal for the purpose since they contain as much as 90% ester and 10% free IAA (3), and it can be determined whether this is an "equilibrium" by determining whether the same equilibrium is reached following introduction of either ^{14}C-labeled free IAA or IAA-myo-inositol into the seedling. "Equilibrium", as here used, connotes convertibility of IAA to IAA conjugate and IAA conjugate to IAA and does not imply a single reversible pathway.

Application of ^{14}C-labeled IAA-myo-inositol to the endosperm results in 90% ester and 10% free IAA in the shoot, and these are the expected concentrations (11, 31). Application of free ^{14}C-IAA to the endosperm results in as much as 70% ester and 30% free IAA in the shoot (18), and since this was an earlier experiment, with insufficient understanding of ester lability, we conclude that the equilibrium between IAA and its conjugates can be approached from either side of the equation:

$$\text{IAA} + \text{myo-inositol} \rightleftharpoons \text{IAA-myo-inositol}$$

The esterification or hydrolysis occurred in the shoot and not in the endosperm since, had synthesis or hydrolysis occurred in the endosperm, dilution by endogenous pools would have precluded detection in the shoot. Thus, it is established that enzymes to make and hydrolyze IAA esters are active in *Zea* shoots. Now, can this equilibrium be perturbed?

Perturbation of the Homeostatic System by Light. Do environmental stimuli change the rate of growth by perturbing the relative amounts of free and conjugated hormone? Two experimental systems have been utilized to test this hypothesis, photoinhibition of growth and geotropic curvature of etiolated plants. These two growth moderating systems are dissimilar in the nature of the stimulus, and if both perturb the IAA-IAA ester equilibrium, there is assurance that the system proposed for growth control has broad significance.

The relationship between free and ester-IAA and growth, following photoinhibition of growth, is summarized in Table 4 (2, 5). A growth-inhibiting light flash of 20 s results in 90 min in a 42% reduction of free IAA and a 34% reduction in growth. Owing to the multiplicity of esters, an exact stoichiometry cannot be established, but the light flash results in increased ester IAA and less free IAA. This supports the postulate that growth is regulated by the enzymes that make and hydrolyze conjugates of growth-promoting hormones (5).

Table 4. Comparative changes in growth rate, free IAA and ester IAA in *Zea* seedlings following a growth-inhibiting light flash

	Dark ($mm \cdot 90\ min^{-1}$)	Light	Δ	%
Growth	3.6	2.6	– 1.1	– 34
		μg/kg fresh weight		
Free IAA	23	13	– 10	– 42
Free + ester IAA	68	77	+ 9	+ 11

An Assay for IAA by Gas Chromatographic-Selected Ion Monitoring-Mass Spectrometry (gc-sim-ms). The photoinhibition studies described above provided large amounts of plant tissue, and thus a ^{14}C-IAA-isotope dilution assay (3,4) was adequate. To assay free and esterified IAA on the upper and lower halves of a geotropically stimulated corn shoot required reduction of sample size to ca. 10 g of tissue comprising about 500 half-shoots. Thus, a more sensitive assay for IAA was required.

All assays for IAA must utilize an internal standard since the planar structure of IAA and high density of π bonding electrons renders it unstable and easily adsorbed (25, 27). An internal standard with deuterium in positions 4, 5, 6 and 7 of the ring is ideal since: (1) the deuterium at all four positions is stable to alkaline hydrolysis, so d_4-IAA may serve as an internal standard for free IAA and the ester- and amide-linked IAA liberated by alkaline hydrolysis (26); (2) the presence of 4 deuterium separates the standard from background caused by the normal abundances of heavy isotopes, and (3) mass spectrometric sensitivity is enhanced by monitoring for broad mass peaks.

Dr. V. Magnus (26) synthesized 4,5,6,7 tetradeutero indole-3-acetic acid by cycliz-ing d_5-aniline with a α-cyanoethylacetoaceticacid ethyl ester (12, 22). Previously pub-lished procedures were modified to permit isolation and subsequent decarboxylation of the 2-carboxyl intermediate without introduction of the fifth, somewhat less stable, deuterium in position 2. The availability of d_4-IAA permitted assays utilizing gc-sim-ms on free IAA and IAA liberated by ester hydrolysis. Previous uses of gc-sim-ms have been described (21).

The procedure is as follows: The harvested tissue is homogenized in acetone con-taining trace amounts of ^{14}C-IAA (to facilitate peak location in subsequent purifica-tion steps) and a known amount of d_4-IAA. The extract, after allowing time for iso-tope equilibration, is filtered and concentrated. For "free" IAA, the concentrate is acidified and IAA extracted into ether or chloroform. For "total" (free plus ester) IAA the concentrate is made one normal with respect to NaOH (4), incubated for 1 h, then acidified and extracted, as for free IAA. The solvent extracts are chromatograph-ed on DEAE-Sephadex and a reverse phase HPLC column (Schulze, Cohen, unpub-lished), and the purified IAA is then methylated and chromatographed on a gc column

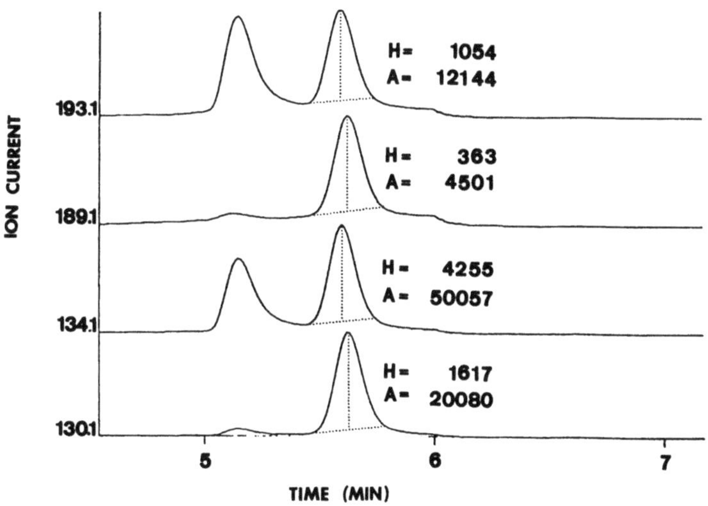

Fig. 3. A plot of the 70 eV ion current at the indicated masses for methyl IAA and $d_4$methyl IAA as a function of gas chromatographic retention time.

Percent d_4-IAA, by area, at masses 130, 134, is $\frac{50057}{50057 + 20080} = 71.4\%$

Percent d_4-IAA, by area, at masses 189, 193 is $\frac{12144}{12144 + 4501} = 73.0\%$

Percent d_4 IAA, by height, at masses 130, 134 is $\frac{4255}{4255 + 1617} = 72.5\%$

Percent d_4-IAA, by height, at masses 189, 193 is $\frac{1054}{1054 + 363} = 74.4\%$

and the mean of 4 measurements = 72.8% d_4-IAA. The peak preceding the methyl IAA at 134 and 193 is a methyl methoxycinnamic fragment ion. Visually the peaks at all 4 masses appear compar-able in size since, for ease of measurement of height and area, the peaks can be increased in size without changing absolute counts in the peak by using sim-ms detection. The molecular ion of the methyl ester of d_4-IAA is at 193 (189 for non-deuterated IAA), while the major fragment ion in a 70 eV spectrum is at 134 (130 for the plants' non-deuterated IAA). Despite the rigorous pre-purification procedure employed, the samples are still contaminated with UV-absorbing and highly fluorescent methoxycinnamic acids. Thus, assay procedures employing UV or fluorescence detection are invalid for at least *Zea.*

Figure 3 shows a gc-sim-ms plot of ion current at the indicated masses as a function of retention time. Since the amount of d_4-IAA added is known, the amount of IAA in the plant can be calculated. With this method, about 4 samples can be assayed in 1 week. It thus remains a laborious and expensive assay, but it is reliable since for any compound to interfere with this assay, the compound must cofractionate with IAA during DEAE, HPLC and gc chromatography and then yield percentages of fragment ions identical to IAA. In addition, the enormous dynamic range of the mass spectrometer permits peaks to be enlarged to any size for purposes of accurately integrating the number of counts in the peak.

Perturbation of the Homeostatic System by Gravity. With the availability of the new gc-sim-ms assay for IAA it was possible to ask if gravity causes upward curvature of a shoot by varying the ratio of free to conjugated growth hormone on the upper and lower sides of the shoot. The data of Table 5 show that free IAA is increased in the lower side of a geotropically stimulated shoot, and by contrast, free IAA is reduced in the slowly growing upper half. This is the first mass spectral confirmation of bioassay data for changes in endogenous IAA in the upper and lower halves during geostimulation. Ester IAA is slightly, but not significantly, lower in the lower half. Preliminary studies indicated a stoichiometry between free IAA increase and ester decrease. However the standard error for determining the relatively large amounts of ester, as compared to free IAA, is almost equal to the change in free IAA. Thus, whether geotropic curvature utilizes the homeostatic system observed for photoinhibition and for transport remains unknown, and a distinction between lateral transport (cf. 45) and conjugate hydrolysis as a mechanism for geo-induced curvature cannot yet be made.

Table 5. Amounts of free IAA and ester in the upper and lower halves of geotropically stimulated *Zea* shoots [a]

	$\mu gm \cdot gm\ dry^{-1}$	$pmol \cdot 1/2\ plant^{-1}$
Free, upper half	0.37 ± 0.03	5.3 ± 1.3
Free, lower half	0.57 ± 0.09	7.6 ± 1.2
Ester, upper half	1.35 ± 0.34	19.2 ± 1.4
Ester, lower half	1.32 ± 0.10	18.9 ± 1.1

[a] Mean of 10 determinations using n − 1 = 9 to compute the standard error of the mean

The following is presented as a working hypothesis (2, 5, 23). As diagrammed in Fig. 4B, internal metabolic processes are regulated by feedback control mechanisms (cf. 29). Such controls gear the rate of metabolism to the rate of utilization of the metabolite. For example, sugar metabolism (A→B→C→D) is feedback controlled and is in turn coupled to organic acid metabolism (1→2→3→4), and this is in turn coupled and *locked in a metabolic grid* to amino acid metabolism (I→II→III→IV). Such controls, like those of a German Autobahn, impose no limits on upper rates of metabolism other than enzyme turnover numbers. Here the environment must impact upon the organism, and as shown in Fig. 4A, we postulate this is done by affecting the enzyme systems that regulate the relative amounts of hormone and hormone conjugates.

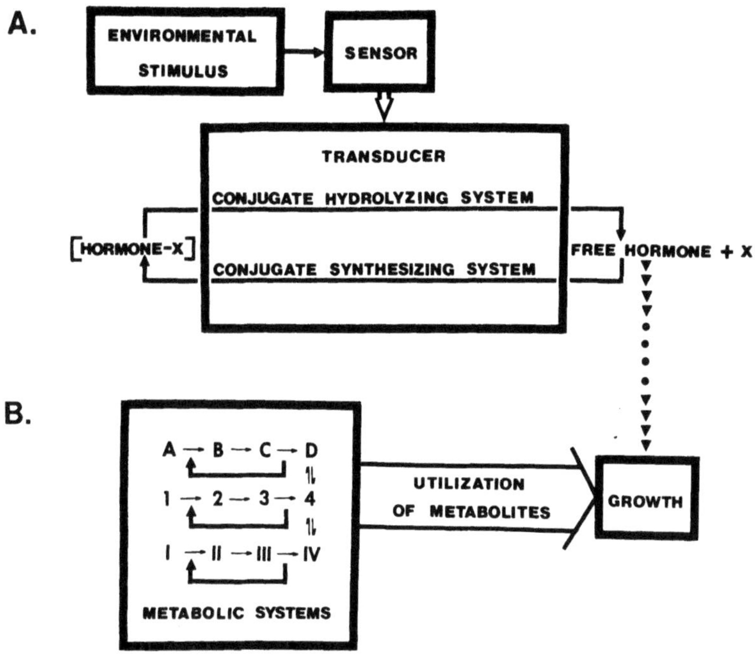

Fig. 4. A working hypothesis for the integration of hormonal and metabolic control systems. In **B** A→B→C→D represents, for example, sugar metabolic interconversions with └──┘↑ showing feedback control. Pathways dependent upon sugar for their carbon skeletons, as for example, organic acid and amino acid metabolism are shown as *1, 2, 3, 4,* and *I, II, III, IV.* Hormonal control, as shown in **A**, has been described (2) and, as discussed above, would control **B** by controlling the overall rate of utilization of the metabolites produced by metabolism

Our overall concept of this control is shown in Figure 4A. The environment affects a sensor and the sensor in turn transfers the stimulus to a transducer (2). We postulate that *the transducers of environmental stimuli are the enzyme systems that make and hydrolyze the hormone conjugates,* and thus, the rate of metabolite utilization is geared to the rate of growth permitted by the environment.

Acknowledgments. The work of J. Cohen, A. Ehmann, E. Epstein, Pat Hall, Prudence Hall, J. Kopcewicz, C. Labarca, V. Magnus, L. Michalczuk, P. Nicholls, J. Nowacki, F. Percival, Z. Piskornik, A. Schulze and M. Ueda provide the data for this report. The Metabolic Biology Section of the U.S. National Science Foundation has provided continued financial support and Drs. E. Romanoff and

J. Shen-Miller provided advice and valuable discussion. Drs. E. Chapman and C.C. Sweeley (NIH RR00480, DOE EY-76-C-02-1338, and Michigan State University) made possible mass spectral studies. Ms. J. Di Lucca Schlub aided in manuscript preparation, and this is journal article 9244 from the MSU Agricultural Experiment Station.

References

1. Andreae, W.A., Good, N.E.: Plant Physiol. *30,* 380–383 (1955)
2. Bandurski, R.S.: In: Cyclitols and Phosphoinositides. Wells, W.W., Eisenberg, F. (eds.), pp. 35–53. London, New York: Academic Press 1978
3. Bandurski, R.S., Schulze, A.: Plant Physiol. *54,* 257–262 (1974)
4. Bandurski, R.S., Schulze, A.: Plant Physiol. *60,* 211–213 (1977)
5. Bandurski, R.S., Schulze, A., Cohen, J.D.: Biochem. Biophys. Res. Commun. *79,* 1219–1223 (1977)
6. Berger, J., Avery, G.S.: Am. J. Bot. *31,* 199–203 (1944)
7. Cholodny, N.: Planta *23,* 289–312 (1935)
8. Cohen, J.D., Bandurski R.S.: Planta *139,* 203–208 (1978)
9. Ehmann, A.: Carbohydr. Res. *34,* 99–114 (1974)
10. Ehmann, A., Bandurski, R.S.: Carbohydr. Res. *36,* 1–12 (1974)
11. Epstein, E., Cohen, J.D., Bandurski, R.S.: Plant Physiol. *65,* 415–421 (1980)
12. Feofilaktov, V.V., Semenova, N.K.: J. Gen. Chem. SSSR *23,* 669–679 (1953)
13. Galston, A.W.: In: Handbuch der Pflanzenphysiologie. Ruhland, W. (ed.). Vol. XVII/1, pp. 492–529. Berlin, Göttingen, Heidelberg: Springer 1959
14. Gillespie, B., Thimann, K.V.: Plant Physiol. *38,* 214–225 (1963)
15. Gustafson, F.G.: Am. J. Bot. *28,* 947–951 (1941)
16. Haagen-Smit, A.J., Leech, W.D., Bergen, W.R.: Am. J. Bot. *29,* 500–506 (1942)
17. Hall, P.J., Bandurski, R.S.: Plant Physiol. *63*(S), 50 (1979)
18. Hall, P.L., Bandurski, R.S.: Plant Physiol. *61,* 425–429 (1978)
19. Hamilton, R.J., Bandurski, R.S., Grigsby, B.H.: Plant Physiol. *36,* 354–359 (1961)
20. Isaiah: The Old Testament 40:6
21. Jenden, D.J., Cho, A.K.: Annu. Rev. Pharmacol. *13,* 371–390 (1973)
22. King, F.E., L'Ecuyer, R.: J. Chem. Soc. 1901–1905 (1934)
23. Kopcewicz, J., Ehmann, A., Bandurski, R.S.: Plant Physiol. *54,* 846–851 (1974)
24. Labarca, C., Nicholls, P.B., Bandurski, R.S.: Biochem. Biophys. Res. Commun. *20,* 641–646 (1966)
25. Little, C.H.H., Heald, J.K., Browning, G.: Planta *139,* 133–138 (1978)
26. Magnus, V., Bandurski, R.S.: Plant Physiol. *61*(S), 63 (1978)
27. Mann, J.D., Jaworski, E.G., Planta *92,* 285–291 (1970)
28. Michalczuk, L., Bandurski, R.S.: Plant Physiol. *63*(S), 50 (1979)
29. Monod, J., Wyman, J., Changeux, J.P.: J. Mol. Biol. *12,* 88 (1965)
30. Nowacki J. Cohen, J.D., Bandurski, R.S.: J. Label. Compd. Radiopharm. *15,* 325–329 (1978)
31. Nowacki, J., Bandurski, R.S.: Plant Physiol. *65,* 422–427 (1980)
32. Percival, F., Bandurski, R.S.: Plant Physiol. *58,* 60–67 (1970)
33. Rittenberg, D. Foster, G.L.: J. Biol. Chem. *133,* 737–744 (1940)
34. Schocken, V.: Arch. Biochem. *23,* 198–204 (1949)
35. Sembdner, G.: In: Biochemistry and Chemistry of Plant Growth Regulators. Schreiber, K., Schütte, H.R., Sembdner, G. (eds.), pp. 283–302. Halle/GDR: Inst. of Plant Biochem. 1974
36. Skoog, F.: J. Gen. Physiol. *20,* 311–334 (1937)
37. Skoog, F. Thimann, K.V.: Science *92,* 64 (1940)
38. Sprinson, D.B., Rittenberg, D.: J. Clin. Invest. *28,* 715 (1949)
39. Sweeley, C.C., Elliott, W.H., Fries, I., Ryhagz, R.: Anal. Chem. *38,* 1549 (1966)

40. Thimann, K.V.: J. Biol. Chem. *109*, 279–291 (1935)
41. Tillberg, E.: Physiol. Plant. *31*, 271–274 (1974)
42. Ueda, M., Bandurski, R.S.: Plant Physiol. *44*, 1175–1181 (1969)
43. Ueda, M., Bandurski, R.S.: Phytochemistry *13*, 243–253 (1974)
44. van Overbeek, J.: Am. J. Bot. *28*, 1–10 (1941)
45. Went F.W. Thimann, K.V.: Phytohormones. New York: Macmillan 1937
46. Zenk, M.H.: Nature (Lond.) *191*, 493–494 (1961)
47. Zilversmit, D.B., Enterman, C., Fishler, M.C.: J. Gen. Physiol. *26*, 325–331 (1943)

The Mechanism of Transmembrane Auxin Transport and Its Relation to the Chemiosmotic Hypothesis of the Polar Transport of Auxin

P.H. RUBERY [1]

The coordinated development of plants requires and reflects a controlled distribution of growth substances which, by interaction with receptors, bring about the biochemical and biophysical changes that culminate in morphogenesis. Transport is a central factor influencing cellular hormone concentration and hence the proportion of occupied receptors. Polar transport in a preferred morphologically defined direction has been most extensively studied and characterized for auxin although abscisic acid, gibberellins and perhaps cytokinins may behave similarly in some instances (1). The "chemiosmotic" hypothesis of polar auxin transport was proposed independently by Rubery and Sheldrake (2) and by Raven (3). It has recently been reviewed by Goldsmith (1). In this paper I shall discuss this new hypothesis together with the theoretical arguments and experimental data that led to its formulation. The key considerations are the mechanism and energetics of transmembrane auxin movement and the basis and maintenance of the cellular asymmetry underlying the polarity of the tissue as a whole.

Transmembrane Auxin Transport

Cell suspension cultures (2, 4–8) and giant algal cells (3, 4) have been used for detailed kinetic studies of transmembrane auxin transport over times ranging from 15 s to several minutes. However, the general features revealed have been shown in complementary experiments to apply also to auxin uptake by segments cut from intact tissues capable of polar auxin movement (8, 9).

The feature which dominates auxin transport is the lipid-soluble weakly acidic nature of both the naturally occurring IAA and of most synthetic auxins such as NAA (1-naphthylacetic acid) and 2,4-D (pK values 4.7, 4.2, and 2.8 respectively). The movement of both the hydrophobic undissociated acid molecules (designated AH for lipophilic weak acids) and of the more polar anions (A^-) must be taken into account. The relative concentrations of these species depend on the strength of the acid and the solution pH. Consider first nonmediated diffusive transport between two buffered aqueous compartments through a biological membrane across which an electrical potential difference is maintained. A convenient and important example is the plasmalemma, membrane potential negative inside, which separates cytoplasm from the cell wall. The *total* auxin concentration in the wall will be lower than in a bulk incubation

1 University of Cambridge, Department of Biochemistry, Tennis Court Road, Cambridge CB2 1QW, United Kingdom

medium if the fixed negative charges of the wall constitute an effective Donnan phase: the [AH] is unaltered but $[A^-]$ is lowered and $[H^+]$ is correspondingly raised with respect to the bulk phase; the product $[H^+][A^-]$ is constant.

Let auxin initially be present only outside the cells where the pH is assumed to be buffered at a more acidic value then the cytoplasmic pH of about 7. AH is by far the more permeant species (2, 3) and will cross the membrane by passive diffusion in response to a concentration (strictly activity) gradient. Making the heuristically useful restriction (to be lifted later) of membrane impermeability to A^-, equilibrium will be attained when the AH concentrations are the same in each compartment. Provided that the relatively alkaline cytoplasm is adequately buffered against the protons released by auxin ionization, the total auxin concentration in the cytoplasm will exceed that outside the cell because of the higher anion concentration needed to satisfy the dissociation constant. The equilibrium condition is expressed most simply by Eq. (1) and more conveniently by Eq. (2):

$$H_iA_i = H_oA_o \tag{1}$$

$$\frac{{}^tA_i}{{}^tA_o} = \frac{1 + 10^{(pH_i - pK)}}{1 + 10^{(pH_o - pK)}} = \frac{(H_i^+ + K)H_o^+}{(H_o^+ + K)H_i^+} \tag{2}$$

where H is hydrogen ion concentration, A is anion concentration, ${}^tA$ is total auxin concentration ($[A^-]$ + [AH]), K is the dissociation constant, and subscripts i and o refer to inner (cytoplasmic) and outer compartments respectively.

The auxin accumulation found experimentally is below that predicted by Eq. (2) which does not allow for anion permeability [see Eq. (3) below]. However, the data show the undissociated acid to be the major permeant species: for auxins [IAA, NAA, and 2,4-D (4, 5)] and other liphophilic weak acids [benzoic acid (5) and abscisic acid – Rubery, unpubl.], the net uptake rate is larger the greater the pH difference between cytoplasm and incubation medium and the pH dependence of influx follows a characteristic "titration" curve, which is typically displaced from the pK of the acid by about plus half a pH unit (3, 4). A reasonable explanation for this alkaline shift has been suggested (10). For highly permeant species such as IAAH, diffusion through the unstirred layer adjoining the membrane can be rate limiting for uptake. But as IAAH is depleted in the boundary layer, its concentration can be topped up by association of IAA^- and H^+. When the pH is increased, uptake of the weak acid does not fall as rapidly as predicted from its dissociation constant because the increased $[IAA^-]$ effectively facilitates IAAH diffusion through the unstirred layer. The most important factor influencing the boundary layer thickness is the vigor of stirring the bulk solution.

Now consider the effect that anion permeability will have on the accumulation ratio ${}^tA_i/{}^tA_o$ (Figs. 1, 2). Auxin anions will cross the membrane in response to the electrochemical potential gradient in A^- since both electrical and chemical (concentration) driving forces operate on the negatively charged ions. Since A^- is less permeant than AH – Raven (3) suggests values of 10^{-6} and 10^{-3} cm/s respectively for the permeability coefficients – its passive efflux from the cell down the electrochemical gradient set up by AH uptake and ionization will act as a partial leak on auxin accumulation.

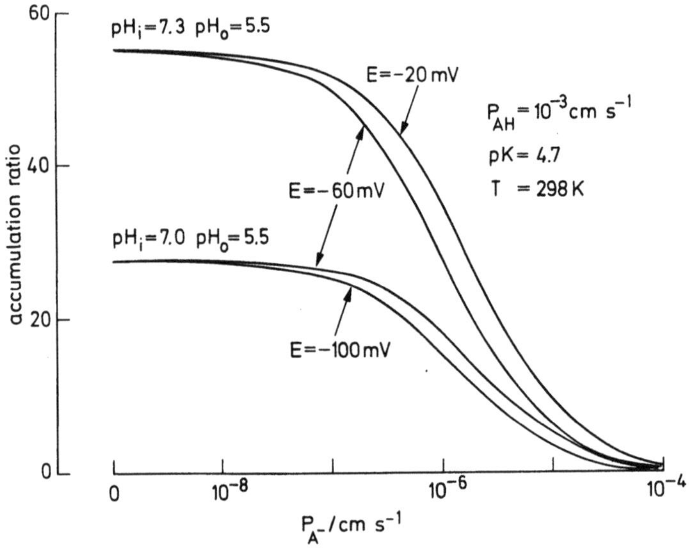

Fig. 1. The behavior of the accumulation ratio for IAA as a function of P_A at different values of E and pH_i, according to Eq. (3)

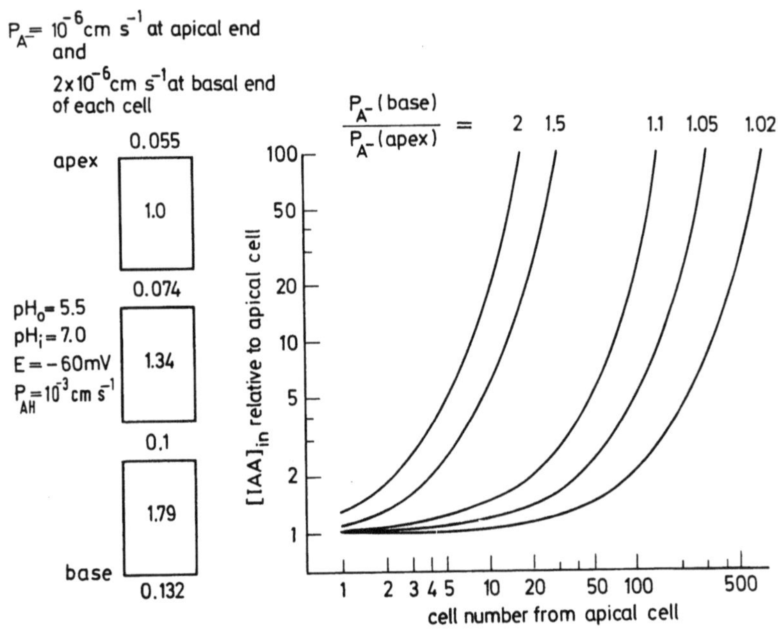

Fig. 2. Simulation of IAA equilibrium distribution [Eq. (3)] between cytoplasm and exterior of a file of cells which are relatively more permeable to IAA^- at their basal ends. The choice of P_{A^-} values is arbitrary: P_{A^-}/P_{AH} could well be $> 10^{-3}$ because of carrier participation. The *numbers inside* and *outside the boxes* ("cells") on the *left-hand side* represent total IAA concentration; the vacuole has been omitted for simplicity

At equilibrium, the *total* influx of auxin species is equal to the *total* efflux, and the general expression for the accumulation ratio [Eq. (3)] includes both the permeability coefficients and the membrane potential.

$$\frac{{}^{t}A_i}{{}^{t}A_o} = \frac{(H_i^+ + K)(P_{AH}\, H_o^+\, RT\,[1 - e^{-FE/RT}] + P_{A^-} KFE)}{(H_o^+ + K)(P_{AH}\, H_i^+\, RT\,[1 - e^{-FE/RT}] + P_A KFEe^{-FE/RT})} \tag{3}$$

where E is the membrane potential, R the gas constant, T the temperature, F the Faraday, P_A and P_{AH} the permeability coefficients for the anion and undissociated acid respectively, and other symbols as in Eq. (2).

If P_{A^-} is zero, Eq. (3) reduces to Eq. (2). But if P_{A^-} increases, the accumulation ratio falls as the "leak effect" of anion permeability comes into play (Fig. 1). The accumulation ratio between incubation medium and cells masks the separate contributions of cytoplasm and vacuole, but under most conditions the buffered alkaline cytoplasm remains a sink for auxin. The extent to which the cytoplasmic auxin concentration exceeds that in the more acidic vacuole will depend on the permeability properties of the tonoplast. The amount of auxin in the vacuole, which occupies the majority of the cell volume, may be comparable to or greater than the amount in the cytoplasm. Indeed, guestimation of vacuolar parameters makes assessment of P_{A^-} for the plasmamembrane an uncertain exercise. However, for the Virginia Creeper crown gall cells used in my experiments, a minimal value of about 2×10^{-4} cm/s for the P_{AH} of IAA can be calculated. Corresponding values for 2,4-D, NAA and benzoic acid are approximately 10^{-3} cm/s, and for abscisic acid about 10^{-5} cm/s.

The effect on the accumulation ratio of systematically altering the other variables and parameters in Eq. (3) can be seen conveniently using a programmable calculator. Also, sensitivity functions may be derived by appropriate partial differentiation of the accumulation ratio. Some of the conclusions may be summarized: (1) the accumulation ratio is raised by increasing the pH gradient and it is more sensitive to pH changes in the more alkaline compartment, (2) hyperpolarization of the membrane decreases the accumulation ratio which is, however, much less sensitive to this than to pH changes for given values of the permeability coefficients – the higher P_{A^-}/P_{AH}, the greater the effect of changes in membrane potential, (3) the dependence of the accumulation ratio on the dissociation constant of the auxin passes through a maximum, which corresponds to a minimum perturbation of the ratio by anion leakage. It is interesting to note that the optimal pK is 4.7 (that of IAA) if $pH_o = 5$, $pH_i = 7$, $E = -0.06$V and $P_{A^-}/P_{AH} = 10^{-3}$.

A further equation of interest gives the time ($t_{1/2}$) for the internal concentration to rise from zero to 50 percent of the equilibrium value:

$$t_{1/2} = \frac{V \ln 2}{L} \left[\frac{(H_i^+ + K)(1 - e^{-FE/RT})}{P_{AH}\, H_i^+ (1 - e^{-FE/RT}) + P_{A^-} K \frac{FE}{RT}\, e^{-FE/RT}} \right] \tag{4}$$

where V/L is the cell volume to surface area ratio. The $t_{1/2}$ value is decreased if either P_{AH} or P_A is raised, if E becomes more negative or if the internal pH is lowered. It is independent of the external pH.

The pH gradient-driven accumulation of IAA was first pointed out several decades ago (11, 12) although the realization of its possible significance for auxin polar transport is more recent (1, 3). However, it is worth explicitly stating that all the arguments so far advanced are generally applicable to the distribution of lipophilic weak acids and bases across all cellular membranes, the bases being excluded from the more alkaline compartment. Thus, not only auxins, but transport inhibitors such as TIBA (2,3,5-triiodobenzoic acid) and NPA, abscisic acid, gibberellins, and cytokinins may move passively across membranes with a facility dependent on their lipid solubility and show pH-related equilibrium partition. Recent experiments in my laboratory have confirmed this expectation for abscisic acid.

A Model for Polar Auxin Transport

It is likely that tissue polarity reflects an underlying asymmetry of individual cells. Polar auxin transport is highly substrate specific and most lipophilic weak acids (e.g., benzoic) have been shown to move slowly and non-directionally through tissues such as young shoot segments. In deciding which of the factors that alter the accumulation ratio [Eq. (3)] might be invoked to account for cellular asymmetry, this substrate specificity focusses attention on P_{A^-} and P_{AH} rather than on the driving forces for transport. Selection between chemically similar potential substrates is likely to be determined by membrane proteins rather than membrane lipids. We have previously suggested (2, 4–7) that the energy stored in an electrochemical gradient set up by the cell's accumulation of auxin from an acidic apoplast could be applied to polar transport if a carrier protein for auxin anions were preferentially located at the basal ends of cells in the transport pathway. In effect there would be a local increase of P_{A^-} relative to P_{AH} allowing the cytoplasmic concentration of auxin to be in equilibrium with a higher external concentration at the basal end than at the apical end of the cell, because of the resulting lower basal accumulation ratio [Eq. (3), Fig. 1]. The argument can be applied to each cell in a linear file to account for the basipetal polarity of the tissue as a whole (Fig. 2). Only a slight polarity of the individual cells is required (13, 14). The model predicts, in accord with observation, that an uphill concentration gradient of auxin can be established along the tissue in the direction of polar transport (Fig. 2), the time needed depending inter alia on the values of P_{A^-} and P_{AH}. Back-diffusion will oppose this to only a small degree because of the high membrane permeability to IAA. These directional and uphill transport characteristics depend on metabolic energy to maintain the pH and potential differences which might otherwise be dissipated by the proton movements accompanying transmembrane auxin transport itself; energy may also be required to maintain asymmetry of carrier distribution (15). This energy expenditure could be channelled through the plasmalemma- and tonoplast-located proton pumps proposed to participate in a biophysical pH-stat mechanism (16). Goldsmith (1) has paid particular attention to the ability of the model (which she felicitously termed the chemiosmotic polar diffusion hypothesis) to predict both quantitative and qualitative features of polar auxin transport in the context of the mathematics of diffusion processes.

The model requires that auxin move from cell to cell by transport across the plasmamembrane and the intervening cell walls rather than primarily through plasmodesmata: symplastic transport is unlikely to be the dominant pathway for highly permeant molecules. This is supported by the observation that polar transport persists in tissues whose plasmodesmatal connections have been largely broken (17, 18) and that auxin in the polar transport stream passes through the free space between cells (17). Parenthetically, the role suggested for plasmadesmata in root geotropism (19) may be questioned on similar grounds – maize roots whose tips have been plasmolyzed with sorbitol are still geotropically competent (Rubery, unpublished).

Evidence for Carrier-Mediated Auxin Transport

Carrier participation in auxin transport by suspension cultured cells of *Parthenocissus tricuspidata* crown gall was suggested by two findings, subsequently extended to pea and tobacco stem segments (2, 5, 8, 9): (1) Auxin (IAA, 2,4-D) uptake had a saturable component superimposed on the linear nonsaturable diffusive uptake due to AH transport. (2) Net IAA uptake was stimulated by TIBA, a noncompetitive inhibitor of polar auxin transport, due to its inhibition of an IAA efflux component. The simplest interpretation of the detailed results, summarized below, is that two separate carrier systems are present; one electrogenic and TIBA-sensitive, and one electroneutral (7) (Fig. 3).

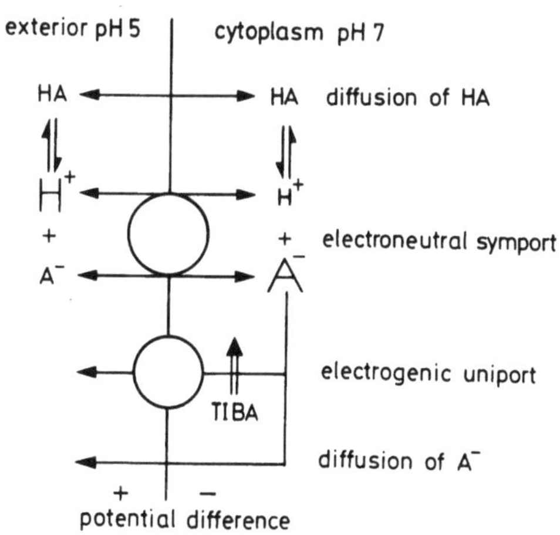

Fig. 3. Diagram to illustrate the proposed mechanisms of carrier-mediated and diffusive IAA transport across the plasmamembrane of a plant cell

The Electroneutral Carrier

The carrier mediating the saturable component of auxin uptake will transport IAA and 2,4-D (K_m values about 1 μM) but not NAA or benzoic acid (2, 5). It shows countertransport behavior [(2); Fig. 4] demonstrating both its reversibility and that carrier-mediated uptake still proceeds in the face of the negative membrane potential as equilibrium is approached. This carrier is thus unlikely to facilitate electrically uncompensated auxin influx. However, an electroneutral mechanism is plausible, entailing a carrier-coupled 1:1 stoichiometric cotransport of the auxin anion with a cation. Uptake showed no dependence on any external cation other than H^+ (6) and the pH optimum of carrier-mediated IAA uptake (pH 5) was higher than that of the stronger acid 2,4-D (pH 4). If the transport-effective species were a protonated form of the carrier bound to an auxin anion, then, on the usual assumptions for transport kinetics, the pH optimum (reflecting the concentration of this species) would be expected to be higher for the weaker acid. Such an auxin anion/H^+ symport would be reversible and accumulative from an acidic external medium, provided translocation of carrier species is rate-limiting for transport. The operation of a kinetically symmetrical and obligatorily coupled symport would be energetically equivalent to diffusive transport of AH (20). More generally, if it were less accumulative than AH diffusion, it could function in a net efflux mode at overall passive flux equilibrium. In any case, its detailed kinetic properties would depend on both the internal and external concentrations of auxin anions and H^+, and on the substrate binding order.

The TIBA-Sensitive Electrogenic Carrier

In cultured crown gall cells, TIBA stimulates the net rate of IAA uptake but has no effect on 2,4-D, NAA, benzoic acid or abscisic acid movement. However, in pea stem segments both IAA and NAA uptake are increased (2, 5, 9). The TIBA inhibition of efflux responsible for these effects can be directly demonstrated (2). Preloading experiments and the dependence of TIBA-sensitivity on the external pH suggest that TIBA can penetrate as (TIBAH) and interact with an IAA efflux component at the inner face of the plasmalemma (2). This may be separate from the electroneutral uptake carrier for which both IAA *and* 2,4-D are substrates. More direct evidence for two carriers comes from the use of 2,4-dinitrophenol (DNP) and of chemical modifying agents (7, 9).

Uptake of 2,4-D is smoothly inhibited by N-ethylmaleimide (NEM, an -SH reagent) due to the gradual abolition of the carrier-mediated component of uptake. In contrast, the net uptake of IAA is unaffected or slightly stimulated by low NEM levels ($< 5\ \mu M$). However, if TIBA is present, IAA uptake is increased in the control and NEM gives a smooth inhibition profile; NEM is able completely to abolish the TIBA-stimulation of uptake at much lower concentrations than needed to inhibit carrier-mediated IAA or 2,4-D uptake. Diethylpyrocarbonate (DEPC, a histidine reagent) gives similar results. Both NEM and DEPC are still effective at low external pH which probably reflects their lipid solubility and consequent carrier inactivation at the inner face of the membrane. p-Chloromercuribenzene-sulfonic acid, a relatively impermeant -SH reagent, inhibits carrier-mediated IAA uptake, with decreasing effectiveness as the pH is lowered, and does not cause a marked differential inhibition of the TIBA-stimulation (7, 9).

The presence of DNP in an incubation medium more acidic than the cytoplasm would be expected to decrease the cytoplasmic pH and depolarize the plasmamembrane because it promotes electrogenic H^+ ion influx. The uptakes of both BA and 2,4-D are smoothly inhibited by DNP due to cytoplasmic acidification. However, its effects on IAA transport are more complex: there is a plateau before substantial inhibition is observed where uptake is relatively insensitive to low DNP concentrations and may even be stimulated; when TIBA is present, IAA uptake now falls sharply as DNP concentration is raised, and the TIBA stimulation relative to controls is thereby abolished. This is probably due to membrane depolarization rather than to concomitant cytoplasmic acidification, since the same pattern is observed when the cells are incubated in imidazole buffer (rather than phosphate/citrate) which is penetrant and alkalinifies the cytoplasm. When the effect of DNP on efflux was examined directly, it was found to stimulate the efflux of 2,4-D, and of IAA in the presence of TIBA; without TIBA, IAA efflux could be inhibited by DNP (7).

Taken together, these data may be interpreted as follows (7, 9): there is an electrogenic carrier, specific for IAA anions (and for NAA^- in pea stem), which is inhibited by TIBA. This is separate from an electroneutral uptake carrier, specific for IAA and 2,4-D, which may be an auxin anion/proton symport (Fig. 3). Both carriers have essential cysteine and histidine residues. Those of the TIBA-sensitive carrier are the more reactive, and also appear to be less accessible at the outer face of the membrane than is the case for the electroneutral carrier. Blocking of these two carriers by NEM or DEPC will have contrasting effects on net IAA uptake; with low inhibitor concentrations, the electrogenic efflux carrier is fully inactivated so that TIBA can no longer have any effect, whereas the uptake carrier is still substantially unaffected. Similarly, a decrease in the electrical component of the driving force brought about by DNP would decrease the contribution of IAA anion efflux to net uptake and thus reduce the stimulatory effect of TIBA by diminishing the magnitude of its target component within the total flux. The plateau is observed before DNP inhibition of net IAA uptake is apparent and could thus result from opposing effects of DNP – on the one hand, to promote diffusive efflux of IAAH and perhaps carrier-mediated efflux on an IAA^-/H^+ symport, and on the other hand to inhibit electrogenic efflux of IAA^-. The smooth inhibition of 2,4-D uptake suggests that any electrogenic efflux of 2,4-D^- is much less than that of IAA^-, in agreement with the lack of effect of TIBA on 2,4-D transport.

In the model for polar auxin transport described earlier, the structural basis of cell polarity was suggested to be an enhanced basal permeability of the cells to auxin anions brought about by asymmetrical distribution of a carrier. The more appropriate carrier for such a role is clearly the electrogenic efflux carrier: it is antagonized by the polar transport inhibitor TIBA; its substrate specificity in pea stem parallels the polar transport specificity of the tissue; and in tobacco stem it is only present in the vascular cylinder where the polar transport pathway is found (Rubery, unpubl.). It is sensitive to both pH and electrical gradients across the membrane and is thus potentially responsive to changes in these parameters due to physiological stimuli.

The auxin uptake carrier does not have an obvious role in view of the ease with which undissociated acid molecules can cross membranes. It could be a relatively significant uptake mechanism at low auxin concentration and extracellular pH near neutrality. Also, its preferential apical localization, so increasing P_{AH}, would tend to favor basipetal polar transport and reinforce the effects of a basally localized anion carrier.

Correlations of Transport Kinetics with Auxin Binding Studies

The predictions made by transport kinetics can be pursued by in vitro study of the binding of auxins and modulating ligands to receptors. It is clear that plant cells possess several distinct auxin binding components which in principle include enzymes and action receptors as well as transport proteins. Attribution of function in the latter cases will depend heavily on reconstituted systems using purified components; at present only circumstantial evidence is available, although solubilization and partial purification of a membrane bound auxin receptor has been achieved (21).

The cellular locations of auxin binding sites have been studied after density gradient centrifugation of tissue homogenates by comparing their distribution with that of probable markers. The majority of sites are intracellular, "site I" occurring on the endoplasmic reticulum and "site II" has been tentatively assigned to the tonoplast (22) – study of isolated vacuoles may strengthen this possibility. Recently, a "site III" has been partially characterized in a plasmamembrane-rich fraction from zucchini hypocotyls and maize coleoptiles (22, 23). All of these sites, whether or not solubilized, have a dissociation constant for IAA of ca. 10^{-6} M and bind subsaturating concentrations optimally at ca. pH 5. This reminiscent of the auxin uptake carrier, although the efflux carrier is currently inaccesible to such kinetic characterization.

The specificity of "site III" resembles that of carrier-mediated auxin transport: 2,4-D binds less strongly than IAA; 1-NAA is only weakly competitive for IAA binding and may even enhance it. But the most compelling reason, apart from its plasmamembrane localization, for believing "site III" to represent a transport protein is its unique response to non-competitive transport inhibitors such as TIBA and NPA which *stimulate* IAA binding (23). It has not been excluded that this stimulation may reflect inhibition of carrier-mediated IAA efflux from the vesicles which largely constitute the membrane fraction. However, NPA and TIBA bind at separate sites from IAA (24) and may modulate its binding via conformational changes: inhibition of IAA transport could occur if they favor a nonproductive mode of IAA binding, perhaps opening up a second site, leading to immobilization of the carrier at the inner face of the membrane. If separate polypeptides are required for inhibitor and substrate binding, the IAA carrier could include catalytic and regulatory subunits, either permanently associated or capable of interaction dependent on the proteins' mobility in the membrane. It is relevant that the TIBA stimulation of IAA binding occurs at 2°C.

The existence of a natural class of transport modulators is suggested by the transport inhibiting and binding properties of "Kartoffelfaktor", first demonstrated in potato tubers but also present in maize coleoptiles (25). Its further characterization is eagerly awaited. "Supernatant factor", which modifies the binding affinity of auxins and inhibits auxin-induced growth (26, 27), has been identified as a mixture of 6-methoxy-2-benzoxazolinone and 6,7-dimethoxy-2-benzoxazolinone (27).

The Relationship Between Auxin Action and Polar Transport

The acid growth theory of rapid auxin-stimulated cell elongation requires that IAA enhance proton secretion in a dose dependent fashion. The IAA-promoted wall acidification is electrogenic, hyperpolarizing the membrane potential, and is probably accompanied by some increase in cytoplasmic pH although this will be limited by buffering mechanisms. The fungal toxin fusicoccin has similar effects exerted via a distinct receptor (28). All these changes will influence the transmembrane distribution of IAA [Eq. (3)] so that IAA action would be expected to interact with IAA transport by autocatalytically increasing the accumulation ratio between cytoplasm and apoplast, the pH changes dominating the contrary effect of membrane hyperpolarization (see earlier). This positive feed-forward increase in intracellular IAA levels has obvious significance for other auxin-dependent phenomena including hormonal control of gene expression.

In an auxin responsive polarly transporting tissue, cells towards the basal end would exhibit increasing electrogenic H^+ efflux because of their greater auxin concentration. Thus the occurrence of polar auxin transport could lead to the development of pH and electrical gradients along the tissue (15). On this basis, the apoplast at the base of the tissue would become both acidic and electrically positive with respect to the apoplast in the apical region, and vice versa for the symplast (15). A number of experimental observations summarized by Goldsmith (1) are in accord with this prediction. Raven (15) has suggested that membrane electrophoresis in response to the induced electrical gradients could maintain or fix the basal distribution of an auxin anion carrier having a net negative charge exposed at the outer membrane surface, or a net positive charge at the inner surface. Current views of transport mechanisms envisage oligomeric carriers which span the membrane forming a selective channel for passage of substrate (29). However, lateral movement of proteins in plant plasmamembranes remains conjectural, especially regarding the influence of the apressed cell wall. [Interestingly, Raven (15) also suggests that heterogeneity of cell wall charge distribution may be responsible for the origin of transport polarity in embryos.]

Taken together, these considerations offer a plausible basis for the known ability of auxin to maintain and stimulate its own transport, and are compatible with the electrical and auxin concentration oscillations that can occur in coleoptiles (1). Moreover, it might be expected that polar gradients of other lipophilic weak acids would arise in response to transcellular pH and potential differences brought about by polar auxin transport or other stimuli such as light or gravity in tropic tissues. Carriers are not necessary because differences in cellular apical and basal accumulation ratios would arise out of the cell's position in the proposed gradients of transport driving forces. Limited simulation studies (Rubery, unpubl.) suggest that the direction and intensity of such induced polarity would depend on such factors as the relative magnitudes of P_{AH} and P_{A^-}, of the electrical gradient, and of the symplastic (base alkaline) and apoplastic (base acidic) pH gradients: if P_{A^-}/P_{AH} is invariant with position (i.e. no carrier asymmetry, including the case $P_{A^-} = 0$), a change in the internal pH is needed to generate a substrate gradient in the symplast.

It is known that IAA will stimulate the basipetal polar transport of 2,4-D (30) which is sluggish alone, and may not involve strong interaction with the auxin anion carrier (5, 9). The scattered experimental observations of the polar transport of non-auxin growth substances (1) are inconsistent and erratic, perhaps partly because of undefined auxin status of the tissues used. Auxin induction of the polar transport of other lipid soluble weak acids such as abscisic acid, gibberellic acid, and benzoic acid is a testable prediction.

If borne out, these speculations could advance our understanding of the interaction of plant hormones in the control of growth and development. However, intracellular and extracellular pH, membrane potentials and hormone metabolism are subject to multiple interdependent control mechanisms; the dynamic situation in growing differentiating plant tissues will of course be exceedingly complex. Its further analysis will require a cooperative interplay of mathematical modelling, computer simulation, measurement of transport driving forces, and of hormone levels and profiles.

References

1. Goldsmith, M.H.M.: Annu. Rev. Plant Physiol. *28,* 439–478 (1977)
2. Rubery, P.H., Sheldrake, A.R.: Planta *118,* 101–121 (1974)
3. Raven, J.A.: New Phytol. *74,* 163–172 (1975)
4. Rubery, P.H., Sheldrake, A.R.: Nature (Lond.) New Biol. *224,* 285–288 (1973)
5. Rubery, P.H.: Planta *135,* 275–283 (1977)
6. Rubery, P.H.: Planta *142,* 203–206 (1978)
7. Rubery, P.H.: Planta *144,* 173–178 (1979)
8. Rubery, P.H.: Plant Sci. Lett. *14,* 365–371 (1979)
9. Davies, P.J., Rubery, P.H.: Planta *142,* 211–219 (1978)
10. Gutknecht, J., Tosteson, D.C.: Science *182,* 1258–1261 (1973)
11. Albaum, H.G., Kaiser, S., Nestler, H.A.: Am. J. Bot. *24,* 513–518 (1937)
12. Sutter, E.: Ber. Schweiz. Bot. Ges. *54,* 197–244 (1944)
13. Leopold, A.C., Hall, O.F.: Plant Physiol. *41,* 1476–1480 (1960)
14. De La Fuente, R.K., Leopold, A.C.: Plant Physiol. *41,* 1481–1484 (1960)
15. Raven, J.A.: New Phytol. *82,* 285–291 (1979)
16. Smith, F.A., Raven, J.A.: Encycl. Plant Physiol. New Ser. A *2,* 317–346 (1976)
17. Cande, W.Z., Ray, P.M.: Planta *129,* 43–52 (1976)
18. Sheldrake, A.R.: Planta *145,* 113–117 (1979)
19. Juniper, B.E.: Annu. Rev. Plant Physiol. *27,* 385–406 (1976)
20. Stein, W.D., Honig, B.: Mol. Cell Biochem. *15,* 27–44 (1977)
21. Cross, J.W., Briggs, W.R.: Plant Physiol. *62,* 152–157 (1978)
22. Dohrmann, U., Hertel, R., Kowalik, H.: Planta *140,* 97–106 (1978)
23. Jacobs, M., Hertel, R.: Planta *142,* 1–10 (1978)
24. Thomson, K.-S., Hertel, R., Muller, S., Tavares, J.E.: Planta *109,* 337–352 (1973)
25. Trillmich, K., Michalke, W.: Planta *145,* 119–127 (1979)
26. Ray, P.M., Dohrmann, U., Hertel, R.: Plant Physiol. *60,* 585–591 (1977)
27. Venis, M.A., Watson, P.J.: Planta *142,* 103–107 (1978)
28. Dohrmann, U., Hertel, R., Pesci, P., Cocucci, S.M., Marré, E., Randazzo, G., Ballio, A.: Plant Sci. Lett. *9,* 291–299 (1977)
29. Ho, M.K., Guidotti, G.: J. Biol. Chem. *250,* 675–683 (1975)
30. Hertel, R., Flory, R.: Planta *82,* 123–140 (1968)

Purification and Properties of Membrane-Bound Auxin Receptors in Corn

M.A. VENIS [1]

Introduction

Binding sites in corn coleoptile membranes with high affinity for auxins were first detected by Hertel et al. (1). Subsequently we reported evidence for two kinetic classes of auxin binding sites (2, 3) and described their solubilization and partial purification (4). Other workers could distinguish only a single kinetic class of binding sites, either in the membrane-bound (5) or solubilized state (6). On the other hand, Dohrmann et al. (7) have suggested that there are three types of auxin-binding sites, two of which have affinities for 1-naphthylacetic acid (NAA) comparable to those reported by ourselves. The third site, which preferentially binds 2,4-dichlorophenoxyacetic acid (2,4-D) and may be an auxin transport site (8), is only revealed under specialized assay conditions.

In attempting to resolve the question of the number of discrete auxin-binding species present in corn membranes, we have sought methods for their further purification and resolution. We also describe evidence bearing on the physiological relevance of the auxin-binding sites and inter-laboratory discrepancies in the molecular weight of solubilized binding proteins (4, 6).

Materials and Methods

Chemicals. 1-naphthyl-1-^{14}C-acetic acid (61 mCi/mmol) and 3-indolyl-1-^{14}C-acetic acid (52 mCi/mmol) were obtained from the Radiochemical Centre, Amersham. MBOA (6-methoxy-2-benzoxazolinone) and DMBOA (6,7-dimethoxy-2-benzoxazolinone) were isolated from corn tissues (9), BOA (2-benzoxazolinone) and 5-chloro-BOA (5-chloro-2-benzoxazolinone) were from Aldrich Chemicals, and DIBOA (2,4-dihydroxy-1,4-benzoxazin-3-one) and HBOA (2-hydroxy-1,4-benzoxazin-3-one) were obtained from Professor P. Sammes, The City University, London.

Membrane Preparation and Solubilization. Membranes of corn coleoptiles (*Zea mays*, cv. Kelvedon 33 or Blizzard) either with (solubilization and purification studies) or without (pellet binding assays) primary leaves were prepared by differential centrifugation between 4000 *g* and 80000 *g*. Procedures and media were as described previously (2), except that in many experiments 0.25 mM PMSF (phenylmethylsulfonyl fluoride)

1 Shell Biosciences Laboratory, Sittingbourne Research Centre, Sittingbourne, Kent, ME9 8AT, United Kingdom

was added to the homogenization buffer. Auxin-binding proteins were solubilized from acetone treated membranes (4) using standard pH 5.5. buffer (2) or, for ion exchange chromatography, this buffer fivefold diluted (DBB). Occasionally, for gel filtration experiments, proteins were solubilized in the appropriate column equilibration buffer (see Table 1). In some cases, designed to mimic the conditions of Cross and Briggs (6) as closely as possible, further minor modifications were introduced: homogenization medium contained 14 mM 2-mercaptoethanol, and tris-HCl rather than tris-acetate; pH of the resuspension medium was adjusted with citric acid in place of acetic acid; acetone treatment was at –15°C rather than 4°C.

Molecular Weight Estimation by Gel Filtration. Solubilized membrane extracts (1.0–1.5 ml, equivalent to 7–10 g tissue) were applied to columns of Sephadex G75 or G100 (Pharmacia) or Bio-Gel A 1.5 m or A 0.5 m (Bio-Rad), dimensions 1.5 cm × 29 cm or 1.5 cm × 85 cm. The columns were equilibrated and eluted either in standard citrate-acetate binding buffer, pH 5.5 (2) or in 10 mM tris-HCl, pH 7.6, with or without 0.1 M NaCl (6). Fractions of 1.1 ml were collected and 1 ml aliquots were assayed for auxin binding by equilibrium dialysis (4). Prior to assay, fractions of pH 7.6 were adjusted to pH 5.5 by addition of 25 μl of 0.2 M citrate-acetate, pH 5.0, 0.2 M $MgSO_4$ buffer. Columns were calibrated with proteins of known molecular weights.

Bioassays. Saturable binding of NAA-^{14}C to membranes (0.15 μM NAA-^{14}C ± 100 μM NAA) was determined by a pelleting assay (2). Solubilized binding proteins were assayed by equilibrium dialysis (4) against NAA-^{14}C (ca. 0.15 μM equilibrium concentration). Auxin-induced growth was determined at 0.3 μM IAA (3-indolyl-acetic acid) by using oat coleoptile sections (9).

Chromatographic Purification Methods. Initial purification of solubilized auxin-binding proteins was generally by step-gradient elution (4) on DEAE Bio-Gel (Bio-Rad). For resolution studies a 70 ml continuous gradient from 0.25–1.0 × standard binding buffer (containing 2.5–10 mM citrate) was applied to a 3 cm × 1 cm column. In all cases, samples were loaded in DBB and columns were initially eluted with the same buffer until the A_{280} (Isco UA-2 monitor) returned to baseline. Several liganded Sepharoses were investigated as aids to purification: (1) Poly(U)-Sepharose (Pharmacia), (2) Heparin-Sepharose, prepared by coupling heparin (Sigma, grade I) to CNBr-activated Sepharose 4B(10), (3) Cu^{2+}-iminodiacetic acid (IDA)-Sepharose, prepared by coupling IDA-Na_2 (Aldrich) to Epoxy-Sepharose (Pharmacia), then saturating ca. 60% of the column bed with $CuSO_4$, (4) Aminohexyl (AH)-Sepharose (Pharmacia). For preliminary studies, crude acetone powder extracts were applied to 4 cm × 1.5 cm columns of the adsorbents in binding buffer. The columns were eluted with binding buffer to baseline A_{280}, then step-wise with NaCl (0.1 M, 0.2 M, 0.5 M, 1.0 M) in binding buffer. With Cu^{2+}-IDA columns, the 1 M NaCl step was followed by 1 M NaCl-EDTA-Na_2. Fractions were collected and assayed for NAA-^{14}C binding and protein content (12). In sequential purification schemes the final step was gel filtration on 85 cm × 1.5 cm columns of Sephadex G100 or G75 or Sephacryl (Pharmacia). Aliquots of the 1.5–2.0 ml fractions were assayed for protein and for NAA-^{14}C binding and the remainder of the fractions with activity were concentrated by centrifugal ultrafiltration with CF25 membrane cones (Amicon). Concentrates were analysed by polyacrylamide gel electrophoresis according to the suppliers' instructions on either 7.5% Biophore gels (Bio-Rad) in tris-glycine or (after treatment with 1% SDS and 2% 2-mercaptoethanol, 90°C, 2 min) on 12% Biophore gels in tris-acetate-SDS. The latter gels were calibrated with bovine serum albumin, ovalbumin, chymotrypsinogen A and cytochrome c as m.w. standards.

Isoelectric Focusing. Solubilized auxin-binding proteins pre-purified by step-gradient elution on DEAE-Bio-Gel were precipitated with ammonium sulphate (80% saturation) and desalted on a 6 cm × 1.5 cm Sephadex G25 (Pharmacia) column equilibrated in DBB-5% v/v glycerol. Aliquots (600–900 μl) were adjusted to 1.5% w/v ampholyte and subjected to isoelectric focusing by the method of O'Brien et al. (13) using a column (18 cm × 1 cm) of Sephadex G15 (Pharmacia) equilibrated in 1.5% w/v ampholyte – 5% v/v glycerol. Experiments were run with ampholytes of various pH ranges, obtained from LKB, Pharmacia and Bio-Rad. After focusing at 400 v for 16 h at 2°C, the column was eluted in ampholyte-glycerol and alternate one-drop and five-drop (235 μl) fractions were collected. The former were used for pH estimation after addition of 1 ml of water, and the latter were assayed for A_{280} and NAA-^{14}C binding after buffering with 850 μl of 50 mM citrate-acetate binding buffer, pH 5.2 or 5.5. The pH range in the buffered fractions was pH 5.3–5.5.

Enzyme Assays. K^+-ATPase was assayed as previously described (3), except that 10 mM citrate-acetate pH 5.5 was used in place of tris-MES. Conjugation of IAA-^{14}C or NAA-^{14}C with glucose or myo-inositol was examined by the procedure of Kopcewicz et al. (14), except that assays were run at a range of pH values (pH 5–8) and for glucose ester assays, inositol was replaced by either UDPG, glucose-1-phosphate, glucose-6-phosphate or glucose (0.2 mM). Assays for auxin-aspartate conjugation were run over the same pH range in 1 ml reaction volumes containing 0.5 μCi IAA-^{14}C or NAA-^{14}C, potassium phosphate (10 mM), aspartic acid (1 mM), ATP (1 mM) and CoA (0.2 mM). After incubation at 30°C for 2 h, the mixture was acidified (pH 3), partitioned to ethyl acetate and examined by silica tlc in iso-propanol:ammonia:water (8:1:1), followed by radioscanning (15). IAA oxidase was assayed by incubation for 30 min at 30°C in 1 ml reaction volumes containing 50 mM citrate-acetate buffer, pH 5.5, 0.1 mM IAA, 0.5 mM hydrogen peroxide and 0.5 mM p-coumaric acid. IAA remaining was determined by adding 2 ml of Salkowski reagent (16) and measuring A_{530} after 30 min.

Results

Molecular Weight of Solubilized Binding Proteins

Auxin-binding proteins solubilized from corn membranes by an acetone powder method have consistently, in our hands, shown m.w. of ca. 40,000 by gel filtration on Sephadex G100 (4). However, Cross and Briggs recently reported a m.w. of 80,000 using our solubilization procedure and implied that the lower value we had observed may result from proteolysis, since the main departure in technique was inclusion of the protease inhibitor PMSF in the initial membrane homogenization medium (6). These workers used a different gel permeation column (Bio-Gel A 1.5 m), run at higher pH and ionic strength, claiming aggregation to ca. 200,000 m.w. in the absence of 0.1 M NaCl. To investigate this discrepancy, we have made m.w. comparisons under many different conditions, using ± PMSF extracts and varying the column material, pH and ionic strength.

Table 1. Gel filtration of crude acetone powder extracts

PMSF in extraction medium	Gel permeation conditions			Apparent m.w. at binding peak
	Column	Buffer	NaCl	
a. None	Sephadex G100	10 mM citrate-acetate, pH 5.5	0	43,800
b. 0.25 mM	Sephadex G100	10 mM citrate-acetate, pH 5.5	0	43,800
c. 0.25 mM	Sephadex G75	10 mM citrate-acetate, pH 5.5	0	39,300
d. 0.25 mM	Sephadex G75	10 mM tris-HCl, pH 7.6	0.1 M	42,000
e. 0.25 mM	Bio-Gel A 0.5 m	10 mM tris-HCl, pH 7.6	0.1 M	40,700
f. 0.25 mM	Bio-Gel A 1.5 m	10 mM tris-HCl, pH 7.6	0.1 M	41,000
g. 0.25 mM	Bio-Gel A 1.5 m	10 mM tris-HCl, pH 7.6	0	40,800

From Table 1, it is clear that irrespective of the conditions used, whether our original ones ± PMSF (a, b), those of Cross and Briggs (f, g) or other combinations (c–e), the auxin-binding proteins have a m.w. around 40,000, with no evidence of aggregation at low ionic strength (a–c, g). Comparable results have been obtained with two different corn varieties and with post-DEAE fractions as well as with crude extracts. Mixing the sample with blue dextran (procedure of Cross and Briggs) was also without effect on the elution behaviour. The source of the m.w. discrepancy thus remains obscure, but does not appear to be dependent on the use of PMSF.

Purification and Resolution

We have described the partial purification of solubilized auxin-binding proteins by step-gradient ion exchange and gel permeation chromatography (4). Investigation of additional means of purification revealed several liganded Sepharose columns that were effective (see Methods section). When extracts were fractionated on columns of poly (U)-, heparin-, or Cu^{2+}-IDA-Sepharose, auxin-binding activity was found exclusively in the non-retained fractions eluted with starting buffer, separated from indifferent proteins eluted subsequently with NaCl. On AH-Sepharose, the binding proteins appear in the 0.1 M NaCl eluate.

By combining two of these chromatographic steps with the previously used ion-exchange (DEAE Bio-Gel) and gel permeation methods (Sephadex G100 or Sephacryl), auxin-binding proteins of high purity have been obtained (Fig. 1, Table 2). The most purified fractions yield two predominant protein bands on both non-dissociating and SDS polyacrylamide gels. The apparent m.w. of the SDS gel bands are 46,000 and 40,000, agreeing with estimates from gel filtration [(4) and Table 1] and indicating that these are monomeric proteins. The highest specific activities obtained in the final

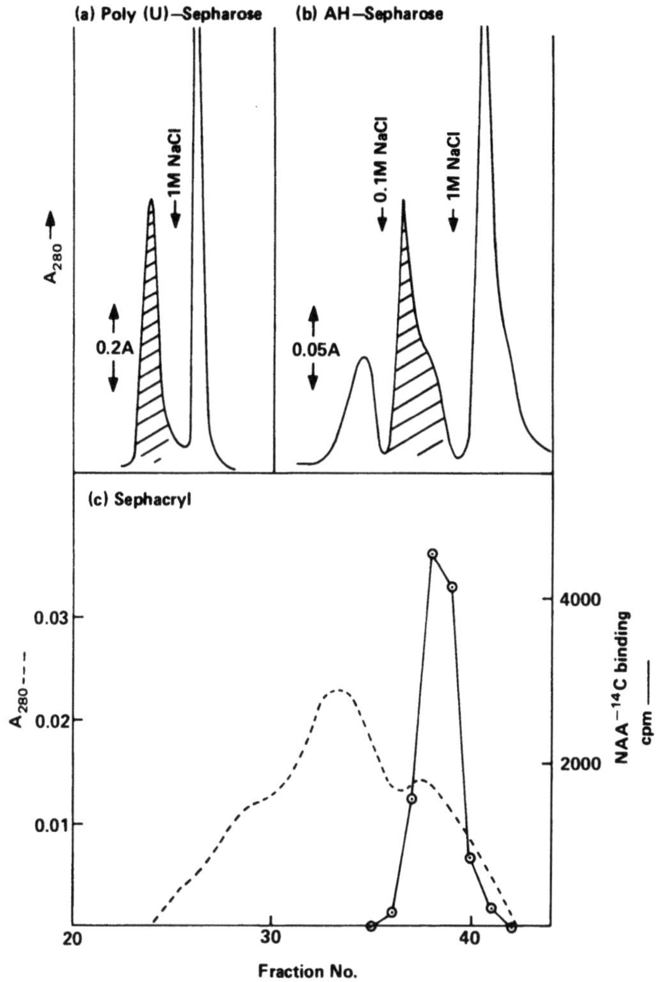

Fig. 1. Purification of post-DEAE auxin-binding proteins. The *shaded peaks* in **a** and **b** denote the auxin-binding fractions that were pooled and applied to the subsequent column

Table 2. Purification of auxin-binding proteins from corn membranes (260 g)

Fraction	Total binding dpm	Total protein mg	Specific activity dpm/mg
Crude extract	559,747	194.50	2,878
DEAE Bio-Gel	303,398	8.21	36,943
Poly(U)-Sepharose	216,000	1.13	192,000
AH-Sepharose	141,704	0.54	262,416
Sephacryl	74,666	0.15	493,823 (mean)
			564,553 (max)

purification step (Table 2) closely approach the theoretical maximum and it is therefore tempting to suggest that both the bands seen on acrylamide gels are auxin-binding proteins. Unfortunately, attempts to resolve this question unambiguously by eluting the bands individually from non-dissociating gels have so far been unsuccessful, in that no binding activity could be recovered.

In an alternative approach to the question of the number of discrete auxin-binding proteins present in corn membranes, we have sought other means of resolution that might permit recovery of biological activity. Since two major proteins could be resolved electrophoretically on acrylamide gels, the O'Brien et al. (13) variant of isoelectric focusing was the method selected initially. This technique uses a column of Sephadex G15 as the anti-convectant instead of a sucrose gradient and permits the analysis and recovery in low volumes of small amounts of protein. Focusing of post-DEAE fractions using a variety of LKB ampholytes showed that the pH 4–6 range yielded good recovery of binding activity (60%–100%) but imperfect resolution – generally one main peak with a shoulder. With the pH 3.5–5 range, two main peaks of auxin-binding activity are consistently seen (Fig. 2), but the overall recovery of activity is always

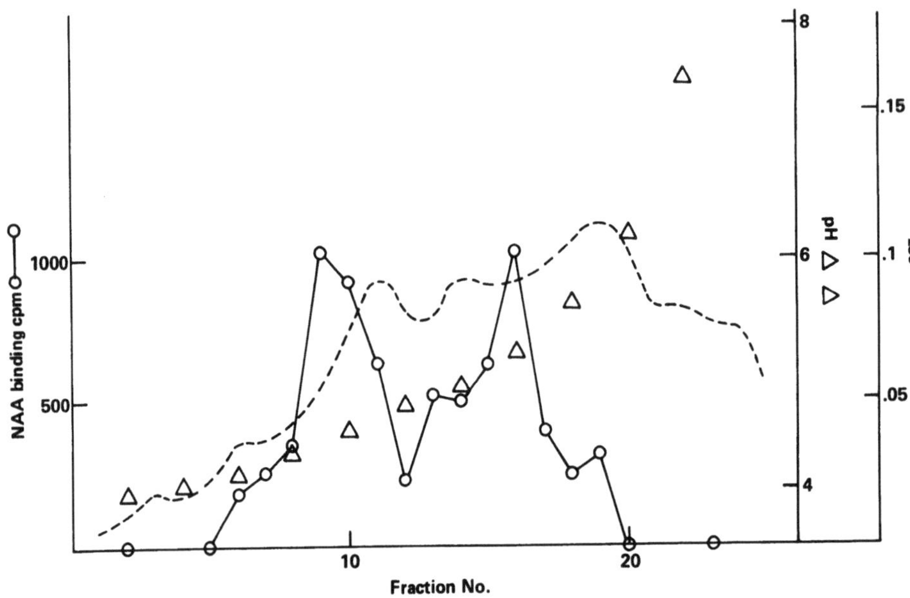

Fig. 2. Isoelectric focusing of auxin-binding proteins

very low (10%–15%). Similar results have been obtained with low pH range ampholytes from other suppliers (Pharmacia, Bio-Rad). The ampholyte mixture does not itself significantly affect auxin binding by a prefocused extract and various modifications in technique have failed to improve recovery. The isoelectric focusing results are thus suggestive of more than one auxin-binding species, but the low recoveries preclude the use of this technique for their preparative resolution.

Step-gradient elution on DEAE Bio-Gel has been our standard initial purification step. Preliminary work suggested that there was no advantage in using continuous gradients, but further investigation has shown that with a shallower gradient, from 0.25–1.0 x standard binding buffer, the auxin-binding activity can be resolved into two main peaks (Fig. 3). We now plan to determine whether these represent distinct binding proteins.

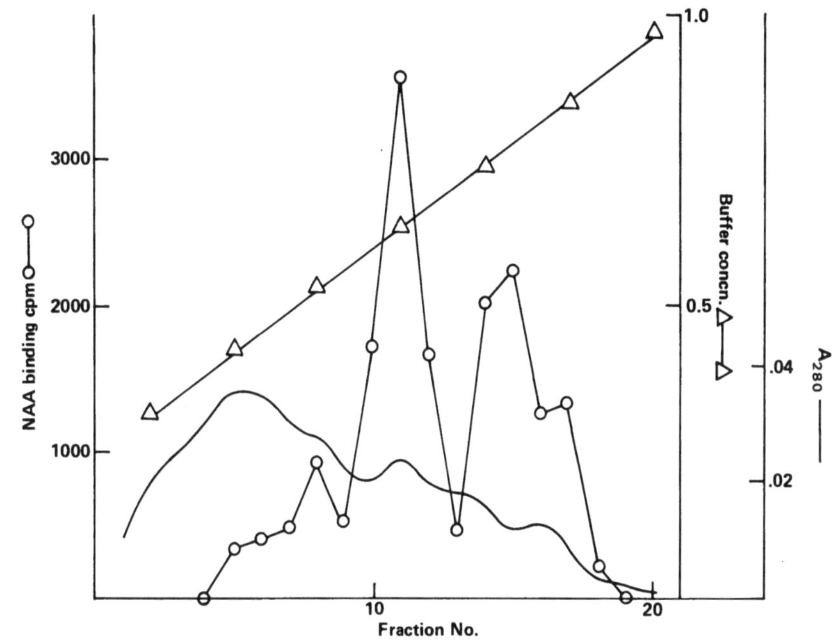

Fig. 3. Gradient chromatography of auxin-binding proteins on DEAE Bio-gel

Enzymic Activities

Could the auxin-binding proteins be auxin-metabolising enzymes rather than receptor proteins? Auxins such as IAA and NAA are known to form conjugates with aspartic acid and sugars; assays for such activities in membranes or membrane extracts have proved negative. Crude membrane extracts do contain IAA oxidase activity (totally peroxide-dependent) and a portion of this activity remains associated with the binding proteins through successive purification on DEAE Bio-Gel, poly(U)-Sepharose and AH-Sepharose. Upon further chromatography on Sephadex G75, the residual IAA oxidase appears predominantly in a symmetrical peak that overlaps, but is clearly non-coincident with the auxin-binding peak (Fig. 4). Furthermore, horseradish peroxidase (Sigma, 200 μg/ml) exhibits no NAA-binding activity.

In the light of proton pump theories of auxin action (17), there is a possibility that the auxin-binding proteins could be associated with ATPase activity. Acetone powder extracts of corn membranes do show such activity, but in agreement with the findings of Cross et al. (18) for Triton extracts, ATPase and auxin-binding activities are resolved from one another on gel permeation columns.

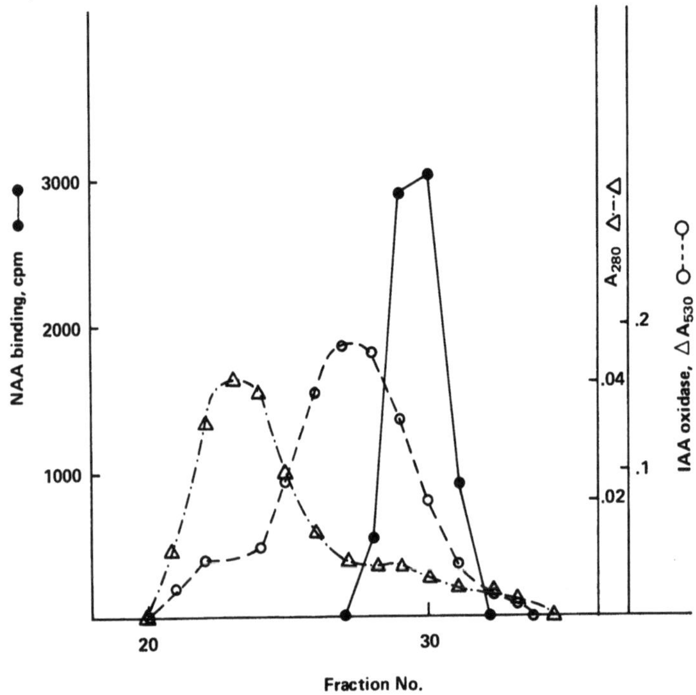

Fig. 4. Distribution of IAA oxidase and auxin-binding activities following chromatography on Sephadex G75

Modifiers of Auxin Binding

Corn coleoptiles contain a low molecular weight modifier of auxin binding, termed supernatant factor (SF), which reduces the binding affinity of the membrane sites for NAA (5). It was reported (19) that the affinity for certain physiologically inactive ligands is lowered to an even greater extent, while the binding affinity for some auxin analogs, e.g., 2,4-D is actually increased by SF. The suggestion has therefore been made (5) that SF may be a natural regulator of auxin activity. We were able to isolate and identify SF (9) as a mixture of 6-methoxy-2-benzoxazolinone (MBOA) and 6,7-dimethoxy-2-benzoxazolinone (DMBOA). The parent compound 2-benzoxazolinone (BOA) was also found, but in much smaller amount. These are known natural products in corn and in certain other Gramineae, and they have been implicated in resistance of cereals to fungi, insects and chloro-s-triazine herbicides [(20) and references therein]. In the intact plant the compounds occur as benzoxazinone glucosides, from which free benzoxazinones are released enzymically upon cell damage, e.g., by pathogen attack or by homogenization. On warming in dilute aqueous solution these aglucones undergo ring contraction to produce the benzoxazolinones.

MBOA and DMBOA were isolated by following inhibition of NAA-membrane binding during purification, a convenient measure of SF activity. Both compounds are active in such assays, DMBOA being several times more active than MBOA; BOA is only weakly active. The compounds were also found to inhibit auxin induced growth of oat coleoptiles, and their relative activity in such assays correlated well with their inhibition of auxin-membrane binding (9). From Table 3 it can be seen that this general correlation extends to the three additional analogs that have been tested: 5-chloro-BOA, DIBOA (the benzoxazinone precursor of BOA) and HBOA (the lactam analog of DIBOA).

Table 3. Inhibitory effects of benzoxazolinone analogs (10 μg/ml) on NAA binding to corn coleoptile membranes and on IAA-induced growth of oat coleoptiles

Compound	DMBOA	DIBOA	MBOA	HBOA	5-Cl-BOA	BOA
% Inhibition of binding	69	48	36	14	12	8
% Inhibition of growth	78	46	44	29	26	17

Discussion

We have been unable to detect any auxin-binding proteins of m.w. in excess of 40,000 –45,000, even when PMSF is included during membrane homogenization (Table 1). Cross and Briggs (6) state that their buffer was saturated before use with PMSF, without specifying the actual concentration. We have used 0.25 mM PMSF, which is probably close to saturation since prolonged stirring at room temperature is required to effect solution. We conclude, therefore, that the m.w. values we consistently observe under a wide variety of conditions (Table 1) are not the result of proteolysis. Since the Cross-Briggs material aggregates at low ionic strength (unlike ours), it may be that their 80,000 m.w. species is itself an aggregate and that aggregation is a feature of the corn variety used.

The suggestion from the acrylamide gel patterns that there are two auxin-binding proteins is supported by the resolution of two auxin-binding peaks either by isoelectric focusing (Fig. 2) or by ion exchange chromatography (Fig. 3). Further purification of the individual peaks obtained by the latter technique should enable this question to be addressed with greater confidence.

The auxin-binding proteins do not appear to be associated with ATPase, IAA oxidase or auxin conjugating enzymes. Negative evidence of acyl aspartate synthesis is equivocal, since no one has been able to demonstrate aspartate conjugation in vitro. It is possible, therefore, that our cell-free conditions were inappropriate for the detection of this activity. However, tissues that are known to conjugate auxins efficiently with aspartate (e.g., peas) have much reduced auxin-binding activity relative to corn, and auxin induction of the aspartate conjugation system (15) does not improve auxin binding by pea tissues.

Identification of supernatant factor as a mixture of compounds of the benzoxazolinone family has enabled a connection to be made between auxin binding and a physiological response. Since the species distribution of these compounds is known to be very restricted (9), they cannot play a general role in auxin physiology. Nevertheless, the correlation found between the activities of benzoxazolinones and analogs in inhibiting auxin-binding site interaction and auxin-induced growth [Table 3 and (9)] provides persuasive circumstantial evidence that these binding sites represent physiological receptor sites for auxin action.

Acknowledgment. I wish to thank Richard Bumpus for able technical assistance.

References

1. Hertel, R., Thomson, K., Russo, V.E.A.: Planta *107*, 325–340 (1972)
2. Batt, S., Wilkins, M.B., Venis, M.A.: Planta *130*, 7–13 (1976)
3. Batt, S., Venis, M.A.: Planta *130*, 15–21 (1976)
4. Venis, M.A.: Nature (Lond.) *266*, 268–269 (1977)
5. Ray, P.M., Dohrmann, U., Hertel, R.: Plant Physiol. *59*, 357–364 (1977)
6. Cross, J.W., Briggs, W.R.: Plant Physiol. *62*, 152–157 (1978)
7. Dohrmann, U., Hertel, R., Kowalik, H.: Planta *140*, 97–106 (1978)
8. Jacobs, M., Hertel, R.: Planta *142*, 1–10 (1978)
9. Venis, M.A., Watson, P.J.: Planta *142*, 103–107 (1978)
10. Molinari, A.M., Medici, N., Moncharmont, B., Puca, G.A.: Proc. Natl. Acad. Sci USA *74*, 4886–4890 (1977)
11. Porath, J., Carlsson, J., Olssen, I., Belfrage, G.: Nature (Lond.) *258*, 598–599 (1975)
12. Lowry, O.H., Rosebrough, J.J., Farr, A.L., Randall, R.J.: J. Biol. Chem. *193*, 265–275 (1951)
13. O'Brien, T.J., Liebke, H.H., Cheung, H.S., Johnson, L.K.: Anal. Biochem. *72*, 38–44 (1976)
14. Kopcewicz, J., Ehmann, A., Bandurski, R.S.: Plant Physiol. *54*, 846–851 (1974)
15. Venis, M.A.: Plant Physiol. *49*, 24–27 (1972)
16. Gordon, S.A., Weber, R.P.: Plant Physiol. *26*, 192–195 (1951)
17. Hager, A., Menzel, H., Krauss, A.: Planta *100*, 47–75 (1971)
18. Cross, J.W., Briggs, W.R., Dohrmann, U.C., Ray, P.M.: Plant Physiol. *61*, 581–584 (1978)
19. Ray, P.M., Dohrmann, U., Hertel, R.: Plant Physiol. *60*, 585–591 (1977)
20. Klun, J.A., Tipton, C.L., Robinson, J.F., Ostrem, D.L., Beroza, M.: J. Agric. Food Chem. *18*, 663–665 (1970)

Auxin and H^+-Excretion: The State of Our Knowledge [1]

R.E. CLELAND [2]

For more than 50 years scientists have been attempting to explain how auxin induces cell elongation in stem and coleoptile cells. Many mechanisms have been proposed only to be discarded with amazing rapidity. The theory receiving the most attention at present is the acid-growth theory which states (1) that auxin causes cells to excrete protons and that the resulting lowered pH of the wall solution activates some enzyme capable of cleaving load-bearing bonds in the cell wall. Since this theory was first formulated in 1970–71 (2, 3), a large body of papers have appeared, both attacking and supporting this theory, with the result that considerable confusion exists. The purpose of this article is to summarize some of the conflicting information, and attempt to show why problems have arisen and how they can be overcome.

To determine whether the acid-growth theory is applicable to a tissue the elongation of which is influenced by auxin, one may use the following four predictions which arise from the theory:

1) auxin must cause the cells to excrete protons; 2) exogenous protons must be able to substitute for auxin in causing cell wall loosening and growth; 3) neutral buffers infiltrated into the walls must inhibit auxin-induced growth; and 4) other agents which induce H^+-excretion, such as the phytotoxin fusicoccin (FC), must cause a parallel promotion of growth. If each of these four predictions applies, one can say with some confidence that the auxin-induced growth occurs via acid-growth.

It may sound easy to test these four predictions, but in fact it is difficult to do so, and contradictory results have often been obtained. To illustrate this, consider the auxin-induced growth of soybean hypocotyl sections. Does auxin cause these cells to excrete protons? Vanderhoef et al. (4) reported that the pH of the solution surrounding soybean hypocotyl sections equilibrated at 5.3, with or without auxin. But David Rayle and I (5), using a different technique, found that in the absence of auxin the pH external to the epidermal walls was 5.8, and that upon addition of auxin the pH began to drop after a lag of about 10–12 min and finally equilibrated at a pH below 4.8. Why were such different results obtained in these two studies? To understand this we must consider the difficulties in measuring proton excretion. If proton excretion is to be measured, the protons must escape the tissue and migrate to the measuring pH electrode. Some of the excreted protons will be absorbed onto the fixed carboxyl groups of the wall and thus never escape. If the walls are prewashed with acid to saturate these carboxyl groups, auxin-induced H^+-excretion is more readily detected (6).

---

1 Abbreviations: CHI, cycloheximide; DCCD, dicyclohexylcarbodiimide; FC, fusicoccin, ER, endoplasmic reticulum; MDMP, 2-(4-methyl-2,6-dinitroanilino)-N-methyl proprionamide

2 Department of Botany, University of Washington, Seattle, WA 98195, USA

Those protons which escape absorption must now diffuse to the electrode; the timing and amount of H^+-excretion will depend on the distance the protons must diffuse and the solution/tissue volume ratio. Thus auxin-induced H^+-excretion is detected more rapidly when microelectrodes are inserted into the tissue (7) or when an electrode is placed in direct contact with the cell surface (8), or if the solution/tissue volume ratio is decreased by increasing the number of sections per ml of solution (9). But the most important factor affecting detection of proton excretion is the cuticle. Sheryl Dreyer (10) has examined the permeability of soybean and sunflower hypocotyl cuticles to protons and has shown that their permeability to protons is extremely low, on the same order as that reported for other cations such as K^+ (11). Thus the intact cuticle cannot help but trap most of the excreted protons inside the tissue, preventing their detection. If the cuticle is abraded with emery powder (scrubbed), the permeability is greatly increased; SEM pictures show that scrubbing results in many minute holes and tears in the cuticle (10). Alternatively, the cuticle of coleoptiles can be removed by peeling the epidermal layer from the coleoptile (peeled).

The reason that Rayle was able to detect auxin-induced H^+-excretion in soybean hypocotyl sections is that he employed conditions which maximized his chances of detecting the proton excretion; namely, scrubbed sections, a surface-applied electrode and a minimal volume of solution. Vanderhoef et al. (4), on the other hand, used unscrubbed sections, an external electrode and a much larger solution/tissue volume ratio.

One might expect to be able to detect auxin-induced H^+-excretion, even when the cuticle is intact, because of the escape of protons from the cut ends. Indeed, Marré et al. (12) have demonstrated auxin-induced proton excretion from intact pea epicotyl tissues. But in many tissues this may not be possible for the following reason. In dicot tissues the control of auxin-induced growth resides in only the outer epidermal and subepidermal cell layers (13, 14); this is the basis for the split-pea test for auxin (15). The cortical cells, which comprise the majority of most stems, apparently have low or no auxin sensitivity. Every cell comes to a pH equilibrium with its external solution at some pH determined by the ratio of proton excretion and proton absorption (16). For auxin-insensitive cells this equilibrium pH is in the 5.5–6.0 range, while for cells undergoing auxin-induced H^+-excretion it is in the 4.5–5.0 range (16). If the auxin-insensitive cells predominate, they may simply reabsorb protons excreted from the auxin-sensitive cells so rapidly that no auxin-induced acidification of the external medium can be detected. The wall solution around the auxin-sensitive cells will still be in the 4.5–5.0 range, however.

Many tissues have now been examined for auxin-induced H^+-excretion; some of the results are summarized in Table 1. In general, rapid auxin-induced proton excretion has been detected whenever the experimenters used scrubbed sections (or for coleoptiles, peeled tissue) and a low solution/tissue volume ratio, while proton excretion was not detected when the sections had an intact cuticle, where a large number of auxin-insensitive cells were exposed to the solution, or where microelectrodes were placed in auxin-insensitive regions of the tissue.

Acid-induced extension is easier to measure. An effective technique (29) is to place freeze-thawed sections under constant stress (10–20 g load) in a pH 7 buffer until the viscoelastic extension is finished (30–60 min), then to change to a more acidic solution

Table 1. Measurements of rapid auxin-induced H^+-excretion

Tissue	Conditions	References
A. Successfully detected:		
Avena coleoptile	Peeled	(17, 18)
	Scrubbed	(19)
Zea mays coleoptile	Intact	(20)
	Cell sheets	(21)
	Microelectrode	(7)
Pisum sativum epicotyl	Intact	(12)
	Microelectrode	(7)
	Scrubbed	(15)
	Protoplasts	(22)
	Peeled	(23)
	Epidermal strip	(24)
Helianthus annuus hypocot.	Scrubbed	(25)
Glycine max hypocot.	Scrubbed	(5)
Phaseolus vulgaris hypocot.	Scrubbed	(25)
Phaseolus aureus hypocot.	Scrubbed	(26)
Cucurbita pepo hypocot.	Scrubbed	(25)
Cucumis sativa hypocot.	Scrubbed	(25)
Ipomoea tricolor hypocot.	Scrubbed	(25)
B. Not detected:		
Avena coleoptile	Microelectrode	(26)
Lupinus angustifolius hypocot.	in xylem	(26)
Pisum sativum epicotyl	Peeled/intact	(27)
Helianthus annuus hypocot.	Bisected	(28)
Glycine max hypocot.	Intact	(4)

and measure the extension rate after some standard time (e.g., 20 min). Since the cuticle is just as effective in preventing proton entry as it is in preventing proton exit, the cuticle must be rendered permeable, or it must be peeled off. Scrubbed soybean hypocotyls show acid-induced extension whenever the pH is below 6.5, with the magnitude of the acid response increasing down to a pH of at least 3.0 (5). A wide variety of tissues have now been examined [e.g., (25)] and all show good acid-induced extension in response to pH of 4.5–5.0, the pH range expected for the wall solution of auxin-treated cells.

Although the ability of neutral buffers to inhibit auxin-induced growth is one of the most diagnostic tests for acid-growth, it has been used only rarely. Since buffers penetrate the cuticle even more poorly than protons (10), and penetrate a section only slowly via the cut ends (30), it is necessary to remove or at least scarify the cuticle in order to allow the buffer to infiltrate the walls of the auxin-sensitive cells. Thus, it should be no surprise that the auxin-induced growth of intact sections is unaffected by neutral buffers (31). In our experience, organic buffers such as MES-tris, and HEPES-NaOH are more effective growth inhibitors than phosphate buffer at the same pH and buffer strength (5). In such studies one must remember to include as a control a salt or osmoticum at the same osmolarity as the buffer being tested.

Durand and Rayle (15) first showed that neutral buffers can inhibit auxin-induced growth of *Avena* coleoptile and pea epicotyl sections, and we have confirmed these results. The results with soybean hypocotyl tissues depend upon whether or not the cuticle is intact; neutral buffers either had no effect or inhibited only during the first hour of growth when intact sections were used (32), but inhibited auxin-induced growth for at least the first 6 h when the cuticle was scrubbed (5). It is clear that buffer inhibition experiments need to be conducted on a much wider range of auxin-sensitive tissues.

The ability of FC to induce parallel levels of H^+-excretion and elongation in auxin-sensitive tissues was first demonstrated by Marré et al. (12) with pea epicotyl sections, and similar results have since been obtained with a variety of tissues including *Avena* coleoptiles (33), soybean hypocotyls (5) and *Phaseolus aureus* hypocotyls (34).

In conclusion, the four predictions concerning the acid-growth theory have been tested on only three tissues; *Avena* coleoptiles, pea epicotyls and soybean hypocotyls. In each case all four predictions were confirmed. More tissues must be examined in detail before we can conclude that auxin-induced growth, in general, involves excretion of protons. Of more significance is the fact that no auxin-sensitive tissue has failed to meet the criteria for acid-growth.

Let us now consider a closely related question: is H^+-excretion the only auxin-mediated process needed for cell elongation? Growth rate curves provide evidence that cell elongation involves more than one step. Following addition of auxin the growth rate rises to a peak, and then often falls before rising again to a second plateau (35, 36). These two peaks differ in sensitivity to inhibition by cytokinins (36) and cycloheximide (37). In addition, exogenous acid mimics only the first peak; the growth rate rises rapidly in response to acid but then falls and no second growth rate peak occurs (32). For these reasons it has been suggested (32) that the first peak may be due to auxin-induced proton excretion, but the second is due to some completely different auxin-mediated process. The fact that neutral buffers inhibit auxin-induced growth during both the first and second peaks (5) indicates that H^+-excretion is required for growth, whenever it occurs. We suggest that the first peak of growth rate occurs in response to auxin-induced H^+-excretion alone, but that the second peak requires both proton excretion and a second, auxin-mediated factor.

What is that second factor? Consideration of the equation governing cell elongation, ($dV/dt = \frac{Lp \cdot WEx \cdot (P-Y)}{Lp + WEx}$, where $WEx = f(\Delta H^+, WLC)$ = wall extensibility, dV/dt = growth rate, Lp = hydraulic conductivity, P = turgor pressure, Y = wall yield stress, ΔH^+ = proton excretion, and WLC = wall loosening capacity) indicates that in addition to proton excretion, four factors could be controlling the rate of cell elongation; namely, the hydraulic conductivity, the turgor pressure, the wall yield stress, and the wall-loosening capacity – the capacity of the wall to be loosened in response to acid. We are currently examining the effect of auxin on each of these four factors in *Avena* coleoptiles with the goal of determining which factor or factors is influenced by auxin so as to cause the second peak in the growth rate curve (38). While the studies are far from complete, certain trends have become obvious. We find no effect of auxin on the wall yield stress. The hydraulic conductivity of coleoptile sections has yet to be measured. Boyer and Wu (39) found that auxin increases the hydraulic conductivity of soybean seedlings and suggest that this may be a major factor in controlling elongation in intact plants. It seems doubtful, however, that the hydraulic conductivity will be as important for sections floating on water, but such measurements certainly must be made. Turgor pressure is primarily determined by the osmotic concentration of the cells (assuming a high hydraulic conductivity). As a cell takes up water during growth, the osmotic solutes become diluted and the turgor pressure falls, unless osmoregulation takes place; i.e., unless sufficient osmotic solutes are taken up or formed in a cell to maintain turgor. Auxin-induced growth of *Avena* coleoptiles requires no osmoregulation for the first 3–4 h, but thereafter if a rapid growth rate is to be maintained, osmoregulation must occur (38). Sections incubated with sugars or salts take up more solutes in the presence of auxin, but this is actually a response to the increased growth rather than to auxin itself as shown by the fact that the extra sugar uptake is inhibited whenever auxin-induced growth is inhibited by calcium ions or by osmotica (38).

The one factor which changes in response to auxin is the wall-loosening capacity (WLC). WLC is determined by freezing-thawing sections after the desired incubation, and then measuring their extension rate when placed under constant tension and acidic (pH 4.0) conditions. In the absence of auxin the WLC of *Avena* coleoptile sections decreases steadily until after 12 h it is only 50% of its initial value. But in the presence of auxin the WLC increases by at least 50% over 4 h, and then remains at this elevated level (38). This is a response to auxin, as the increase in WLC does not occur when growth is promoted by exogenous protons or by FC. We would suggest that the growth rate during the second peak is controlled by a combination of factors: by auxin-induced H^+-excretion, by auxin-induced increase in WLC, and by a growth-associated osmoregulation.

Finally, let us consider the mechanism by which auxin causes cells to excrete protons. A number of possible mechanisms have been proposed involving respiratory CO_2, organic acid excretion, electron transport pathway, H^+/K^+ antiport, PM-associated ATPase, or the bucket-brigade, but most can be discarded. For example, auxin causes an increase in respiration in many cells (40), and the increased respiratory CO_2 must contribute to the acidity of the wall solution. The fact that after auxin-induced acidification, removal of CO_2 from the medium does not increase the pH significantly indicates that most of the acidification occurs by some other pathway (41). A second possibility is that proton excretion occurs by means of a plasma membrane-associated electron transport pathway; components of such a pathway have been demonstrated, for example, in a plasma membrane preparation from corn coleoptiles (42). However, salicylhydroxamic acid, a potent inhibitor of this electron transport pathway, is relatively ineffective as an inhibitor of auxin-induced H^+-excretion (38). A third possibility is that H^+-excretion occurs by an electroneutral H^+/K^+ antiport. This seems unlikely because auxin causes a hyperpolarization of the membrane potential (43), and because the effect of auxin on K^+ uptake is poorly correlated with its effect on H^+-excretion, at least during the first hour (44).

Hager et al. (3) suggested that auxin activates a plasma membrane-associated, electrogenic ATPase. This is still the most widely considered mechanism. It is supported by some reports of in vitro stimulation of ATPases by auxin (45,46). The responses, however, tend to be small and difficult to reproduce. In addition, the auxin-binding sites are clearly different from the ATPases of corn coleoptiles (47). This mechanism is also supported by the fact that auxin-induced H^+-excretion is inhibited by inhibitors of ATP synthesis (17, 18) and by ATPase inhibitors such as DES and DCCD (48). But this theory fails to account for two important characteristics of the auxin response. First, there is a 8–12 min lag between addition of auxin to a tissue and the start of rapid cell elongation (49). Only under exceptional circumstances can this lag be reduced significantly. The lag is not a function of the time needed for auxin to penetrate cells, but apparently reflects the time needed for some event to occur within the cells. There is no apparent reason why such a lag would occur if the principal action of auxin is on a plasma membrane-associated ATPase. Second, auxin-induced H^+-excretion and growth require the continued synthesis of new proteins (50); inhibition of protein synthesis by CHI or MDMP stops auxin-induced growth after lags of only 5–15 min (51). Bates and Cleland (52) have examined the effects of auxin on the patterns of protein synthesis in *Avena* coleoptiles and have concluded that the proteins required for cell elongation are being synthesized both in the presence and absence of auxin. Again, there is no apparent reason why auxin activation of an ATPase would require protein synthesis. Thus, some modification of the ATPase mechanism is needed to account for these two characteristics.

One possibility is the "bucket-brigade" mechanism (53). This assumes that the ER is producing vesicles (buckets) which are moving to and fusing with the plasma membrane, where they empty their contents into the wall solution. It is proposed that the ER vesicles contain the ATPases which, when activated by auxin, pump protons into the vesicles. The contents of the vesicles, in the absence of auxin, would be around pH 7, but in the presence of auxin the contents would be strongly acidic.

The attractiveness of this mechanism is that it accounts for both the requirement for protein synthesis and for the lag. Continued production of new vesicles would certainly require protein synthesis, since any vesicle could be used only once. The lag would reflect the time required to fill a vesicle with protons and then move it to the plasma membrane. This theory accounts for the fact that auxin binding sites in corn coleoptiles are primarily localized at the ER (53), and is compatible with auxin binding sites at the plasma membrane (54), as these would presumedly be remnants from already-fused vesicles.

However, the theory is not without its problems. First, in order for sufficient protons to be transported, the pH within each vesicle would have to be $\leqslant 2$. This is possible, as stomach mucosa vesicles have a pH at least this low (55). Second, inhibition of protein synthesis would have to stop the formation and movement of ER vesicles. Inhibition of protein synthesis does not appear to prevent wall matrix synthesis in pea stems (56) which presumedly involves movement of Golgi vesicles to the plasma membrane, but the situation for ER vesicles may be different. Finally, the ability of auxin to induce H^+-excretion appears to be influenced by the pH of the external medium, in that no auxin-induced proton excretion can be detected when the pH is 4.5 or below (16). It is difficult to see how the external pH could regulate the rate at which protons are pumped into ER vesicles, but it could alter the rate at which vesicles fuse with the plasma membrane or open to the wall solution.

The bucket-brigade is a viable and interesting theory, but it lacks concrete evidence in its support. What is needed now is an EM study of ER vesicles, and histochemical determination of their pH in the presence and absence of auxin. Such measurements could determine whether this theory can still be retained, or discarded as have so many of its predecessors.

In summary, we must note that after 50 years we still do not know the mechanism of auxin action. The current theory, the acid-growth theory, has already enjoyed a longer than usual lifetime and seems to be getting stronger each year. But it does not explain all aspects of auxin-induced cell elongation, and the mechanism by which auxin causes proton excretion is far from understood.

Acknowledgment. Unpublished research described in this article was supported by Contract EY-76-06-2225-T19 from the U.S. Department of Energy.

References

1. Rayle, D.L., Cleland, R.E.: Curr. Top. Dev. Biol. *11*, 187–214 (1977)
2. Rayle, D.L., Cleland, R.E.: Plant Physiol. *46*, 250–253 (1970)
3. Hager, A., Menzel, H., Krauss, A.: Planta *100*, 47–75 (1971)
4. Vanderhoef, L.N., Findley, J.S., Burke, J.J., Blizzard, W.E.: Plant Physiol. *59*, 1000–1003 (1977)
5. Cleland, R.E., Rayle, D.L.: Plant Physiol. *61*, 43 (1978)
6. Cohen, J.D., Nadler, K.D.: Plant Physiol. *57*, 347–350 (1975)
7. Jacobs, M., Ray, P.M.: Plant Physiol. *58*, 203–209 (1976)
8. Cleland, R.E.: Plant Physiol. *58*, 210–213 (1976)
9. Rubinstein, B.: Unpublished data
10. Dreyer, S.A., Seymour, V., Cleland, R.E.: Submitted for publication (1980)
11. McFarlane, J.C., Berry, W.L.: Plant Physiol. *53*, 723–727 (1974)
12. Marré, E., Lado, P., Rasi Caldogno, F., Colombo, R.: Plant Sci. Lett. *1*, 179–184 (1973)
13. Diehl, J.M., Gorter, C.J., van Itersen Jr., G., Kleinhoonte, A.: Rec. Trav. Bot. Neerl. *36*, 709–798 (1939)
14. Masuda, Y., Yamamoto, R.: Plant Cell Physiol. *27*, 109–115 (1972)
15. Durand, H., Rayle, D.L.: Planta *114*, 185–193 (1973)
16. Cleland, R.E.: Planta *127*, 233–242 (1975)
17. Rayle, D.L.: Planta *114*, 63–73 (1973)
18. Cleland, R.E.: Proc. Natl. Acad. Sci. USA *70*, 3092–3093 (1973)
19. Pike, C.S., Richardson, A.E.: Plant Physiol. *57*, 615–617 (1977)
20. Polevoy, V.V., Salamatova, T.S.: Fiziol. Rast. *22*, 519–522 (1975)
21. Darvill, A.G., Smith, C.J., Hall, M.A.: New Phytol. *80*, 503–516 (1978)
22. Böttger, M., Engvild, K.C., Soll, H.: Planta *140*, 89–92 (1978)
23. Yamamoto, R., Maki, K., Yamagata, Y., Masuda, Y.: Plant Cell Physiol. *15*, 823–831 (1974)
24. Yamagata, Y., Masuda, Y.: Plant Cell Physiol. *17*, 1235–1242 (1976)
25. Mentze, J., Raymond, B., Cohen, J.D., Rayle, D.L.: Plant Physiol. *60*, 509–512 (1977)
26. Penny, P., Dunlop, J., Perley, J.E., Penny, D.: Plant Sci. Lett. *4*, 35–40 (1975)
27. Parrish, D.J., Davies, P.J.: Plant Physiol. *59*, 574–578 (1977)
28. Ilan, I.: Physiol. Plant. *28*, 146–148 (1973)
29. Rayle, D.L., Haughton, P.M., Cleland, R.E.: Proc. Natl. Acad. Sci. USA *67*, 1814–1817 (1970)
30. Kuraishi, S.: Plant Cell Physiol. *14*, 689–718 (1973)
31. Nitsch, J.P., Nitsch, C.: Plant Physiol. *31*, 94–111 (1956)
32. Vanderhoef, L.N., Shen Lu, T-Y., Williams, C.A.: Plant Physiol. *59*, 1004–1007 (1977)
33. Cleland, R.E.: Planta *128*, 201–206 (1976)
34. Goldberg, R., Prat, R.: Physiol. Veg. *17*, 83–94 (1979)

35. Penny, P., Penny, D., Marshall, D., Heyes, J.K.: J. Exp. Bot. *23*, 23–36 (1972)
36. Vanderhoef, L.N., Stahl, C.A.: Proc. Natl. Acad. Sci. USA *72*, 1822–1825 (1975)
37. Vanderhoef, L.N., Stahl, C.A., Lu, T.Y.S.: Plant Physiol. *58*, 402–404 (1976)
38. Cleland, R.E.: Unpublished data
39. Boyer, J.S., Wu, G.: Planta *138*, 227–237 (1978)
40. Sloane, M., Sadava, D.: Nature (Lond.) *257*, 68–69 (1975)
41. Oberbauer, S., Rayle, D., Johnson, K., Cleland, R.E.: Plant Sci. Lett. *11*, 317–322 (1978)
42. Goldsmith, M.H.M., Caubergs, R., Briggs, W.R.: Plant Physiol. *63*, 155 (1979)
43. Cleland, R.E., Prins, H.B.A., Harper, J.R., Higinbotham, N.: Plant Physiol. *59*, 395–397 (1977)
44. Cleland, R.E.: Biochem. Biophys. Res. Commun. *69*, 333–338 (1976)
45. Kasamo, K., Yamaki, T.: Plant Cell Physiol. *15*, 965–970 (1974)
46. Scherer, G.F.E., Morré, D.J.: Biochem. Biophys. Res. Commun. *84*, 238–247 (1978)
47. Cross, J.W., Briggs, W.R., Dohrmann, U.C., Ray, P.M.: Plant Physiol. *61*, 581–584 (1978)
48. Marré, E., Lado, P., Rasi-Caldogno, F., Colombo, R., De Michelis, M.I.: Plant Sci. Lett. *3*, 365–379 (1974)
49. Evans, M.L.: Annu. Rev. Plant Physiol. *25*, 195–223 (1974)
50. Nooden, L.D., Thimann, K.V.: Proc. Natl. Acad. Sci. USA *50*, 194–200 (1963)
51. Bates, G.W., Cleland, R.E.: Planta *145*, 437–443 (1979)
52. Bates, G.W., Cleland, R.E.: Plant Physiol. *59*, 40 (1977)
53. Ray, P.M.: Plant Physiol. *59*, 594–599 (1977)
54. Batt, S. Venis, M.A.: Planta *130*, 15–21 (1976)
55. Rabon, E., Chang, H., Sachs, G.: Biochemistry *17*, 3345–3353 (1978)
56. Ray, P.M.: Plant Physiol. *51*, 609–614 (1973)

Auxin-Induced Changes in Noncellulosic Polysaccharides of Cell Walls of Monocot Coleoptiles and Dicot Stems [1]

Y. MASUDA [2]

Introduction

Auxin-induced cell wall loosening is considered to be brought about by biochemical modifications in cell wall components. There have been a number of studies on the changes in cell wall components during auxin-induced cell extension, and in some cases a good correlation between changes in mechanical properties and biochemical changes of the cell wall has been reported (5, 6).

Also, extensive studies on the chemical structure of the primary cell wall (1) offer a basis for understanding auxin-induced cell wall loosening. However, there are some indications that the chemical structure of the cell wall differs with the plant. The noncellulosic polysaccharide fraction of the monocot cell wall is rich in Ara, Xyl and Glc, whereas that of the dicot cell wall contains a large amount of Gal (6).

The purpose of the present study was (a) to compare changes in the cell wall composition of etiolated oat coleoptile segments and light-grown azuki epicotyl segments in response to added auxin, (b) to analyze the detailed molecular modifications in hemicellulose composition of oat coleoptile segments when auxin induces cell extension, and (c) to observe changes in the cell wall components in azuki epicotyl segments due to synthesis which occurs in the presence of added sucrose in a medium which enhances auxin-induced stem extension.

Comparison of Noncellulosic Polysaccharides in Monocotyledonous and Dicotyledonous Plants

Coleoptiles or epicotyls which had been starved (6 h for oat, 3 h for azuki) and incubated, were homogenized and centrifuged. Then the residue was treated with pronase and pancreatic amylase to obtain holocellulose (see Fig. 3). The holocellulose was extracted with 20 mM ammonium oxalate (pH 4.0) at 65°C to obtain PC, then with 4% KOH to obtain HC I and finally with 24% KOH to obtain HC II, the residue being α-cellulose.

---

1 *Abbreviations:* IAA, indole-3-acetic acid; HPLC, high-pressure liquid chromatography; GLC, gas-liquid chromatography; PC, pectic fraction; HC I, hemicellulose I, HC II, hemicellulose II; HC A, hemicellulose A; HC B, hemicellulose B; UA, uronic acid; Ara, arabinose; Xyl, xylose, Man, mannose; Gal, galactose; Glc, glucose; Rha, rhamnose

2 Department of Biology, Faculty of Science, Osaka City University, Sumiyoshi-ku, Osaka 558, Japan

Noncellulosic neutral sugars are considered to be contained mostly in the HC fractions in the monocot cell wall, which has a very small amount of pectic substances (see below). However, the PC of the dicot cell wall contains a large amount of pectic substances and neutral sugars. The differences in the compositions of the cell walls can be seen in Fig. 1. In oat coleoptiles, PC accounts for a small part of the cell wall and HC I is more abundant than HC II. HC I is relatively rich in UA (presumably glucuronic acid). In azuki epicotyls, PC accounts for a large portion of the wall and contains large amounts of UA (presumably galacturonic acids) and neutral sugars. In contrast to findings in oat coleoptiles, the amount of HC II is much greater than that of HC I. These results suggest that the cell wall structure is quite different in oat and azuki.

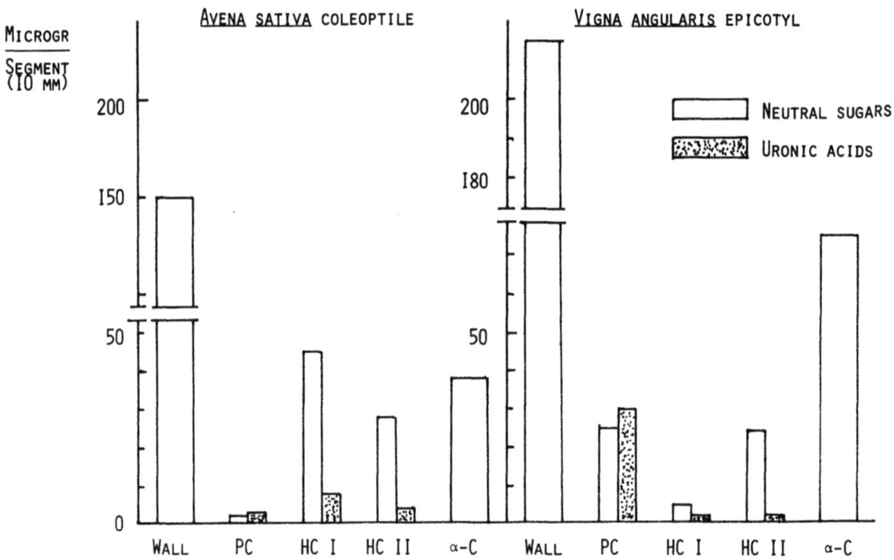

Fig. 1. Amounts of different fractions of the cell wall of 4-day old etiolated oat coleoptiles and 6-day old light-grown azuki epicotyls. Each fraction was extracted and its neutral sugars estimated by the phenol-sulfuric acid method and the uronic acid content estimated by the carbazole method. The sugar content is expressed as μg per 10 mm segment

The neutral sugar composition of each fraction of the cell wall from oat coleoptiles and azuki epicotyls is shown in Fig. 2. PC of the oat cell wall contains mainly Ara, Xyl and Glc, whereas that of the azuki cell wall contains a large amount of Gal and lesser amounts of Ara and Rha but not Xyl. The composition of HC I is different in oat and azuki, the former containing Ara, Xyl, and Glc but the latter Xyl, Gal and lesser amounts of Ara and Glc. The composition of HC II appears to be similar in oat and azuki, being rich in Xyl and Glc, although oat also contains a significant amount of Ara.

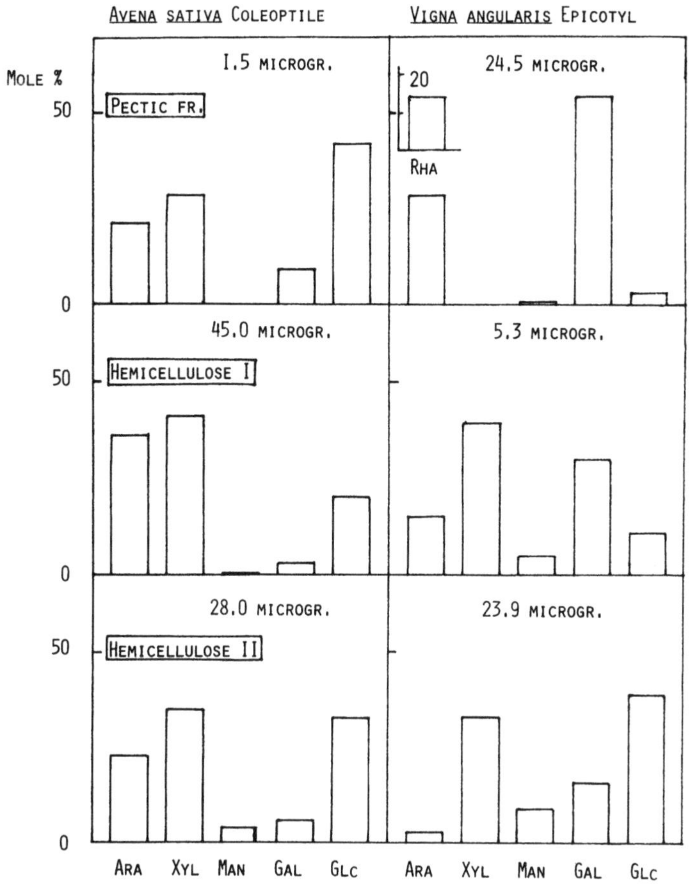

Fig. 2. Neutral sugar composition of different fractions of the cell walls of oat coleoptiles and azuki epicotyls. Each fraction was hydrolyzed with trifluoroacetic acid, reduced and acetylated, then the neutral sugars were determined by GLC. The amount of sugars per 10 mm segment is also given

Changes in the Cell Wall Composition of Oat Coleoptile Segments

HCs are classified in different ways depending on the manner of their extraction. As indicated in Fig. 3, HC I is the fraction extracted with 4% KOH from holocellulose, and HC II is that extracted with 24% KOH from the residue after the above extraction. HC A is the precipitate obtained when the 17.5% NaOH extract of the holocellulose is neutralized. The alcohol precipitate of the supernatant is HC B. Recently, Wada and Ray (14) analyzed the oat coleoptile cell wall and found HC A (9.3%) to consist of xyloglucans and HC B (38.8%) of glucuronoarabinoxylans and β-glucans, leaving α-cellulose (29.1%). We obtained similar values, i.e., 2.8%, 38.9%, and 23.2% respectively. This extraction procedure is unsuitable for certain analyses of wall polysaccharides, such as determinations of molecular weight distribution by HPLC or column chromatography, since HC A, which is soluble only in alkaline solution, cannot be chromatographed for lack of column materials stable to alkali. We thus extracted HC I and II, although small amounts of polysaccharide precipitate after neutralization of the alkali extracts, especially in HC II. Although the amounts of HC A and HC II are rather small, the neutral sugar compositions are similar except for Ara in HC II, possibly due to contamination of HC I in HC II.

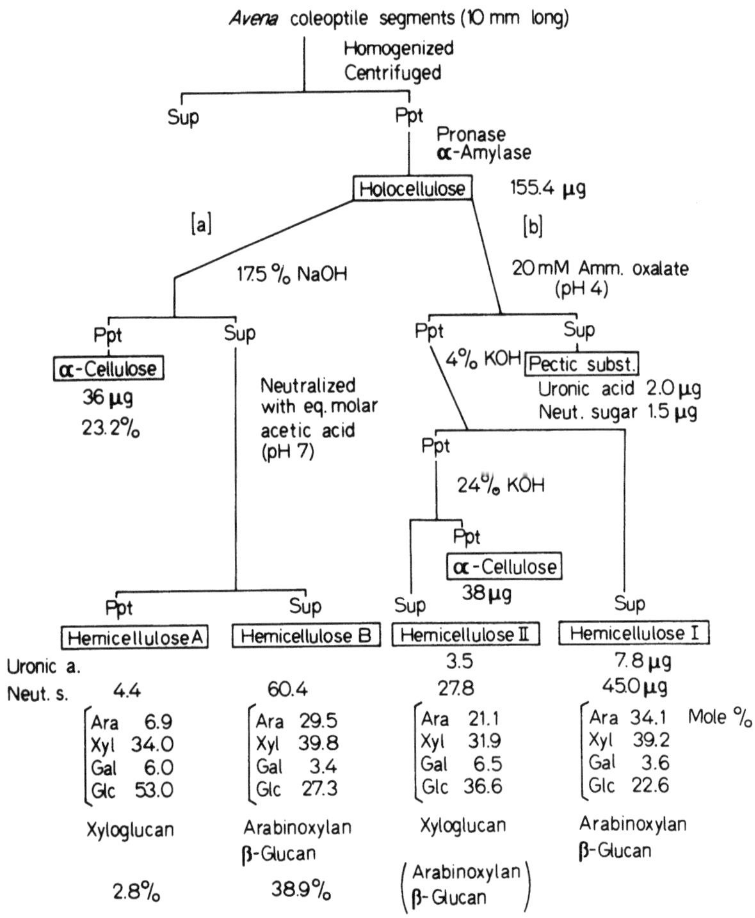

Fig. 3. Extraction procedures for the separation of different fractions of the cell wall of oat coleoptiles. Values given in μg are the amounts of the holocellulose, cellulose, neutral sugars or UA per 10 mm coleoptile segment. Neutral sugar compositions in mol % of HC A, HC B, HC I and HC II are also given

Oat coleoptile segments were treated with or without 10^{-5} M IAA for 5 h, then extracted and the neutral sugars were analyzed (Fig. 4). IAA caused a significant decrease in the mol % of Glc and a small increase in that of Ara and Xyl in HC B and HC I. On the other hand, it caused little change in the mol % of Glc, but a very small decrease in that of Ara and Xyl in HC A and II. Previous reports (3, 4, 11, 12) have presented evi-

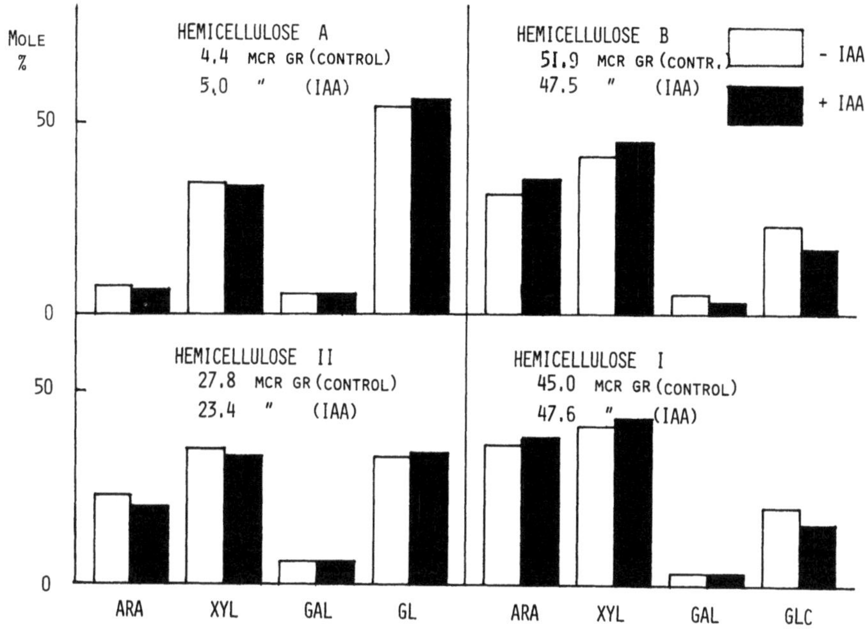

Fig. 4. Effect of IAA on the neutral sugar composition, mol %, of HC A, HC B, HC I and HC II of the oat coleoptile cell wall. Coleoptile segments were incubated for 5 h in buffer solutions with or without 10^{-5} M IAA and extracted by the procedures shown in Fig. 3. The amount of sugars in each fraction is given as μg per segment (initial length: 10 mm)

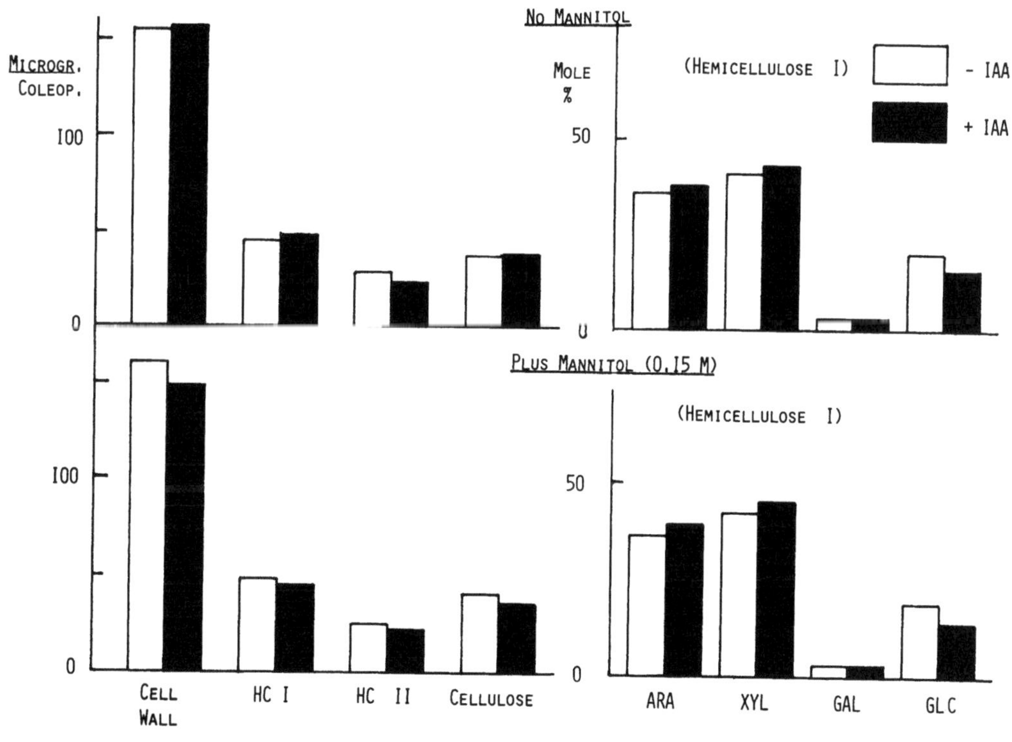

Fig. 5. Effect of IAA on the amount of different cell wall fractions and the neutral sugar composition of HC I in oat coleoptiles. Segments were incubated for 5 h in buffer solutions in the presence or absence of 0.15 M mannitol, with or without 10^{-5} M IAA, and extracted by the procedures shown in Fig. 3

dence that IAA causes a specific decrease in the amount of noncellulosic (trifluoroacetic acid-hydrolyzable) Glc but little changes in the amounts of other sugars. This glucan which is degraded during IAA treatment appears to originate in HC I or HC B; the increase of Ara and Xyl in HC B or HC I and the decrease in HC II and A may cancel each other when all noncellulosic neutral sugars are hydrolyzed together.

To determine whether or not these changes result from IAA-induced cell extension, segments were incubated in the presence of 0.15 M mannitol (Fig. 5). As previously reported (3), the amount of cell wall (holocellulose) per segment changed very little during IAA-induced growth and decreased when segments were treated with IAA in the presence of mannitol. The net amounts of HC I, II and cellulose per segment also decreased in response to IAA when segments were incubated in the presence of mannitol. These data suggest that IAA does not stimulate the net synthesis of cell wall which accompanies cell extension. Neutral sugar analysis showed that IAA also caused a specific decrease in the mol % of Glc even in the presence of mannitol. This confirms that the decrease in the amount of Glc in HC I is not the result of IAA-induced cell extension but is a direct effect of IAA.

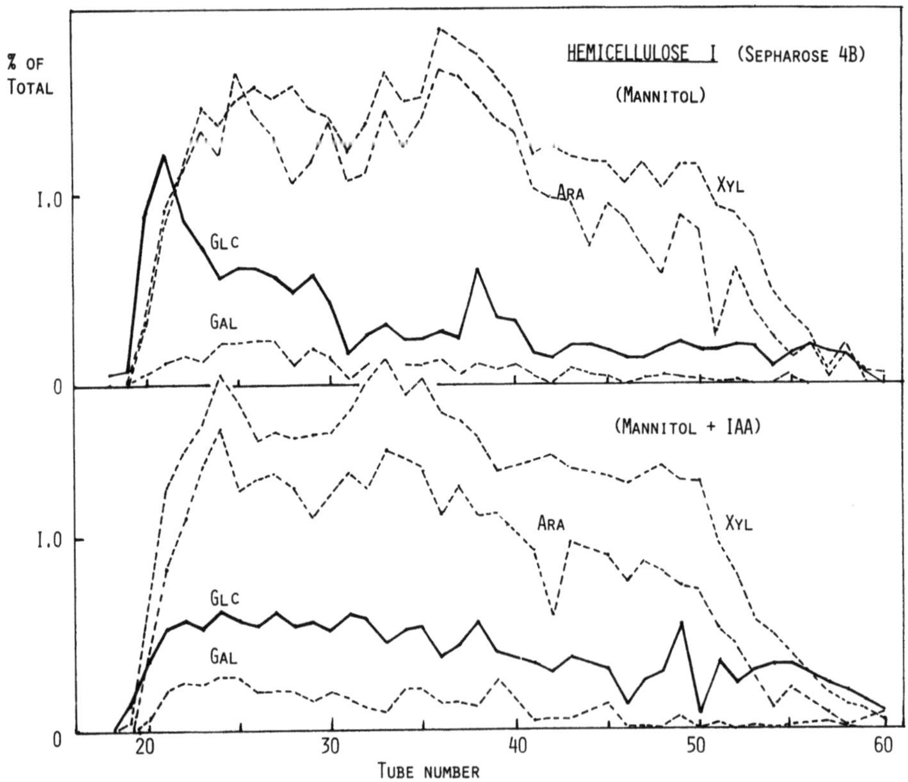

Fig. 6. Sepharose 4B column chromatogram of HC I extracted from oat coleoptile cell wall. Segments were incubated for 5 h with or without 10^{-5} M IAA in the presence of 0.15 M mannitol, then HC I was extracted. The alkaline HC I fraction was neutralized, desalted on Bio Gel P-2, then chromatographed on Sepharose 4B. Polysaccharides in each tube were hydrolyzed and subjected to GLC. The relative amounts of the constituting neutral sugars *(ordinate)* are given as % of the total

The HC I fractions from these experiments were neutralized with acetic acid to pH 7, desalted on a Bio-Gel P-2 column, concentrated and then chromatographed through Sepharose 4B. The elution pattern with 50 mM acetate buffer (pH 5.0) is shown in Fig. 6. Arabinoxylans (possibly glucuronoarabinoxylans) with a wide range of molecular weights seem to be present together with small amounts of Gal. These polysaccharides show very small differences due to IAA. In the control HC I, there is a rather sharp peak of glucans with relatively high molecular weights (> 4 million). The amount of these glucans decreased in the HC I from IAA-treated coleoptile cell wall, which appeared to correspond to the decrease in the level of Glc found earlier. Yamamoto and Nevins (15) have reported that the oat coleoptile contains β-1,3:1,4-mixed glucans. By analogy, HC I also contains these mixed glucans which may be degraded during IAA treatment. Similar chromatography was carried out on HC II, and the levels of certain glucans were found to have decreased due to IAA. These glucans may be like the glucans in HC I which are affected by IAA. Methylation analyses (data not shown) indicate that the glucans in HC I and II which are affected by IAA are similar to the mixed glucans of Yamamoto and Nevins (15).

Changes in the Cell Wall Composition of Azuki Epicotyl Segments During Extension

Azuki epicotyl segments were incubated for 20 h in 10^{-4} M IAA in the presence or absence of 50 mM sucrose which enhances IAA-induced growth (10, 13). As shown in Fig. 7, IAA stimulated the net synthesis of UA in the PC and of cellulose in the absence

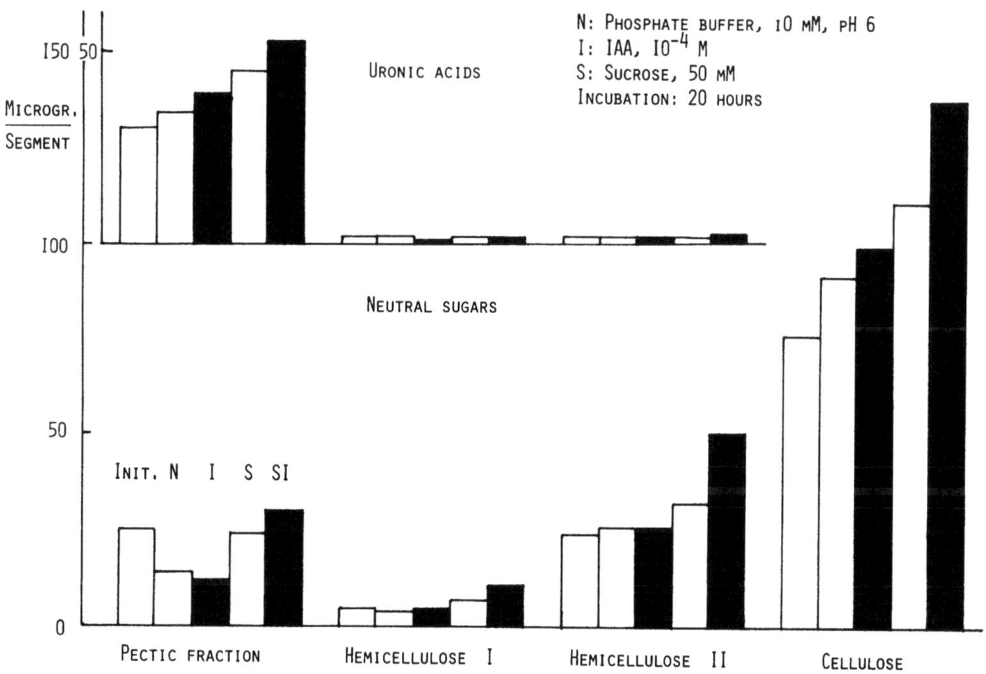

Fig. 7. Effect of IAA and/or sucrose on different fractions of the azuki epicotyl cell wall. Segments were incubated for 20 h in buffer, 10^{-4} M IAA, 50 mM sucrose, or IAA + sucrose, then each fraction was extracted by the procedures shown in Fig. 3. The amount of sugars was determined as for Fig. 1

of sucrose. IAA appears to cause a decrease in the amount of neutral sugars in the PC. GLC analysis of the fraction clearly showed that IAA caused a decrease in the relative amount of Gal; 33 mol % in the control and 24 mol % in the IAA-treatment (Fig. 8A).

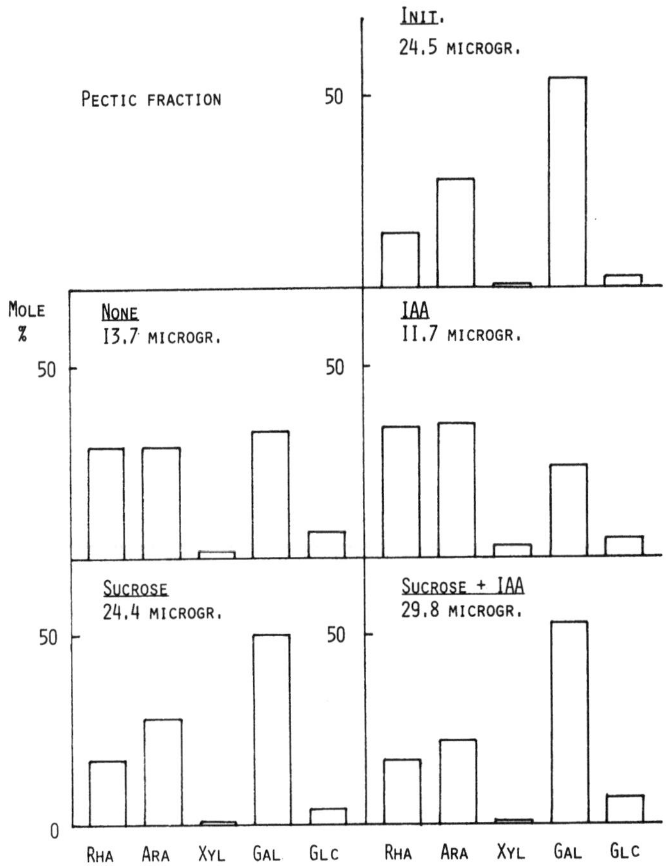

Fig. 8 A, B. Effect of IAA and/or sucrose on neutral sugar compositions of PC and HC I in azuki epicotyl segments. Segments were incubated as shown in Fig. 7. Each fraction was extracted, hydrolyzed and subjected to GLC. Neutral sugar composition is expressed as mol %, and the net amounts of neutral sugars in each fraction per segment (initial length: 10 mm) are given. **A** PC

In the presence of sucrose, the amounts of UA in the PC, neutral sugars in HC I and II, and cellulose increased, and these increases were further promoted by IAA (Fig. 7). GLC analysis showed that the sugar composition of these fractions changed very little due to the presence of sucrose with and without IAA, except for HC I wherein the relative amount of Gal increased in response to IAA in the presence of sucrose; 27 mol % for sucrose alone, 34 mol % for IAA + sucrose-treated (Fig. 8B).

The PC and HC I + II were extracted as before and chromatographed on Sepharose 4B. UA in PC appeared in two main peaks. In both the presence and absence of sucrose, IAA caused an increase in the amount of low molecular weight polyuronides. This increase was greater in the presence of sucrose. This fraction had a class of galactans with relatively high molecular weights; the amount tended to decrease in the absence of sucrose and the decrease was accelerated by IAA. Changes in HC I + II were conspic-

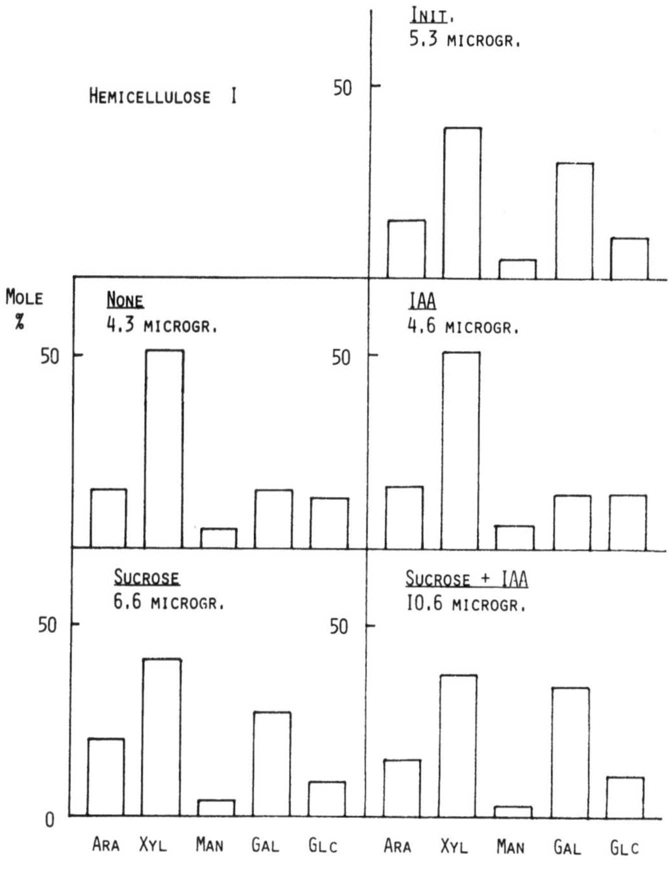

Fig. 8B. HC I

uous if the segments were incubated in the presence of sucrose, i.e., IAA caused the appearance of high molecular weight xyloglucans and galactans.

During a 20-h incubation period, particularly in the presence of sucrose, active cell wall synthesis occurred in azuki epicotyl segments. This wall synthesis may be accompanied by, but may not be the cause of, IAA-induced cell extension. However, the specific decrease in the amount of Gal in the PC due to IAA in the absence of sucrose may be significant in IAA-induced cell extension. This point needs further investigation.

We did not investigate the change in the cell wall composition associated with IAA-induced cell extension during a relatively short incubation period which is not affected by sucrose (10, 13). If any changes do occur, they should provide information on cell wall loosening which is not affected by added sucrose in azuki stems (9).

Consideration

Our findings on the structure of the primary cell wall of oat coleoptiles generally agree with those of Ray and co-workers (2, 14). Although we have not characterized the polyuronides in HC I, they are considered to be parts of glucuronoarabinoxylan. UA found in HC I is considered to be quite labile, since most of it (> 65%) disappears on dialysis. The chromatography patterns and results of methylation analysis show that the oat coleoptile cell wall contains a small amount of xyloglucan in HC II or HC A. Albersheim (1) reported that β-glucan is not a structural component but a storage polysaccharide in barley and also in oat cell walls. This erroneous notion was based on experiments with bacterial α-amylase (Sigma type III-A), which showed that the barley cell wall completely lacks 3-linked glucosyl residues (7). This enzyme contains a β-D-glucanase which has been shown to hydrolyze β-1,3:1,4-mixed glucans of the oat coleoptile cell wall (15). It is now confirmed that the cell wall not only of oat coleoptiles but also of other grasses (coleoptiles of *Zea mays, Hordeum vulgare, Triticum vulgare* and *Secale cereale*, and mesocotyls of *Sorghum bicolor*) contain the β-1,3:1,4-mixed glucans. All contain approximately 40–100 μg β-glucans per mg cell wall, of which about 30% are 1,3-linkage glucans (8).

In the oat coleoptile cell wall, the most conspicuous change due to auxin, judging from methylation analysis, is the specific decrease in the amount of β-1,3:1,4-glucans. In addition, there is a small increase due to auxin in the amounts of Ara and Xyl which may constitute glucuronoarabinoxylan. These changes should be closely related to auxin-induced cell wall loosening (6). Auxin-induced rapid cell extension is considered to be the consequence of the breaking of some hemicellulose-mediated interconnections between cellulose microfibrils. We believe that the changes we found in hemicellulosic glucans are associated with the breaking of these interconnections.

The situation in the dicot cell wall appears more complex, since growing stems are generally supposed to actively synthesize cell wall components. This is clearly shown by the data in Fig. 7. Nevertheless, even in dicot stems, as in monocot coleoptiles, the initiation of cell wall extension caused by auxin should be a consequence of breaking some interconnections between cellulose microfibrils. According to the Albersheim et al. model (1), in the dicot cell wall, approximately 80% of the xyloglucan chains hydrogen-bonded with cellulose microfibrils are attached to rhamnogalacturonans through arabinogalactan. Our data on azuki cell wall show that the major constituents of HC I + II are xyloglucans and galactans, while rhamnose is located only in PC which also contains UA and galactans. Chromatography resolved xyloglucans, xylans, and galactans, indicating that they may be separate polysaccharides. We feel that Albersheim's model for the primary cell wall should be revised and does not represent the structure of the monocot cell wall, which is simpler than that of the dicot cell wall; for example, the monocot wall contains no rhamnogalacturonans.

Under conditions where only limited wall synthesis occurs (in absence of sucrose), the only conspicuous change caused by auxin is the decrease in the amount of galactans in PC, although short-term experiments are still needed. The other change caused by auxin is the increase in the amount of low molecular weight polyuronides also in PC. These findings suggest that in the dicot cell wall, modifications of galactans and

UA are important changes related to auxin-induced cell extension as exemplified in azuki epicotyls.

Acknowledgments. I wish to thank my colleagues, Drs. S. Kamisaka, R. Yamamoto, N. Sakurai and K. Nishitani, for their cooperation and discussions. Parts of the present study were supported by grants from the Ministry of Education.

References

1. Albersheim, P.: In: Plant Growth Regulation. Pilet, P.E. (ed.), pp. 1–12. Berlin-Heidelberg-New York: Springer 1977
2. Labavitch, J.M., Ray, P.M.: Phytochemistry *17*, 933–937 (1978)
3. Loescher, W. Nevins, D.J.: Plant Physiol. *50*, 556–563 (1972)
4. Loescher, W., Nevins, D.J.: Plant Physiol. *52*, 248–251 (1973)
5. Masuda, Y.: In: Plant Growth Regulation. Pilet, P.E. (ed.), pp. 21–26. Berlin-Heidelberg-New York: Springer 1977
6. Masuda, Y.: Bot Mag. Spec. Issue *1*, 103–123 (1978)
7. McNeil, M., Albersheim, P., Taiz, L., Jones, R.L.: Plant Physiol. *55*, 64–68 (1975)
8. Nevins, D.J., Yamamoto, R., Huber, D.J.: Phytochemistry *17*, 1503–1505 (1978)
9. Nishitani, K., Masuda, Y.: Plant Cell Physiol. *20*, 63–74 (1979)
10. Nishitani, K., Shibaoka, H., Masuda, Y.: Plant Cell Physiol. *20*, 463–472 (1979)
11. Sakurai, N., Nevins, D.J., Masuda, Y.: Plant Cell Physiol. *18*, 371–380 (1977)
12. Sakurai, N. Masuda, Y.: Plant Cell Physiol. *18*, 587–594 (1977)
13. Shibaoka, H.: Plant Cell Physiol. *13*, 461–469 (1972)
14. Wada, S., Ray, P.M.: Phytochemistry *17*, 923–931 (1978)
15. Yamamoto, R., Nevins, D.J.: Carbohydr. Res. *67*, 275–280 (1978)

Auxin-Regulated Elongation: A Summary Hypothesis

L.N. VANDERHOEF [1]

Introduction

Auxin has been studied for fifty years as a regulating hormone in many developmental phenomena (1). Much of this work has been devoted to its effect on cell enlargement (2–6); however, our accumulated knowledge of auxin-regulated cell enlargement does not include an understanding of its primary mode of action. In the 1960's the "gene expression" hypothesis of Key, and others (4) was popularly accepted. In the early 1970's, however, the "wall acidification" (7) hypothesis of Rayle and Cleland (5) and Hager et al. (8) was proposed. It has been supported in numerous reports (5).

Discovery and study of two separate elongation responses to auxin (9–11) and recent studies of elongation-associated gene activity (12) lead me to propose that the hypotheses are not incompatible. While neither accounts for all of our relevant knowledge, together they can account for all aspects of auxin-controlled cell enlargement.

The Gene Expression Hypothesis

Many studies of hormone-induced elongation were based on the simple straight-growth bioassay which consisted of the addition of exogenous auxin to an elongating segment which had been excised from its endogenous auxin supply. Accumulated growth after several hours was measured. The gene expression hypothesis was supported by experiments similar to the growth bioassay, but with the additional use of transcription and translation inhibitors such as actinomycin D and cycloheximide. In the absence of RNA and/or protein synthesis, auxin was not able to induce the higher rate of elongation. Thus, the hypothesis was supported (4).

These experiments and their many sophisticated variations were commonly reported and accepted for both plant and animal hormone studies. It is especially clear in hindsight, though, that interpretation of these experiments was difficult, since it was always possible that the inhibition of cell growth in the presence of actinomycin D and/or cycloheximide was caused by events which were not related to auxin action.

The gene expression hypothesis was challenged in 1969 by Evans and Ray (13). These authors presented an unusually inventive technique for measuring minute-by-minute growth, experimental results that measured a short (10 min) lag time for auxin action, and an eloquent discussion of the logic which led them to rule out gene ex-

1 Department of Botany, University of Illinois, Urbana, IL 61801, USA

pression in auxin-promoted cell enlargment based on growth kinetics. While the fast response to auxin had been known for many years (14), its presentation and discussion in this new light promoted it from a curiousity to an important contribution to our understanding of auxin action.

The Wall Acidification Hypothesis

The Evans and Ray experiments initiated a series of debates regarding the validity of the gene expression hypothesis, and its popularity waned. The resulting void was filled in 1971 when Cleland (2) and Hager et al. (8) independently suggested that acidification of the cell wall may mediate in auxin action. This hypothesis, attractive for its simplicity, stated that auxin maintains the cell wall pH near 5, causing the increased rate of elongation. Two kinds of data supported the hypothesis. First, in most systems where exogenous auxin could induce a higher rate of cell elongation, lowering the pH from 6 to about 4 could also induce a higher rate of cell elongation (9, 15–17). Second, in certain circumstances auxin could be shown to induce acidification of external medium by auxin-responsive tissue (18). Also, in one case (19) auxin was shown to induce responsive cells to acidify the free space of pea stem segments and maize coleoptile segments.

The wall acidification hypothesis is currently being tested by the research of many auxin physiologists. Two problems have emerged. First, auxin-induced cell wall acidification was not demonstrable in all auxin-responsive systems [(20) and references therein]; this, however, may be a problem of methods and cannot yet be assumed to constitute a major criticism of the hypothesis (21). Second, acid-induced elongation does not, in fact, mimic auxin-induced elongation. Cells elongating in response to lowered pH never attain a steady-state rate. The H^+-induced increase in elongation rate is transient and over within 60 to 100 min, the actual time depending on the species and experimental conditions (5, 9, 16, 20, 22). This is different from auxin-induced elongation, where a steady-state rate is attained and can be retained for several hours. Furthermore, acid damage is not responsible for the transient nature of acid-induced elongation; segments which have previously experienced the acid-induced burst of elongation are fully responsive to subsequent auxin addition (9, 20).

This major problem with the wall-acidification hypothesis suggested that it should be expanded to include additional components. Experiments which have distinguished two separate elongation responses to auxin also indicate that wall loosening alone does not account for all components of auxin-regulated cell enlargement.

The Separation of Two Responses to Auxin

When auxin was added to auxin-depleted elongating soybean hypocotyl segments, the results were unusual (9). Rather than simply rise to a higher steady-state rate, the rate rose, then dropped, then finally rose again to a steady-state rate (Fig. 1). Others (23, 24) had also seen unusual early kinetics. While they were not identical to our determinations, they were similar. We now know that the common explanation for these early fluctuations in the growth rate was that auxin induced elongation in two separate ways, i.e., auxin induced a rapid (but transient) wall-loosening response, as well as a more permanent response that began after a longer lag time and resulted in steady-state growth. It seemed quite likely (and has been confirmed) that the chemical and physical processes that constitute the two responses were not identical (11, 12).

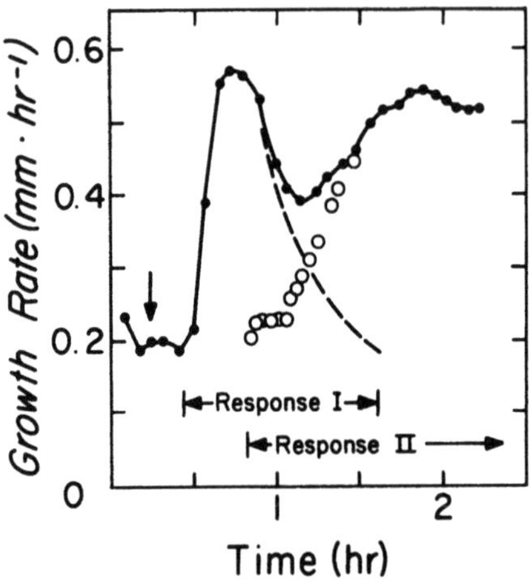

Fig. 1. Rate kinetics show two separate elongation responses to auxin by soybean hypocotyl segments. The *arrow* marks auxin addition to auxin-depleted excised hypocotyl segments

The supportive data for two separate elongation responses have appeared elsewhere, and are briefly summarized as follows: Cytokinin inhibits the second response, but not the first (9, 25); acid-induced elongation mimics only the first response, i.e., the growth rate goes up and comes down without ever attaining a steady-state rate (9, 20); auxin analogs can temporally separate the first response from the second (10, 24); the limiting proteins for the two responses are different (11); the confusing early elongation kinetics in response to auxin that varied from species to species were explained by the concept of two elongation responses to auxin (9, 10).

Subsequent independent experiments in other laboratories have confirmed the existence of two separate elongation responses to auxin, and proton-independent growth has been demonstrated in corn (26).

A Reappraisal of the Hypothesis of Auxin Action in Cell Elongation

The discovery that auxin induces two separate elongation responses makes it clear that the "gene expression" and "wall acidification" hypotheses are not incompatible. In fact, all available data can be accommodated if one assumes that auxin regulates both wall loosening and the supply of wall materials, as shown in Fig. 2. Thus the reintroduction of auxin to excised, auxin-depleted segments causes wall loosening (very likely mediated by wall acidification) and a subsequent burst of turgor-driven elongation. This elongation is transient; it lasts 30 to 90 min, depending on conditions and species. Elongation cannot continue by the first wall-loosening auxin action alone. This very

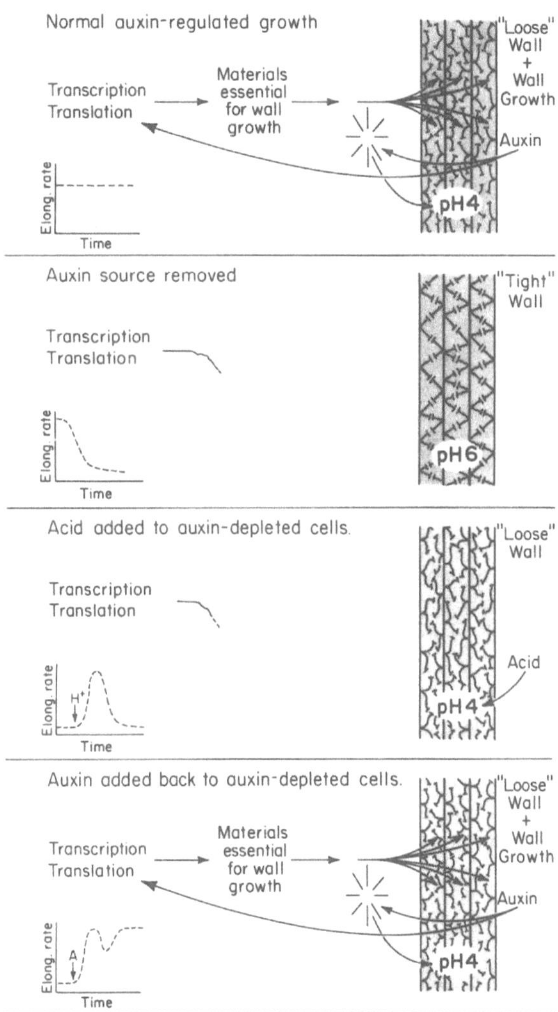

Fig. 2. Events in cell elongation. Auxin is postulated to regulate both wall loosening and wall synthesis

important point was first established with experiments which employed cytokinin, an elongation inhibitor (9). Auxin must have another activity in its regulation of cell elongation, very likely the production of compounds necessary for sustained cell elongation.

There is very little direct evidence for gene expression regulation by hormones in an eukaryotic system. However, this kind of auxin action must be considered most likely, in light of several lines of indirect and circumstantial evidence. First, it is now clear that the "fast response" experiments do not, after all, refute the tenets of the gene activation hypothesis. The arguments of Evans and Ray (13) are applicable to the first elongation response only. This is best demonstrated by replotting some data that were submitted in evidence of the contention that actinomycin D did not inhibit auxin-induced elongation. While the first elongation response was not affected by actinomycin D, the second elongation response was strongly affected by the inhibitor

(Fig. 3). The experiments of Key et al. and others (4) which supported the gene expression hypothesis can not, after all, be dismissed simply on the basis of the fast response.

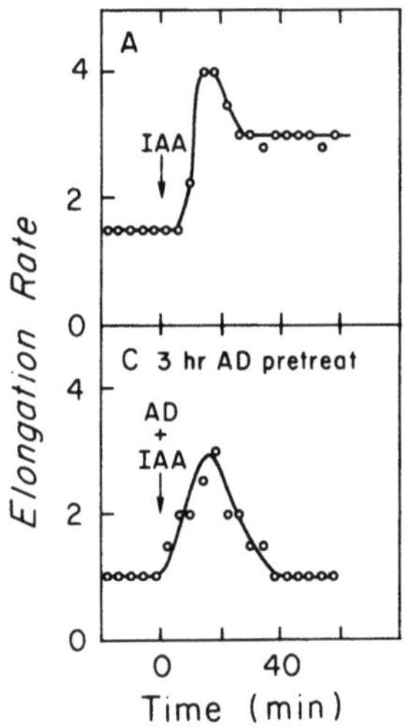

Fig. 3. Actinomycin D inhibits the second, but not the first elongation response in pea stem segments. Redrawn from Fig. 4 of (23) (Barkley and Evans)

A second line of evidence which supports a mediating role of gene activation in auxin-induced elongation comes from studies of auxin-affected protein synthesis in elongating cells. Recently auxin has been shown to induce specific changes in the pattern of protein synthesis in elongating soybean hypocotyl segments (12). Certainly auxin activity at gene expression in sustained cell elongation appears to be a real possibility.

Conclusions

Most of the experiments cited above utilized a common technique. Rapidly elongating stem (or coleoptile) segments were surgically removed from their endogenous auxin supply and incubated in an auxin-free pH 6 medium until the elongation rate was low. Exogenous auxin was then added and the responses monitored. This experimental technique, though not initially foreseen, was absolutely necessary to distinguish the two major events of auxin-regulated elongation. The first auxin-induced burst of elongation is the result of the wall "breaking loose." The first response, therefore, can be observed only if the cells are allowed to incubate for a period of time in the absence of auxin. The loosened wall begins to show the first signs of irreversible elongation about thirty-five minutes after auxin addition. A steady-state rate of elongation is established by 60 min after auxin addition.

The events of elongation as shown in Fig. 2 allow an interesting prediction. If exogenously-supplied auxin causes wall loosening by causing wall acidification, then a segment which is kept at pH 4 and under tension from the moment of excision should show, after depletion of endogenous auxin, only the second response when exogenous auxin is applied. This experiment was performed and this precise result was obtained, i.e., when the wall was kept loose, auxin induced only the second response (Fig. 4). These results are especially interesting because they also rule out an alternative explanation for the first elongation response to auxin. The first elongation response might represent the utilization of a backlog of wall materials that could not be inserted while the wall was temporarily, in the absence of auxin, at a high pH. However, if this alternative explanation were correct, the 0 to 40 min growth rate kinetics should have been measurably different at pH 4 and 6, i.e., those segments at pH 4 should have maintained a higher growth rate while utilizing the proposed backlog of materials. They did not.

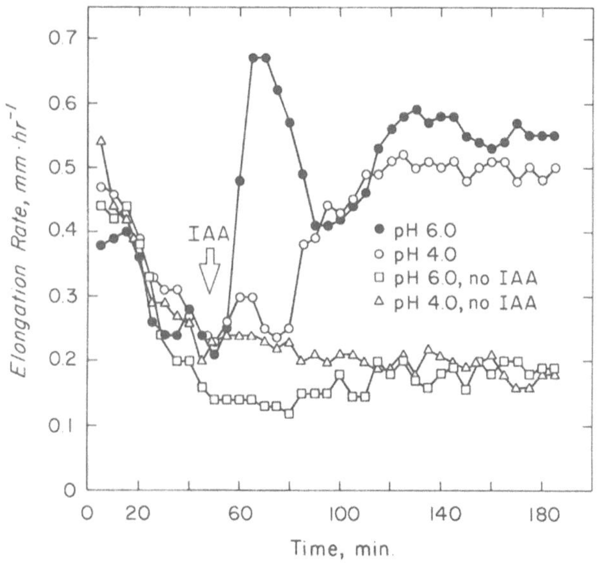

Fig. 4. Auxin induces only the second response in acid-pretreated segments. The excised elongation segment was transferred immediately to pH 4 or pH 6 medium in the growth-measuring apparatus. Auxin, final concentration 45 μM, was added when the elongation rate decreased to near 0.2 mm per h

Figure 2 is necessarily non-specific on several points, e.g., on the actual mechanisms of proton-mediated wall loosening and gene expression regulation; the available data give us no substantial hints. On the other hand, the hypothesis is quite specific in omitting any consideration of osmoregulation. The experiments of Boyer and Wu (27) have made it clear, after years of debate, that auxin-induced changes in cell enlargement can not be attributed to changes in cell osmotic potentials or turgor.

Acknowledgments. Work supported by grants NSF BMS72-02496, NSF PCM77-14175, HEW PHS 07030, and by the University of Illinois Graduate Research Board.

References and Notes

1. Thimann, K.: Hormone Action in the Whole Life of Plants. Amherst: U. Mass. Press 1977
2. Cleland, R.: Annu. Rev. Plant Physiol. *22*, 197 (1971)
3. Ray, P.: Adv. Phytochem. *7*, 93 (1974); Trewavas, A.: Progr. Phytochem. *1*, 113 (1968)
4. Key, J.: Annu. Rev. Plant Physiol. *20*, 449 (1969)
5. Rayle, D., Cleland, R.: Curr. Top. Dev. Biol. *34*, 187 (1977)
6. Evans M.: Annu. Rev. Plant Physiol. *25*, 195 (1974)
7. The hypothesis has various names, e.g., the "acid growth" hypothesis, the "proton extrusion" hypothesis. I propose "wall acidification" hypothesis because it is more accurately descriptive of the hypothesis, and it reflects the supporting data
8. Hager, A. Menzel, H., Krauss, A.: Planta *100*, 47 (1971)
9. Vanderhoef, L., Stahl, C.: Proc. Natl. Acad. Sci. USA *72*, 1822 (1975)
10. Vanderhoef, L., Stahl, C., Williams, C., Brinkman, K., Greenfield, J.: Plant Physiol. *57*, 817 (1976)
11. Vanderhoef, L, Stahl, C., Lu, T-Y.: Plant Physiol. *58*, 402 (1976)
12. Zurfluh, L., Guilfoyle, T.: Plant Physiol. *63*, S143 (1979)
13. Evans M. Ray, P.: J. Gen. Physiol. *53*, 1 (1969)
14. Yamaki, T.: Sci. Pap. Coll. Gen. Educ. Univ. Tokyo *4*, 129 (1954); Köhler, K.: Planta *47*, 159 (1956); Ray, P., Ruesink, A.: Dev. Biol. *4*, 377 (1962)
15. Bonner, J.: Protoplasma *21*, 406 (1934)
16. Rayle, D., Cleland, R.: Plant Physiol. *46*, 250 (1970); Evans, M., Ray, P., Reinhold, L.: Plant Physiol. *47*, 335 (1971)
17. Rayle, D., Haughton, P., Cleland R.: Proc. Natl. Acad. Sci. USA *67*, 1814 (1970); Rayle, D., Cleland, R.: Planta *104*, 282 (1972)
18. Cleland, R.: Proc. Natl. Acad. Sci USA *70*, 3092 (1973); Rayle, D.: Planta *114*, 63 (1973); Mentze, J., Raymond, B., Cohen, J., Rayle, D.: Plant Physiol. *60*, 509 (1977)
19. Jacobs, M., Ray, P.: Plant Physiol. *58*, 203 (1976)
20. Vanderhoef, L., Findley, J., Burke, J., Blizzard, W.: Plant Physiol. *59*, 1000 (1977); Vanderhoef, L., Lu, T-Y., Williams, C.: Plant Physiol. *59*, 1004 (1977); Parrish, D., Davies, P.: Plant Physiol. *60*, 509 (1977)
21. The cuticle presents a special problem for workers intending to measure auxin-modified wall pH. If it is left intact it may slow the H^+ equilibrium between the wall and external medium. If it is removed by abrasion or stripping, extensive cell wounding occurs, which may cause artifactual pH changes. Additionally, pH microelectrodes to directly and accurately measure the wall pH are not available.
22. Kazama, H., Katsumi, M.: Plant Cell Physiol. *17*, 467 (1976)
23. Barkley, G., Evans, M.: Plant Physiol. *45*, 143 (1970); Penny, D., Penny, P., Munro, J., Bailey, P.: In: Plant Growth Substances. Carr, D. (ed.), pp. 52–61. Berlin-Heidelberg-New York: Springer 1970; Barkley, G., Leopold, A.: Plant Physiol. *52*, 6 (1973)
24. Penny, P., Penny, D., Marshall, D., Heyes, J.: J. Exp. Bot. *23*, 23 (1972)
25. Vanderhoef, L., Stahl, C., Seigel, N., Zeigler, R.: Physiol. Plant. *29*, 22 (1973)
26. Kazama, H., Katsumi, M.: Plant Cell Physiol. *17*, 467 (1976); Sakurai, N., Nevins, D., Masuda, Y.: Plant Cell Physiol. *18*, 371 (1977); Vesper, M., Evans, M., Cline, M.: Plant Physiol. *63*, S21 (1979)
27. Boyer, J., Wu, G.: Planta *139*, 227 (1978)

Auxin-Induced Specific Changes in the Pattern of Protein Synthesis in Soybean Hypocotyl Sections

L. ZURFLUH and T. GUILFOYLE [1]

Introduction

It is generally agreed that continued protein synthesis is required for the expression of auxin-induced cell elongation, but controversy has existed over whether this response involves the synthesis of any specific "growth-limiting" proteins. To help resolve this controversy, we have employed the high-resolution two-dimensional gel electrophoresis system described by O'Farrell (1) to examine the spectrum of polypeptides synthesized in elongating soybean hypocotyl sections which were incubated in a medium containing 2,4-dichlorophenoxyacetic acid (2,4-D) (auxin-treated) or lacking 2,4-D (untreated). We have also utilized this technique to analyze the polypeptides synthesized in untreated and auxin-treated sections of mature (basal) soybean hypocotyls where auxin enhances the level of RNA and protein synthesis relative to that observed in untreated sections (2). Our results indicate that auxin treatment alters the pattern of protein synthesis and/or causes a charge modification of proteins in both elongating and basal sections of soybean hypocotyl.

Materials and Methods

After 72 h of germination at 30°C in the dark, 1.2-cm sections were excised from the elongating region (approximately 0.5 to 1.7 cm below the cotyledon) and from the basal region (the 1.2-cm region just above the root-shoot transition zone) of soybean (*Glycine max* var. Wayne) hypocotyls. Untreated sections (5 sections/ml) were incubated at 30°C with continuous shaking in a solution containing 2% sucrose, 50 μg/ml chloramphenicol, 10 mM sodium phosphate (pH 6) or potassium phosphate (pH 6) buffer. Auxin-treated sections were incubated in an identical medium with the exception that 2,4-D was present at 5×10^{-5} *M* for elongating sections and 5×10^{-4} *M* for basal sections. After incubating the sections for 2 h, ^{35}S-methionine (100 μCi/ml) was added to the medium and incubation was continued for 3 h. During the 5-h incubation period, untreated and auxin-treated elongating sections increased in length by 8% and 28%, respectively. Following incubation, the terminal 1-mm segments of each section were excised, and the terminal and internal segments were separated. Twelve or 25 unlabeled 0.8-cm sections were combined with the 1-mm terminal segments to facilitate preparation of the labeled proteins in a manner similar to that for internal

1 Department of Botany, University of Minnesota, St. Paul, Minnesota 55108, USA

segments. The tissue was homogenized in 50 mM Tris-HCl (pH 7.9 at 4°C), 1 mM EDTA, 15 mM 2-mercaptoethanol, and 500 mM ammonium sulfate. The homogenate was filtered through Miracloth and centrifuged at 8000 *g* for 15 min. Proteins from the supernatant were concentrated by ammonium sulfate precipitation (approximately 90% saturation with ammonium sulfate), and the precipitate was collected by centrifugation at 5000 *g* for 5 min. The precipitate in one case (Sample Buffer A) was solubilized in 0.5% sodium dodecyl sulfate (SDS), 9.5 *M* urea, 5% 2-mercaptoethanol, 0.2% Ampholines (pH 3.5–10), and 10% glycerol. In the other case (Sample Buffer B), the precipitate was solubilized in 10 mM Tris-HCl (pH 7.9 at 4°C), 1 mM EDTA, and 15 mM 2-mercaptoethanol at 2°C and dialyzed against several changes of the same buffer at 2°C; the sample was then made 1.7% SDS and 8% 2-mercaptoethanol and boiled for 3 min; after the sample was cooled to 20°C, urea and Ampholines (pH 3.5–10) were added to 9.5 *M* and 0.2%, respectively. The method described for Sample Buffer B was preferred (and utilized in all cases except for Fig. 3), since greater resolution and less polypeptide aggregation and streaking are observed on two-dimensional gels compared to solubilization with Sample Buffer A.

One-dimensional gel electrophoresis in the presence of SDS was performed on 0.75-mm thick slab gels by the method of Laemmli (3). Two-dimensional gel electrophoresis with isoelectric focusing in the first dimension and SDS in the second dimension was performed essentially as described by O'Farrell (1, 4). Sample treatment and loading on first dimension gels was similar to the SDS protocol described by O'Farrell (5). Fluorography was performed according to the method of Bonner and Laskey (6).

Results

Elongating Sections

Because of the possible wounding effects on the pattern of protein synthesis in excised sections of hypocotyl tissue, we removed the terminal 1-mm segments of each section after incubation. Approximately 70% of the ^{35}S-methionine incorporated into protein was recovered in the 1-mm terminal segments of each section, and the ^{35}S-labeled polypeptides observed in terminal 1-mm segments were significantly different from those observed in internal segments (Fig. 1). Electrophoresis in one dimension (SDS) revealed no obvious differences in the polypeptides synthesized in untreated and auxin-treated terminal segments (Fig. 1) or in whole sections without their termini removed (data not shown); however, several differences were observed in the polypeptide patterns of internal segments from untreated and auxin-treated elongating hypocotyl sections (Fig. 1).

Most of the ^{35}S-methionine was incorporated into polypeptides in the range of 30,000 to 60,000 daltons, and we utilized two-dimensional gel electrophoresis (1) to enhance the resolution of these polypeptides. Polypeptide patterns displayed on two-dimensional gels from untreated and auxin-treated terminal 1-mm segments are highly similar although minor differences in intensity of some spots are evident (Fig. 2). The overall patterns of polypeptides synthesized in untreated and auxin-treated internal segments are similar, but differences in the intensity of some polypeptides are more

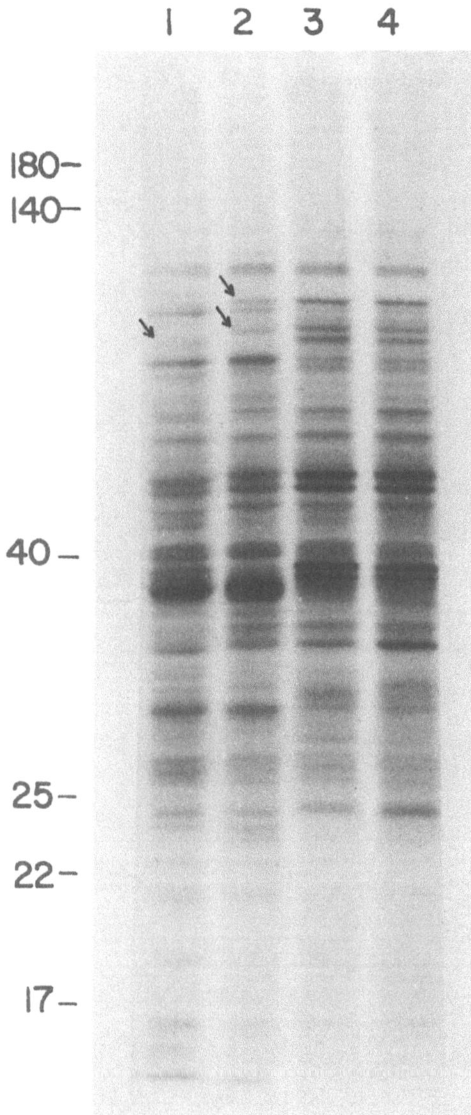

Fig. 1. Fluorograph of ^{35}S-labeled polypeptides from elongating sections resolved by electrophoresis on concave exponential gradient (10%–16%) polyacrylamide gels in the presence of SDS. To each lane, 28,000 cpm were applied. *Lane 1* untreated internal segments; *Lane 2* auxin-treated internal segments; *Lane 3* untreated terminal segments; *Lane 4* auxin-treated terminal segments. Numbers adjacent to the figure indicate molecular weights in kilodaltons as determined with cauliflower RNA polymerase IIB subunits and bovine serum albumin as molecular weight standards. *Arrows* indicate examples of differences in polypeptides synthesized in untreated and auxin-treated sections

striking than those observed with terminal segments, and the synthesis of several polypeptides appears to be strongly activated or repressed by auxin treatment (Fig. 3). The most significant differences in untreated and auxin-treated elongating internal segments observed on two-dimensional gels are the appearance of a polypeptide of approximately 100,000 daltons with an isoelectric point of about 6.1, the appearance of two polypeptides of approximately 40,000 daltons with isoelectric points of about 5.9 and 6.1, and the disappearance of a polypeptide of approximately 40,000 daltons with an isoelectric point of about 6.3 in auxin-treated tissue.

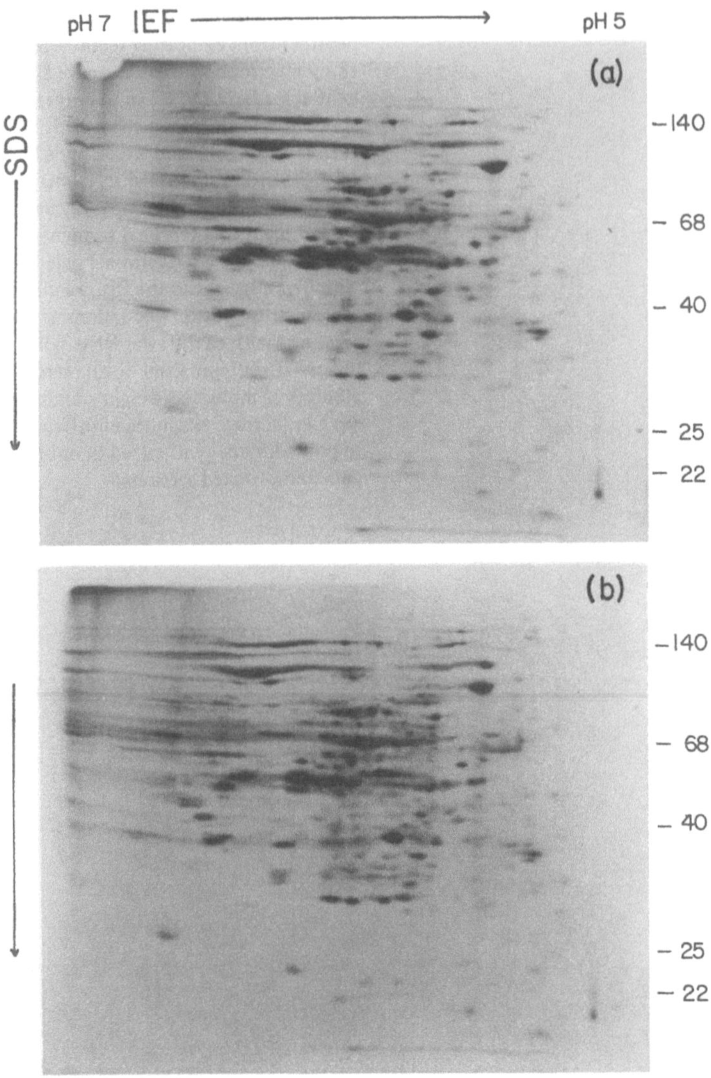

Fig. 2a, b. Fluorographs of ^{35}S-labeled polypeptides from elongating terminal segments resolved by electrophoresis on two-dimensional polyacrylamide gels (1). Samples loaded on first dimension gels were 54 μg protein and 272,000 cpm for both untreated and auxin-treated tissue. Second dimension gels are 10% polyacrylamide. **a** untreated segments; **b** auxin-treated segments. *IEF* isoelectric focusing

Mature Sections

In contrast to the elongating sections, terminal 1-mm and internal segments of basal sections displayed similar patterns of polypeptide synthesis as revealed by one-dimensional gel electrophoresis (Fig. 4), and most of the ^{35}S-methionine was incorporated

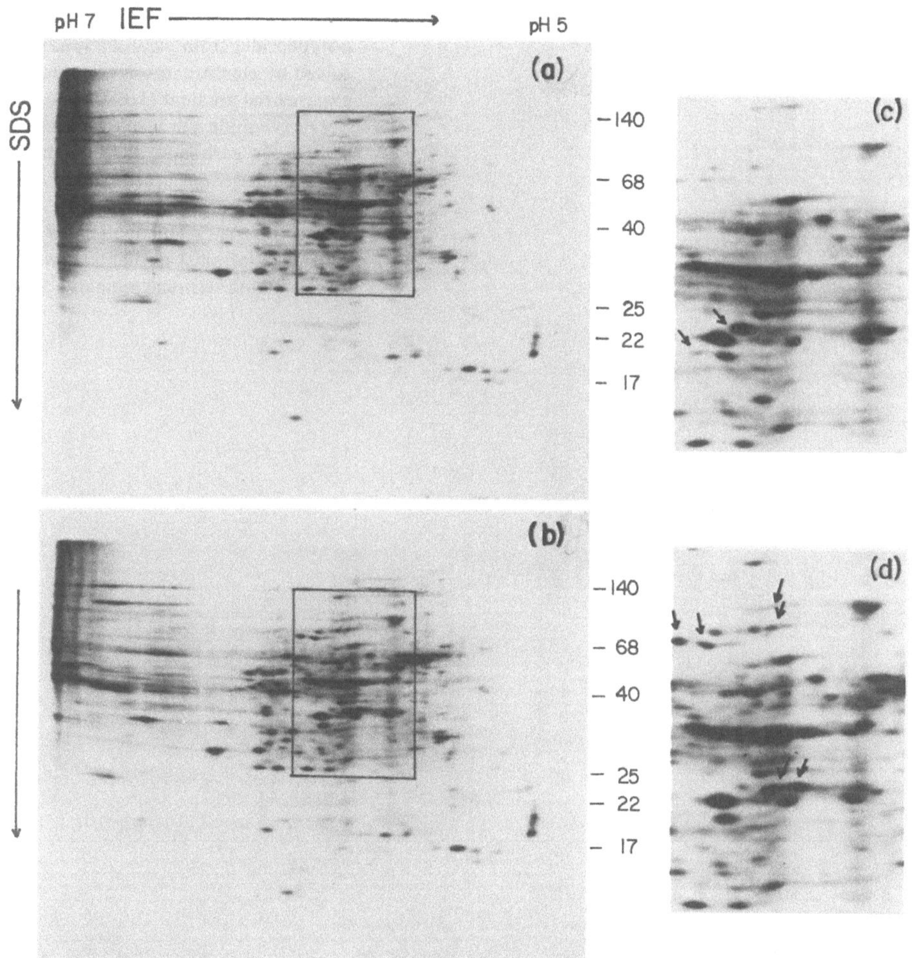

Fig. 3a–d. Fluorographs of ^{35}S-labeled polypeptides from elongating internal segments resolved by electrophoresis on two-dimensional polyacrylamide gels. Samples loaded on first dimension gels were 52 μg protein in each case; 78,000 cpm were applied in the case of untreated and 118,000 cpm were applied in the case of auxin-treated tissue. Second dimension gels are concave exponential gradient (10%–16%) polyacrylamide gels. **a** untreated segments; **b** auxin-treated segments; **c** enlargement of region outlined in **a**; **d** enlargement of region outlined in **b**

into polypeptides of internal segments rather than terminal 1-mm segments. One-dimensional gel analysis is sufficient to resolve many significant differences in the patterns of protein synthesis in untreated and auxin-treated basal hypocotyl sections (Fig. 4). Synthesis of a large spectrum of polypeptides is promoted or repressed in response to auxin treatment as evidenced by the polypeptide patterns observed on two-dimensional gels (Fig. 5).

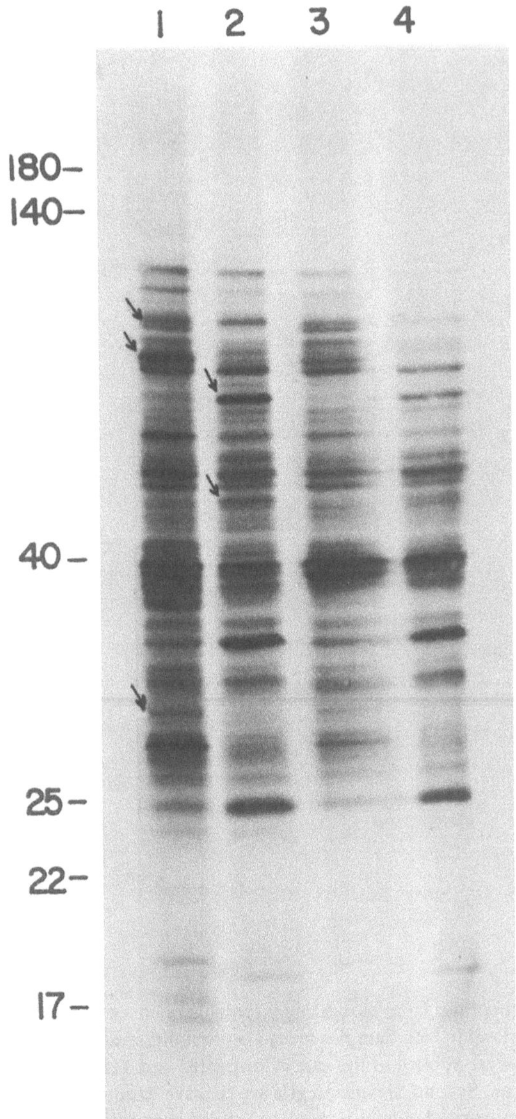

Fig. 4. Fluorograph of ^{35}S-labeled polypeptides from basal sections resolved by electrophoresis on concave exponential gradient (10%–16%) polyacrylamide gels in the presence of SDS. To each lane, 28,000 cpm were applied. *Lane 1* untreated internal segments; *Lane 2* auxin-treated internal segments; *Lane 3* untreated terminal segments; *Lane 4* auxin-treated terminal segments

Discussion

In order to clearly observe auxin-induced changes in the spectrum of polypeptides synthesized in soybean hypocotyl sections, it is necessary to employ high-resolution techniques, such as one-dimensional exponential gradient polyacrylamide gels and two-dimensional polyacrylamide gels. With elongating sections, it is also important to remove the wounded portions (e.g., 1-mm terminal segments) since polypeptides synthesized in response to wounding can mask the changes in polypeptide synthesis induced by auxin.

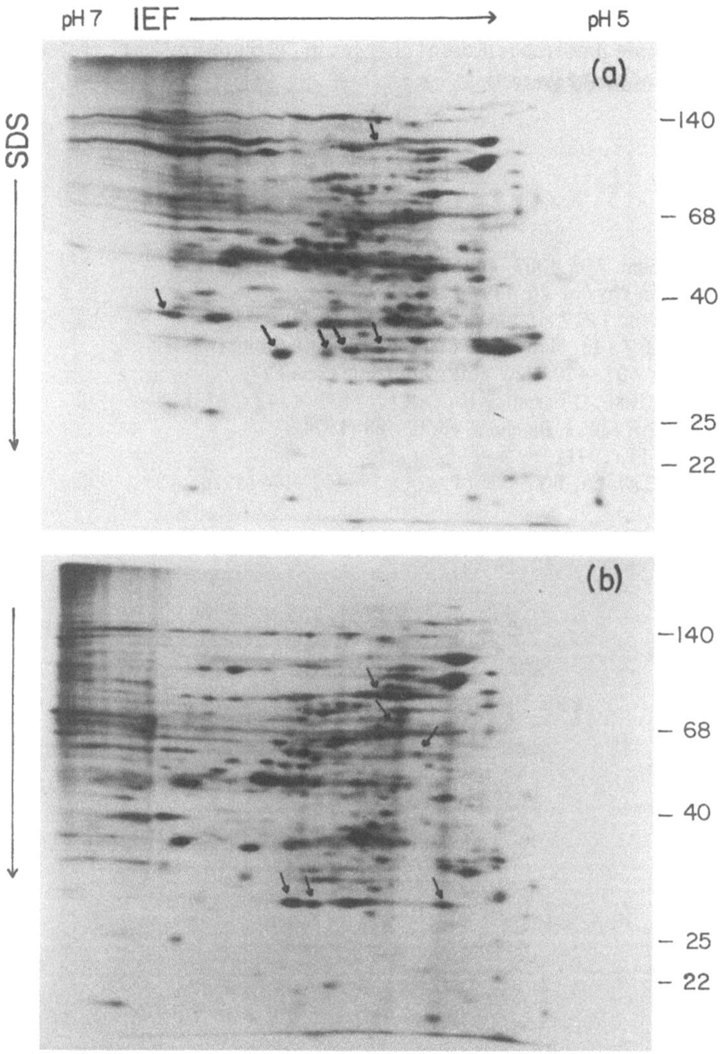

Fig. 5a, b. Fluorographs of ^{35}S-labeled polypeptides from basal internal segments resolved by electrophoresis on two-dimensional polyacrylamide gels. Samples loaded on first dimension gels were 18 μg protein in each case; 424,000 cpm were applied in the case of untreated and 409,000 cpm were applied in the case of auxin-treated tissue. Second dimension gels are 10% polyacrylamide. **a** untreated segments; **b** auxin-treated segments

Our results show that auxin treatment alters the pattern of protein synthesis and/ or causes charge modification of polypeptides in incubated sections of soybean hypocotyl. The changes in ^{35}S-methionine-labeled protein patterns induced by auxin are highly different in elongating and mature hypocotyl sections. It remains to be determined (a) whether auxin-induced changes in the pattern of polypeptide synthesis result from changes in the population of specific messenger RNAs, (b) what specific enzymatic activities or structural roles these polypeptides represent and (c) whether these changes in polypeptide patterns represent changes in "growth-limiting" proteins (7, 8) required for auxin-induced growth.

References

1. O'Farrell, P.H.: J. Biol. Chem. *250*, 4007–4021 (1975)
2. Key, J.L.: Annu. Rev. Plant Physiol. *20*, 449–474 (1969)
3. Laemmli, U.K.: Nature (Lond.) *227*, 680–685 (1970)
4. O'Farrell, P.H., O'Farrell, P.Z.: In: Methods in Cell Biology, XVI. Stein, G., Stein, J., Kleinsmith, L.J. (eds.), pp. 407–420. New York: Academic Press 1977
5. O'Farrell, P.Z., Goodman, H.M., O'Farrell, P.H.: Cell *12*, 1133–1142 (1977)
6. Bonner, W.M., Laskey, R.A.: Eur. J. Biochem. *46*, 83–88 (1974)
7. Cleland, R.: Planta *99*, 1–11 (1971)
8. Vanderhoef, L.N., Stahl, C.A., Lu, T-Y.: Plant Physiol. *58*, 402–404 (1976)

Auxins – Summary of Other Reports

L. TAIZ [1]

The "acid growth hypothesis" continues to influence research on auxin-induced elongation. Indirect evidence in support of acid growth in corn root segments was presented by Mueller and Scott, who found that growth inhibition caused by anaerobiosis and low temperature (15°) could be reversed by 10^{-4} M HCl. However, Abu-Tabikh and Vanderhoef pointed out that measurements of cell wall pH must take into account the intrinsic buffering capacity of the wall, which tends to maintain the free space solution at about pH 5.4. In soybean hypocotyl, the wall-buffering capacity is independent of auxin, but is subject to inhibition by metabolic poisons. In the Characean alga, *Nitella,* Metraux et al. provided evidence that light-induced proton extrusion, which occurs in localized bands along a single giant cell, promotes elongation in this region by enhancing wall extensibility. Since labeling studies indicated that glucose incorporation occurs equally in growing and nongrowing regions of the wall, it was concluded that proton extrusion, rather than wall deposition, regulates cell elongation.

The mechanism of acid-induced wall extensibility in plasmolyzed or frozen-thawed *Helianthus* stem segments was explored by Böttger and Soll. Evaluation of the pH and temperature curves for wall extension under an externally applied force led them to conclude that a physical process such as the dissolution of calcium pectate bridges, rather than a wall enzyme, may be responsible for acid-induced extensibility. Whether or not wall enzymes are involved in acid growth, Huber and Nevins have isolated endo- and exoglucanases from corn coleoptile walls. These glucanases appear to be responsible for both the autolytic behavior in isolated wall preparations and the decrease in hemicellulosic β-D-glucan associated with auxin-induced growth.

In addition to wall acidification, auxin has more long-term effects on the cell. Dute and Vanderhoef provided additional support for the "dual mechanism" model of elongation in the corn mesocotyl, based on the biphasic growth response. Bown and Smith reported on the pH-stat mechanism which operates to maintain cytoplasmic pH during proton secretion. The role of PEP carboxylase and malic enzyme as a pH-stat system was studied using metabolities with modulator functions at various pH values. The results were, in general, consistent with the pH-stat mechanism, although not all the criteria for this function were fulfilled.

The effect of auxin on structural gene expression in soybean hypocotyls was investigated by Baulcombe et al. Associated with a 2,4-D-induced shift from cell elongation to cell division there were increases and decreases in a few mRNA sequences, as judged

1 Thimann Laboratories, University of California, Santa Cruz, California 95064, USA

by in vitro translation products using polyadenylated RNA. Hybridization analysis indicated that the major effect of auxin was to decrease some abundant RNA species.

Auxin control of the cell cycle was also investigated in several tissue culture systems. Minocha found that ABA increased the percentage of cell divisions in auxin-treated Jerusalem artichoke tissue slices. Everett presented evidence that 2,4-D promotes cell division indirectly in sycamore cells by maintaining some essential metabolic process. Leguay found that a threshold level of 2,4-D was necessary to maintain cell division in sycamore cells.

Finally, there were two papers on auxin-binding to receptors. Pooviah and Mudge were able to show that the specificity of several auxin analogs, with regard to growth promotion of strawberry receptacles, paralleled their binding affinities to membranes isolated from the same tissue, which suggests that the binding has physiological significance. Oostrom et al. reported partial purification of a cytoplasmic auxin receptor in tobacco callus cells. This receptor, in the presence of auxin, promoted transcription in isolated nuclei.

Cytokinins

Chairmen: N. J. LEONARD and D. J. ARMSTRONG

Metabolites of Cytokinins

B. ENTSCH, D.S. LETHAM, C.W. PARKER, R.E. SUMMONS, and B.I. GOLLNOW [1]

Introduction

The view that root-produced cytokinin moves in the xylem to the shoot where it regulates development and senescence is now a widely accepted concept in developmental botany. However, little is known of the regulation of cytokinin movement to the shoot or of the distribution and metabolism of cytokinin in shoot organs. To provide a chemical basis for worthwhile physiological studies in this area, we have studied the metabolites formed from cytokinins supplied to shoot tissues and have characterized two of the relevant enzyme systems. Finally, the identification and quantitation of cytokinins in plant tissues by mass spectrometry is discussed.

Results and Discussion

Metabolite Structure, Occurrence and Significance

When supplied exogenously to shoot and other plant tissues, zeatin is converted to a complex of metabolites (Table 1; abbreviations used for metabolites listed therein). Recently identified metabolites are discussed in this paper and include glucose conjugates with a β-D-glucopyranosyl moiety at positions 7, 9, or *O*. Other zeatin metabolites are β-[6-(4-hydroxy-3-methylbut-trans-2-enylamino)purin-9-yl]alanine, a compound termed lupinic acid, and dihydrolupinic acid. Metabolites of 6-benzylaminopurine (BAP) identified recently are the 3-, 7- and 9-glucopyranosides, termed 3G-BAP, 7G-BAP and 9G-BAP respectively, and the BAP analogue of lupinic acid. All the above metabolites have been identified unequivocally by comparison with unambiguously synthesized compounds (4, 6, 7, 20–23, 25, 30, 31). Many of these metabolites have unusual structures. Thus, the only previously known purines from natural sources with a sugar at position 7 are a few compounds related to vitamin B_{12}, and in these the sugar is ribose. The only other known natural purines with a sugar at position 3 are 3-ribosyluric acid and its 5′-phosphate, both obtained from beef blood. Purines with an amino acid moiety attached to a ring N atom have not been found previously in plants.

Formation of metabolites of exogenously supplied zeatin in various plant tissues is summarized in Table 1. The compilation is confined to studies in which the identification of glucoside metabolites involved critical cochromatography with authentic syn-

1 Research School of Biological Sciences, Australian National University, Canberra, A.C.T., Australia

Table 1. Metabolites of zeatin formed in plant tissues. [^{3}H]Zeatin was supplied to all tissues except sweet corn kernels which received [^{3}H]zeatin riboside. These cytokinins were metabolized for over 20 h before the tissue was extracted. A major metabolite is denoted by +++, while + indicates a minor metabolite; a metabolite present in intermediate amounts is designated ++. A dash indicates the metabolite was not detected after a critical examination of the extract. A blank indicates the metabolite, if present, was at a level considerably below that denoted by one +

Identified Metabolites [a]	Derooted radish seedlings (29)	Radish roots (11)	Derooted sweet corn seedlings (30)	Sweet corn roots (30)	Sweet corn kernels [b]	Lupin shoots (31)	Immature lupin seeds [b]	Poplar leaves (6)	Immature apple seeds [b]
Simple Derivatives of Zeatin									
Dihydrozeatin	–	–			+	++		+	+++
Zeatin riboside (ZR)	+	+		+		+	++	+	+
Dihydrozeatin riboside (DZR)						+	+++		+
Zeatin Nucleotides	+	+	–	++		+++		–	
Dihydrozeatin Nucleotides	–	–			} +	++	} ++	–	} +
Glucoside Metabolites									
7-Glucosylzeatin (7GZ)	+++	+++	++	+		+		+	
9-Glucosylzeatin (9GZ)				+++		+		+	+
O-Glucosylzeatin (OGZ)						+++	++	+++	+
O-Glucosyldihydrozeatin (OGDZ)					+	+ [c]	++	+++	+
O-Glucosylzeatin riboside (OGZR)						+ [c]	++	++ [d]	
O-Glucosyldihydrozeatin riboside (OGDZR)					++	+ [c]	+++	+++	
Amino Acid Metabolites									
Lupinic acid	–					+++	++	–	++
Dihydrolupinic acid	–					+ [c]	+	–	
Products of N^6 side-chain cleavage									
Adenine	+		+++	++	+++	++		+	++
Adenosine	+		+++	++	++	++		++	+++
Adenine nucleotides	+	+	++	++	+++	++	++	+	++

[a] Identification of metabolites was based on chromatographic and electrophoretic studies. In the case of *O*-glucoside and nucleotide metabolites and lupinic acid, enzymic degradation was also used to establish identity

[b] Unpublished work [c] Metabolite in *Lupinus luteus*; the other data apply to *L. angustifolius* [d] Metabolite of zeatin riboside in *Populus nigra*; the other data apply to zeatin metabolism in *P. alba*

thetic glucosides. In some tissues (e.g., derooted radish seedlings, radish roots, *Zea mays* roots), only one major metabolite was formed and in this the zeatin side-chain was conserved. In other tissues (e.g., derooted sweet corn seedlings, and sweet corn kernels), adenine and its derivatives were the major metabolites, and hence isoprenoid side-chain cleavage was the dominant form of metabolism. In lupin shoots and seeds and in poplar leaves, the metabolism of zeatin was very complex. In the case of lupin shoots, it is likely that all 16 listed metabolites were formed plus three partially characterized compounds (31).

Although 7GZ is listed (Table 1) as the dominant metabolite in derooted radish seedlings and in radish roots, zeatin riboside 5′-phosphate was the dominant metabolite after a short period of metabolism (1–3 h).

No definite function has been established for 7- and 9-glucosyl metabolites or for 3G-BAP, the 7-glucosides being much less active than the parent cytokinin. However formation of these metabolites is not simply a mechanism for inactivation of unphysiological excesses of cytokinin in tissue. Thus, when BAP was supplied to excised radish cotyledons at a concentration (0.30 μM) far below the optimum (25 μM) for a growth response, the percentage of the metabolite radioactivity due to 3G-, 7G- and 9G-BAP was almost the same as when BAP was supplied at 120 μM, a supraoptimal level. Gawer et al. (10) detected marked and almost immediate 7G-BAP formation in tobacco cell cultures when BAP was supplied at a concentration of only 6×10^{-3} μM.

All the glucoside metabolites of exogenous zeatin mentioned above have now been found to occur naturally in plants. New free naturally occurring cytokinins to which structures were assigned during the period 1976–1979, and the sources of these compounds, are listed below: OGZ and OGZR, from crown-gall tumor tissue (27, 32) and sweet corn kernels (35); OGDZ, from *Phaseolus vulgaris* leaves [(39), tentative identification] and sweet corn kernels (35); OGDZR, sweet corn kernels (35); DZR, bean leaves (38) and sweet corn kernels (35); 7GZ, radish seeds (36); 9GZ and dihydro-9GZ, sweet corn kernels (35); cis-zeatin riboside, cones of hops (40); 2-methylthio-cis-zeatin, from *Corynebacterium fascians* (2); 2-methylthiozeatin riboside (trans/cis?) and N^6-(isopentenyl)-2-methylthioadenosine, from *Agrobacterium tumefaciens* (18); and 3-(3-amino-3-carboxypropyl)-6-(3-methylbut-2-enylamino)purine (discadenine), *Dictyostelium discoideum* (1). cis-Zeatin riboside may also occur in potato tubers (26) and in *Mercurialis ambigua* (5). This cytokinin had previously been isolated only from tRNA hydrolysates. The number of free naturally occurring cytokinins now identified is 25.

Identification of Cytokinins by Mass Spectrometry

In our recent studies of sweet corn cytokinins, which identified OGZ, OGZR, OGDZ, OGDZR, 9GZ, dihydro-9GZ and DZR, chemical-ionization mass spectrometry was used for the first time to elucidate cytokinin structure, and the technique was applied to TMS derivatives. Our results suggest that use of the chemical-ionization technique for cytokinins merits further development, since not only do we observe the expected enhancement of abundances for molecular ion species but also prominent structure-related fragmentations often not evident in the electron-impact mass spectra (EI-MS). In the chemical-ionization mass spectra (CI-MS) recorded, the MH^+ peak was the most intense above *m/e* 400.

The chemical-ionization technique would appear to be of particular value for cytokinins having a sugar moiety linked through the side-chain oxygen atom. Thus, under electron-impact conditions, underivatized OGZR (MW. 513) does not show a M^+, and the only prominent fragment ions are derived from the zeatin moiety. The EI-MS of hepta-TMS OGZR exhibits an extremely weak M^+ (*m/e* 1017) and a weak M^+-CH_3 (*m/e* 1002) which could easily be concealed if background was high. The base peak at *m/e* 550, which arises by allylic cleavage, dominates the extremely complex EI-MS, while other structurally significant ions, including those derived from the sugar moieties, are all of very low abundance. In contrast, the high-mass region of the relatively simple CI-MS showed the following peaks: (*m/e* values > 400 with rel. int. in parentheses): 1020 (75), 1019 (95), 1018 (100), 1003 (2), 552 (variable int., 15–42), 550 (45), 484 (62), 467 (17), 439 (15). The peak at *m/e* 1018 is due to the MH^+ ion, while that at *m/e* 484 is attributable to cleavage of the complete side chain from N^6 with charge retention on the protonated base portion. This fragmentation is a prominent feature of the CI-MS we have recorded for other zeatin derivatives and is not evident or is insignificant in the corresponding EI-MS. Peaks at *m/e* 439 and 467 are due to fragment ions derived from the glucose moiety. These peaks, which are very intense (about 70% of MH^+) in the CI-MS of TMS OGZ, could arise from a population of MH^+ ions in which the glucose ring oxygen is protonated. Allylic cleavage with elimination of the C-1′ hydrogen could give the ion of *m/e* 467; elimination of CO from this could then yield the ion of *m/e* 439.

An unusual feature of many of the isobutane chemical-ionization spectra we have determined is the presence of anomalous ions of variable intensity at one atomic mass unit above the expected MH^+ peak. The origin of these ions is being studied.

Enzymes of Cytokinin Metabolism

The only enzymes which show specificity for cytokinins and which have been purified from plants are cytokinin oxidase (41), and cytokinin 7-glucosyltransferase and β-(9-cytokinin)alanine synthase, both characterized in our laboratory. Some properties of the two latter enzymes are discussed below.

From radish cotyledons, we extracted two distinct fractions of enzyme activity capable of forming 7GZ and traces of 9GZ. Both appeared to be necessary to account for the proportion of 7- and 9-glucose conjugates formed when BAP was fed to cotyledons (8). So far, one of these trace enzymes has been studied in detail (9) and named cytokinin-7-glucosyltransferase. The enzyme preparation was purified to a stage where kinetic behavior and substrate specificity could be studied without any interfering reactions. By HPLC on μBondapak C_{18} columns, glucoside conjugates could be quantitated accurately in less than n mole amounts in the presence of a 100-fold or more of purine substrate. The enzyme was found to be specific for UDP-glucose and TDP-glucose, but UDP-glucose is the most likely substrate in vivo. The transferase exhibited a strong preference for an adenine ring with a side chain of at least three carbon atoms attached to position N^6. Some aromatic substituents are also acceptable, and there is reasonable correlation between the general requirements for substrate acceptance and hormone activity. Compounds with aliphatic side chains are converted to 7-glucosides plus traces of 9-glucosides, but some aromatic side chains result in 7- and 9-glucosides in similar amounts from a single compound.

Some potential inhibitors of the enzyme with respect to zeatin as substrate have been examined, and it is clear that the most effective are compounds which are structurally very similar to the most active cytokinins. Two compounds which are competitive inhibitors of zeatin are worth special attention; 3-methyl-7-(*n*-pentylamino)pyrazolo[4,3-*d*]-pyrimidine (I) has a K_i of 2.2×10^{-5} M, and 4-(cyclopentylamino)-2-methylthiopyrrolo[2,3-*d*]pyrimidine has a K_i of 2.7×10^{-5} M. Skoog et al. (33) obtained evidence that I was a specific competitive inhibitor of cytokinins in induction of growth in tissue cultures, a result not substantiated by Helgeson et al. (14). Inhibition of glucosylation by I provides the first convincing evidence that I does compete with cytokinin at a site exhibiting considerable specificity for cytokinin-active molecules.

A metabolite of zeatin recently discovered is lupinic acid (Table 1), obtained from lupin shoots and seed and apple seed after supplying them with zeatin. Lupinic acid is very weakly active in bioassays. Preliminary work with a cell-free extract from lupin seed indicated that *O*-acetylserine was the source of the alanine moiety in lupinic acid (28). We have now further purified the enzyme activity from developing lupin seeds. The preparation catalyses the following reaction:

O-acetylserine + zeatin → acetate + lupinic acid.

No other products were formed by the purified preparation and mass spectrometry confirmed the structure of the lupinic acid produced. A full investigation of this enzyme was possible when paired-ion chromatography was used in conjunction with HPLC to accurately quantitate substrate and product in the enzyme reaction. Analysis of the dependence of the reaction on the concentration of substrates established that

Table 2. Specificity towards the acceptor of an alanine residue exhibited by β-(9-cytokinin)alanine synthase. The enzyme was obtained by purification of an extract from developing seed of *Lupinus luteus.* Results are confined to N^6 derivatives of adenine. Rates of formation of product are given relative to zeatin, for 1.5 mM substrate, in standardized assays, with a saturating concentration of *O*-acetylserine. Extensive results were collected for three compounds, with kinetic parameters as listed below

Compound	Relative rate	Km (mM)	V_{max} (μmol h^{-1} mg^{-1} protein)
Adenine	7.5	26.00	1.47
Methyladenine	25.0		
Hydroxyethyladenine	20.0		
Propyladenine	57.0		
Isopentenyladenine	290.0		
Isopentyladenine	99.0		
Zeatin (trans)	100.0	0.88	1.03
Dihydrozeatin	30.0		
cis-Zeatin	< 5.0		
Hydroxyhexyladenine	64.0		
O-β-D-Glucopyranosylzeatin	34.0	2.5	0.55

the mechanism was probably of the *ping pong bi bi* type (nomenclature of Cleland), with the K_m for zeatin, 8.8×10^{-4} M, and for *O*-acetylserine, 4.7×10^{-5} M.

The enzyme required an adenine ring for activity, but reached optimum activity when adenine was substituted with an alkyl group at N^6 (Table 2). The requirement was rather similar to that for the enzyme forming glucosides, except for a very strong preference for the trans rather than the cis isomer of zeatin. The kinetic parameters in Table 2 show that the discrimination on the basis of side chain is probably associated with the binding affinity of the enzyme for each compound. It was also found that kinetin and 6-benzylaminopurine were excellent substrates, at least as good as zeatin.

The two enzymes discussed above and cytokinin oxidase could be involved in hormone control by inactivation of existing molecules in a tissue. Unlike some other substituted forms of zeatin, the products formed by the transferase and synthase probably do not readily revert to zeatin in vivo (29, 31).

Purification and Quantitation of Cytokinins in Tissues

The limitations of cytokinin bioassays and the desirability of chemical quantitation methods have been discussed recently (19). Extraction and a degree of purification are necessary first steps in both chemical and biological quantitation of tissue levels of cytokinin. The procedures used have been reviewed recently (16, 19). HPLC undoubtedly holds great promise for the future. With the development of reverse-phase column packings (e.g., C_{18}- and C_8-substituted porous silica) with high capacity and high resolution, HPLC became particularly valuable for purification of cytokinins (3, 15, 17, 34, 35).

We have developed procedures for purifying cytokinins from plant tissues which are capable of resolving all naturally occurring cytokinin bases, ribosides, and glucosides, resulting in products of high purity (34, 35). Decomposition of metabolites is negligible due to the mild conditions used. When ^{3}H-labeled zeatin riboside (ZR) is used as an indicator during extraction and purification, overall recoveries of 60% and more are obtained. Extract equivalent to a maximum of 120 g of tissue (fresh weight) can be fractionated without any repetition of steps. The first step, chromatography on cellulose-phosphate, gives high recovery of cytokinins from crude extracts with removal of most of the unwanted solid matter. The great bulk of remaining solid matter is then eliminated by HPLC of the evaporated cellulose-phosphate eluate on Bondapak C_{18}/Porasil B, resulting in only mg or sub-mg quantities for later chromatographic steps. The next stage, chromatography on a column of μBondapak Phenyl, is most significant, since it results in another substantial purification, and appearance of individual peaks of cytokinins, if these are present in μg quantities. The final stages of purification involve the use of high-resolution analytical HPLC columns (μBondapak C_{18}, μBondapak Phenyl) to yield compounds sufficiently pure for direct-probe introduction into the mass spectrometer, a GC step being unnecessary (34, 35). Such purity was not demonstrated in isolations of cytokinins by HPLC reported in the literature. The use of the procedure is illustrated by the rapid isolation of known and new endogenous cytokinins from developing *Zea mays* seed (35).

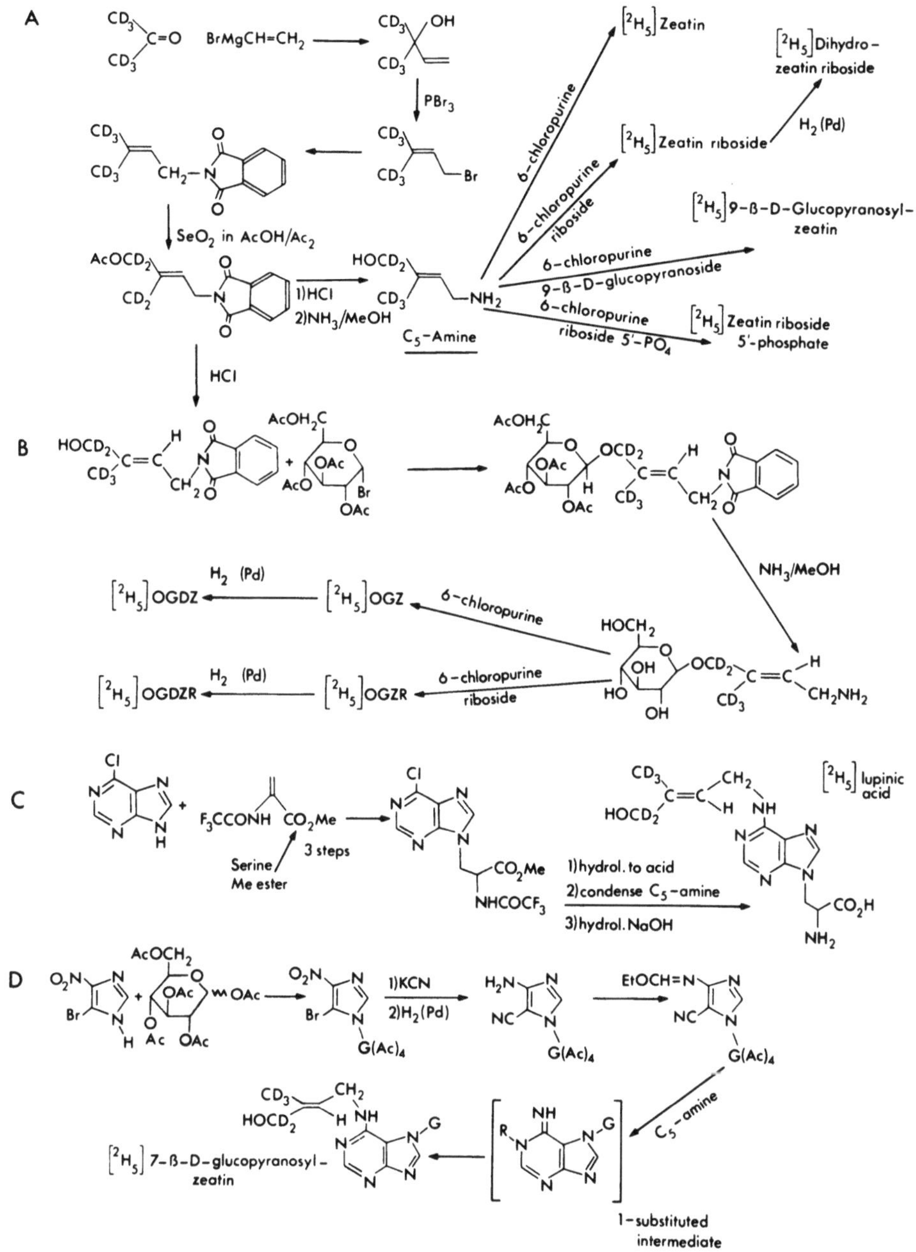

Fig. 1A–D. An outline of the synthesis of deuterium-labeled cytokinins. Each cytokinin carries five deuterium (D) atoms derived from hexadeutero-acetone. **A** the synthesis of 9-glycosides of zeatin. **B** the synthesis of *O*-β-D-glucopyranosylzeatin (OGZ), *O*-β-D-glucopyranosyldihydrozeatin (OGDZ) and the corresponding 9-ribosides (OGZR and OGDZR, respectively). **C** the synthesis of lupinic acid. **D** the synthesis of 7-β-D-glucopyranosylzeatin

Methods based on mass spectrometry are the best for reliable quantitation of cytokinins. Isotopically labeled cytokinins are required as internal standards for addition to the extracting solvent in order to accurately correct for losses in extraction and subsequent purification. Previously we have quantitated cytokinins in radish seed using dideuterium-labeled standards, the extracts being purified by TLC prior to mass spectral quantitation based on both complete EI-MS and SICM (36). The practices of quantitating cytokinins by mass spectrometry with no internal standard (5, 24), or with an isomeric compound as standard (37) could yield misleading results. The same applies to single ion current monitoring at low resolution (24, 37) and to addition of internal standards just prior to mass spectrometry, instead of to the extracting solvent (13).

To provide internal standards, we have now synthesized all major naturally occurring cytokinins with five deuterium atoms at specific non-exchangeable positions in the isoprenoid side chain [see Fig. 1, (34)]. These were added to the extracting solvent, recovered in a purified form by HPLC as outlined above and converted to TMS derivatives for determination of complete mass spectra by direct probe methods. Quantitation was based on the ratio of endogenous cytokinin (D_0) to $^2H_5$-analog (D_5) as evidenced by peak heights of selected ion pairs [see (34) for discussion of suitable ions]. Alternatively, the TMS derivatives of $^2H_5$-labeled zeatin, zeatin riboside and OGZ, and the corresponding dihydro compounds, may be subjected to GC-MS analysis. In this method, less stringent purification by HPLC is acceptable. It should be noted that OGZR and OGDZR are not amenable to routine GC-MS analysis, and direct probe methods are essential for these glucosides. Using the above methods, 100 ng of cytokinin can be quantitated very accurately with simultaneous unequivocal identification

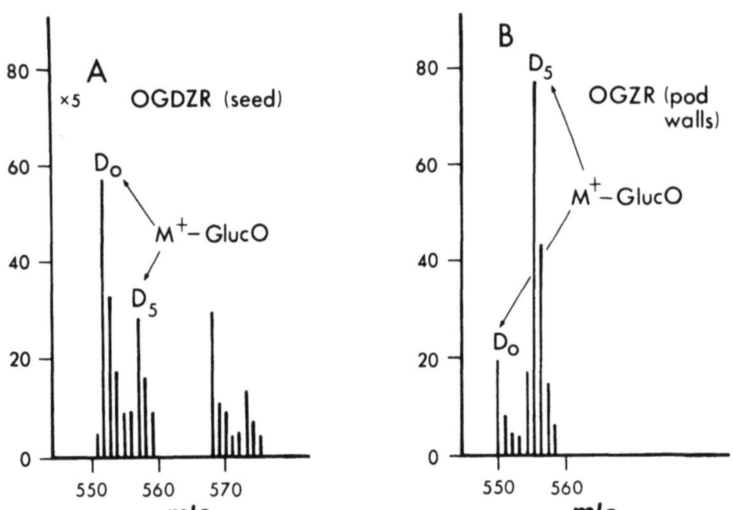

Fig. 2A, B. Regions of mass spectra of TMS derivatives of OGDZR (A) and OGZR (B) showing the ions used to quantitate levels of these cytokinins in lupin pods. D_0 and D_5 refer to the endogenous cytokinin and $^2H_5$-analog (internal standard) respectively. GlucO denotes a tetra-TMS glucosyl moiety; in both spectra, the quantifying ions were M$^+$-GlucO

provided by the complete spectra; with SICM, a few ng are sufficient [cf. (13, 36)] but unequivocal identifications are not obtained.

The levels of cytokinins in lupin (*Lupinus luteus,* plants about 14 days after petal fall) seed and pod walls were determined by the above method (direct probe technique) and were as follows (μg/100 g; values for pod walls in parentheses): OGZ, 0.68 (0.60); OGDZ, 1.7 (8.1); OGZR, 3.7 (3.9); OGDZR, 37 (110); ZR, 24 (0.0); DZR, 20 (16). Regions of two typical mass spectra used in the above quantitation are presented in Fig. 2. Developing seeds are frequently a rich source of cytokinin activity. Hence the high levels of ZR, DZR and OGDZR observed in lupin seed were not unexpected, the content of ZR being very similar to that found by mass spectrometry in immature sweet corn kernels (34). However, the presence of high levels of some cytokinins (e.g., OGDZR, DZR) in lupin pod walls was not anticipated. These results have physiological relevance. The cytokinin content of pea seeds increases markedly when the intact pods are cultured in vitro (12), and this increase has been used as evidence that pea seeds synthesize cytokinins (12). Because lupin pod walls contain DZR and a very high level (1 μg/g) of OGDZR, which is potentially capable of enzymic conversion to more active compounds, the observed increase in cytokinin content of pea seeds (12) may only reflect import of cytokinins from pod walls. Hence the origin of the cytokinins in pea seeds requires more critical study.

The new techniques for quantitating cytokinin and other plant hormone levels which are now being developed herald a new phase in plant hormone physiology. Bioassays have provided useful clues concerning correlations between cytokinin levels and plant development. It is important to confirm these correlations and look for new relationships by mass spectrometry techniques which, if properly employed, have the precision and certainty characteristic of chemical analysis.

References

1. Abe, H., Uchiyama, M., Tanaka, Y., Saito, H.: Tetrahedron Lett. *1976,* 3807 (1976)
2. Armstrong, D.J., Scarbrough, E., Skoog, F., Cole, D.L., Leonard, N.J.: Plant Physiol. *58,* 749 (1976)
3. Carnes, M.G., Brenner, M.L., Andersen, C.R.: J. Chromatogr. *108,* 95 (1975)
4. Cowley, D.E., Duke, C.C., Liepa, A.J., MacLeod, J.K., Letham, D.S.: Aust. J. Chem. *31,* 1095 (1978)
5. Dauphin, B., Teller, G., Durand, B.: Planta *144,* 113 (1979)
6. Duke, C.C., Letham, D.S., Parker, C.W., MacLeod, J.K., Summons, R.E.: Phytochemistry *18,* 819 (1979)
7. Duke, C.C., MacLeod, J.K., Summons, R.E., Letham, D.S., Parker, C.W.: Aust. J. Chem. *31,* 1291 (1978)
8. Entsch, B., Letham, D.S.: Plant Sci. Lett. *14,* 205 (1979)
9. Entsch, B., Parker, C.W., Letham, D.S., Summons, R.E.: Biochim. Biophys. Acta *570,* 124 (1979)
10. Gawer, M., Laloue, M., Terrine, C., Guern, J.: Plant Sci. Lett. *8,* 267 (1977)
11. Gordon, M.E., Letham, D.S., Parker, C.W.: Ann. Bot. *38,* 809 (1974)
12. Hahn, H., de Zacks R., Kende, H.: Naturwissenschaften *61,* 170 (1974)
13. Hashizume, T., Sugiyama, T., Imura, M., Cory, H.T., Scott, M.F., McCloskey, J.A.: Anal. Biochem. *92,* 111 (1979)

14. Helgeson, J.P., Haberlack, G.T., Hecht, S.M.: In: Plant Growth Substances 1973, p. 485. Tokyo: Hirokawa Publishing Co. 1974
15. Holland, J.A., McKerrell, E.H., Fuell, K.J., Burrows, W.J.: J. Chromatogr. *166*, 545 (1978)
16. Horgan, R.: In: Isolation of Plant Growth Substances. Hillman, J.R. (ed.), p. 98. Cambridge: Cambridge University Press 1978
17. Kannangara, T., Durley, R.C., Simpson, G.M.: Physiol. Plant. *44*, 295 (1978)
18. Kaiss-Chapman, R.W., Morris, R.O.: Biochem. Biophys. Res. Commun. *76*, 453 (1977)
19. Letham, D.S.: In: Phytohormones and Related Compounds: a Comprehensive Treatise. Letham, D.S., Goodwin, P.B., Higgins, T.J.V. (eds.), Vol. 1, p. 205. Amsterdam: Elsevier/ North-Holland 1978 (see also p. 18 of Chapter 1)
20. Letham, D.S., Gollnow, B.I., Parker, C.W.: Plant Sci. Lett. *15*, 217 (1979)
21. Letham, D.S., Parker, C.W., Duke, C.C., Summons, R.E., MacLeod, J.K.: Ann. Bot. *41*, 261 (1977)
22. Letham, D.S., Summons, R.E., Parker, C.W., MacLeod, J.K.: Planta *146*, 71 (1979)
23. Letham, D.S., Wilson, M.M., Parker, C.W., Jenkins, I.D., MacLeod, J.K., Summons, R.E.: Biochim. Biophys. Acta *399*, 61 (1975)
24. Lorenzi, R., Horgan, R., Wareing, P.F.: Biochem. Physiol. Pflanz. *168*, 333 (1975)
25. MacLeod, J.K., Summons, R.E., Parker, C.W., Letham, D.S.: J. Chem. Soc. Chem. Commun. *1975*, 809 (1975)
26. Mauk, C.S., Langille, A.R.: Plant Physiol. *62*, 438 (1978)
27. Morris, R.O.: Plant Physiol. *59*, 1029 (1977)
28. Murakoshi, I., Ikegami, F., Ookawa, K., Haginiwa, J., Letham, D.S.: Chem. Pharm. Bull. *25*, 520 (1977)
29. Parker, C.W., Letham, D.S.: Planta *114*, 199 (1973)
30. Parker, C.W., Letham, D.S.: Planta *115*, 337 (1974)
31. Parker, C.W., Letham, D.S., Gollnow, B.I., Summons, R.E., Duke, C.C., MacLeod, J.K.: Planta *142*, 239 (1978)
32. Peterson, J.B., Miller, C.O.: Plant Physiol. *59*, 1026 (1977)
33. Skoog, F., Schmitz, R.Y., Bock, R.M., Hecht, S.M.: Phytochemistry *12*, 25 (1973)
34. Summons, R.E., Duke, C.C., Eichholzer, J.V., Entsch, B., Letham, D.S., MacLeod, J.K., Parker, C.W.: Biomed. Mass Spectrometry (in press, 1980)
35. Summons, R.E., Entsch, B., Letham, D.S., Gollnow, B.I., MacLeod, J.K.: Planta *147*, 422 (1980)
36. Summons R.E., MacLeod, J.K., Parker, C.W., Letham, D.S.: FEBS Lett. *82*, 211 (1977)
37. Thompson, A.G., Horgan, R., Heald, J.K.: Planta *124*, 207 (1975)
38. Wang, T.L., Horgan, R.: Planta *140*, 151 (1978)
39. Wang, T.L., Thompson, A.G., Horgan, R.: Planta *135*, 285 (1977)
40. Watanabe, N., Yokota, T., Takahashi, N.: Agric. Biol. Chem. *42*, 2415 (1978)
41. Whitty, C.D., Hall, R.H.: Can. J. Biochem. *52*, 789 (1974)

Cytokinin Action on Enzyme Activities in Plants

O.N. KULAEVA [1]

Introduction

Many data have been collected in recent years dealing with cytokinin effects on enzyme activities in plants (2, 3, 5, 10, 14). In some cases cytokinin was shown to enhance the synthesis of an enzyme (3, 10) or to induce synthesis of a new protein (6). In some plant tissues, cytokinin-induced increase in enzyme activity is preceded by activation of RNA synthesis (27). However, it was shown that the change in the polysome : monosome ratio effected by cytokinin treatment was not blocked by inhibition of transcription (6).

It is too early to generalize these experimental data obtained on different tissues because two questions are still open:

1. Is the molecular mechanism of cytokinin action one and the same in controlling different physiological processes in plants?
2. Does cytokinin act at various independent parallel levels in the plant cell or does it affect one primary mechanism?

To answer these questions it is necessary first to study cytokinin effects on various processes in the same plant material and then to compare the action of cytokinin in different plant tissues. In particular it is of interest to compare the action of cytokinin in the systems in which it enhances enzyme activities and in which it induces de novo synthesis of enzyme, because activation and induction by cytokinin are thought to be carried out via different molecular mechanisms, albeit these may be concurrent.

We have investigated cytokinin promotion of enzyme activities in excised pumpkin cotyledons (1, 22) and induction of enzyme synthesis in excised embryos of *Agrostemma githago* L., in which Borriss (2, 3) and Kende et al. (9) have shown cytokinin-induced de novo synthesis of nitrate reductase (NR).

Materials and Methods

Excised cotyledons from dry pumpkin seeds (cv. Mozolevskaja) or from etiolated 3- to 4-day-old seedlings (8, 22), were incubated in 6-benzylaminopurine (BA) solution (10 mg/l) in Petri dishes in growth chambers at 28°C under continuous illumination with fluorescent lamps (29000 ergs/cm^2). Procedures for measuring enzyme activities in cotyledons have been described (7, 8, 11). We determined incorporation of label

1 K.A. Timiriazev Institute of Plant Physiology, Moscow, USSR

into proteins (13), rate of protein synthesis in vitro by isolated ribosomes (31), polysome : monosome ratios (31), incorporation of label into RNA (22), the effect of BA on rRNA accumulation in cytoplasm and chloroplasts (23) and on incorporation of label into nuclear and chloroplast RNAs.

Barley (cv. Viner) was grown in soil in a greenhouse. The first leaves were used for isolation of chromatin (26), nuclei and chloroplasts (17). These preparations were used for assaying RNA polymerase activitiy (17, 26).

The experiments with excised *A. githago* embryos were performed according to the procedure developed by Borriss (2). NR activity (18), and incorporation of label into proteins (18) and RNA (18, 19) were measured.

Results and Discussion

Cytokinin Action on Enzyme Activities in Excised Pumpkin Cotyledons

Excised pumpkin cotyledons are very sensitive to exogenous cytokinin, because their content of endogenous cytokinins decreases sharply after isolation (4). Treatment with BA stimulated their growth (1, 22) and led to development of subcellular structure

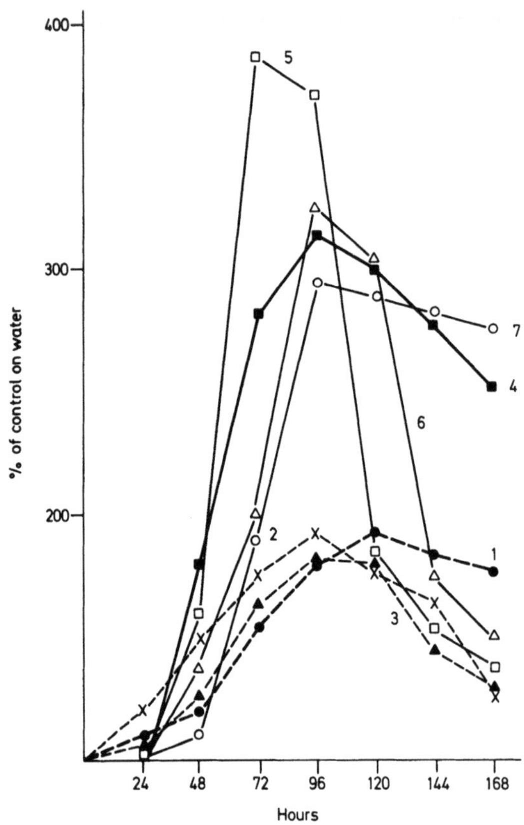

Fig. 1. Effects of BA on growth, chlorophyll content and enzyme activities of excised pumpkin cotyledons. Data are presented as percent of control (water-grown) cotyledons: *1* cotyledon growth; *2* endopeptidase; *3* acid pyrophosphatase; *4* chlorophyll content; *5* ribulosediphosphate carboxylase; *6* alkaline pyrophosphatase; *7* malic enzyme

characteristic of cotyledon transformation from a storage organ to a green cotyledonary leaf (11, 22, 14). Cytokinin promoted the mobilization of protein and fat, the development of endoplasmic reticulum and mitochondria, and the maturation of proplastids into chloroplasts (11, 22, 14). Accordingly, cytokinin activated a number of enzymes, participating in a wide range of metabolic reactions in the cotyledons (7, 8, 11). We have shown this for endopeptidases which degrade aleurone proteins (7), for pyrophosphatases which catalyze various syntheses (8), for malic enzyme participating in organic acid metabolism (8) and for ribulosediphosphate carboxylase and phosphoenolpyruvate carboxylase, two enzymes which function in photosynthesis (11). The enzymes investigated may be divided into two groups according to their response to cytokinin treatment (Fig. 1). The first group is characterized by a pronounced response to cytokinin treatment; their activities and chlorophyll accumulation in cotyledons being enhanced to the same extent. The enzymes in the second group show responses to cytokinin which are correlated with the growth response of the cotyledons. The first group of enzymes can be said to relate to chloroplast differentiation, while the second group relates to cytokinin-stimulated growth of the cotyledons. These data support the assumption that cytokinin affects individual enzyme activities through its action in the regulation of programmed development, the realization of which depends on coordinated functioning of numerous enzymes.

Stimulation by cytokinin of enzyme activities in pumpkin cotyledons is associated with protein synthesis, because cycloheximide eliminates this effect and inhibits the concomitant stimulatory effects on cotyledon growth and chloroplast differentiation (7, 8, 11). Accordingly, cytokinin promotes incorporation of label into protein in excised pumpkin cotyledons (13) and accelerates changes which occur in the antigenic protein spectrum in the course of growth and greening of cotyledons. Cytokinin has been found to increase the rate of in vitro protein synthesis by ribosomes isolated from cytokinin-treated cotyledons as compared with untreated controls [(31; Fig. 2]. This effect is accompanied by an increase in the polysome : monosome ratio in the cells of the cotyledons (12) (Fig. 3). Hence the cytokinin effect on enzyme activities in excised pumpkin cotyledons is associated with an increase in total protein synthesis.

In considering at what level of regulation cytokinin intervenes to effect this increase, the following two findings are of interest. First, the polysome : monosome

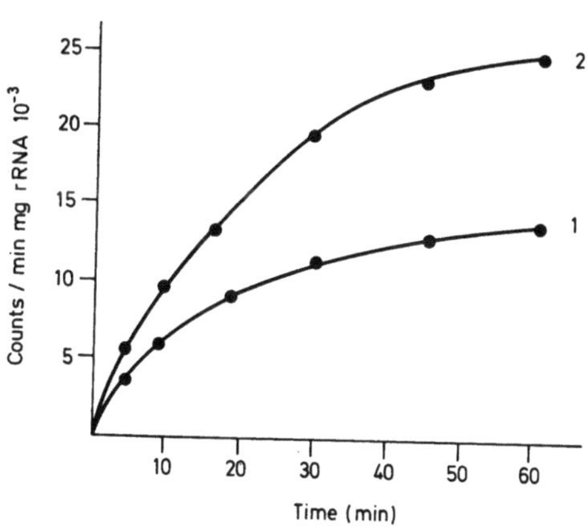

Fig. 2. ^{14}C-leucine incorporation into protein in vitro by polysomes from excised pumpkin cotyledons. Cotyledons excised from etiolated 4-day-old seedlings were preincubated for 24 h in water in the dark, then incubated under illumination for 24 h, *1* in water, or *2* in BA (10 mg/l)

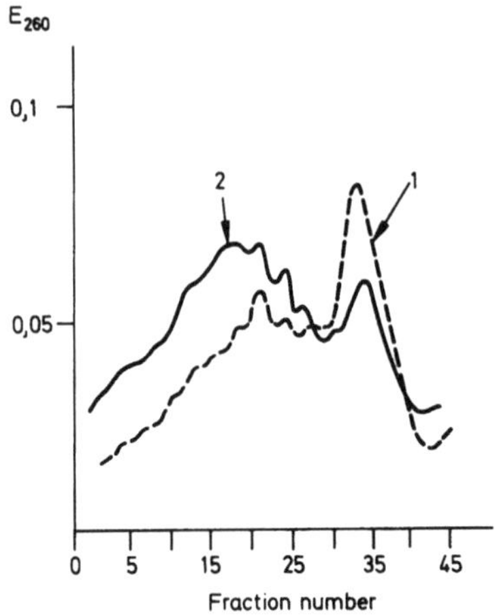

Fig. 3. Sucrose density gradient analysis of polysomes from pumpkin cotyledons held for 24 h *1* in water, or *2* in BA (10 mg/l) solution. Cotyledons were excised from 4-day-old etiolated seedlings and preincubated for 24 h in the dark in water. [For details see (12)]

ratio rose throughout the first 2 h of cytokinin treatment of the cotyledons, and the rise was not counteracted by cordycepin, which inhibited RNA labeling by 80% in this experiment. This observation confirms the proposal by Fosket et al. (6) of a posttranscriptional effect of cytokinin on translation. Second, 2 h after the cytokinin treatment, when the polysome fraction had increased, activation of transcription had already occurred, a fact which would also account for enhancement of translation. Hence, an action of cytokinin on protein synthesis at the transcriptional and posttranscriptional levels both remain equally probable.

The promotion of transcription by cytokinin can be revealed much earlier than its effect on the growth of cotyledons and on enzyme development. Cytokinin activates both nuclear and chloroplast RNA synthesis (23). It is evident that such activation could result from modified template activity of chromatin or from increased activity of RNA polymerases. Unfortunately, excised pumpkin cotyledons contain too much reserve material to be suitable for isolation of chromatin and RNA polymerase preparations.

In vitro Cytokinin-Activation of Chromatin-Bound RNA Polymerase from Barley Leaves

As to mechanisms whereby cytokinin may activate transcription, consideration of its effect on RNA polymerase in barley leaves is pertinent (17, 25, 26). Cytokinin is known to activate the synthesis of all types of RNA in detached barley leaves (16, 30) and to enhance chromatin-bound RNA polymerase I activity in leaves (25, 28). This latter effect may result from either activation of preformed enzyme or promotion of RNA polymerase synthesis. The first possibility is supported by our data (Table 1) that the activity of chromatin-bound RNA polymerase can be increased by cytokinin added to the leaf homogenization medium in the course of chromatin isolation (26). This effect was strictly dependent on the cytokinin concentration (Fig. 4). The optimal BA concentration was much lower in this case than in experiments with intact

Table 1. Effect of BA added to barley leaf homogenate on chromatin-bound and solubilized RNA polymerase activity

Treatment	Chromatin-bound enzyme		Solubilized enzyme	
	c.p.m./100 μg DNA	%	c.p.m./100 μg of protein	%
Control	1145	100	1140	100
BA (0.1 mg/l)	1633	142	2681	235

BA was added to homogenization medium. Ten-day-old barley leaves were homogenized to obtain chromatin preparations containing RNA polymerase. The reaction mixture for RNA polymerase estimation contained 33.5 mM Tris-HCl buffer, pH 8.0, 6.7 mM $MgCl_2$, 6.7 mM 2-mercaptoethanol, 1.34 mM nucleoside triphosphates including ^{14}C-ATP and chromatin (20–100 μg DNA). Incubation period 15 min at 30°C. [Details in (26)]

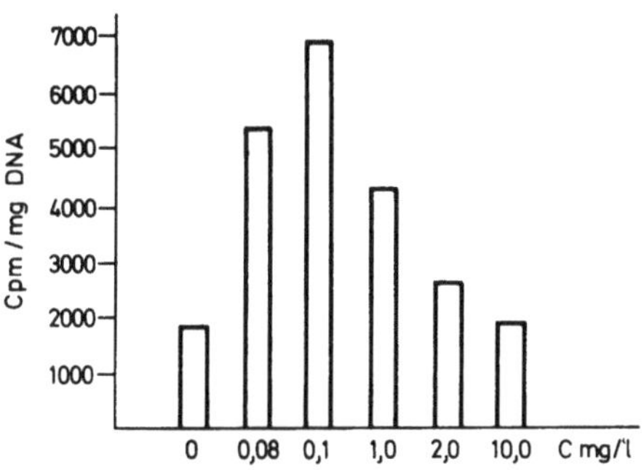

Fig. 4. Effect of BA added to the leaf homogenization medium on the activity of chromatin-bound RNA polymerase (see legend to Table 1)

leaves (26). The effectiveness of cytokinin added to the leaf homogenate excludes the possibility of hormone action on RNA polymerase synthesis in these experiments and points rather to a cytokinin effect on RNA polymerase activity in this case.

But this kind of experiment does not reveal whether or not cytokinin affected RNA polymerase directly or through its influence on chromatin to which RNA polymerase is bound. In any case, cytokinin added to the homogenate did increase the activity of chromatin-bound RNA polymerase, and the increase persisted after enzyme solubilization (Table 1).

BA addition at successive stages of chromatin purification or directly to the incubation medium for RNA synthesis had no effect. The data support the assumption that some nonchromatin cytokinin acceptor(s) or cofactor(s) are necessary for its action on chromatin-bound RNA polymerase.

Confirmation of this assumption came from the experiments with leaves of different ages. Cytokinin added to a leaf homogenate stimulated chromatin-bound RNA polymerase only in fully expanded, 8- to 13-day-old leaves (Table 2). No stimulation was observed either in young, expanding leaves (3- to 5-day-old) or in old, yellow leaves (15-day-old). Chromatin-bound RNA polymerase of 3-day-old leaves was sensitive to cytokinin only after the following procedure: The homogenate of 10-day-old leaves, after cytokinin addition, was filtered through two layers of cheesecloth and

Miracloth, chromatin-containing precipitate was removed by low speed centrifugation (4000 rpm) and the BA-containing supernatant was used as a homogenization medium for isolation of chromatin-bound RNA polymerase from 3-day-old leaves (Table 3).

All this evidence is consistent with the above-mentioned assumption that some non-chromatin factors present in the supernatant of cytokinin-sensitive 10-day-old leaves are necessary for cytokinin promotion of chromatin-bound RNA polymerase activity.

Table 2. Effect of BA added to the leaf homogenate on activity of chromatin-bound RNA polymerase prepared from barley leaves of different ages. Change in activity is expressed as percent of activity in absence of BAP

Age of leaves (days)	6	8	10	15
Changes in activity (%)	– 24	+ 160	+ 66	– 27

BA (0.1 mg/l) was added to homogenization medium (26). The incubation medium for estimating RNA polymerase activity is specified in legend to Table 1

Table 3. Effect of supernatant from 10-day-old barley leaf homogenate on activity of chromatin-bound RNA polymerase obtained from 3-day-old leaves

Age of leaves (days)	BA (0.1 mg/l)	Homogenization medium	^{14}C-AMP incorporation	
			c.p.m./100 μg DNA	%
3	–	Buffer	25,622	100
	+		16,415	64
10	–	Buffer	1,632	100
	+		14,466	886
3	–	Supernatant from 10-day-old leaves	9,630	100
	+		15,652	162

For details see legend to Table 1

Cytokinin Action on Nuclear RNA Polymerase I and II and on Chloroplast RNA Polymerase Isolated from Barley Leaves

As the above effect of BA on chromatin-bound RNA polymerase in leaf homogenates can differ from the effect in intact plants, our next aim was to test the effect of BA in a more structured system, i.e., in isolated nuclei. Also in these experiments cytokinin was added just at the start of leaf homogenization. The isolated nuclei contained active RNA polymerase I and II (Table 4). The incorporation of label into nuclear RNA was markedly increased by cytokinin (17).

Cytokinin stimulation of nuclear RNA polymerases, as that of chromatin-bound RNA polymerase was obtained only with fully expanded, 9- to 10-day-old leaves. It was also obtained with nuclei from isolated protoplasts from barley leaves.

Table 4. Effect of BA added to leaf homogenization medium on RNA polymerase activity of nuclei and chloroplasts isolated from 10-day-old barley leaves

RNA polymerase	BA (0.1 mg/l) addition	^{14}C-AMP incorporation	
		c.p.m./100 μg DNA	%
RNA-P-I	–	2,902	100
	+	5,524	190
RNA-P-II	–	3,420	100
	+	12,374	362
RNA-P of chloroplasts	–	27,180	100
	+	48,240	175

Leaf homogenization medium: 0.05 M Tris-HCl (pH 7.8); 0.1 M sucrose; 0.01 M KCl; 0.01 M $MgCl_2$; 0.004 M mercaptoethanol; 25 μg/ml chloramphenicol. Homogenate was filtered through two layers of cheesecloth, two layers of flannel, one layer of nylon and centrifuged for 10 min at 1000 *g*. The nuclei and chloroplasts were purified by sucrose density gradient centrifugation. Nuclei were assayed for RNA polymerase I and II activities. Incubation medium (0.5 ml) for RNA polymerase I: tris-HCl buffer 100 mM (pH 8.3); $MgCl_2$ – 16 mM; GTP, CTP, UTP – 0.2 mM; ^{14}C-ATP – 0.006 mM (321 μCi/μmol); nuclei 20–150 μg DNA. Incubation medium for RNA polymerase II: the same composition as for RNA polymerase I, but pH 7.8, $MgCl_2$ was excluded and 4 mM $MnCl_2$ and 0.25 M $(NH_4)_2SO_4$ were added. Nuclei were incubated at 32°C for 25 min. For the estimation of RNA polymerase activity of chloroplasts the same incubation medium was used as for RNA polymerase I. Chloroplasts were incubated at 25°C for 25 min

BA addition to the leaf homogenates also resulted in increased RNA polymerase activity in chloroplast suspensions (17) purified by sucrose density gradient centrifugation. In the course of chloroplast isolation, the integrity of their envelopes was disturbed. This facilitated the penetration of labeled triphosphate to the membrane-bound DNA-RNA-polymerase complex in the chloroplasts. Hence, the activation of chloroplast RNA labeling by cytokinin does not seem to be ascribable to a cytokinin effect on the permeability of the envelope to labeled precursors. As chloroplast RNA polymerase is coded for in the nucleus and formed in the cytoplasm, its content in chloroplasts cannot have been changed by cytokinin added just prior to homogenization of the leaves. Hence cytokinin probably enhanced the activity of preformed chloroplast RNA polymerase.

The failure of cytokinin to affect the Hill reaction in isolated chloroplasts is consistent with its specific action being on the RNA polymerases. Thus the results obtained with isolated nuclei and chloroplasts are in good agreement with the data on chromatin-bound RNA polymerase.

It may be assumed, therefore, that the in vitro activation of chromatin-bound RNA polymerase by cytokinin must not be an artifact due to the procedure of chromatin isolation but represents a bona fide hormone effect exerted on the nuclear RNA polymerases in the cells. It is probable that this mechanism is operative in activating synthesis of all types of RNA in the detached barley leaves (16, 30) as well as in the excised pumpkin cotyledons (22, 23) where the hormone was found to stimulate total RNA and protein synthesis.

Cytokinin Induction of Nitrate Reductase in Excised Embryos of *Agrostemma githago*

A significantly different mode of cytokinin action is involved in the case of enzyme induction not accompanied by activation of total protein and RNA synthesis, as demonstrated in excised embryos of *Agrostemma githago* L.

Borriss (2, 3) was the first to demonstrate that cytokinin, as well as nitrate, can induce NR activity in excised embryos of *A. githago.* According to data obtained by Kende et al. (10) and Borriss (3), cytokinin induces de novo synthesis of the enzyme. Both the hormone and nitrate induce synthesis of the same enzyme, but the hormone does so independently of the nitrate (9, 21). For this reason excised *A. githago* embryos are very suitable for investigating the action of cytokinin as an inducer. As our experiments with cycloheximide added at various intervals throughout the enzyme induction period have shown (18), NR is a very labile protein, readily inactivated after inhibition of protein synthesis by cycloheximide (Fig. 5). BA does not stabilize NR, and BA-induced increase in NR activity depends on protein synthesis continuing throughout the whole period of induction. Similarly, experiments with actinomycin D have shown that cytokinin-induced NR activity depends on continuous RNA synthesis throughout the period of induction (Fig. 5). Blocking RNA synthesis with this

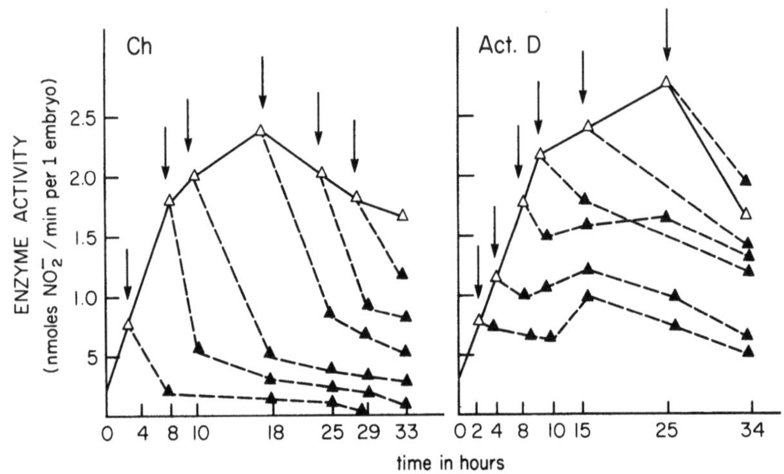

Fig. 5. Time course of cycloheximide and actinomycin D action on cytokinin-induced NR activity in isolated embryos of *A. githago.* Procedures for excision and incubation of embryos are described in the legend to Table 5. *Arrow* addition of inhibitor. *Solid line* enzyme activity in the presence of BA *dotted line* the combined effects of BA and inhibitor. Concentration of BA – 2 mg/l, cycloheximide – 10 mg/l, actinomycin D – 15 mg/l. Assay conditions have been described (18)

antibiotic stopped the increase in enzyme activity. However, throughout the initial 2–4 h of BA treatment, enzyme activity remained constant at the level achieved at the moment of actinomycin D addition. As mentioned above, NR is highly unstable. The fact that the initial level of enzyme activity is retained following treatment with actinomycin D can only be explained as a dynamic equilibrium between the rates of biosynthesis and breakdown of the enzyme. Hence mRNA for NR is stable at the beginning of embryo growth, and BA has increased the enzyme activity via synthesis of mRNA specific for NR or some other proteins responsible for NR activity in the embryos.

An essential difference between excised *A. githago* embryos and excised pumpkin cotyledons is the absence of BA-induced activation either of total RNA or of rRNA and tRNA synthesis in the embryos. This lack is related to the failure of cytokinin to activate total protein synthesis (9) in embryos. It is evident that the molecular mechanism of cytokinin action in NR synthesis in *A. githago* embryos differs from its mechanism of action in stimulating total protein synthesis in pumpkin cotyledons. The possibility is not excluded that cytokinin can act at the level of gene control in this case.

The excised *A. githago* embryos is a suitable system for investigating interaction of the hormone with substrate and metabolite (glucose) in the regulation of enzyme activity.

It was first demonstrated by Kende et al. (9), and we (21) have confirmed, that cytokinin and nitrate act independently in inducing NR activity in excised *A. githago* embryos. We have shown, furthermore, that glucose can induce NR activity in this system (20). We have tested a cytokinin antagonist, 4-cyclopentylamino-2-methylthiopyrrolo(2,3*d*)pyrimidine, which effectively inhibits cytokinin-induced growth of tissue cultures (29), for its effect on NR induction. This compound did not affect NR induction by nitrate, but inhibited enzyme induction by either glucose or cytokinin to the same extent (Table 5). It should be reemphasized that the mechanism of NR induction by nitrate and by cytokinin are independent. Our results indicate that glucose may induce NR through an influence on synthesis or activity of endogenous cyto-

Table 5. Effect of anticytokinin on the induction of NR activity in isolated embryos of *A. githago* by BA, NO_3^- or glucose

Treatment	nmol NO_2^- formed / min embryo	Inhibition %
BA	1.92 ± 0.05	
BA + anticytokinin	1.17 ± 0.02	39
NO_3^-	1.54 ± 0.02	
NO_3^- + anticytokinin	1.61 ± 0.01	0
Glucose	1.80 ± 0.04	
Glucose + anticytokinin	1.08 ± 0.11	40

The seeds of *A. githago* were imbibed for 12 h at 30°C in a medium containing 0.01 M $CaCl_2$: 0.01 M KCl=1:1. The seeds were sterilized with 30% H_2O_2 for 6 min and subjected to the following treatments under aseptic conditions: Isolated embryos were preincubated for 10 h in medium with or without the anticytokinin, 4-cyclopentylamino-2-methylpyrrolo(2,3-d)pyrimidine (40.3 μM). Then either BA (4 × 10^{-2} μM); KNO_3 (50 mM) or glucose (50 mM) was added to the incubation medium. Embryos were incubated with the test substance for 24 h at 30°C in the dark. NR activity was measured as described (18). NR activity of control embryos was 0.11 nmol NO_2^- formed/min/embryo

kinins (20). The data obtained reveal a complex system of intracellular control of enzyme activities in plants, involving interactions between hormone, substrate and other metabolites.

Acknowledgment. I thank F. Skoog for a sample of anticytokinin.

References

1. Banerji, D., Laloraya, M.M.: Naturwissenschaften *52,* 349–350 (1965)
2. Borriss, H.: Wiss. Z. Univ. Rostock., Math. Naturwiss. Reihe *16,* 629–639 (1967)
3. Borriss, H.: In: Regulation of Developmental Processes in Plants. Schütte, H.R., Gross, D. (eds), pp. 98–110. Jena: Fischer 1977
4. Engelbrecht, L., Rybicka, H., Kulaeva, O.: Biochem. Physiol. Pflanz. *169,* 317–320 (1976)
5. Feierabend, J.: Planta *94,* 1–15 (1970)
6. Fosket, D.E., Tepfer, D.A.: In Vitro *14,* 63–75 (1978)
7. Karavaiko, N.N., Krawiaiz, K., Khokhlova, V.A., Kulaeva, O.N.: Fiziol. Rast. *25,* 803–809 (1978)
8. Karavaiko, N.N., Ohmann, E., Kulaeva, O.N.: Fiziol. Rast. *22,* 903–908 (1975)
9. Kende, H., Hahn, H., Kays, E.: Plant Physiol. *48;* 702–706 (1971)
10. Kende, H., Schen, T.C.: Biochim. Biophys. Acta *286,* 118–125 (1972)
11. Khokhlova, V.A., Karavaiko, N.N., Podergina, T.A., Kulaeva, O.N.: Cytologia *XX,* 1033–1039 (1978)
12. Klyachko, N.L., Ananiev, E.D., Kulaeva, O.N.: Doklady (Moscow) *243,* 1334–1336 (1978)
13. Klyachko, N.L., Yakovleva, L.A., Kulaeva, O.N.: Fiziol. Rast. *20,* 1219–1223 (1973)
14. Kulaeva, O.N.: Cytokinins, their structure and function. 264 pp. Moscow: Nauka 1973
15. Kulaeva, O.N.: Fiziol. Rast. *25,* 990–1008 (1978)
16. Kulaeva, O.N., Selivankina, S. Yu., Kuroyedov, V.A.: Fiziol. Rast. *18,* 746–753 (1971)
17. Kulaeva, O.N., Selivankina, S.Yu., Romanko, E.G., Nikolaeva, M.K., Nichiporovich, A.A.: Fiziol. Rast. *26,* 1016–1027 (1979)
18. Kulaeva, O.N., Kuznetsov, Vl.V., Kuznetsov, V.V.: Fiziol. Rast. *23,* 1225–1263 (1976)
19. Kuznetsov, Vl.V., Kuznetsov, V.V., Kulaeva, O.N.: Fiziol. Rast. *26,* 309–317 (1979)
20. Kuznetsov, V.V., Kuznetsov, Vl.V., Kulaeva, O.N.: Fiziol. Rast. *26,* 728–736 (1979)
21. Kuznetsov, V.V., Kuznetsov, Vl.V., Kulaeva, O.N.: Biochimia *44,* 684–692 (1979)
22. Mikulovich, T.P., Khokhlova, V.A., Kulaeva, O.N., Sveschnikova, I.N.: Fiziol. Rast. *18,* 98–106 (1971)
23. Mikulovich, T.P., Wollgiehn, R., Khokhlova, W.A., Neumann, D., Kulaeva, O.N.: Biochem. Physiol. Pflanz. *172,* 101–110 (1978)
24. Rauser, W.E.: Can. J. Bot. *49,* 311–316 (1971)
25. Romanko, E.G., Selivankina, S.Yu., Kuroyedov, V.A.: Fiziol. Rast. *25,* 1199–1205 (1978)
26. Selivankina, S.Yu., Romanko, E.G., Kuroyedov, V.A., Kulaeva, O.N.: Fiziol. Rast. *26,* 41–47 (1979)
27. Selivankina, S.Yu., Romanko, E.G., Kuroyedov, V.A., Ohmann, E.: Fiziol. Rast. *23,* 1011–1017 (1976)
28. Szweykowska, A.: In: Int. Conf. Plant Growth Substances. Abstr. p. 62. Prague, 72 pp. (1978)
29. Skoog, F., Schmitz, R.Y., Hecht, S.M., Frye, R.B.: Proc. Natl. Acad. Sci. USA *72,* 3508–3512 (1975)
30. Srivastava, B.J.S.: In: Biochemistry and Physiology of Plant Growth Substances. Wightman, F., Setterfield, G. (eds.), pp. 1479–1494. Ottawa: Runge Press 1968
31. Yakovleva, L.A., Klyachko, N.L., Kulaeva, O.N.: Mol. Biol. *11,* 868–876 (1977)

Presence and Possible Functions of Cytokinins in RNA

C. PÉAUD-LENOËL and J-P. JOUANNEAU [1]

Introduction

Nowadays, four lines of research seem promising as guides to significant molecular events following the application of a cytokinin to a responsive biological system. The first is the study of the stepwise metabolic reactions in which the free cytokinin bases or their nucleosides are involved after they are absorbed by the cells. Another line is expected to elucidate the regulatory role of cytokinin-receptor proteins with a high specific affinity for their ligand. A third problem is to identify specific and early markers in plant material treated by cytokinins. A fourth line of experiments stems from the fact that cytokinins are purine bases or analogs, leaving aside, N,N′-diphenyl- or N-phenyl,N′-pyridyl-ureas and their derivatives. Thirteen years ago, it was observed that nucleosides with a cytokinin potency were present in tRNA molecules. This discovery led to the stimulating suggestion that exogenous cytokinins might control the effectiveness of tRNA in some key functions, presumably by replacing the natural hypermodified base at its anticodon proximal position. The experimental attempts to prove this theory did not reconcile the facts, but it remains reasonable to assume that cytokinins may act physiologically through a regulatory role in some reaction(s) involving RNA functions. We have kept this idea in mind and tried to connect and clarify the results of experiments to investigate whether free cytokinins are active in this way. See cited reviews (1–7) for additional information on cytokinin activity at the molecular level.

tRNAs Contain Natural Hypermodified Nucleosides That Are Active Cytokinins in the Free State

Hypermodified Bases in the Polynucleotide Chains of tRNA

N^6-(Δ^2-isopentenyl) adenine[2] and its trans-4-hydroxy analog, zeatin, were recognized as active cytokinins in the free state shortly before they were characterized in tRNA:

---

1 Laboratoire de Biochimie Fonctionnelle des Plantes, Faculté des Sciences de Luminy, 13288 Marseille, Cedex 2, France

2 *Abbreviations:* bzl^6Ade: N^6-benzyladenine; bzl^6Ado: N^6-benzyladenosine; fr^6Ade: N^6-furfuryladenine or kinetin; fr^6A: N^6-furfuryladenosine or kinetin riboside; i^6A: N^6-(Δ^2-isopentenyl)adenosine: io^6A: ribosylzeatin; ms^2i^6A: N^6-(Δ^2-isopentenyl)-2-methylthio-adenosine; ms^2io^6A: 2-methylthioribosylzeatin; t^6A: N-(9-β-D-ribofuranosylpurin-6yl-carbamoyl)threonine. Other abbreviations as recommended by IUPAC

in 1966, i^6A was identified as a hypermodified nucleoside in the primary structure of $tRNA^{Ser}_{Yeast}$ I and II. The presence of cytokinins was further detected in tRNA of various origins by their biological activities. As reviewed in 1–4, i^6A, io^6A, ms^2i^6A, ms^2io^6A are the cytokinins which have been found as hypermodified nucleosides in tRNA. The most frequent cytokinins in prokaryotic tRNA are i^6A and ms^2i^6A; io^6A or ms^2io^6A have been described in *Pseudomonas, Corynebacterium, Rhizobium* and *Agrobacterium* (8, 9, 69). To date i^6A is the only cytokinin nucleoside characterized in animal tRNA, whereas all the four nucleosides are present in plant tRNA. It was thought that cis-io^6A occurred in tRNA, whereas free zeatin occurred as the trans-isomer. However, trans-zeatin has been reported in some tRNA (9, 10). Determinations of the amount of cytokinins in tRNA (Table 1) rely on biological activity tests. In spite of some uncertainties, the results indicate that defined species of isoacceptor tRNA contain one cytokinin nucleoside per molecule, whereas other isoacceptors may contain another hypermodified base or no hypermodification at all. Moreover, if the molecular population of a tRNA issued from the same transcript is considered, some molecules are hydroxy-pentenylated, whereas others may remain unsubstituted in the mature tRNA. It is noteworthy that tobacco callus contains tRNA-linked cytokinin nucleosides, whether or not this tissue requires exogenous cytokinin for growth (13, 14).

Table 1. Cytokinin content of tRNA

Reference	Source of tRNA	Mol of cytokinin-containing tRNA / mol tRNA
(1)	Yeast total tRNA	1/22
(11)	Yeast total tRNA	1/10
(1)	Pea total tRNA	1/167
(11)	*E. coli* total tRNA	1/30
(12)	*E. coli* $tRNA^{Phe}$	1 i^6A / 8.4; 1 ms^2i^6A / 75
(12)	*E. coli* $tRNA^{Cys}$	1 i^6A / 60; 1 ms^2i^6A / 1.6

In eukaryotic organisms, the organelles, mitochondria and plastids, contain their own specific tRNA equipment (15), distinct from the cytoplasmic tRNA. Comparative analyses provide excellent evidence that the hypermodified bases of isoacceptor tRNAs are indeed different according to their organelle or cytoplasmic origin. From the standpoint of cytokinin content, as well as from the comparison of whole primary structures, a wide homology is found between prokaryotic and organelle tRNA (Table 2). A good illustration of this homology is provided by the compared aminoacylation of cytoplasmic or organelle tRNA with homologous or heterologous tRNA synthetases: *E. coli* aminoacyl-tRNA synthetases recognize the isoacceptor tRNA of plastids and vice versa, whereas the cytoplasmic tRNAs are recognized only by cytoplasmic synthetases (19, 20).

Table 2. Cytokinins in cytoplasmic and chloroplast tRNA

Cytokinin	*Euglena*				*Phaseolus vulgaris*	
	Total tRNA (16)		$tRNA^{Phe}$ (17)		$tRNA^{Phe}$ (18)	
	cyto	chloro	cyto	chloro	cyto	chloro
i^6A	+	+				
ms^2i^6A				+		+
$c\text{-}io^6A$	mainly					
ms^2io^6A		only				
Y base			+		+	

Codon-Anticodon Specificity and tRNA-linked Cytokinin Nucleosides

A well-documented relationship exists between the presence of a cytokinin in the tRNA polynucleotide chain and the codon to which it pairs. In prokaryotes and in monocellular eukaryotes, the rule is that these hypermodified bases are specific to most of the isoacceptors responding to codons of initial base U (1–4). In multicellular eukaryotes, the cytokinin nucleosides are located in only one or a few isoaccepting tRNA species answering to the same triplets with a U_1 base (Table 3). From the known primary structures of many tRNA species [see (26)], it became obvious that the cytokinin nucleosides always occurs at a position 3′-adjacent to the anticodon triplet. On the other hand, many tRNA species are known which recognize the $U_1X_2X_3$ codon family and in which the cytokinin is replaced either by A or m^1G or the ψ base. The situation is parallel in the tRNA family pairing to codons of initial base A: some tRNA species contain t^6A in the position 3′-adjacent to the anticodon.

Table 3. Distribution of cytokinins in isoacceptor tRNAs responding to codons $U\,X_1X_2$

References	Source of tRNA	Isoacceptor tRNA					
		Phe	Leu	Ser	Tyr	Cys	Trp
(4, 12)	*E. coli*	+	+	+	+	+	+
(21)	Yeast	0 or Y	?	+	+	+	0
(22)	*Drosophila*			$Ser_{4,6,7}$			
(23)	Wheat germ	0 or Y	Leu_6	2 major 2 minor	0	0	0
(24)	Pea seeds		Leu_6				
(25)	Soybean cotyledons		$Leu_{5,6}$				

Mechanism of Isopentenylation

It is well established that the biosynthesis of cytokinin nucleosides in tRNA proceeds through a post-transcriptional modification of the precursor. The enzyme responsible for this modification, Δ^2-isopentenyl-pyrophosphate: tRNA Δ^2-isopentenyltransferase

(EC 2.5.), was discovered by Kline et al., and its properties have been clarified by many contributions [see (1)]. The specificity of the acceptor structure in the Δ^2-isopentenyltransferase reaction was investigated by the use of native or $KMnO_4$-treated tRNA populations or, more recently, with synthetic polynucleotide acceptors (27, 28): it was found that the isopentenylation was specific for A adjacent to the 3′ end of the anticodon. Natural acceptors could be replaced by sequences A A A ψ C or A A A A C, where the second base was the only one isopentenylated. Oligo A chains of 3 to 7 nucleotides also were good acceptors and might represent the minimal structural requirement for the enzyme activity. Synthetic poly A was an acceptor recognized by the corn root enzyme; the natural rRNA was not isopentenylated (27, 28). Additional evidence is necessary to be sure that the use of heterologous acceptors for in vitro reactions faithfully reproduced the situation prevailing in vivo. Isopentenylation in vitro did not require the prior methiolation of adenine when ms^2i^6A was the final product. These results strongly suggest that i^6A and ms^2i^6A are step-wise maturation products of the precursor tRNA. It has been shown that isopentenylation is a late event, compared to the substitution U → T in tRNA precursors (29).

Impact of the Isopentenyl (or Analog) Substitutions of Adenine on the Molecular Functions of tRNA

Many experiments have been undertaken to unravel the effect(s) of the hypermodified nucleosides 3′-adjacent to some tRNA anticodons. It has been recognized that the aminoacylation reactions are not controlled by these nucleotides. At least, the recognition of tRNA by the amino acid-specific synthetase was not affected (1, 5, 29–33). It seems that the specific CTP: or ATP: tRNA nucleotidyltransferases binding pC and pA to the 3′-end of tRNA also were unaffected by the isopentenylation of A in the anticodon loop (32, 33). In one report, it was mentioned that another function of a tRNA species, the binding to the translation complex Ile-$tRNA^{Ile}_{E.\ coli}$-T_{fl}-GTP, was not affected by the hypermodified base t^6A (33). On the other hand, hypermodified nucleosides are required for the binding of some aminoacyl-tRNAs to the ribosome-mRNA complex and for normal translation (1, 5, 32, 33). A good example is the experiment of Gefter and Russell (30) using the infection of an *E. coli* amber mutant by phage φ 80, an inducer of the synthesis of $tRNA^{Tyr}_{Su_3^+}$. In a reconstructed translation system, the suppression of the amber mutation is observed only if $tRNA^{Tyr}_{Su}$ contains i^6A or, better, ms^2i^6A: both substituents allow the binding of the tRNA to the ribosome-messenger complex, and thereby the recognition of the UAG codon, normally a terminator. According to Miller et al. (33), things are not so clear and the stimulating effect of the hypermodified bases upon the binding of aminoacyl-tRNA to the ribosome-messenger complex is controlled by the Mg^{2+} concentration. Moreover, as previously discussed, hypermodified bases are present in only a few iso acceptor tRNAs. Therefore, the in vivo relationship between cytokinin nucleosides and translational events must be evaluated with caution.

Possible Relationship Between the Biological Activity of Cytokinins and Their Presence in tRNA

When cells differentiate, new genes are transcribed, including genes coding for new tRNA species. Evidence has been provided that during the course of organ differentiation, the transcription of isoacceptor tRNA was adapted in such a way that the frequency of a particular anticodon fitted with the frequency of the corresponding codon in the mRNA translated into a specific protein. This was demonstrated for fibroin biosynthesis by the silk worm (34) and for zein biosynthesis in immature corn kernels (35). Correlated with the nature of the prevailing amino acids in these proteins, the cytokinin content of the tRNA set is likely modified. According to other reports, the observed variations in tRNA may also reflect post-transcriptional modifications; as for example the changes in the tRNA map of *Dictyostelium* in the course of its differentiation (36). In *Bacillus subtilis,* the observation of a vegetative phase i^6A-$tRNA^{Tyr}$ and a sporulation phase ms^2i^6A-$tRNA^{Tyr}$ led to the analysis of their nucleotide composition which, apart from the hypermodification, is identical (37). Such discrete changes in the anticodon region may have some impact at the translational level (30).

A noteworthy example of tRNA map change was observed during the light-promoted transformation of etioplasts to chloroplasts: an increase in $tRNA_5^{Leu}$ and and $tRNA_6^{Leu}$ content was noted on reverse phase chromatograms (38). These tRNAs originate from plastids (20, 39); they recognize the triplet UUG in the presence of 70S ribosomes (40) and are probably coded by the same plastid gene. They are likely to differ only by post-transcriptional modifications (20) and probably contain cytokinin nucleosides; this has been recently shown in the case of $tRNA_6^{Leu}$ in which io^6A was identified (24). In the foregoing examples, the molecular adaptation of tRNA increases the efficiency of protein synthesis. Nevertheless, the logical link between these adaptations and the impact of free cytokinins is not in sight. Some correlation was described between the biological activity of applied free cytokinins and the observed changes of the tRNA maps: an example is the change of the tRNA chromatographic profile observed during the flowering of *Mercurialis annua,* a dioecious plant. In this plant, female flowering was obtained by treatment of genetically male individuals with exogenous cytokinins. Consequently, a female-type tRNA map and female-specific aminoacyl-tRNA synthetases were observed (41). This phenomenon can be compared to the observed increase of $tRNA_5^{Leu}$ and $tRNA_6^{Leu}$ after cytokinin treatment of soybean cotyledons (42). As mentioned above, these tRNA species are probably plastidial: this suggestion is strengthened by the comparison of the results of Burkard et al. (38) and of the RPC fractionation of soybean tRNA by Cherry and Anderson (42) (Fig. 1). The physiological activity of cytokinins on plastid differentiation in cotyledons is well documented (43). It seems reasonable to suggest that the $tRNA_5^{Leu}$ and $tRNA_6^{Leu}$, and possibly other tRNAs, may be molecular markers of the biological response of plastids to exogenous or endogenous cytokinins.

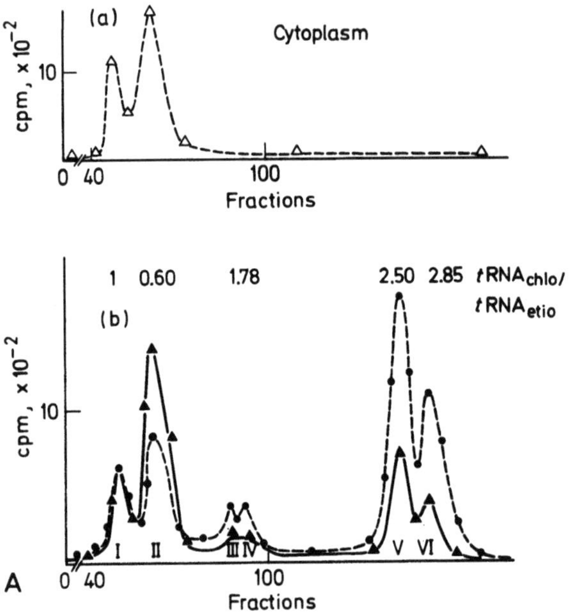

Fig. 1A, B. RPC analysis of leucyl-tRNA isoacceptors: **A** Leucyl-tRNAs from bean leaves, aminoacylated with their homologous transferases: **a** tRNA from cytoplasm; **b** tRNA from chloroplasts *(dashed line)* or from etioplasts *(solid line)* (38)

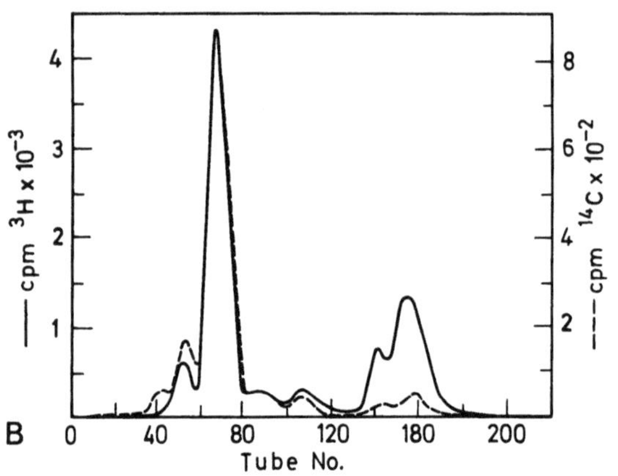

Fig. 1B. Leucyl-tRNAs from soybean hypocotyls treated by bzl^6 Ade *(solid line)* or not treated *(dashed line)* and aminoacylated with cotyledon transferases (42)

tRNA as a Source of Free Natural Cytokinins

A number of authors have suggested that tRNA, by natural degradation in the cells, might be the source of endogenous cytokinins in the free state. Such a hypothesis requires measurements of tRNA half-life, related to the cytokinin requirement of the considered organism. In *Lactobacillus* (44), the half-life of bulk tRNA was estimated to be 3–4 h, i.e., several cell generations. The average half-life of corn root tip tRNA is of the order of 3 days and varies according to the position of the sampled cells (45). This kind of estimation is difficult on account of the stringency of the labeled precursor pool from which the nucleotides are rapidly incorporated into newborn RNA molecules (46). According to our results obtained with tobacco cell suspension cultures, some unidentified tRNA species might turn over more rapidly than others. The average turnover was estimated to be 2–3 cell generations; this is too slow to meet the cytokinin requirement of these cells (46). Other evidence against tRNA as a candidate for the origin of free cytokinins is that the commonest cytokinin nucleoside in plant tRNA is the cis-zeatin riboside whereas the free isomer is trans-zeatin [see (4)]. Consequently, a direct biosynthetic pathway for cytokinins has been sought. Taya et al (47) identified 5'AMP as a direct precursor of i^6A by in vitro experiments with *Dictyostelium* extracts. Burrows (48) has shown that adenine was a precursor of trans-zeatin in tobacco callus tissue, and Chen et al. (49) reported evidence that periodate-oxidized adenosine was N^6-isopentenylated by autonomous tobacco tissue, although this compound could not be directly phosphorylated. The evidence is strong for a direct biosynthetic pathway of free natural cytokinins without excluding degradation of tRNA as contributing to the cytokinin pool.

Synthetic Cytokinins Incorporated into RNA Without Intramolecular Modification

Are Exogenous Cytokinins Linked into RNA Chains in Plant Cells?

The question of the incorporation of cytokinins into RNA was raised even before i^6A was discovered in tRNA. The incorporation of bzl^6-[8-^{14}C] Ade into RNA of tobacco and soybean tissues was first investigated by Fox et al. (50–52): during the incubation period, much randomization of the label occurred; most of the ^{14}C was recovered from AMP and GMP in the RNA digests. However, the intact labeled precursor was traced both into tRNA and into rRNA. Analytical problems raised questions about the actual existence of a covalent linkage between the precursor and the polyribonucleotides. With elaborate techniques of tRNA purification and cytokinin identification (mass spectrometry), Burrows et al. (13) studied the cytokinins linked to tRNA of cytokinin-dependent tobacco callus cultured on medium containing bzl^6Ade. These authors were able to identify in the tRNA fraction, in addition to the natural compounds i^6A, io^6A and ms^2io^6A, a small amount of bzl^6Ado.

On the other hand, Kende and Tavares [see (4)] argued against the possible incorporation of cytokinins into RNA by pointing out that 9-methyl-bzl^6Ade, was not incorporated in soybean callus RNA, but was still a potent cytokinin. Elliott and Murray estimated the maximal amount of bzl^6Ade incorporated into RNA of soybean tissue to be low and suggested that bzl^6Ade recovered from the tRNA fraction might have been synthesized by transbenzylation from the exogenous precursor onto transcribed adenine nucleosides (53).

To clarify this problem, we incubated tobacco cell suspensions with precursor bzl^6-[^{3}H] Ade; the cold perchloric acid-insoluble fraction was prepared, and the nature of the radioactive compounds tenaciously bound to this fraction was investigated (54):

a significant amount of the radioactive material was contained in polyribonucleotide-linked bzl^{6}Ade. The largest part of the label was recovered in other macromolecular fractions in which no evidence of covalently linked precursor was found. Hence, it appeared essential to demonstrate the covalent nature of the linkage between the cytokinin precursor and RNA.

Exogenous Cytokinins Linked to Purified RNA Fractions from Tobacco Callus

Two lines of experiments carried out independently with different methodology but with convergent quantitative results gave evidence of direct incorporation. In several papers (13, 55–58), the Madison group reported the incorporation of bzl^6 [8-^{14}C] Ade or fr^6 [8-^{14}C] Ade or double-labeled [^{3}H]bzl^6 - [8-^{14}C] Ade into cytokinin-dependent tobacco callus. The specific radioactivity of the precursors (20–30 mCi/mmol) justi-

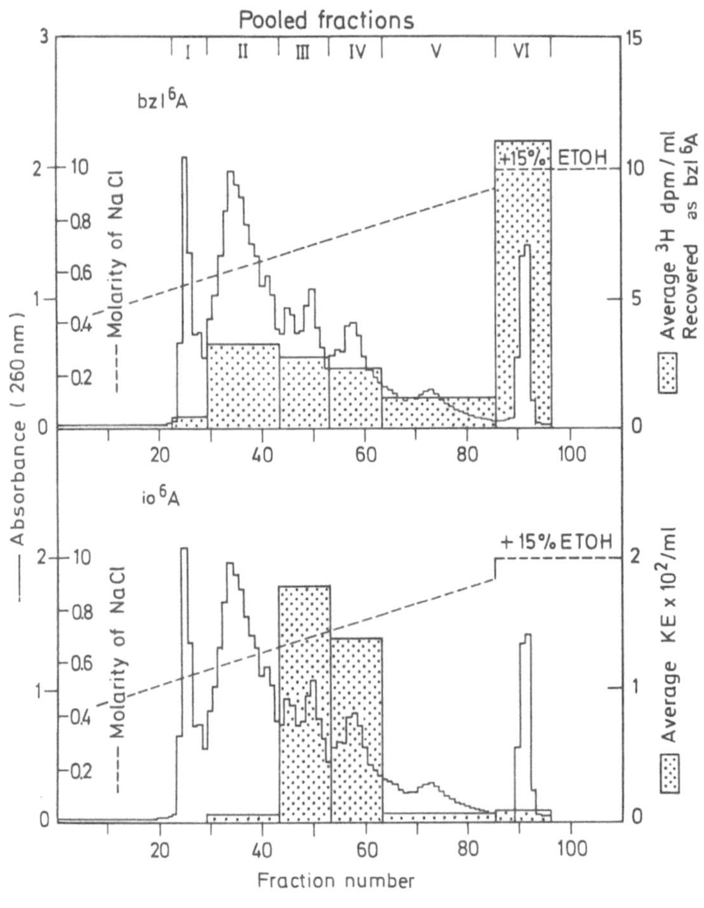

Fig. 2. BD-cellulose chromatography of tobacco callus tRNA labeled by [^{3}H]bzl^6 Ade. Distribution of bzl^6 Ado and io^6 Ado in tRNA fractions. (56)

fied incubations for two to three weeks before RNA was extracted. The purified RNA was fractionated and provided a 4S + 5S fraction, 18S and 25S rRNA species. The purity of these fractions was carefully tested, especially by polyacrylamide gel electrophoresis. The 4S + 5S fraction contained radioactivity which survived all fractionation procedures and RNA denaturation (55, 56). BD-cellulose fractionation showed that bzl^{6}Ade-containing tRNA fractions were different from the c-io^{6}A-containing fractions (Fig. 2). The different RNA fractions were digested by RNases and alkaline phosphatase: the nucleoside digest was then enriched in hydrophobic components by ethyl acetate extraction. The nucleosides were further identified by Sephadex LH-20 co-chromatography with authentic compounds or by GLC and MS (55, 57).

When the rRNA was analyzed (56–58), the radioactivity of the labeled precursor was found to be coincident with the 18S and 25S species which were both labeled.

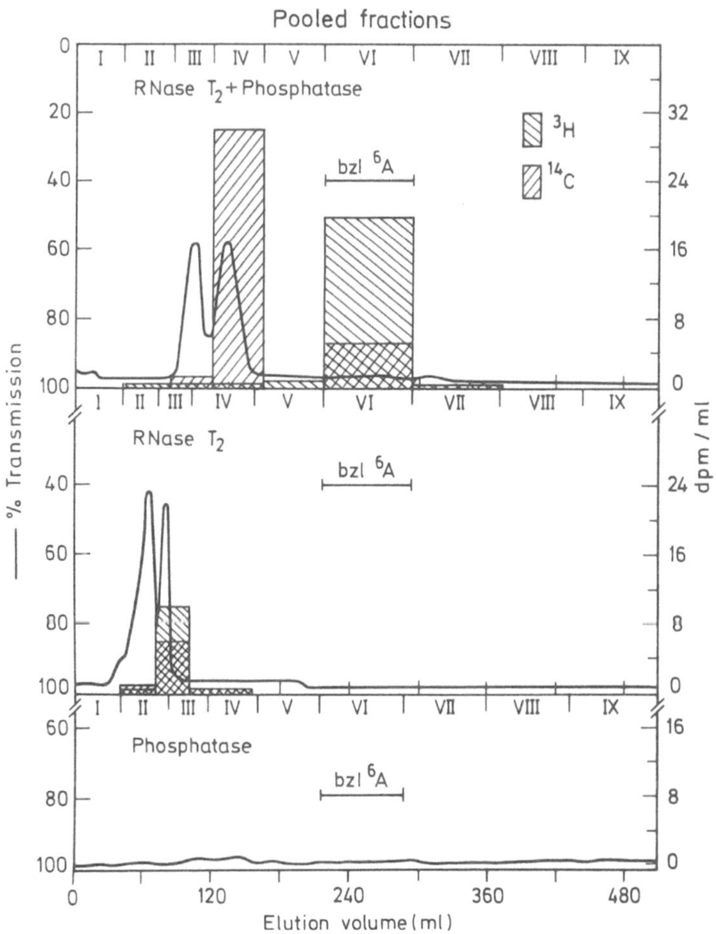

Fig. 3. Recovery of bzl^{6}Ado from [^{3}H]bzl^6-[^{14}C]Ade labeled tobacco callus rRNA submitted to different enzymatic digestions. Ethyl acetate soluble nucleosides from each digest were chromatographed on LH-20 Sephadex. (56)

This label also survived RNA denaturation. The radioactive nucleosides were not recovered by T_2 RNase treatment alone (Fig. 3), nor by treatment with a mixture of RNase T_1 plus RNase A. After any of these RNase treatments, an additional phosphatase digestion was required to obtain radioactive nucleosides: this experiment provided evidence against the possibility that the radioactive precursor was tenaciously bound but not covalently linked to RNA. In the presence of carrier poly A, the radioactive material was solubilized by the combination of RNases T_1 + A. Therefore, it was unlikely that the precursor-labeled cytokinin was interspersed in poly A chains. Murai et al. (58) labeled rRNA by incubation of the tissue with fr^6-[8-^{14}C] Ade; rRNA was purified, digested by RNases T_1 + A and analyzed by ion exchange chromatography. A significant amount of the ^{14}C was contained in fr^6A recovered from oligonucleotides of the structure A^nPy or A^nG. Assuming that fr^6-[8-^{14}C] Ade replaced A in the oligonucleotides, fr^6A was estimated to be four-fold more frequent in APy or A^2Py than in AG or A^2G.

Incorporation of Bzl^{6}Ade Into tRNA and rRNA of Cell Suspensions

In our laboratory, the labeling of RNA by exogenous cytokinin was obtained in a different way (54, 59). Tracer bzl^6-[2-^{3}H] Ade and, later, [p-^{3}H] bzl^6Ade of high specific activity (about 20 Ci/mmol) were synthesized. It was possible to obtain enough labeled RNA from a few grams of tobacco cells incubated with a physiological concentration of these tracers during 10–16 h. This period, during which no morphogenetic change was observed, was previously shown to trigger a subsequent mitotic wave in

Table 4. Incorporation of N^6-benzyladenine into RNA of tobacco cell suspensions (59)

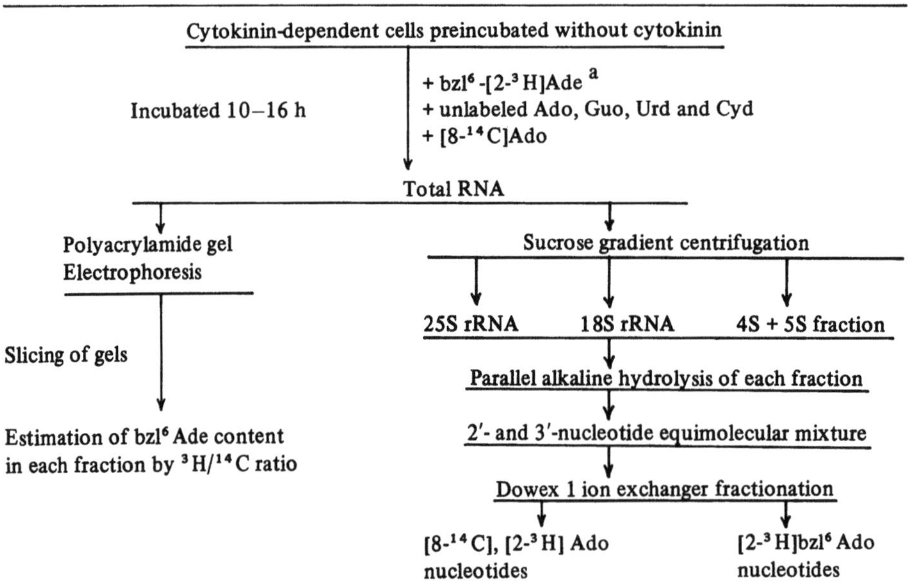

[a] Also incubated with [p-^{3}H]bzl^6Ade: in this case, little ^{3}H is recovered in Ado-nucleotides

control cultures [see (7)]. The RNA was extracted, purified and fractionated: each RNA fraction was found to contain radioactivity. We chose to identify the radioactive nucleotides after alkaline digestion of RNA: this technique provided a way to separate the potential 5′-phosphate nucleosides, issued from the soluble pool of the cells, from the equimolecular 2′- and 3′-phosphate nucleosides produced from the RNA via intermediary nucleoside 2′,3′-cyclic phosphodiesters during the alkaline attack on the polyribonucleotides. During incubation of the cells with the precursor cytokinin, [8-^{14}C] Ado was added as a tracer of the rate of RNA biosynthesis and as a competitor of the [2-^{3}H] Ade produced by debenzylation of bzl^6-[2-^{3}H] Ade in the tobacco cells. [8-^{14}C] Ado was also useful as a tracer of possible transbenzylation reactions which would be recorded on the chromatogram by the presence of [^{14}C]-labeled-bzl^6 Ado 2′- and 3′-phosphate nucleosides (Fig. 4). Table 4 outlines the analysis of RNA by this method (59). The ^{3}H/^{14}C ratio of the fractions collected from polyacrylamide gels provided an estimate of the bzl^6 Ade content of the fractions. The amounts of cytokinin nucleotide linked to the 25S and 18S species of rRNA and to the 4S + 5S fraction were determined by nucleotide ion exchange chromatography and counting of the fractions.

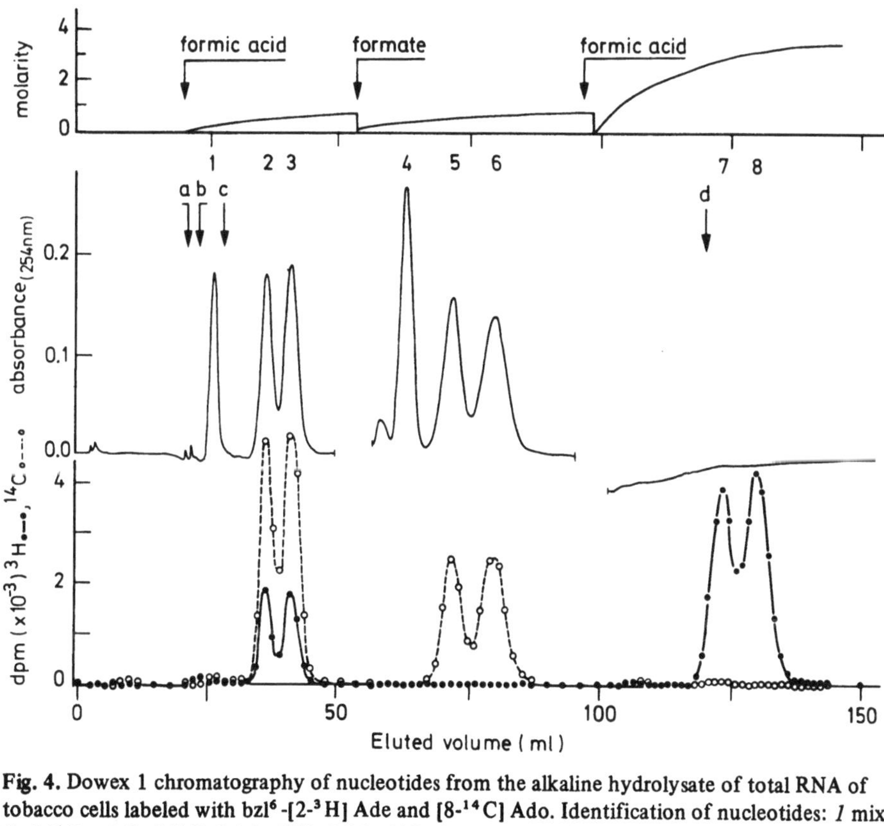

Fig. 4. Dowex 1 chromatography of nucleotides from the alkaline hydrolysate of total RNA of tobacco cells labeled with bzl^6-[2-^{3}H] Ade and [8-^{14}C] Ado. Identification of nucleotides: *1* mixed Cyd-2′-P and Cyd-3′-P; *2* Ado-2′-P; *3* Ado-3′-P; *4* mixed Urd-2′-P and Urd-3′-P; *5* Guo-2′-P; *6* Guo-3′-P; *7* bzl^6 Ado-2′-P; *8* bzl^6 Ado-3′-P. Chromatographic location of control bzl^6 Ade metabolites: *a* bzl^6 Ade-7-glucoside; *b* bzl^6 Ade; *c* bzl^6 Ado; *d* bzl^6 Ado-5′-P. (59)

Significance of Cytokinin Incorporation in RNA

The conclusions drawn from experiments of both groups are in close agreement. When N^6-substituted adenine precursors were labeled on the adenine moiety, a large part of the radioactivity was recovered from Ado or Guo in RNA digests. The labeling of Ado or Guo was extremely small when sidechain-labeled precursors were used. These results suggest that Ado and Guo are produced by biological splitting of the side chains from the precursors and not by randomization of the label. From purine ring-labeled cytokinins, the incorporation into Ado or Guo was strongly inhibited by the addition of the four nucleosides Ado, Guo, Urd, Cyd to the incubation medium of the cells (60). Whatever the position of the label in the precursor, RNA was labeled by direct incorporation of the cytokinin: bzl^6Ado was recovered from the RNA without intramolecular modification, as shown by double labeling of bzl^6Ade both on the side chain by [^{3}H] and on the purine ring by [^{14}C] (55). Separate evidence was obtained from the measurements of respective isotopic dilution of [^{3}H]-purine-labeled bzl^6Ade incorporated into Ado and bzl^6Ado nucleotides of RNA: no evidence was obtained from our experiments that any transbenzylation reaction took place (59).

The frequencies of incorporation of labeled cytokinins in RNA are of the same order of magnitude in different reports: tRNA contained one labeled cytokinin per 10^5 to 7.5×10^5 bases, i.e., $1/10^3$ to $1/10^4$ tRNA molecules: this is a very low level of incorporation. Figure 2 shows that there is no clear relationship between the presence of natural cytokinin nucleosides in some tRNA species and the incorporation of substituted adenine precursors into the tRNA fraction (55). In 18S and 25S rRNA species, the observed incorporation of cytokinins is three to four fold more frequent than in tRNA on a nucleotide basis, i.e., 1 for 100 to 1 for 25 rRNA molecules. This is the minimum order of magnitude since the ratio of the newly synthesized rRNA to the total rRNA is difficult to estimate. The 18S species was always found a little more labeled than its 25S counterpart; 5.8S RNA was probably labeled at the same rate as the 25S species. Incorporation of precursor bzl^6Ade in polydisperse RNA fractions was very small or insignificant (59).

No experimental evidence is available on the mechanism by which this incorporation takes place. It is reasonable to postulate that the origin of the RNA-linked bzl^6Ado nucleotide is the 5′-triphosphate bzl^6Ado, a compound identified in the tobacco tissue pool (61). It is hazardous to speculate any further: we do not know whether this incorporation takes place at random, as a result of mismatches of complementary bases during transcription or if cytokinin nucleotides are located in particular areas of RNA molecules, presumably by post-transcriptional events. The question of the incorporation of natural free cytokinins is also unanswered.

Concluding Remarks

The low level of cytokinin incorporation into rRNA does not preclude any physiological significance; if the cytokinin-bearing rRNAs are associated with ribosomes, the presence of these bases may change the functions of such ribosomes (for instance their binding to membranes) without change in the ribosome population. A cytokinin-binding protein has been reported to be, in part, bound to ribosomes (62). The affinity of this protein for ribosomes and its functions might be affected by the presence of an rRNA-linked cytokinin nucleotide. These suggestions form a basis for future experiments.

The problem raised by the presence of natural cytokinin nucleotides in tRNA is entirely different. As discussed above, such hypermodified bases have been detected in all organisms so far studied, although there has been no direct evidence that their presence or structure is connected with the activity of the free cytokinins. In some cases, exogenous cytokinins induced changes of the tRNA maps of plants, and there is evidence that some of these changes resulted in the synthesis of tRNA species belonging to differentiated plastids. Differentiation of plastids promoted by exogenous cytokinins is well documented at the physiological level. Few molecular markers of this effect have been described (63). The stimulation of the tRNA-translated chloroplast genes and other plastid-specific markers may become useful tools for investigating the molecular impact of cytokinins.

It is now generally recognized that the presence of hypermodified bases in tRNA has a bearing of the binding of aminoacyl-tRNA to the ribosome-messenger complex: this, in turn, may affect the translation process. This is the most important molecular effect observed so far for tRNA-linked cytokinin structures. It may appear surprising that, by virtue of only one hydrophobic side chain in a 75 nucleotide molecule, the key property of tRNA was thereby changed. We must take into account the cloverleaf structure hiding the paired polynucleotide stems with prominent unpaired loops: therefore, the presence of a cytokinin substituent in the anticodon loop may markedly change its conformation. Regarding cytokinins in tRNA, another interesting situation has been described in two missense suppressor $tRNA^{Gly}$ *E. coli* mutants. In the wild type, these tRNAs recognize codons GGG or GGA. The mutants (64) alternatively bear the anticodon mutation $C_1 \rightarrow U_1$, accompanied by the post-transcriptional change of the 3′-adjacent A to t^6A, or the mutation $C_1 \rightarrow A_1$ at the same locus accompanied by a parallel change of the 3′-adjacent A to ms^2i^6A. This situation suggests that the mutation $C_1 \rightarrow A_1$ brings about the poly A structure of the anticodon loop which is recognized by the proper isopentenyltransferase and methiolase, as previously discussed.

The hypothesis of transcriptional control by cytokinins remains attractive. In spite of some pertinent reports [see (6)], the stimulation of the biosynthesis of RNA by these factors is not well understood. As recently discussed by several authors (7, 65), it is likely that only a few genes might be activated and their transcripts would not appear as a large increase of RNA synthesis, except as a secondary consequence. Specific transcription products should be sought rather than a mass effect of cytokinins upon transcription rate. If gene activation is considered, it is useful to mention the experiments of Cortese et al. (66) and Rizzino et al. (67) with HisT mutants of *E. coli.* In these mutants, $tRNA^{His}$, $tRNA^{Leu}$ and $tRNA^{Ile}$ are altered in post-transcriptionally modified bases. The mutant phenotypes lack *attenuation* of *His, Leu* and *Val* operons (attenuation is the partial termination of the transcription by RNA polymerase in a particular region of an operon) with, as a consequence, an increase in the production of the relevant mRNA coded by these operons. A similar regulation of *Trp* operon transcription by attenuation and its mutant modifications were studied by Stauffer et al. (68): this attenuation implies the association of Trp-tRNA^{Trp}, or uncharged tRNA^{Trp}, plus a specific protein with the specific attenuator segment of the *Trp* operon DNA. The presence or the absence of the hypermodified base in the attenuating tRNA might be important for the efficiency of attenuation.

Acknowledgments. Authors' research reported in this paper was supported by the Centre National de la Recherche Scientifique (E.R. 104) and the Délégation à la Recherche Scientifique et Technique, France (Contract No. 79.7.059).

References

1. Hall, R.H.: Prog. Nucleic Acid Res. Mol. Biol. *10,* 57 (1970)
2. Hall, R.H.: Annu. Rev. Plant Physiol. *24,* 415 (1973)
3. Kende, H.: Int. Rev. Cytol. *31,* 301 (1971)
4. Skoog, F., Armstrong, D.J.: Annu. Rev. Plant Physiol. *21,* 359 (1970)
5. Nishimura, S.: Prog. Nucleic Acid Res. Mol. Biol. *12,* 49 (1972)
6. Burrows, W.J.: Curr. Adv. Plant Sci. *7,* 837 (1975)
7. Péaud-Lenoël, C.: In: Plant Growth Regulation. 9th Int. Conf. Plant Growth Substances 1976. Pilet, P.E. (ed.), p. 240. Berlin-Heidelberg-New York: Springer 1977
8. Thimmappaya, B., Cherayil, J.D.: Biochem. Biophys. Res. Commun. *60,* 665 (1974)
9. Chapman, M.W., Morris, R.O., Zaerr, J.B.: Nature (Lond.) *262,* 153 (1976)
10. Vreman, H.J., Skoog, F., Frihart, C.R., Leonard, N.J.: Plant Physiol. *49,* 848 (1972)
11. Babcock, D.F., Morris, R.O.: Biochemistry *9,* 3701 (1970)
12. Bartz, J., Söll, D., Burrows, W.J., Skoog, F.: Proc. Natl. Acad. Sci. USA *67,* 1148 (1970)
13. Burrows, W.J., Skoog, F., Leonard, N.J.: Biochemistry *10,* 2189 (1971)
14. Dyson, W.H., Hall, R.H.: Plant Physiol. *50,* 616 (1972)
15. Barnett, W.E., Schwartzbach, S.D., Hecker, L.I.: Prog. Nucleic Acid Res. Mol. Biol. *21,* 143 (1978)
16. Swaminathan, S., Bock, R.M., Skoog, F.: Plant Physiol. *59,* 558 (1977)
17. Hecker, L.I., Uziel, M., Barnett, W.E.: Nucleic Acid Res. *3,* 371 (1976)
18. Guillemaut, P., Martin, R., Weil, H.J.: FEBS Lett. *63,* 273 (1976)
19. Jeannin, G., Burkard, G., Weil, J.H.: Plant Sci. Lett. *13,* 75 (1978)
20. Weil, J.H., Burkard, G., Guillemaut, P., Jeannin, G., Martin, R., Steinmetz, A.: In: Nucleic Acids and Protein Synthesis in Plants. Bogorad, L., Weil, H.J. (eds.), p. 97. New York-London: Plenum Press 1977
21. Armstrong, D.J., Skoog, F., Kirkegaard, L.H., Hampel, A.E., Bock, R.M., Gillam, I., Tener, G.M.: Proc. Natl. Acad. Sci. USA *63,* 504 (1969)
22. White, B.N., Dunn, R., Gillam, I., Tener, G.M., Armstrong, D.J., Skoog, F., Frihart, C.R., Leonard, N.J.: J. Biol. Chem. *250,* 515 (1975)
23. Struxness, L.A., Armstrong, D.J., Gillam, I., Tener, G.M., Burrows, W.J., Skoog, F.: Plant Physiol. *63,* 35 (1979)
24. Einset, J.W., Swaminathan, S., Skoog, F.: Plant Physiol. *58,* 140 (1976)
25. Lester, B.R., Morris, R.O., Cherry, J.H.: Plant Physiol. *63,* 87 (1979)
26. Jukes, T.H.: Adv. Enzymol. *47,* 375 (1978)
27. Holtz, J., Klämbt, D.: Z. Physiol. Chem. *359,* 89 (1978)
28. Bartz, J.K., Söll, D.: Biochimie *54,* 31 (1972)
29. Blatt, B., Feldmann, H.: FEBS Lett. *37,* 129 (1973)
30. Gefter, M.L., Russell, R.L.: J. Mol. Biol. *39,* 145 (1969)
31. Krauss, G., Peters, F., Maass, G.: Nucleic Acid Res. *3,* 631 (1976)

32. Thiebe, R., Zachau, H.G.: Eur. J. Biochem. *5*, 546 (1968)
33. Miller, J.P., Hussain, Z., Schweizer, P.: Nucleic Acid Res. *3*, 1185 (1976)
34. Garel J.P.: Nature (Lond.) *260*, 805 (1976)
35. Viotti, A., Balducci, C., Weil, J.H.: Biochim. Biophys. Acta *517*, 125 (1978)
36. Palatnik, C.M., Katz, E.R., Brenner, M.: J. Biol. Chem. *252*, 694 (1977)
37. Keith, G., Rogg, H., Dirheimer, G., Menichi, B., Heymant, T.: FEBS Lett. *61*, 120 (1976)
38. Burkard, G., Vaultier, J.P., Weil, H.J.: Phytochemistry *11*, 1351 (1972)
39. Guderian, R.H., Pulliam, R.L., Gordon, M.P.: Biochim. Biophys. Acta *262*, 50 (1972)
40. Ramasia, J., Guillemaut, P., Weil, J.H.: FEBS Lett. *75*, 128 (1977)
41. Bazin, M., Chabin, A., Durand, R.: Dev. Biol. *44*, 288 (1975)
42. Cherry, J.H., Anderson, M.B.: In: Plant Growth Substances 1970. Carr, D.J. (ed.), p. 181. Berlin-Heidelberg-New York: Springer 1972
43. Harvey, M.N.R., Lu, B.C., Fletcher, R.A.: Can. J. Bot. *52*, 2581 (1974)
44. Leineweber, M., Klämbt, D.: Physiol. Plant. *30*, 327 (1974)
45. Klemen, F., Klämbt, D.: Physiol. Plant. *31*, 186 (1974)
46. Teyssendier de la Serve, B., Péaud-Lenoël, C.: Physiol. Veg. *16*, 279 (1978)
47. Taya, Y., Tanaka, Y., Nishimura, S.: Nature (Lond.) *271*, 545 (1978)
48. Burrows, W.J.: Biochem. Biophys. Res. Commun. *84*, 743 (1978)
49. Chen, C.M., Eckert, R.L., McChesney, J.D.: FEBS Lett. *64*, 429 (1976)
50. Fox, J.E.: Plant Physiol. *41*, 75 (1966)
51. Fox, J.E., Chen, C.M.: J. Biol. Chem. *242*, 4490 (1967)
52. Dyson, W.H.: Quoted in ref. 55
53. Elliott, D.C., Murray, A.W.: Biochem. J. *130*, 1157 (1972)
54. Jouanneau, J.P., Gandar, J.C., Péaud-Lenoël, C.: Plant Sci. Lett. *9*, 77 (1977)
55. Walker, G.C., Leonard, N.J., Armstrong, D.J., Murai, N., Skoog, F.: Plant Physiol. *54*, 737 (1974)
56. Armstrong, D.J., Murai, N., Taller, B.J., Skoog, F.: Plant Physiol. *57*, 15 (1976)
57. Murai, N., Taller, B.J., Armstrong, D.J., Skoog, F., Micke, M.A., Schnoes, H.K.: Plant Physiol. *60*, 197 (1977)
58. Murai, N., Armstrong, D.J., Taller, B.J., Skoog, F.: Plant Physiol. *61*, 318 (1978)
59. Teyssendier de la Serve, B., Jouanneau, J.P.: Biochimie *61*, 913 (1979)
60. Jouanneau, J.P., Gandar, J.C.: C.R. Acad. Sci. *288*, 611 (1979)
61. Laloue, M., Terrine, C., Gawer, M.: FEBS Lett. *46*, 45 (1974)
62. Fox, J.E.: In: Plant Growth Regulation, 9th Int. Conf. on Plant Growth Substances 1976. Pilet, P.E. (ed.), p. 139. Berlin-Heidelberg-New York: Springer 1977
63. Lescure, A.M.: Cell Differ. *7*, 139 (1978)
64. Roberts, J.W., Carbon, J.: Nature (Lond.) *250*, 412 (1974)
65. Fosket, D.E., Volk, M.J., Goldsmith, M.R.: Plant Physiol. *60*, 554 (1977)
66. Cortese, R., Landsberg, R., Von der Haar, R.A., Umbarger, H.E., Ames, B.N.: Proc. Natl. Acad. Sci. USA *71*, 1857 (1974)
67. Rizzino, A.A., Bresalier, R.S., Frendlich, M.: J. Bacteriol. *117*, 449 (1974)
68. Stauffer, G.V., Zurawski, G., Yanofsky, C.: Proc. Natl. Acad. Sci. USA *75*, 4833 (1978)
69. Cherayil, J.D., Lipsett, M.N.: J. Bacteriol. *131*, 741 (1977)

Probing the Cytokinin Receptor Site(s)

S.M. HECHT [1]

Our present understanding of the nature of the molecules that promote cytokinin activity is based on the finding by Miller et al. (1–3) of the growth factor kinetin (*1*; 6-furfurylaminopurine) in old and heated samples of DNA. The first active analog, 6-benzylaminopurine *(2)* (4–6), was prepared within days and presaged the synthesis of large numbers of compounds that have helped to define the structure-activity relationships for such species with considerable precision [see, e.g., (7–12)]. Although kinetin is presumably an artifact formed by rearrangement of 2′-deoxyadenosine, structurally related species such as 6-(3-methyl-2-butenylamino)purine *(3)* were subsequently identified as the growth factors in certain plant pathogens (13–15) and have now been shown to occur more generally in plants at the purine (16–25) and purine ribonucleos(t)ide levels (26–28). As discussed in the preceding report, cytokinin-active nucleosides also occur in transfer and ribosomal RNAs, although it is not clear that their presence in such RNAs is at all related to their function(s) as plant hormones.

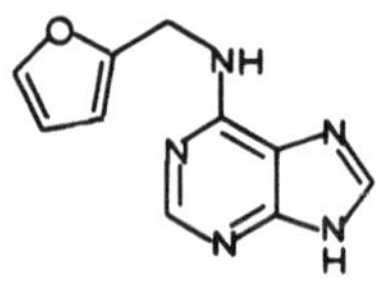

1

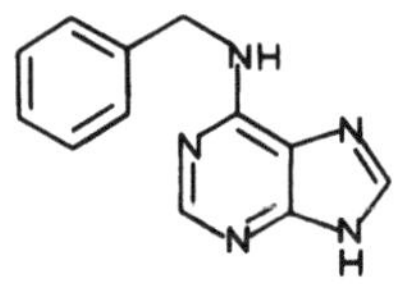

2

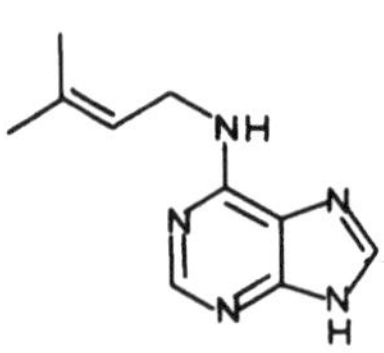

3

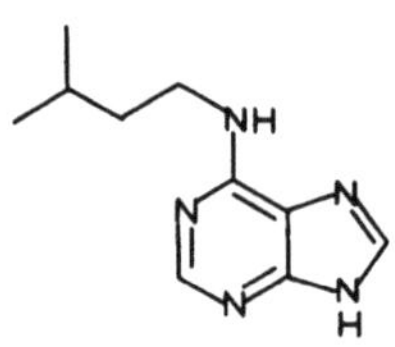

4

Structural Requirements for Cytokinin Activity

One of the interesting features revealed by structural studies of cytokinins is the extent to which the expression of cytokinin activity is influenced by structural variations in the cytokinin. Not only was activity in the purine series found to be limited to those compounds with substituents on C-6 (7–9), but activity was also shown to be greatest for those analogs with substituents having 4–7 carbon atoms (10) attached to the purine nucleus through a nitrogen atom (10, 29). Additional substitution at the N-1, N-3, N^6, N-7 or N-9-positions resulted in compounds having substantially less

1 Departments of Chemistry and Biology, University of Virginia, Charlottesville, Virginia 22901, USA

activity (10), while substituents at C-2 or C-8 had less influence on cytokinin activity (30, 31). Importantly, alteration of the heterocyclic nucleus typically effected a drastic diminution in activity (32).

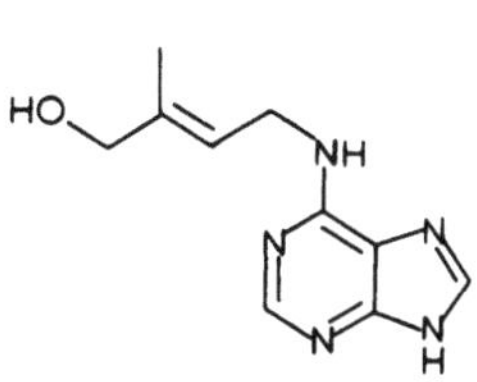

5

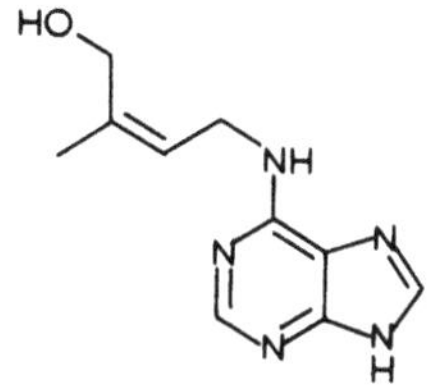

6

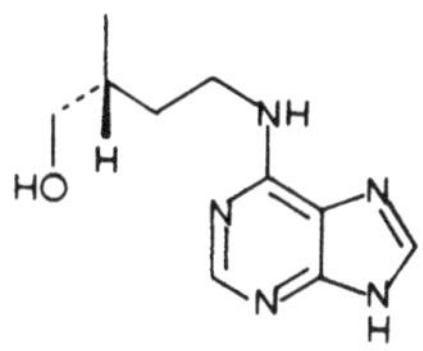

7

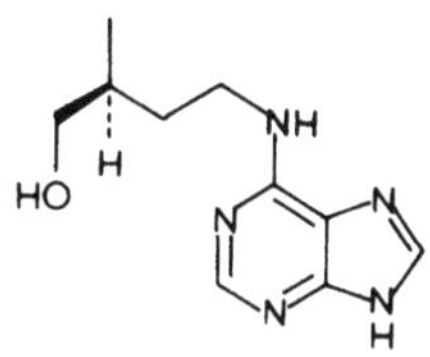

8

Even among purine derivatives having N^6-substituents of optimal size, the nature of individual substituents was important. For example, the potent cytokinin 6-(3-methyl-2-butenylamino)purine *(3)* was ten times as active as the respective saturated analog [6-(3-methylbutylamino)purine *(4)*] in the tobacco bioassay (10). Formal introduction of a 4-OH group onto *3* increased activity approximately two-fold for the trans-isomer *(5)*, but about 100-fold for the respective cis-isomer *(6)* (33); this finding was consistent with those obtained for other isomeric pairs of cytokinin analogs (33–36). Although R-(+) and S-(–) dihydrozeatins (*7* and *8*) were found to have similar activities in the tobacco bioassay, the dextrorotatory species was more effective in the lettuce seed germination assay, in promoting increased fresh weight of excised radish cotyledons and also as an inhibitor of chlorophyll degradation in senescing radish cotyledons (37). Thus, the expression of cytokinin activity is dependent on the spatial arrangement, as well as on the type of atoms present in the N^6-substituent.

Design of a Cytokinin Antagonist

Although all plants are thought to utilize cytokinins as growth hormones, relatively few species respond to exogenous cytokinins, presumably because the majority make their own. In this context, it seemed of interest to prepare a specific anticytokinin, as a compound of this type could potentially extend cytokinin studies to plants unresponsive to exogenous cytokinins. The dependence of cytokinin activity on precisely defined structural parameters, as noted above, and the ability of potent cytokinins to elicit detectable growth responses (in the tobacco bioassay) at concentrations as low as 10^{-11} M, suggested the existence of a high affinity cellular receptor site(s) for the cytokinins. It was hoped that the expression of cytokinin activity depended on more than the ability of the active species to bind to such sites, i.e., that it would be possible to design compounds that were not cytokinins per se but nonetheless bound to the cytokinin receptor site(s) (and could thus block cytokinin utilization). In the design process it seemed reasonable to begin with 6-(3-methyl-2-butenylamine)purine *(3)* as a "prototype" species having good affinity for the postulated receptor site(s) and to alter its structure systematically in a way that would diminish its efficiency as a cytokinin, but not necessarily affect its binding properties. As discussed above, *3* is an extremely active cytokinin and almost all types of structural alterations diminish that activity; moreover, the effects of such alterations tend to be additive (38). Therefore, compounds *9*–*12* were prepared for testing as potential anticytokinins. It may be noted that all of these compounds have a pyrazolo[4,3-*d*]pyrimidine nucleus, as compared with the purine nucleus in *3*, and were therefore expected to have greatly diminished cytokinin activity. Moreover, compounds *10* and *12* are additionally substituted on C-3 (a feature that might be expected to reduce activity severalfold, based on analogy with purine-derived cytokinins); the isopentyl substituents on N^7 in *11* and *12* were also expected to diminish activity substantially, in parallel with the tenfold reduction observed for the formal conversion *3* → *4*. Thus the activities of *9*–*12* as cytokinins were expected to diminish in the order *9* → *10* → *11* → *12*.

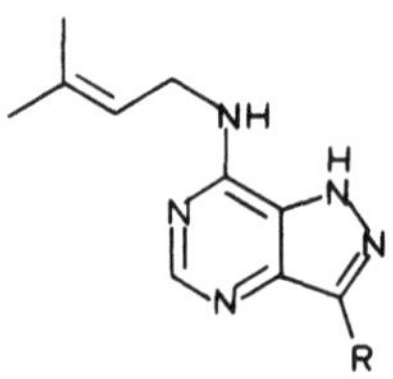

9 R=H
10 R=CH_3

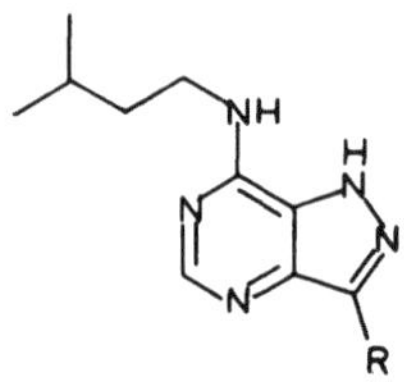

11 R=H
12 R=CH_3

Testing of the Potential Anticytokinins in the Tobacco Bioassay

Each of the test compounds was assayed as a cytokinin in the standard assay system (10, 39) in replicate cultures at each of several concentrations. After five weeks of growth in continuous diffuse light the tobacco callus was harvested for measurement of fresh weight yields. As shown in Table 1, all of the substituted pyrazolo[4,3-*d*]pyrimidines were much less active than *3*, but did have the anticipated order of activities among themselves. Thus *9* was the most active, producing detectable growth when applied at 0.24 μM concentration and eliciting maximal response at 0.73 μM (as compared with 3×10^{-4} μM and 3×10^{-3} μM, respectively, for *3* in this particular experiment).

Table 1. Cytokinin activity of substituted 7-aminopyrazolo[4,3-*d*]pyrimidines [a]

Compound	Range of concentrations tested	Minimum concentration for	
		Detection	Maximum growth
3	0.0003–0.027	0.0003	0.003
9	0.081 –6.6	0.24	0.73
10	0.24 –20	0.73	6.6
11	0.24 –20	6.6	NR [b]
12	0.24 –20	NA [c]	

[a] In comparison with compound *3*
[b] NR, not reached
[c] NA, not active

Analog *10* produced maximal growth response only at 6.6 μM concentration, while *11* did not elicit maximal growth at any tested concentration (up to 20 μM). Compound *12*, containing three formal structural alterations relative to *3*, was totally without cytokinin activity.

Although compound *12* was without activity as a cytokinin, when added to media containing potent cytokinins it diminished the apparent utilization of such species; this is illustrated in Fig. 1 for 6-benzylaminopurine *(2)*. As shown in the figure, admixture of *12* to cultures containing optimal concentrations of the cytokinin completely

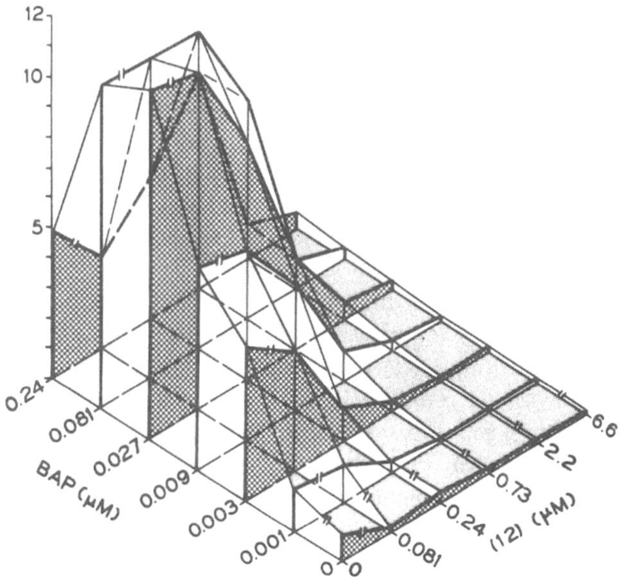

Fig. 1. Effect of 6-benzylaminopurine (BAP; *2*) and 3-methyl-7-(3-methylbutylamino)pyrazolo [4,3-*d*]pyrimidine *(12)* on the fresh weight yield of tobacco callus

eliminated cytokinin-induced growth when the inhibitor was present in 100-fold molar excess. As would be expected if *12* were a specific anticytokinin, the addition of greater amounts of *2* to the culture medium reversed the inhibition observed after a five-week growth period. As indicated in the figure, at moderate concentrations of *2* and *12*, the inclusion of more *2* in the culture medium enhanced growth while greater amounts of *12* gave further inhibition. Moreover, when tobacco tissue was maintained on a medium containing compound *3* (1×10^{-2} μM), and transferred on and off a medium containing *12* (0.1 μM) in addition to *3*, the inhibition of callus growth (measured in terms of fresh weight yield as a function of time) was found to be reversible (40).

Another criterion utilized to determine whether *12* was a specific anticytokinin involved its ability to inhibit the growth induced by cytokinins of widely varying potency. Compound *12* was found to inhibit tobacco callus growth induced by *2* and *3*, although greater concentrations were required to effect inhibition of *3*, which is tenfold more active as a cytokinin than *2*. When *2, 3* and diphenylurea (which is about

1/1000 as active as *3*) were tested for their ability to reverse inhibition of growth induced by compound *12*, compound *2* was found to be only about 1/3 as effective as *3* and diphenylurea only about 1/500 as effective. Thus the ability of *12* to inhibit cytokinin-induced growth was proportional to the activity of the individual cytokinins, as would be expected for a specific anticytokinin.

Implicit in the rationale for the preparation of structural analogs of *3* (such as *12*) as potential cytokinin antagonists is the assumption that the anticytokinin activity of such compounds should be dependent on their structural similarity to the cytokinins. Preparation and testing of additional analogs with a variety of substituents on N^7 (analogous to N^6 in the purine series) demonstrated that those species having substituents with 4–7 carbon atoms were the most active as anticytokinins, while those with much larger or smaller substituents were much less active. Thus the structural features associated with intense cytokinin activity in the purine series also correlated with strong anticytokinin activity in the pyrazolo[4,3-*d*]pyrimidine series (41).

Also investigated as potential anticytokinins were several 2-substituted 4-alkylaminopyrrolo[2,3-*d*]pyrimidines (42, 43); the test results for certain (2-methylthio)-4-alkylaminopyrrolo[2,3-*d*]pyrimidines are shown in Table 2. As indicated in the table, the derivatives lacking the 2-methylthio substituent were found to be weak cytokinins in the tobacco bioassay, whereas those having the 2-substituent lacked cytokinin activity, but were potent anticytokinins. 4-Cyclopentylamino-2-methylthiopyrrolo[2,3-*d*]-pyrimidine was the best of these, eliciting a detectable response at 9×10^{-3} μM concentration when tested against cytokinin *3* (at 3×10^{-3} μM concentration). This antagonist was thus severalfold more active than any of the pyrazolo[4,3-*d*]pyrimidine derivatives tested. Consistent with the belief that the 4-alkylamino-2-methylthiopyrrolo[2,3-*d*]pyrimidines were specific anticytokinins were the observations that compound *3* effected reversal of the inhibition of cytokinin-mediated growth caused by these compounds and that the structurally related analog lacking the N^7-substituent was inactive as a cytokinin or anticytokinin (42).

Similar results were obtained by Iwamura et al. (43) in their study of 4-alkylamino-2-methylpyrrolo[2,3-*d*]pyrimidines. It is of interest that these workers analyzed quantitatively the structure-activity relationship for those compounds having cytokinin or anticytokinin activity and were able to correlate the type and intensity of activity with the maximum width of the N^4-substituent as measured from the bond between N^4 and C^{α}. Also of interest is the observation that 4-cyclobutylamino-2-methylpyrrolo[2,3-*d*]-pyrimidine was the most active anticytokinin in this structural series. This finding prompted the preparation of 4-cyclobutylamino-2-methylthiopyrrolo[2,3-*d*]pyrimidine (*13*) in collaboration with Dr. David Evans. Testing of the compound by Drs. Folke Skoog and Ruth Schmitz has shown it to have intense activity as an anticytokinin in the tobacco bioassay.

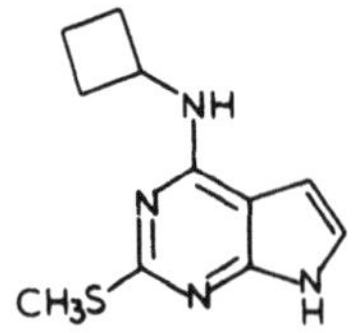

13

Table 2. Biological activity of substituted 4-aminopyrrolo[2,3-*d*]pyrimidines [a]

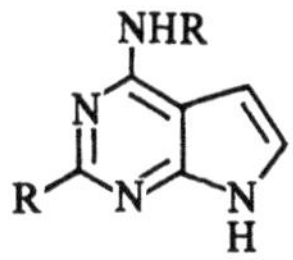

R	R′	Range of conc tested (μM)	Cytokinin activity min conc (μM) for: Detection	Cytokinin activity min conc (μM) for: Maximum growth	Growth inhibition min conc (μM) for: Detection	Growth inhibition min conc (μM) for: Complete inhibition [b]
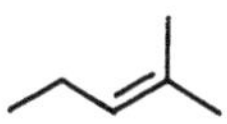	H	0.24–20	0.62	6.6	NA	–
	H	0.24–20	5.8	> 20	NA	–
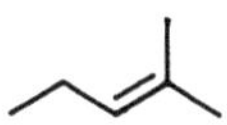	SCH_3	0.24–20	NA	–	NA	–
	SCH_3	0.009–20	NA	–	0.24	2.2
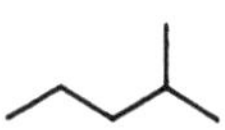	SCH_3	0.27–20	NA	–	0.40	2.0
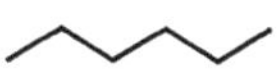	SCH_3	0.08–20	NA	–	0.1	1
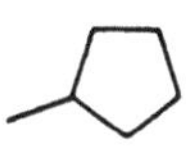	SCH_3	0.009–20	NA	–	0.009	0.05
	SCH_3	0.24–20	NA	–	6.6	> 20
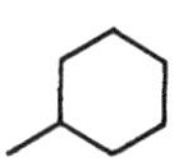	SCH_3	0.003–20	NA	–	0.05	0.60
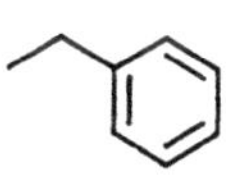	SCH_3	0.24–20	NA	–	10	> 20

[a] Abbreviations: NA, nonactive

[b] In presence of 0.003 μM i^6 Ade

There May Be Multiple Cytokinin Receptor Sites

Although certain of the substituted pyrazolo[4,3-*d*]pyrimidines and pyrrolo[2,3-*d*]-pyrimidines discussed above were shown to oppose the effects of the cytokinins in the tobacco bioassay, all of these compounds were prepared as structural analogs of cytokinins, and it was of interest to determine whether any of them was capable of *reinforcing* other cytokinin-mediated processes.

The first indication that such effects may actually obtain derived from observations made for 4-cyclopentylamino- and 4-cyclohexylamino-2-methylthiopyrrolo-[2,3-*d*]pyrimidines in the tobacco bioassay. In the presence of 11.4 μM indole-3-acetic acid both compounds promoted bud formation, an effect normally associated with high cytokinin-auxin ratios (42). Although the effect was less apparent at lower auxin concentrations, it was also shown that the anticytokinin 2-methyl-3-[(2-ethyl)hexyl-amino]pyrrolo[2,3-*d*]pyrimidine at 40 μM would promote bud formation when applied in the absence of added cytokinin (43). One possible interpretation of these findings is that there are different cytokinin receptor sites for bud formation as compared with cell division and growth and that the cytokinin requirements of these sites (in structural terms) may differ somewhat. Consistent with this interpretation were the observations that 2-methyl-4-phenylaminopyrrolo[2,3-*d*]pyrimidine, an antagonist in the tobacco bioassay, stimulated lettuce seed germination (a cytokinin-mediated process), but had no effect on betacyanin production by *Amaranthus caudatus*. Promotion of betacyanin synthesis was effected by several other 4-alkylamino-2-methylpyrrolo[2,3-*d*]pyrimidine derivatives that had anticytokinin activity in the tobacco bioassay. Although without activity in the tobacco bioassay, 4-hydroxyethylamino-2-methylpyrrolo[2,3-*d*]pyrimidine suppressed cytokinin- (*3*; 1 μM) mediated betacyanin synthesis (43). It is anticipated that the development of quantitative structure-activity relationships for individual cytokinin-dependent functions might eventually make it possible to design compounds that could permit the manipulation of single cytokinin receptor sites.

Biological Effects of the Anticytokinins on Plants

As the preparation of a cytokinin antagonist was originally undertaken in an effort to extend the study of cytokinins to cytokinin-autonomous species, one of the first biological studies carried out with the anticytokinins involved a strain of tobacco callus that grew without exogenous cytokinin. Antagonist *12* was found to inhibit the growth of this tissue in the same fashion as had been observed for the cytokinin-dependent strain, and the inhibition of the autonomous strain could also be reversed by added cytokinin (41). This provided strong evidence that the cytokinin-independent callus produced its own cytokinin, a conclusion reinforced by the isolation of zeatin (4-hydroxy-3-methyl-trans-2-butenylaminopurine) and two other cytokinins from this strain.

As discussed by Dr. Kulaeva in this volume, an additional effect noted for a cytokinin antagonist (4-cyclopentylamino-2-methylthiopyrrolo[2,3-*d*]pyrimidine, 40 μM) is the inhibition of cytokinin-induced nitrate reductase in isolated embryos of *Agro-*

stemma githago (4 x 10^{-2} μM *2*; 39% inhibition). Also of interest are the effects of the antagonists on ethylene production in apple slices. Cytokinins *1* and *3* diminished the production of ethylene in apple and avocado slices; the antagonists opposed this effect (M. Lieberman, personal communication).

Unlike certain of the 4-alkylamino-2-methylpyrrolo[2,3-*d*]pyrimidines discussed above, antagonist *12* had no effect on germination per se, but high (180 μM) concentrations of the compound did affect root initiation and development of *Coleus* cuttings, as well as wheat and radish seedlings. The same compound caused severe wilting of tomato seedlings, but had no effect on mature tomato, sweet corn or tobacco plants.

Tests of two anticytokinins (7-*n*-hexylamino-3-methylpyrazolo[4,3-*d*]pyrimidine and 4-cyclopentylamino-2-methylthiopyrrolo[2,3-*d*]pyrimidine) were also carried out at Dow Chemical Company (Walnut Creek, California). When applied preemergence (10 lb/acre) or postemergence (4000 ppm) both compounds were generally ineffective as herbicides, although the pyrazolo[4,3-*d*]pyrimidine did control crabgrass in the preemergence test. Testing of the compounds as insecticides also showed them to be without activity. The compounds were tested extensively as potential fungicides. The pyrazolo[4,3-*d*]pyrimidine derivative was found to control wheat leaf rust when applied to the seed at 100 ppm and both barley mildew and apple scab at higher (400 ppm) concentration. Control of grape downy mildew was achieved by the use of the pyrrolo-[2,3-*d*]pyrimidine derivative at 100 ppm.

Perhaps the most interesting effect noted at Dow Chemical was that on water uptake. When utilized a level of 15–20 ppm in the soil, two 4-alkylamino-2-methylthiopyrrolo[2,3-*d*]pyrimidines were found to decrease the water requirement for the growth of wheat, corn, soybeans and cotton plants.

At this stage of development, none of the cytokinin antagonists has been found to have activity on an intact plant that could lead to its commercial use. It may be noted, however, that this may simply reflect the nature of the assay system used for development of the anticytokinins and that structural optimization using more appropriate assays might provide compounds with greatly improved activities at the level of whole plants.

Biological Effects of the Anticytokinins on Mammalian Cells

Although originally of interest as a naturally occurring plant growth regulator, N^6-(3-methyl-2-butenyl)adenosine *(14)* enjoys more widespread distribution as one of the cytokinin-active nucleosides which occur as a component of transfer RNAs derived from virtually all forms of life (45). The possible biological importance of the nucleoside in mammalian cells was suggested by the observation that tRNAs from human lymphosarcoma cells had a fourfold greater concentration of this compound than tRNAs from normal lymphocytes (46); in fact, exogenous N^6-(3-methyl-2-butenyl)adenosine has been shown to affect the growth of cultured mouse cells (47–49) and the extent of transformation, growth and mitosis

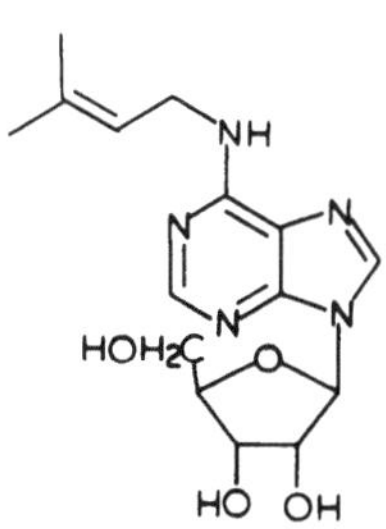

14

of rat and human T lymphocytes pretreated with phytohemagglutinin (PHA) (46, 50–52). Therefore, the additional finding by Mittelman et al. (53) that N^6-(3-methyl-2-butenyl)adenosine may have clinical utility in the treatment of some leukemias promp-prompted us to investigate the physiological effects of certain structural analogs of *14*.

Gallo and his coworkers (46, 51, 52) found that ribonucleoside *14* stimulated cellular transformation and DNA synthesis when added in low concentration to human T lymphocytes well after PHA addition, but inhibited these processes when added soon after PHA or at higher concentration; analogous results were obtained with a line (6410) of human myeloblastic leukemia cells (47–49). While the inhibitory effects obtained by incubation of PHA-treated lymphocytes with N^6-(3-methyl-2-butenyl)adenosine were observed only with certain other N^6-alkyladenosines (52), and no other nucleoside has been reported to elicit the stimulatory effects, we have discovered that several 7-alkylamino-3-methylpyrazolo[4,3-*d*]pyrimidines affect the percent transformation and DNA synthesis of PHA-treated lymphocytes in the same manner as *14*. This is illustrated in Fig. 2 for *14* and 3-methyl-7-*n*-pentylaminopyrazolo[4,3-*d*]pyrimidine, the latter of which was severalfold more potent than compound *14* in this assay system.

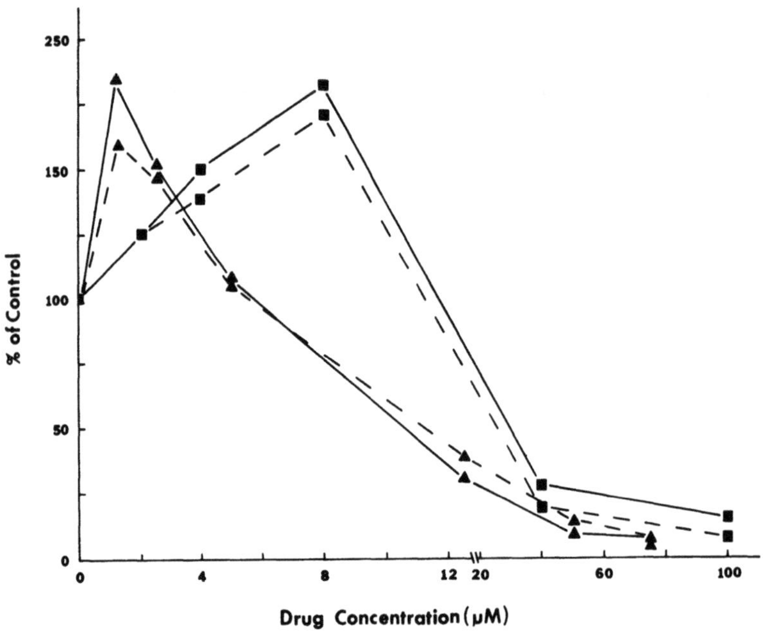

Fig. 2. Dose-response curves indicating the extent of stimulation and depression of transformation (– – –) and DNA synthesis (——) obtained with phytohemagglutinin-treated human T-lymphocytes in the presence of *14* (■) and 3-methyl-7-*n*-pentylaminopyrazolo[4,3-*d*]pyrimidine (▲). Lymphocytes (2×10^6/tube) were incubated in 600 μl of Eagle's minimum essential medium in the presence of 0.15 mg of phytohemagglutinin M for 24 h, then treated with varying concentrations of the drugs (in triplicate) and maintained at 37°C for an additional 24 h. DNA synthesis was measured as incorporation of [^{3}H]-thymidine into 5% perchloric acid-insoluble material; the extent of morphological transformation was determined after the cells were stained with a giemsa-based stain

As noted previously for Sarcoma 180 and carcinoma TA-3 cells (47–49), moderate concentrations of *14* were found to inhibit the growth of 3T3 and 3T6 cells as well as simian virus-transformed 3T3 cells. In analogy with the results obtained with human lymphocytes, in some experiments very low (< 1 μg/ml) concentrations of *14* resulted in growth stimulation. The efficacy of the compound in inhibiting growth was directly proportional to the extent of morphological transformation of the cell line; thus total inhibition of SV3T3 cells was achieved in 72 h with 15 μM *14*, but in a parallel experiment a 600 μM concentration was required for comparable inhibition of 3T3 cells. Moreover, in the presence of moderate concentrations of added *14*, 3T6 and SV3T3 cells were altered morphologically to resemble 3T3 cells. Also tested in comparison with *14* was 3-methyl-7-*n*-pentylaminopyrazolo[4,3-*d*]pyrimidine. Consistent with its greater activity as a regulator of lymphocyte growth, incubation of 3T6 cells in the presence of this compound (3 μM; 48 h) resulted in 45% fewer viable cells than were found in drug-free controls; in the presence of 3 μM *14* the reduction in cell population was < 5%.

In view of the considerable biological activity of the 7-*n*-pentylamino- and other 7-alkylamino-3-methylpyrazolo[4,3-*d*]pyrimidines, it seemed of interest to determine whether species of this type functioned by the same mechanism as N^6-(3-methyl-2-butenyl)adenosine *(14)*. This was done initially by taking advantage of the observation that at very low concentrations of added *14* or 7-alkylamino-3-methylpyrazolo[4,3-*d*]-pyrimidines, little or no inhibition was obtained. As suggested by Fig. 3, however,

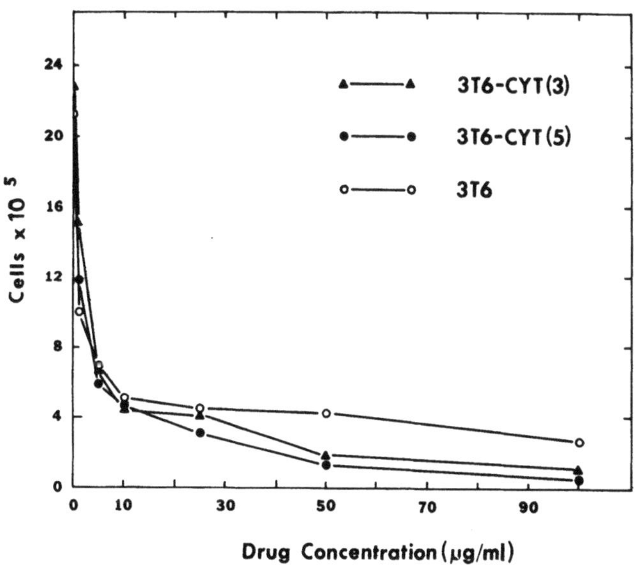

Fig. 3. The inhibitory effect of various concentrations of 3-methyl-7-*n*-pentylaminopyrazolo[4,3-*d*]-pyrimidine on mouse fibroblast (3T6) cells and on two N^6-(3-methyl-2-butenyl)adenosine-resistant lines derived therefrom [3T6-CYT(3), 3T6-CYT(5)]. Mouse fibroblast cells (2×10^4/dish) were grown in 4 ml of Dulbecco's modified Eagle medium (supplemented with penicillin G and streptomycin) and horse serum in the presence of varying amounts (in triplicate) of 3-methyl-7-*n*-pentylaminopyrazolo[4,3-*d*]pyrimidine. After incubation at 37°C for 96 h, the attached cells in each of the dishes were harvested and counted

small increments of these compounds beyond their maximum tolerated concentration caused substantial inhibition of cell growth. Therefore, after careful determination of dose-response curves of low concentrations of *14* and 3-methyl-7-(3-methylbutylamino)pyrazolo[4,3-*d*]pyrimidine, the compounds were added singly and in combination to replicate cultures of 3T6 cells at the maximum tolerated concentration measured for each. Had we failed to observe an enhancement of inhibition in the cultures containing both compounds, we would have concluded that they effected inhibition of the mouse cells by unrelated mechanisms. The actual experimental observation, though, was that cultures grown in the presence of 0.3 μM *14* and 3 μM 3-methyl-7-(3-methylbutylamino)pyrazolo[4,3-*d*]pyrimidine for 4 days had 8% and 21% fewer cells, respectively, than drug-free controls, while in combination the drugs diminished the number of cells by 47%. The synergistic effect obtained with the two types of compounds is consistent with the interpretation that they inhibit the growth of 3T6 cells by a related mechanism. (It has recently been shown that adenosine is cytotoxic to cultured cells if not deaminated prior to conversion to 5′-AMP, the latter of which inhibits uridine biosynthesis (54). The effects obtained with compounds *14* and the pyrazolo[4,3-*d*]pyrimidine derivative were not spared by added uridine.)

As part of their study of the effects of *14* on PHA-transformed lymphocytes, Gallo and his coworkers (46, 51, 52) investigated the mechanism of action of the ribonucleoside. N^6-(3-methyl-2-butenyl)adenosine had no effect on phytohemagglutinin itself, nor did the compound compete with PHA for lymphocyte receptor sites. The additional observation that lymphocytes that had entered S phase were unaffected by *14* led to the conclusion that the nucleoside acted after interaction of PHA with the lymphocyte cell membrane, but before the initiation of DNA synthesis. Since the transformation and growth of mammalian cells is known to be associated with changes in the intracellular concentration of cyclic AMP (55–58), the possibility was considered that *14* might function at the level of cyclic nucleotide metabolism. In fact, exogenous N^6-$O^{2'}$-dibutyryl cyclic AMP had virtually the same effect on transformation and DNA synthesis of PHA-treated lymphocytes as *14*; the latter species was also observed to alter the dose-response curves obtained with the cyclic AMP analog (46, 52). The possible activity of *14* as an inhibitor of cyclic AMP phyosphodiesterase activities was subsequently studied using the high K_m phosphodiesterase from beef heart (59). Compound *14* was found to be a competitive inhibitor of the enzyme (apparent K_m 70 μM) with an apparent K_i of 109 μM. Not unexpectedly, 3-methyl-7-*n*-pentylaminopyrazolo[4,3-*d*]pyrimidine was found to be a better competitive inhibitor (apparent K_i 48 μM).

If the regulatory effects of *14* were mediated via control of cyclic AMP turnover, then the compound should also inhibit the cyclic AMP phosphodiesterase activity in cells known to be physiologically responsive to the nucleoside. Therefore, we have studied the appropriate phosphodiesterase activities isolated from 3T6 cells. The ability of cell-free extracts to convert [^{3}H]-cyclic AMP to AMP was monitored by modification of the procedure of Gilman (60); compound *14* was inhibitory to both the high and low K_m activities. Fractionation of the crude cell extract on Sephadex G-100 afforded a sample of the low K_m activity not contaminated with the low affinity enzyme, and the initial velocity of cyclic AMP hydrolysis by this activity was measured in the presence and absence of *14*. The apparent K_m for the conversion was 2 μM, while the K_i was measured as 6 μM. In analogy with its greater potency toward intact 3T6 cells, 3-methyl-7-*n*-pentylaminopyrazolo[4,3-*d*]pyrimidine was more inhibitory to the low K_m phosphodiesterase activity than *14*. Analysis of the high and low K_m cyclic AMP phosphodiesterase activities from PHA-treated human leukocytes gave results consistent with those obtained for the analogous enzyme activities from 3T6 cells and beef heart. Both activities were inhibited by *14*, and to a greater extent by 3-methyl-7-*n*-pentylaminopyrazolo[4,3-*d*]pyrimidine. Interpretation of this result is complicated by the fact that ~80% of the leukocyte fraction employed consisted of cells other than T-lymphocytes (e.g., B-lymphocytes, monocytes), which were not transformed but still undoubtedly contained cyclic AMP phosphodiesterase activities. Preliminary results obtained with the low K_m activity from purified human T lymphocytes also indicated that the pyrazolo[4,3-*d*]pyrimidine derivative was the better inhibitor (61).

The utility of many potentially interesting chemotherapeutic agents is limited by the development of resistance to the agents. It seemed reasonable to attempt to derive mouse fibroblast cells resistant to the test compounds as a model system. After several passages on nutrient media containing sublethal doses of N^6-(3-methyl-2-butenyl)adenosine, it was possible to effect subsequent transfer of 3T6 cells to similar media containing increasing concentrations of compound *14*. In this fashion we obtained cells that could be cultured in the presence of 600 μM *14* and that maintained their resistance through 15 passages in the absence of the drug. Several clones were derived from the resistant cells and three of these, which had the same morphological characteristics as 3T6 cells, also grew with the same generation time (16 h) and to about the same density. As shown in Fig. 3 for two of these N^6-(3-methyl-2-butenyl)adenosine-resistant clones, growth was still inhibited by exogenous 3-methyl-7-*n*-pentylaminopyrazolo[4,3-*d*]pyrimidine, and to essentially the same extent as that observed for 3T6 cells. Interestingly, to date we have been unable to isolate 3T6 cells resistant to the pyrazolo[4,3-*d*]pyrimidine. One of the clones of 3T6 cells resistant to *14* was analyzed for cyclic AMP phosphodiesterase content, in the belief that a comparison with the corresponding activities from the parent cells might prove instructive. Consistent with the hypothesis that *14* raises the intracellular level of cyclic AMP, the resistant cells had about 1/3 more low K_m activity than 3T6 cells (40 vs 30 pmol/mg/min) and 2–3 times as much high K_m activity.

It has been reported previously that Sarcoma 180 cells resistant to adenosine analogs (including *14*) had increased adenosine deaminase activity, which could contribute to the development of resistance via detoxification of the adenosine analogs (62). In fact, analysis of the N^6-(3-methyl-2-butenyl)adenosine-resistant 3T6 cells showed that they had 75% more adenosine deaminase activity than the sensitive line from which they were derived, as judged by the ability of the cells to deaminate (50 μM) adenosine. In an effort to identify an analog of 3-methyl-7-*n*-pentylaminopyrazolo[4,3-*d*]-pyrimidine that would not be a substrate for adenosine deaminase (and would thus be refractory to this potential mode of inactivation), we considered the results of two published studied (63, 64). By the use of deaza analogs of adenosine, it was established that the presence of N-1, N-3 and N-7 nitrogen atoms was required for deamination, but that the 1-deaza analog of adenosine was bound efficiently by the deaminase from calf intestinal mucosa and inhibited adenosine deamination. Since Rogozinska et al. (65) have shown that the 1-deaza analog of *14* retains considerable cytokinin activity, it was of interest to prepare the corresponding deaza analog of 3-methyl-7-*n*-pentylaminopyrazolo[4,3-*d*]pyrimidine (3-methyl-7-*n*-pentylaminopyrazolo[4,3-*b*]pyridine; *15*). As shown in Scheme 1, this compound was prepared from 3-carboethoxyisoxazole

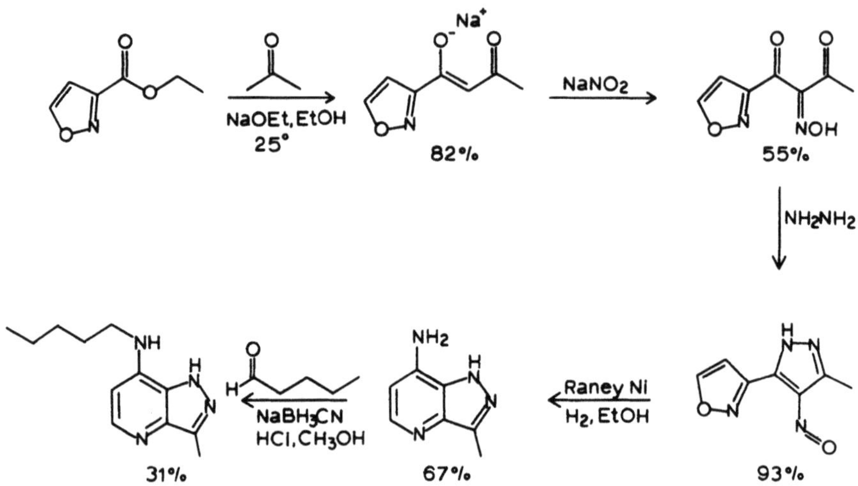

15

(66) in analogy with the procedure of Ajello for the preparation of 7-amino-3,5-dimethylpyrazolo[4,3-*b*]pyridine (67). Introduction of the C-5 substituent to N[7] was accomplished via reductive amination (68) of valeraldehyde (69). Preliminary testing of compound *15* in Prof. Skoog's laboratory has shown it to have weak anticytokinin activity in the tobacco bioassay. It will be of interest to measure its activity in the aforementioned assay systems involving mammalian cells.

Acknowledgments. The plant bioassay results discussed in this contribution were obtained as part of a collaborative effort with Prof. Folke Skoog and Dr. Ruth Schmitz, Institute of Plant Development, University of Wisconsin. Thanks are also due to Dow Chemical Company for permission to cite the results of their studies with certain synthetic cytokinin analogs. I would also like to thank my coworkers for their contributions to this project, especially Drs. R. Bruce Frye, Hector Juarez and David Evans. This work was supported in part by research grants from the National Cancer Institute (CA 14896) and the donors of the Petroleum Research Fund, administered by the American Chemical Society.

References

1. Miller, C.O., Skoog, F., von Saltza, M.H., Strong, F.M.: J. Am. Chem. Soc. *77*, 1392 (1955)
2. Miller, C.O., Skoog, F., Okumura, F.S., von Saltza, M.H., Strong, F.M.: J. Am. Chem. Soc. *77*, 2662 (1955)
3. Miller, C.O., Skoog, F., Okumura, F.S., von Saltza, M.H., Strong, F.M.: J. Am. Chem. Soc. *78*, 1375 (1956)

4. Skinner, C.G., Shive, W.: J. Am. Chem. Soc. *77*, 6692 (1955)
5. Daly, J.W., Christensen, B.E.: J. Org. Chem. *21*, 177 (1956)
6. Bullock, M.W., Hand, J.J.: J. Org. Chem. *22*, 568 (1957)
7. Strong, F.M.: In: Topics in Microbial Chemistry, Chapter III. New York: Wiley 1958
8. Okumura, F.S., Enishi, N., Itoh, H., Masumura, M., Kuraishi, S.: Bull. Chem. Soc. Jpn. *32*, 886 (1959)
9. Kuraishi, S.: Sci. Pap. Coll. Gen. Educ. Univ. Tokyo *9*, 67 (1959)
10. Skoog F., Hamzi, H.Q., Szweykowska, A.M., Leonard, N.J., Carraway, K.L., Fujii, T., Helgeson, J.P., Loeppky, R.N.: Phytochemistry *6*, 1169 (1967)
11. Skoog, F., Armstrong, D.J.: Annu. Rev. Plant Physiol. *21*, 359 (1970)
12. Leonard, N.J.: Recent Adv. Phytochem. *7*, 21 (1974)
13. Klämbt, D., Thies, G., Skoog, F.: Proc. Natl. Acad. Sci. USA *56*, 52 (1966)
14. Helgeson, J.P., Leonard, N.J.: Proc. Natl. Acad. Sci. USA *56*, 60 (1966)
15. Upper, C.D., Helgeson, J.P., Kemp, J.D., Schmidt, C.J.: Plant Physiol. *45*, 543 (1970)
16. Miller, C.O.: Proc. Natl. Acad. Sci. USA *47*, 170 (1961)
17. Letham, D.S., Shannon, J.S., McDonald, I.R.: Proc. Chem. Soc. 230 (1964)
18. Letham, D.S., Miller, C.O.: Plant Cell Physiol. *6*, 355 (1965)
19. Witham, F.H., Miller, C.O.: Physiol. Plant. *18*, 1007 (1965)
20. Letham, D.S.: Phytochemistry *5*, 269 (1966)
21. Miller, C.O.: Science *157*, 1055 (1967)
22. Koshimizu, K., Kusaki, T., Mitsui, T., Matsubara, S.: Tetrahedron Lett. 1317 (1967)
23. Koshimizu, K., Matsubara, S., Kusaki, T., Mitsui, T.: Agric. Biol. Chem. *31*, 795 (1967)
24. Scarbrough, E., Armstrong, D.J., Skoog, F., Frihart, C.R., Leonard, N.J.: Proc. Natl. Acad. Sci. USA *70*, 3825 (1973)
25. Armstrong, D.J., Scarbrough, E., Skoog, F., Cole, D.L., Leonard, N.J.: Plant Physiol. *58*, 749 (1976)
26. Miller, C.O.: Proc. Natl. Acad. Sci. USA *54*, 1052 (1964)
27. Letham, D.S.: Life Sci. *5*, 551 (1966)
28. Letham, D.S.: Phytochemistry *12*, 2445 (1973)
29. Henderson, T.R., Frihart, C.R., Leonard, N.J., Schmitz, R.Y., Skoog, F.: Phytochemistry *14*, 1687 (1975)
30. Hecht, S.M., Leonard, N.J., Schmitz, R.Y., Skoog, F.: Phytochemistry *9*, 1173 (1970)
31. Dammann, L.G., Leonard, N.J., Schmitz, R.Y., Skoog, F.: Phytochemistry *13*, 329 (1974)
32. Skinner, C.G., Shive, W.: Plant Physiol. *32*, 500 (1957)
33. Schmitz, R.Y., Skoog, F., Playtis, A.J., Leonard, N.J.: Plant Physiol. *50*, 702 (1972)
34. Hecht, S.M., Leonard, N.J., Schmitz, R.Y., Skoog, F.: Phytochemistry *9*, 1907 (1970)
35. Hall, R.H., Srivastava, B.I.S.: Life Sci. *7*, 7 (1968)
36. Leonard, N.J., Hecht, S.M., Skoog, F., Schmitz, R.Y.: Proc. Natl. Acad. Sci. USA *63*, 175 (1969)
37. Matsubara, S., Shiojiri, S., Fujii, T., Ogawa, N., Imamura, K., Yamagishi, K., Koshimizu, K.: Phytochemistry *16*, 933 (1977)
38. Schmitz, R.Y., Skoog, F., Hecht, S.M., Bock, R.M., Leonard, N.J.: Phytochemistry *11*, 1603 (1972)
39. Hecht, S.M., Bock, R.M., Schmitz, R.Y., Skoog, F., Leonard, N.J.: Proc. Natl. Acad. Sci. USA *68*, 2608 (1971)
40. Helgeson, J.P., Haberlach, G.T., Hecht, S.M.: In: Plant Growth Substances 1973, pp. 485 ff. Tokyo: Hirokawa 1974
41. Skoog, F., Schmitz, R.Y., Bock, R.M., Hecht, S.M.: Phytochemistry *12*, 25 (1973)
42. Skoog, F., Schmitz, R.Y., Hecht, S.M., Frye, R.B.: Proc. Natl. Acad. Sci. USA *72*, 3508 (1975)
43. Iwamura, H., Masuda, N., Koshimizu, K., Matsubara, S.: Phytochemistry *18*, 217 (1979)
44. Einset, J.W., Skoog, F.: Proc. Natl. Acad. Sci. USA *70*, 658 (1973)
45. Skoog, F., Armstrong, D.J.: Annu. Rev. Plant Physiol. *21*, 359 (1970)
46. Gallo, R.C., Whang-Peng, J.: In: Biological Effects of Polynucleotides. Beers, R.F. Jr., Braun, W. (eds.), p. 303. Berlin-Heidelberg-New York: Springer 1971
47. Fleysher, M.H., Hakala, M.T., Bloch, A., Hall, R.H.: J. Med. Chem. *11*, 717 (1968)

48. Fleysher, M.H., Bloch, A., Hakala, M.T., Nichol, C.A.: J. Med. Chem. *12,* 1056 (1969)
49. Fleysher, M.H.: J. Med. Chem. *15,* 187 (1972)
50. Hacker, B., Feldbush, T.L.: Biochem. Pharmacol. *18,* 847 (1969)
51. Gallo, R.C., Whang-Peng, J., Perry, S.: Science *165,* 400 (1969)
52. Gallo, R.C., Hecht, S.M., Whang-Peng, J., O'Hopp, S.: Biochim. Biophys. Acta *281,* 488 (1972)
53. Mittelman, A., Evans, J.T., Chheda, G.B.: Ann. N.Y. Acad. Sci. *255,* 225 (1975)
54. Ishii, K., Green, H.: J. Cell Sci. *13,* 429 (1973)
55. See, e.g., Heidrick, M.L., Ryan, W.L.: Cancer Res. *31,* 1313 (1971)
56. Sheppard, J.R. Nature (Lond.) New Biol. *236,* 14 (1972)
57. Otten, J., Johnson, G.S., Pastan, I.: Biochem. Biophys. Res. Commun. *44,* 1192 (1971)
58. Seifert, W., Paul, D.: Nature (Lond.) New Biol. *240,* 281 (1972)
59. Hecht, S.M., Faulkner, R.D., Hawrelak, S.D.: Proc. Natl. Acad. Sci. USA *71,* 4670 (1974)
60. Gilman, A.G.: Proc. Natl. Acad. Sci. USA *67,* 305 (1970)
61. Juarez, H., Catsimpoolas, N., Hecht, S.M.: unpublished results
62. Divekar, A.Y., Fleysher, M.H., Slocum, H.K., Kenny, L.N., Hakala, M.T.: Cancer Res. *32,* 2530 (1972)
63. Ikehara, M., Fukii, T.: Biochim. Biophys. Acta *338,* 512 (1974)
64. Kitano, S., Mizuno, Y., Ueyama, M., Tori, K.: Kamisaku, M., Ajisaka, K.: Biochem. Biophys. Res. Commun. *64,* 996 (1975)
65. Rogozinska, J.H., Kroon, C., Salemink, C.A.: Phytochemistry *12,* 2087 (1973)
66. Micetich, R.G.: Can. J. Chem. *48,* 467 (1970)
67. Ajello, E.: J. Heterocycl. Chem. *8,* 1035 (1971)
68. Borch, R.F., Bernstein, M.D., Durst, H.D.: J. Am. Chem. Soc. *93,* 2897 (1971)
69. Evans, D.L.: Ph.D. Thesis, Massachusetts Institute of Technology (1979)

Gibberellins

Chairman: B. O. PHINNEY

Partial Syntheses of Isotopically Labelled Gibberellins

J. MACMILLAN [1]

Introduction

The general chemistry of the gibberellins (GAs) has now been reasonably well explored. Most of the current chemical interest is centered on the total or partial synthesis of the GAs. Part of this interest lies in the challenge to synthetic chemists of devising total syntheses of these complex structures. More germane to plant hormonologists is the interest in the acquisition of isotopically labelled GAs for metabolic studies. In our own work we are also interested in the confirmation, by synthesis, of structures deduced for new GAs from GC-MS data, and in the preparation of GA-derivatives in which the positions of hydroxylation are blocked in order to explore the relationship between hydroxylation and biological action. This paper is, however, confined to a discussion of the chemical preparation of isotopically labelled GAs and their precursors for metabolic studies.

Although the total syntheses of several GAs have been achieved (1, 2), the synthetic routes are long and specifically devised for particular GAs. Several partial syntheses of GAs from other naturally occurring compounds have also been published (2), but they are also lengthy and not generally applicable to other GAs. The most convenient methods are those based upon the accessible fungal GAs, such as GA_3, $GA_{4/7}$ mixtures, and, to a lesser extent, GA_9. Only these partial syntheses are considered here. In the following section, published methods for the preparation of isotopically labelled GAs are briefly reviewed. Then, in the third section, some recent and largely unpublished methods from our own studies are presented.

Review of Previous Work

Wittig Reaction

The most frequently used method (Fig. 1) has been the oxidation of the 16,17-double bond of GAs and of their ent-kaurenoid precursors to give the 17-nor-16-ketone which is then reacted with a ^{14}C- or ^{3}H-methyltriphenylphosphonium halide in the presence of a strong base.

With [^{14}C-methyl]triphenylphosphonium iodide and n-butyl lithium as the base, the following [17-^{14}C]-labelled compounds have been prepared: ent-kaur-16-ene (3);

1 School of Chemistry, The University, Bristol, United Kingdom

ent-kaur-16-en-19-oic acid (4, 5); ent-7α-hydroxy-kaur-16-en-19-oic acid (6); GA_{12}-alcohol (7); GA_{12}-aldehyde (8); GA_{12} (8); GA_9 (3); GA_{13} (8); and GA_{14} (8).

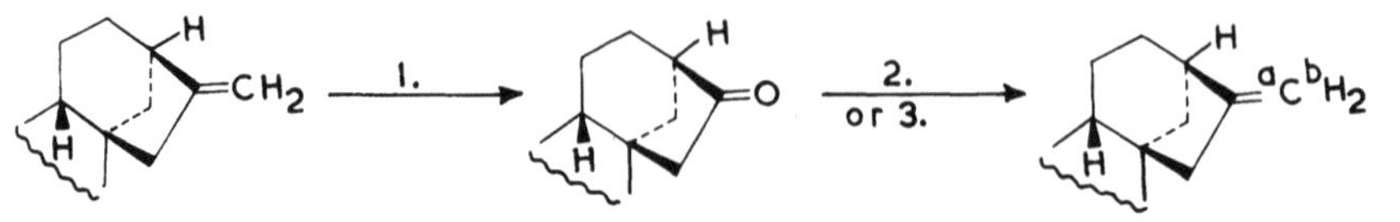

Fig. 1. 17-Labelling by the Wittig reaction; (a) = 12, 13 or 14, and (b) = 1, 2 or 3. Reagents: *1* OsO_4-$NaIO_4$; *2* $Ph_3\overset{+}{P}{}^aC^bH_3\bar{X}$ and base; *3* $Ph_3P = {}^aC^bH_2$ in tetrahydrofuran or benzene

Using n-butyl lithium the yields are generally low, and it is surprising that this base is commonly used since it has been known for many years (9) that lithium ions stabilize the betaine intermediate in Wittig reactions. Higher yields are obtained using either potassium t-butoxide or salt-free methylene triphenylphosphorane, prepared in tetrahydrofuran from methyltriphenylphosphonium bromide (more soluble than the iodide) and an excess of sodium hydride. Examples are provided by Bearder et al. (10, 11).

[^{14}C-Methyl]triphenylphosphonium halides are more expensive and less convenient to use at reasonably high specific activity than the [^{3}H-methyl] salts. Bearder et al. (10) have described a method of exchanging the methyl protons in methyltriphenylphosphonium bromide which is more convenient than the previously published methods (12). The procedure involves reaction of the phosphonium bromide with ^3H_2O (or ^2H_2O) in tetrahydrofuran containing triethylamine at room temperature. Using [^{3}H-methyl]triphenylphosphonium bromide prepared in this way and potassium t-butoxide, Bearder et al. (11) prepared [^{3}H]GA_9 from GA_9-norketone in 80% yield and showed that the ^{3}H-label was scrambled between the 15- and 17-positions.

In earlier Wittig reactions, it was deemed necessary to protect any carboxyl group in the ketone as the ester which was hydrolyzed after the Wittig reaction. However this is not necessary. Neither is it necessary to protect any hydroxyl groups unless they are at position-3, in which case the base used in the reaction causes epimerization, or at position-13 in which case base-catalyzed rearrangement of rings C and D occurs. For example, in the reaction of steviol norketone with [^{3}H-methyl]triphenylphosphonium bromide and potassium t-butoxide, the carboxyl group was not derivatized (10). However, rearrangement of rings C and D occurred, and both [^{3}H]-steviol (36% yield) and epi-[^{3}H]steviol (30%) were obtained; scrambling of the label also occurred (10). Hydroxyl groups can be protected as trimethylsilyl ethers (13) which are stable to base but which can be hydrolyzed by mild acid treatment after the Wittig reaction.

Catalytic Reduction

Selective catalytic reduction of the 1,2-double bond in ring A of GA_3 has been used by several groups to prepare [1,2-$^3H_2$]GA_1. Kende (14) and Musgrave et al. (15), adapting the original method (16) for the partial hydrogenation of GA_3 to GA_1, reduced GA_3 with a tritium-hydrogen mixture to obtain a 30% yield of [^{3}H]GA_1 (1.3 Ci

$mmol^{-1}$) which was purified by TLC and co-crystallized with unlabelled GA_1 to a constant activity of 88.1 mCi $mmol^{-1}$. Pitel and Vining (17) modified the method to prepare [^3H]GA_1 in 17%–20% yield and with a specific activity of 487 mCi $mmol^{-1}$; the [^3H]GA_1 was purified by Sephadex partition chromatography (18). Using carrier-free tritium Nadeau and Rappaport (19) obtained a complex mixture of products from which, inter alia, [^3H]GA_1 (12.6%, 43 Ci $mmol^{-1}$) and [^3H]GA_3 (3.2%, 13 Ci $mmol^{-1}$) were isolated and characterized. [^3H]GA_4 (1.87 Ci $mmol^{-1}$) has been prepared (20) from GA_7 in an analogous manner.

Using the procedure of MacMillan and Pryce (21), Musgrave and Kende (15) and Durley and Pharis (20) have prepared [^3H]GA_5 from [^3H]GA_1 with respective specific activities of 113 and 129 mCi $mmol^{-1}$. Adopting a different and ingenious approach, Murofushi et al. (22) found that the 3-methane sulphonate of GA_3 methyl ester was cleanly reduced with tritium gas to [1-$^3H_1$]GA_5 methyl ester from which [1-$^3H_1$]GA_5 (5.3 Ci $mmol^{-1}$) and, hence, [1-$^3H_1$]GA_8 (0.44 Ci $mmol^{-1}$) were prepared. Murofushi et al. (22) also prepared [2,3-$^3H_2$]GA_{20} (3.3 Ci $mmol^{-1}$) from GA_5 by reduction of the 16,17-epoxide of the methyl ester with tritium, then removing the epoxide followed by hydrolysis of the methyl ester. By an analogous route Yokota et al. (23) have prepared [2,3-$^3H_2$]GA_9 from GA_7.

Exchange Reactions

Bearder et al. (24) have prepared [6-$^2H_1$]GA_{12}-aldehyde (0.9 atoms deuterium per molecule) and [6-$^3H_1$]GA_{12}-aldehyde (0.89 mCi $mmol^{-1}$) by exchanging the 6-proton in GA_{12}-aldehyde with either MeO^2H or ^2H_2O and MeO^3H or ^3H_2O in the presence of base. In a similar way Hedden et al. (25) prepared [6-$^3H_1$]GA_{14}-aldehyde (0.3 mCi $mmol^{-1}$).

Bearder et al. (26) have described the exchange of the 15- and 17-protons in ent-kaur-16-ene with $CF_3CO_2{}^2H$ and $CF_3CO_2{}^3H$ to provide a mixture of ent-kaur-16-ene and ent-kaur-15-ene, appropriately labelled in the 15- and 17-positions and separable by argentation-TLC.

Recent Partial Syntheses from GA_3 and GA_7

Recently several methods have been developed in our laboratory. Three of them are discussed in this section which also includes a partial synthesis of a catabolite of GA_{29}.

Metal Hydride Reduction

A preliminary account of this work has been presented (27). The method is based upon the reduction of 3-didehydroGA_3 methyl ester to 3-epiGA_1 methyl ester with sodium borohydride which was initially described by Gurvich et al. (28). Beale and MacMillan (29) have investigated this conjugate reduction in some detail and have shown (Fig. 2) that the saturated alcohols contain hydrogen atoms at the 1β- and 3α (or β)-positions which are derived from the borohydride and a 2ξ-hydrogen which

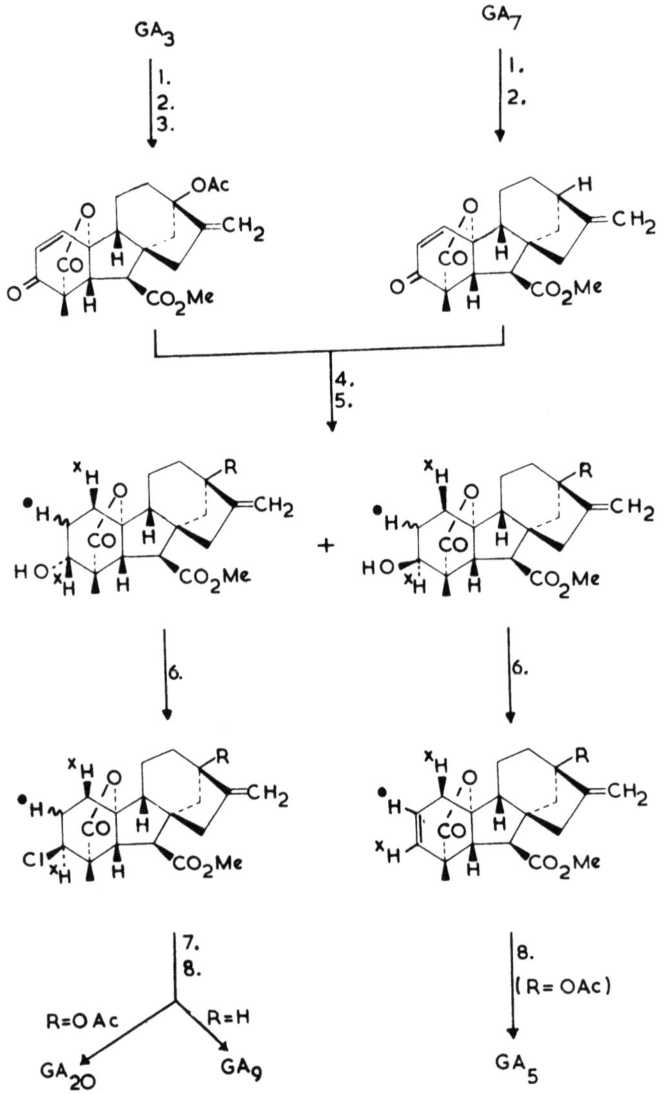

Fig. 2. Preparation of 1-, 2- and 3-labelled GAs from GA_3 and GA_7. Reagents: *1* CH_2N_2 ; *2* MnO_2 ; *3* Ac_2O-C_5H_5N; *4* NaB^XH_4 ; *5* • H_2O; *6* $POCl_3$; *7* n-Bu_3SnH; *8* OH^-

comes from the aqueous acid used to decompose the reduction intermediate. Thus by a selection of borohydride, borodeuteride, or borotriide and $^1H^+$, $^2H^+$, or $^3H^+$, it is possible to prepare 1β,3α(or β)-, 1β,2ξ,3α(or β)-, and 2ξ-labelled alcohols. These results have been exploited by Beale et al. (30) to prepare (Fig. 2) the following GAs from GA_3: GA_{20}, [1β,3α-$^2H_2$]GA_{20}, [1β,3α-$^3H_2$]GA_{20}, GA_5, [1β,3-$^2H_2$]GA_5, and [1β,3-$^3H_2$]GA_5; and the following GAs from GA_7: GA_4, [1β-$^2H_1$]GA_4, [1β,2ξ-$^2H_2$]-GA_4, [1β,3α-$^2H_2$]GA_4, GA_9, [3β-$^2H_1$]GA_9 and [3α-$^2H_1$]GA_9. The route can be adapted to the preparation of GA_1 and 1-, 2-, and 3-labelled GA_1 and is therefore a versatile one.

Chlorination of GA_3 and GA_7

This method, illustrated in Fig. 3, is the best route so far developed for the preparation of GA_5 from GA_3 (P.S. Kirkwood and J. MacMillan, unpublished work). It is illustrated in Fig. 3 for GA_3 methyl ester, and by using tri-n-butyl[^{2}H]stannane in the reduction step, provides [1β-$^2H_1$]GA_5 in excellent overall yield from GA_3. Hanson (unpublished) has also used these chlorination-dechlorination methods with similar results.

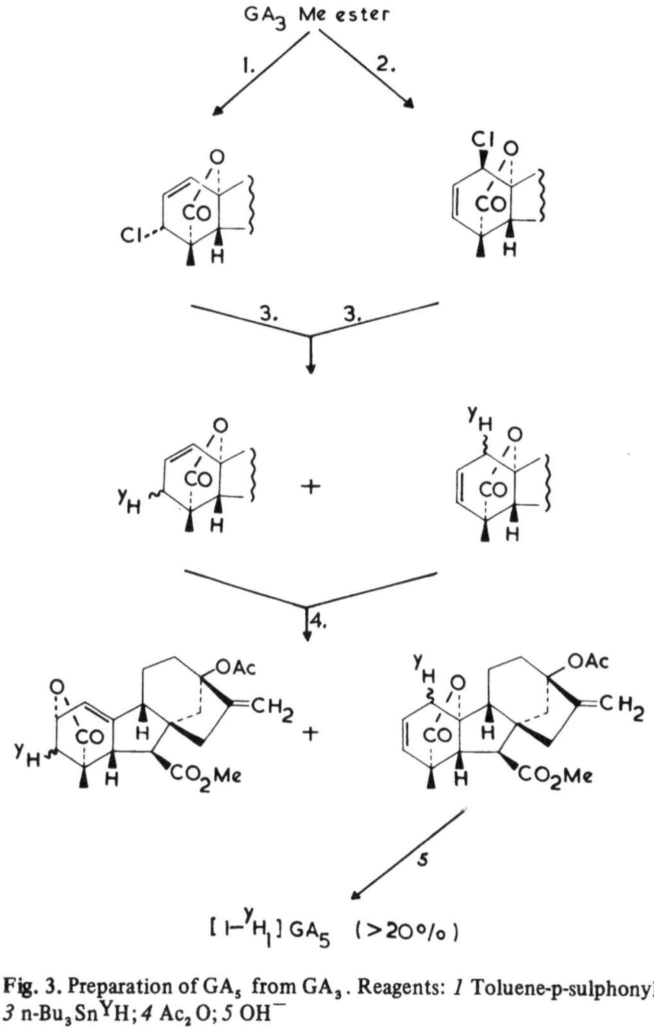

Fig. 3. Preparation of GA_5 from GA_3. Reagents: *1* Toluene-p-sulphonyl chloride, or – LiCl; *2* $SOCl_2$; *3* n-Bu_3Sn^YH; *4* Ac_2O; *5* OH^-

1β-ChloroGA_5 and the 3α-chloro-1-ene have been prepared from GA_5. In the lettuce hypocotyl bioassay, 1β-chloroGA_5 is less active than GA_5, and the 3α-chloro-isomer has about the same activity as GA_5.

Preparation of Labelled 2β-HydroxyGAs

2β-Hydroxylation of GAs occurs in plants. It results in biologically inactive metabolites and may be a process which controls the levels of biologically active hormones. We have examined methods of preparing labelled 2β-hydroxy GAs to investigate their further metabolism.

Using a route essentially the same as that described by Beeley and MacMillan (13) for the preparation of GA_{40} from GA_7, Beale and MacMillan (unpublished) have prepared $[2\alpha\text{-}^2H_1, 17\text{-}^3H_2]GA_{51}$ from GA_7 and $[2\alpha\text{-}^2H_1]GA_{29}$ from GA_3. By modifying this method Kirkwood and MacMillan (unpublished) have synthesized $[17\text{-}^{13}C]$-GA_{29} from GA_3 in good overall yield by the route shown in Fig. 4.

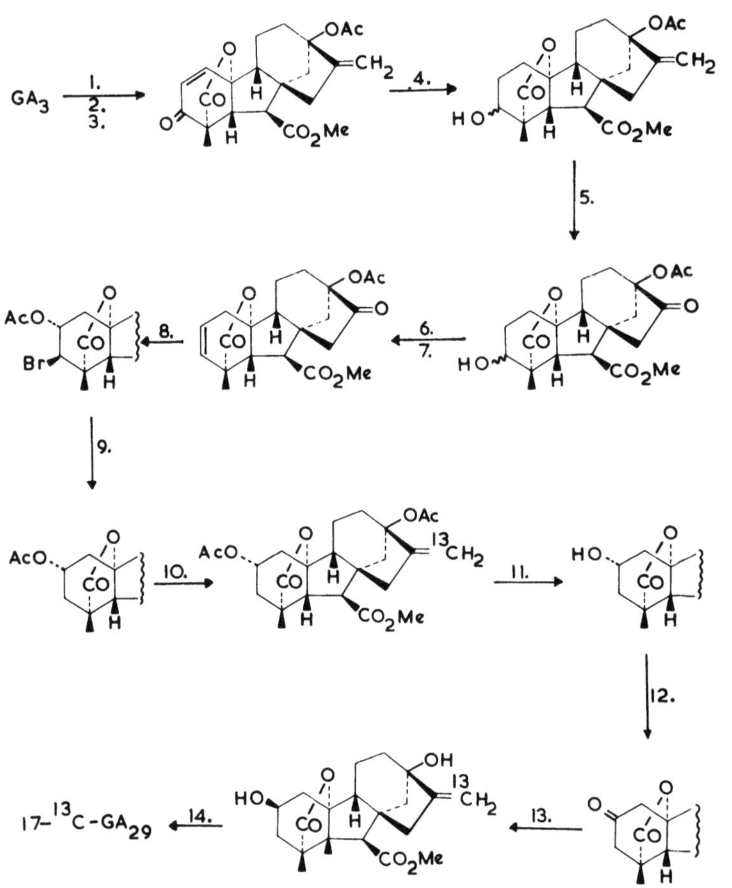

Fig. 4. Preparation of $[17\text{-}^{13}C]GA_{29}$ from GA_3. Reagents: *1* CH_2N_2; *2* MnO_2; *3* Ac_2O; *4* $NaBH_4$; *5* OsO_4-$NaIO_4$; *6* $POBr_3$; *7* 1,5-Diazabricyclo[5,4,0]undec-5-ene; *8* LiAc.$2H_2O$, MeCONHBr; *9* n-Bu_3SnH; *10* $Ph_3P{=}^{13}CH_2$ in C_6H_6; *11* K_2CO_3; *12* $C_5H_5N.CrO_3.HCl$; *13* $NaBH_4$; *14* OH^-

Synthesis of a GA_{29}-Catabolite

A GA_{29}-catabolite in *Pisum sativum* has been prepared from GA_3 by the method shown in Fig. 5 (Gaskin, Kirkwood and MacMillan, unpublished). This synthesis takes advantage of the fact that GA_3 is isomerized by dilute aqueous alkali to the isomeric 19,2-lactone which is further hydrolyzed by stronger alkali to the corresponding tri-hydroxy-dicarboxylic acid. By using ^{18}O-labelled alkali the catabolite was obtained with 0.68 atoms ^{18}O per molecule in the 19-oic acid.

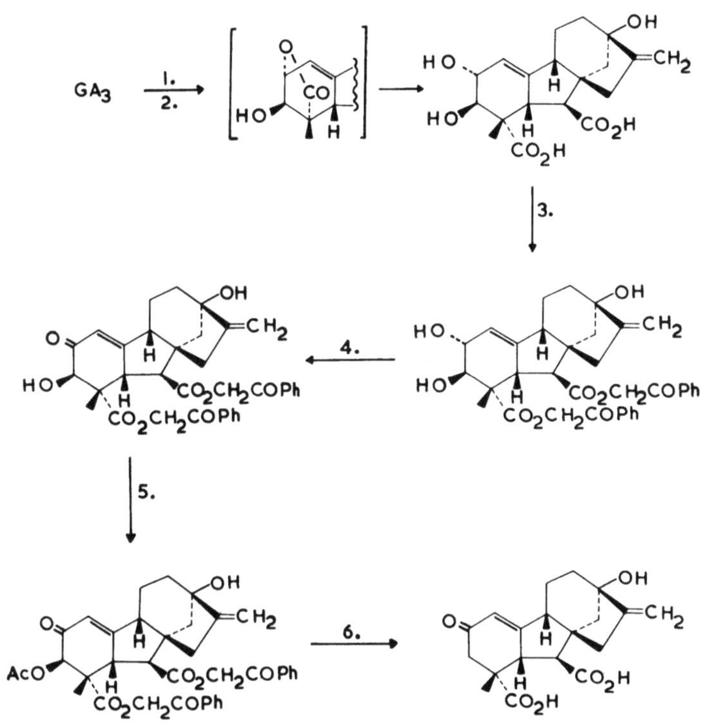

Fig. 5. Partial synthesis of a GA_{29}-catabolite from GA_3. Reagents: *1* 0.8M OH^-, *2* H^+ to pH 9.0; *3* $PhCOCH_2Br$ and 18-crown-6 ether; *4* $C_5H_5N.CrO_3.HCl$; *5* Ac_2O-C_5H_5N; *6* Zn and $MeCO_2H$

Concluding Remarks

In this very brief review of methods of preparing isotopically labelled GAs and their precursors, only chemical syntheses have been discussed. However, enzymatic methods are also available. Cultures of the fungus, *Gibberella fujikuroi,* and enzyme preparations from higher plants have been used to convert radiolabelled precursors into radiolabelled GAs. Dilution of the label can be minimized by dialysis of the enzyme preparations and by blocking GA-biosynthesis in the fungus either by chemical inhibitors or by using mutants. The fungal enzymes show a remarkable degree of nonspecificity, so that unnatural precursors may be converted into unnatural GAs by fungal cultures. The plant enzymes may also show low substrate specificity. These biological methods are, however, more restrictive than the chemical methods described in the two previous sections. They have not been used to prepare GAs labelled with stable isotopes, although the specific activity of [^{14}C]-labelled GAs, prepared from the *C. maxima* enzyme preparations, is high enough for the [^{14}C]-label to be used both as a radioisotope and as a heavy isotope (31) in the cell-free studies reviewed by Graebe in this volume.

A wide selection of chemical methods are now available for the preparation of radiolabelled GAs and GAs labelled with stable isotopes. The advantages of using stable isotopes in conjunction with GC-MS have been discussed (27, 32), and they are further illustrated in the paper by Sponsel in this volume.

Acknowledgments. The financial support of the A.R.C. and S.R.C. is grateful acknowledged.

References

1. Corey, E.J., Danheiser, R.L., Chandrasekaran, S., Keck, G.E., Gopolan, B., Larsen, S.D., Siret, P., Gras, S.-J.: J. Am. Chem. Soc. *100,* 8030 (1978)
2. Fujita, E., Node, M.: Heterocycles *7,* 710 (1977)
3. Cross, B.E., Galt, R.H.B., Hanson, J.R.: J. Chem. Soc. *295* (1964)
4. Geissman, T.A., Verbiscar, A.J., Phinney, B.O., Cragg, B.: Phytochemistry *5,* 933 (1966)
5. Murphy, P.J., West, C.A.: Arch. Biochem. Biophys. *133,* 395 (1969)
6. Hanson, J.R., Hawker, J., White, A.F.: J. Chem. Soc., Perkin *I,* 1892 (1972)
7. Evans, R., Hanson, J.R.: J.Chem. Soc. Perkin *I,* 663 (1975)
8. Cross, B.E., Norton, K., Stewart, J.C.: J. Chem. Soc. *C,* 1054 (1968)
9. Schlosser, M., Christmann, K.F.: Angew. Chem. Int. Ed. Engl. *3,* 636 (1964)
10. Bearder, J.R., Frydman, V.M., Gaskin, P., MacMillan, J., Wels, C.M., Phinney, B.O.: J. Chem. Soc. Perkin *I,* 173 (1976)
11. Bearder, J.R., Frydman, V.M., Gaskin, P., Hatton, I.K., Harvey, W.E., MacMillan, J.: J. Chem. Soc. Perkin *I,* 178 (1976)
12. Kong, P.W.: B. Sci. Thesis, University of Western Australia (1969); Barton, D.H.R., Harrison, D.M., Moss, G.P., Widdowson, D.A.: J. Chem. Soc. *C,* 975 (1970); Bestmann, H.J., Kratzer, O., Simon, H.: Chem. Ber. *95,* 2750 (1962)
13. Beeley, L.J., MacMillan, J.: J. Chem. Soc. Perkin *I,* 1002 (1976)
14. Kende, H.: Plant Physiol. *42,* 1612 (1976)
15. Musgrave, A., Kende, H.: Plant Physiol. *45,* 53 (1970)
16. Jones, D.F., McCloskey, P.: J. Appl. Chem. *13,* 324 (1963)
17. Pitel, D.W., Vining, L.C.: Can. J. Biochem. *48,* 259 (1970)
18. Vining, L.C.: J. Chromatogr. *60,* 141 (1971)
19. Nadeau, R., Rappaport, L.: Phytochemistry *13,* 1537 (1974)
20. Durley, R.C., Pharis, R.P.: Planta *109,* 357 (1973)
21. MacMillan, J., Pryce, R.J.: J. Chem. Soc. *C,* 550 (1967)
22. Murofushi, N., Durley, R.C., Pharis, R.P.: Agric. Biol. Chem. *38,* 475 (1974)
23. Yokota, T., Reeve, D.R., Crozier, A.: Agric. Biol. Chem. *40,* 2091 (1976)
24. Bearder, J.R., MacMillan, J., Phinney, B.O.: Phytochemistry *12,* 2173 (1973)
25. Hedden, P., MacMillan, J., Phinney, B.O.: J. Chem. Soc. Perkin *I,* 587 (1974)
26. Bearder, J.R., MacMillan, J., Wels, C.M., Chaffey, M.B., Phinney, B.O.: Phytochemistry *13,* 911 (1974)
27. MacMillan, J.: Pure Appl. Chem. *50,* 995 (1978)
28. Gurvich, I.A., Kobrina, N.S., Kucherov, V.F.: Bull. Acad. Sci. USSR 1668 (1969)
29. Beale, M.J., MacMillan, J.: J. Chem. Soc. Perkin *I,* 877 (1980)

30. Beale, M.J., Gaskin, P., Kirkwood, P.S., MacMillan, J.: J. Chem. Soc. Perkin *I*, 885 (1980)
31. Bowen, D.H., MacMillan, J., Graebe, J.E.: Phytochemistry *11*, 2253 (1972)
32. Sponsel, V.M., MacMillan, J.: Planta *144*, 69 (1978)

Metabolism of Gibberellins in Immature Seeds of *Pisum sativum*

V.M. SPONSEL [1]

Introduction

Gibberellin (GA) metabolism in higher plants has been studied for the past 15 years with mixed success (1, 2). The entire biosynthetic pathway from mevalonic acid to a C-19 GA has been conclusively demonstrated in cell-free systems from *Cucurbita maxima* (3). However, the pathway from mevalonic acid to GA_{12}-aldehyde has not been convincingly demonstrated in cellular systems from higher plants. The bulk of GA metabolic work with intact higher plants has involved a study of the conversion of one GA into another. This work is almost exclusively confined to the C-19 GAs. Thus our current knowledge of GA metabolism in intact higher plants is fragmentary.

In *Gibberella fujikuroi* there is a remarkable lack of substrate specificity of enzymes for the GA biosynthetic pathway. In higher plants too, a similar lack of substrate specificity may exist, since applied GAs are invariably metabolized. A rational choice of GA and material for metabolic studies must therefore be made if the results are to be meaningful. This is now possible, since a variety of isotopically labelled GAs can be produced by chemical means (4). Rigorous identification of metabolites is also required if the results are to be significant.

In an extensive study of GA metabolism, Pharis and co-workers have applied a variety of GAs to several plants which are routinely used for the bioassay of GAs, namely rice, lettuce, pea (5–7). With time course studies, this approach allows GA metabolism to be correlated with biological response. The work of Silk et al. extends this strategy (8). Another approach taken by Takahashi and co-workers (9), and by ourselves, has been to apply to a given plant material only those GAs which are known to be endogenous to it. This facilitates a study of the metabolism of native GAs. The metabolic routes which operate in a given plant tissue must be well understood before GA metabolism can be correlated with factors such as growth and the environment (10).

This paper provides a chronological review of our work on GA metabolism in immature seeds of *Pisum sativum* cv. Progress No. 9.

Endogenous GAs

The GAs endogenous to immature seeds of *P. sativum* cv. Progress No. 9 have been identified by combined gas chromatography-mass spectrometry (GC-MS) (11) and are are shown in Fig. 1. Quantitation of four of these GAs by selected ion current monitor-

1 School of Chemistry, The University, Bristol, United Kingdom

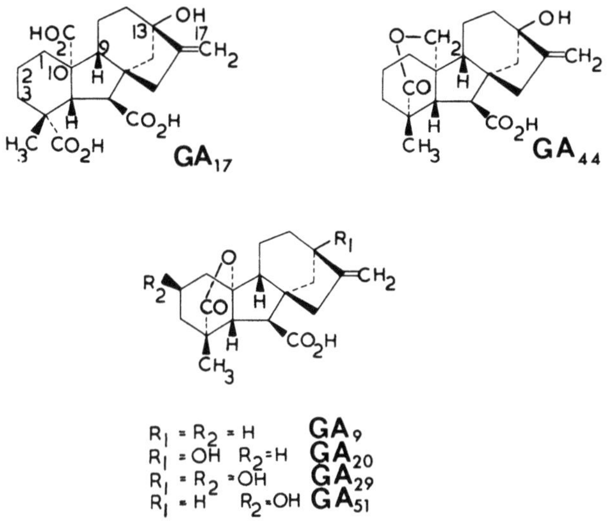

Fig. 1. GAs identified in immature seeds of *Pisum sativum* cv. Progress No. 9

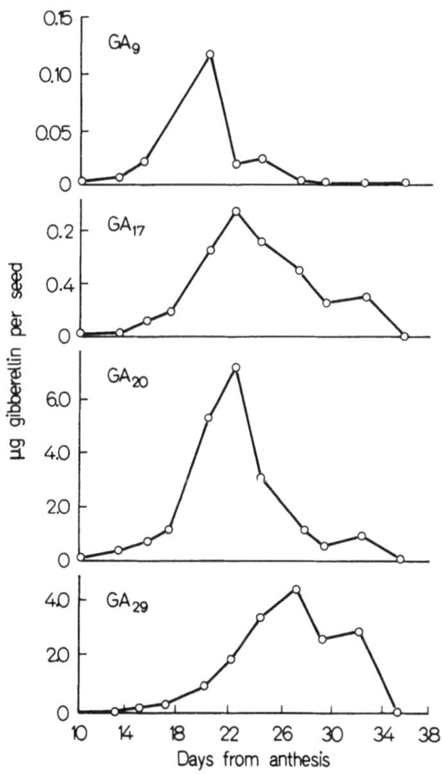

Fig. 2. The levels of GA_9, GA_{17}, GA_{20} and GA_{29} throughout seed maturation (note differing scales on ordinates). From (11)

ing (11) revealed that each GA increased to a maximal level at a defined stage of seed development then decreased to zero or an almost undetectable level in mature seeds (Fig. 2). The growth of seeds in greenhouses during defined "summer" and "winter" conditions (12) is shown in Fig. 3. From this basic ground-work, studies on GA metabolism can be planned with the following criteria in mind: (a) the GA fed should be

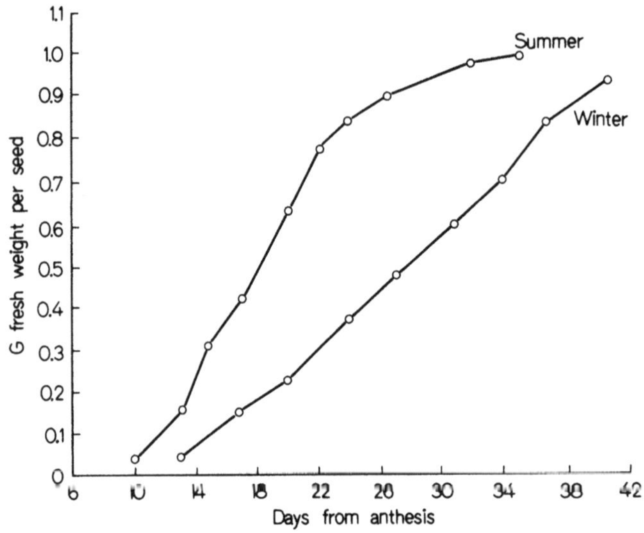

Fig. 3. The growth of pea seeds maturing under "summer" and "winter" growing conditions. From (12)

endogenous, (b) it should be fed at the defined stage of development when the endogenous GA is present, (c) it should be fed in an amount close to the natural level, and (d) the incubation time should correspond to the time interval between the maximal levels of the endogenous GA and its putative metabolite. Although it is difficult to adhere to all these criteria, the following results demonstrate their importance when attempting to establish the native GA pathways operating in pea.

Metabolism of GA_9

The distinct sequential series of maxima suggested the route $GA_9 \rightarrow GA_{20} \rightarrow GA_{29}$. Feeds of [17-$^3H_2$]$GA_9$ to seeds cultured in vitro revealed that [17-$^3H_2$]GA_9 was metabolized to a number of products (Fig. 4) and that these varied with the stage of development of the seeds (12). (The preparation of isotopically labelled GAs is reviewed by MacMillan in this volume.)

Quantitation was previously conducted with "summer"-grown tissue in which the levels of endogenous GA_9 were shown to peak at day 20, and endogenous GA_{20} levels peaked at day 22 (11). Therefore feeds were conducted using 20 day old seeds (weighing ca. 0.7 g/seed) and the incubation period was 2 days. To achieve GC-MS identification of metabolites, [^3H]GA_9 was fed at 10–50 times the endogenous level.

Under these conditions [17-$^3H_2$]GA_9 was metabolized to [^{3}H]GA_{51}, [^{3}H]H_2-GA_{31} and a conjugate of [^{3}H]H_2-GA_{31} (12, 13). It is thought that GA_{51} is the natural metabolite of GA_9. This metabolic step represents 2β-hydroxylation of GA_9. H_2-GA_{31} and its conjugate are probably artefacts of feeding an elevated level of [17-$^3H_2$]GA_9, since they are not endogenous to pea.

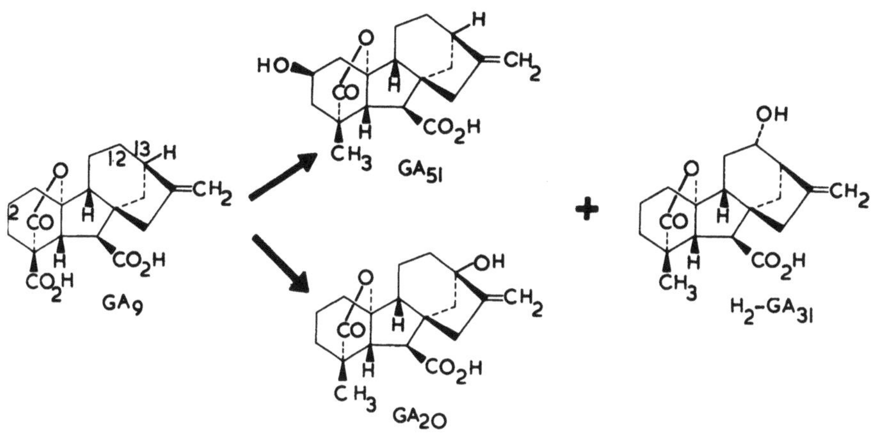

Fig. 4. Metabolites of [17-$^3H_2$]GA_9 in immature pea seeds

Our metabolic studies continued into the "winter" growing season. If [17-$^3H_2$]-GA_9 was fed to 20 day old "winter"-grown seeds (weighing ca. 0.3 g/seed) it was metabolized to [^{3}H]GA_{20}, *not* [^{3}H]GA_{51}. In 35 day old "winter"-grown seeds (weighing 0.7 g/seed and being equivalent to 20 day old "summer"-grown seeds) [17-$^3H_2$]-GA_9 was metabolized instead to [^{3}H]GA_{51}. (All feeds also contained [^{3}H]H_2-GA_{31} and its conjugate.) (12). Hence [17-$^3H_2$]GA_9 was 13-hydroxylated to give [^{3}H]GA_{20} only in small seeds in which endogenous GA_9 and GA_{20} had not accumulated. It is possible that this is due to turn-over rates so rapid that neither GA can accumulate. However, it is questionable whether $GA_9 \rightarrow GA_{20}$ is a natural metabolic sequence in developing pea seeds. Alternatively it may be the result of nonspecific 13-hydroxylating activity. The presence of 13-hydroxylated C-20 GAs in pea (Fig. 1), and cell-free work of Ropers et al. (14), lend support to the idea that GA_{20} is formed from a 13-hydroxylated C-20 precursor. This would be a parallel pathway to the non-13-hydroxylated pathway which would give rise to GA_9 and GA_{51}. The 13-hydroxylating enzyme may be restricted to the earlier stages of seed development, with the 2β-hydroxylating enzyme (catalyzing the conversion of $GA_9 \rightarrow GA_{51}$) being present later. However this suggestion has yet to be confirmed.

Metabolism of GA_{20}

$[17\text{-}^3H_2]GA_{20}$, when fed to seeds cultured in vitro is converted to $[^3H]GA_{29}$ in high yield (13). The greatest conversion was obtained when $[17\text{-}^3H_2]GA_{20}$ was fed and $[^3H]GA_{29}$ was isolated at times coinciding with the maximal levels of endogenous GA_{20} and GA_{29} respectively (13). This is another 2β-hydroxylating step, and on the basis of cell-free work, it and the conversion of $GA_9 \rightarrow GA_{51}$ are probably catalyzed by the same enzyme.

When $[17\text{-}^3H_2]GA_{29}$ was isolated from seeds incubated with $[17\text{-}^3H_2]GA_{20}$ and was fed to 27 day old seeds for 4 or 8 days, it did not appear to be metabolized (13). If $[17\text{-}^3H_2]GA_{20}$ was fed to 22 day old seeds for 4 and 10 days, $[17\text{-}^3H_2]GA_{29}$ was formed but $[^3H]GA_{29}$ was not metabolized. Conjugates of $[^3H]GA_{20}$ and $[^3H]GA_{29}$ were formed, but in small quantity. Thus $[^3H]GA_{29}$ persisted in the seed although the levels of endogenous GA_{29} apparently declined. [It is now known (15) that this decline is slower than that initially reported, and that these feeds should have been conducted for a longer time period.] The use of GAs labelled with stable isotopes, as described below, has provided a means of comparing rates of metabolism of exogenous and endogenous substrates. This technique has been particularly useful for the study of GA_{29} metabolism (15).

Techniques

The feeds so far described were analyzed using gas chromatography-radiocounting. By this method the effluent from the flame ionization detector (FID) of the gas chromatograph is trapped at 0.5 min intervals throughout a GC run, and radioactivity in fractions is determined by liquid scintillation counting (16). GC-MS is then performed under identical GC conditions, so that the total ion current and FID traces can be compared. Identity of a mass peak is correlated with radioactivity detection for identification of metabolites. However, when feeding a GA which is also endogenous, it is necessary to identify a radioactive metabolite in the presence of the same nonradioactive component. Thus during the analysis of $[^3H]GA_9$ feeds, $[^3H]GA_{20}$ or $[^3H]GA_{51}$ and $[^3H]H_2\text{-}GA_{31}$ must be identified in the presence of endogenous GA_{20} and GA_{51}. (These GAs are isomers and their similar retention times on GC are a further complication).

The specific radioactivity of $[^3H]$GAs is normally too low for GC-MS to detect the molecules which contain $[^3H]$. Our current approach is to use compounds labelled with stable isotopes (e.g., $[^2H]$, $[^{13}C]$, $[^{18}O]$), with sufficiently high isotope content that they can be quantitated by mass spectrometry. When these compounds are also labelled with a small quantity of radioactive isotope, they can be used very successfully for metabolic studies. The radiolabel is used as a tracer to follow the process. The stable isotope allows the metabolite to be distinguished from the same endogenous compound and quantitated by GC-MS (14). In our own work we no longer accept that a metabolic step has been unequivocally proven until we can demonstrate *by mass spectrometry* that a stable isotope is incorporated into it from the putative precursor. [GC-MS identification and quantitation of $[^{14}C]$GAs has been described

(17). This is practicable since the proportion of labelled to unlabelled molecules for [^{14}C] and [^{3}H] compounds of the same specific radioactivity is ca. 10^3 times greater for the [^{14}C] compound than for the [^{3}H] compound (17).]

Metabolism of GA_{29}

[2α-$^2H_1$][2α-$^3H_1$]GA_{29} (see Fig. 5 for structure) was fed to 27 day old seeds by direct injection through the pod wall (so-called in vivo feeding). Seeds were harvested immediately and at 2 or 3 day intervals for 13 days. Radioactivity was retained in the seeds, and on extraction it was located in the TLC zone for GA_{29}. In each extract [$^2H_1$][$^3H_1$]GA_{29} was identified by GC-MS. Ratios of nondeuterated GA_{29} to deuterated [$^2H_1$]GA_{29} were computed from the molecular ion clusters in mass spectra of the recovered GA_{29}. Throughout the feed, endogenous nondeuterated GA_{29} declined relative to the deuterated GA_{29}. Thus endogenous nondeuterated GA_{29} was metabolized

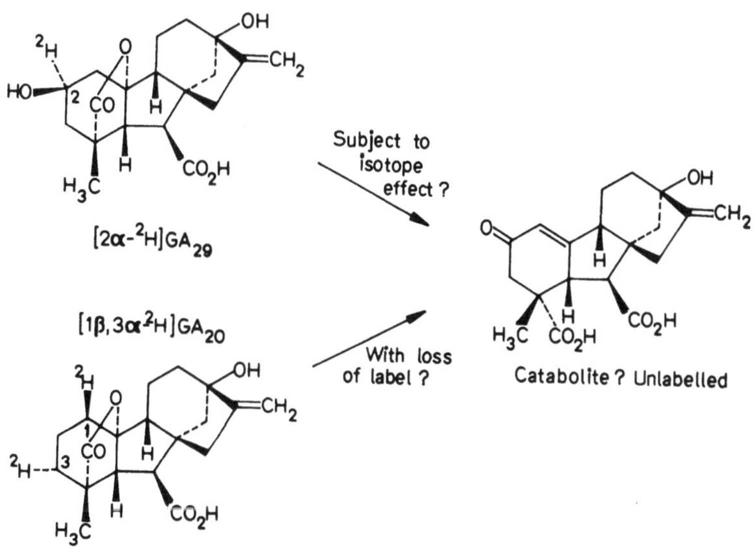

Fig. 5. The structure of a GA catabolite in maturing pea seeds, and its possible formation from [2α-$^2H_1$]GA_{29} and [1β,3α-$^2H_2$]GA_{20}

in preference to the exogenous deuterated substrate. The rate of decline of endogenous GA_{29} was indeed comparable with that in untreated seeds, whilst ca. 50% of the exogenous deuterated GA_{29} fed was unmetabolized after 13 days. Two possible reasons for this non-equivalence of exogenous and endogenous GA_{29} were postulated (15), namely (a) the inability of the exogenous labelled substrate to reach the correct site within the seed for metabolism to occur, or (b) the presence of a primary isotope effect due to metabolism of GA_{29} involving the cleavage of the carbon-deuterium and carbon-tritium bond. Further results discussed below indicated (b) to be the more likely possibility.

A similar feed of $[1\beta,3\alpha\text{-}^2H_2][1\beta,3\alpha\text{-}^3H_2]GA_{20}$ (Fig. 5) to 22 day old seeds was conducted. It was initially metabolized to $[^2H_2][^3H_2]GA_{29}$ (identified by GC-MS), but there was no further accumulation of other radioactive metabolites. Rather, some radioactivity appeared to be lost during metabolism and/or work-up. Ratios of non-deuterated GA_{20} and GA_{29} to mono- and di-deuterated GA_{20} and GA_{29} were again computed from the molecular ion clusters in mass spectra of recovered GA_{20} and GA_{29}. The isotope ratios stabilized by day 5 and remained constant thereafter. Thus endogenous and exogenous GA_{20} and GA_{29} appeared to be metabolized in an equivalent manner, but it was inferred that no metabolite of $[1\beta,3\alpha\text{-}^2H_2][1\beta,3\alpha\text{-}^3H_2]GA_{29}$ accumulated due to loss of radioactive label.

The results of both these feeds have been published in detail (15).

The Identity of a New GA Catabolite

A rational conclusion from the above feeds was that GA_{29} was further metabolized by oxidation at C-2. We have identified a GA catabolite endogenous to pea seeds (15, 18), whose structure is consistent with it being the untraced metabolite of GA_{29}. Formation of this α-β unsaturated ketone (Fig. 5) from $[2\alpha\text{-}^2H_1][2\alpha\text{-}^3H_1]GA_{29}$ would involve breaking a C-$[^2H]$ or C-$[^3H]$ bond and would therefore be subjected to a primary isotope effect, i.e., the endogenous nondeuterated GA would be metabolized faster. Formation of this catabolite from $[1\beta,3\alpha\text{-}^2H_2][1\beta,3\alpha\text{-}^3H_2]GA_{20}$ (Fig. 5), via $[1\beta,3\alpha\text{-}^2H_2][1\beta,3\alpha\text{-}^3H_2]GA_{29}$, would lead to loss of $[^2H]$ or $[^3H]$ from C-1 in formation of the double bond. $[^2H]$ and $[^3H]$ could also be lost from C-3 during work-up due to enolization. Thus the metabolite formed from $[1\beta,3\alpha\text{-}^2H_2][1\beta,3\alpha\text{-}^3H_2]$-$GA_{20}$ would be unlabelled (Fig. 5). This GA catabolite (Fig. 5) was identified in maturing and germinating seeds of pea (15) and was suggested to be the hitherto untraced metabolite of GA_{29}.

Durley et al. (19) have shown that when $[2,3\text{-}^3H_2]GA_{20}$ was fed to pea fruits, tritium is incorporated into a compound with a similar mass spectrum to that of this GA catabolite. They therefore suggested this compound was a metabolite of GA_{20}, probably being formed via GA_{29}. $[17\text{-}^{13}C_1]GA_{29}$ has been prepared (18) to demonstrate that the catabolite is indeed a metabolite of GA_{29}. Our most recent results show that $[17\text{-}^{13}C_1]GA_{29}$ is converted in good yield to $[^{13}C]$-labelled GA catabolite during in vivo feeds. We have not yet been able to prepare $[17\text{-}^3H_2]GA_{29}$ for doubly labelled time course studies. However $[19\text{-}^{18}O_1]$-labelled GA catabolite has been prepared (18). When added to extracts it can be used as an internal standard to quantitate both endogenous and exogenous $[^{13}C]$-labelled catabolite. Preliminary results from an in vivo feed of $[17\text{-}^{13}C_1]GA_{29}$ to 29 day old pea seeds show that 40 day old seeds contain ca. 14 μg endogenous catabolite, and ca. 11 μg $[^{13}C]$-labelled catabolite. Feeds of $[17\text{-}^{13}C_1][17\text{-}^3H_2]GA_{29}$ are planned for the near future.

There are two possible intermediates between GA_{29} and the GA_{29} catabolite (Fig. 6). One, GA_{29} ketone, has not been shown to be endogenous to pea, possibly

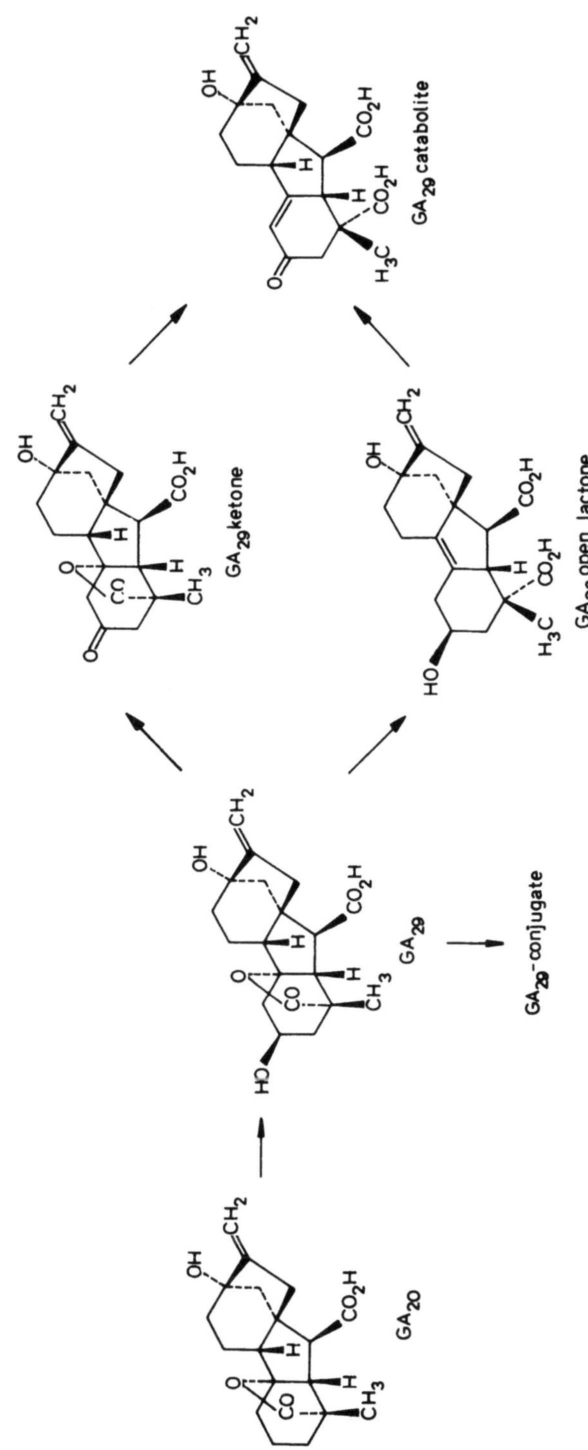

Fig. 6. Possible metabolic sequences in maturing pea seeds. All compounds other than GA_{29} ketone are endogenous to pea

due to extraction and derivatization causing it to be chemically converted to the GA_{29} catabolite. This now seems unlikely on two counts. Firstly if a methanolic extract of pea which has not been subjected to extremes of pH is chemically reduced with sodium borohydride or borodeuteride, the reduction products are consistent with the GA_{29} catabolite, not GA_{29} ketone, being originally present in the extract. Secondly if a methanolic extract, again which has not been exposed to extremes of pH, is derivatized with methoxyamine hydrochloride, the methoxime derivative of the GA_{29} catabolite not of GA_{29} ketone is formed.

The second potential intermediate between GA_{29} and the GA_{29} catabolite is GA_{29} open lactone (Fig. 6). This compound is endogenous to pea (18). Conversion of it to the GA_{29} catabolite would involve oxidation and a shift of the double bond. With the information at hand, this appears to be the more likely intermediate, although we have not yet been able to detect it labelled with [^{13}C] in feeds of [17-$^{13}C_1$]GA_{29}.

Catabolism Versus Conjugation?

In many species seed maturation is accompanied by a conversion of free, acidic GAs to conjugated forms (20). This does not appear to be the case for pea. Only a small quantity of GA_{29} conjugate is formed (15), and yet seemingly unnatural GAs such as [^{3}H]H_2-GA_{31} and a metabolite of [^{3}H]GA_{51} (unpublished) are readily conjugated. Instead it appears that in pea, natural GAs are preferentially catabolized, with high levels of the GA catabolite being readily observed in mature seeds and seedlings. In the closely related legume *Vicia faba*, a strictly analogous situation exists with high levels of this GA_{29} catabolite and low levels of GA conjugates being observed (21). Of possible taxonomic significance is the existence of the converse situation in *Phaseolus* species. Thus high levels of GA conjugates and mere traces of a 3-hydroxylated analogue of the GA_{29} catabolite discussed here are present (21). Thus whilst both catabolic and conjugative pathways for GAs appear to operate in all these legumes, the catabolic pathway may predominate in *Pisum* and *Vicia* and the conjugative pathway may be favored in *Phaseolus*. The situation in other plants remains to be examined.

Acknowledgments. The continuing involvement of Professor J. MacMillan in this work is appreciated. Dr. M.H. Beale, Dr. J.R. Bearder, and Mr. P.S. Kirkwood are thanked for the preparation of isotopically labelled GAs, Mr. P. Gaskin for GC-MS, and the Agricultural Research Council for financial support.

References

1. Hedden, P., MacMillan, J., Phinney, B.O.: Annu. Rev. Plant Physiol. *29*, 149–192 (1978)
2. Graebe, J.E., Ropers, H.-J.: The gibberellins. In: Plant Hormones and Related Compounds. Goodwin, P.B., Higgins, T.J.V. (eds.). Amsterdam: ASP Biol. Med. 1978
3. Graebe, J.E.: This volume
4. MacMillan, J.: This volume
5. Durley, R.C., Pharis, R.P.: Planta *109*, 357–361 (1973)

6. Durley, R.C., Railton, I.D., Pharis, R.P.: Phytochemistry *13*, 547–551 (1974)
7. Durley, R.C., Bewley, J.D., Railton, I.D., Pharis, R.P.: Plant Physiol. *57*, 699–703 (1976)
8. Silk, W.K., Jones, R.L., Stoddart, J.L.: Plant Physiol. *59*, 211–216 (1977)
9. Yamane, H., Murofushi, N., Osada, H., Takahashi, N.: Phytochemistry *16*, 831–835 (1977)
10. Bown, A.W., Reeve, D.R., Crozier, A.: Planta *126*, 83–91 (1975)
11. Frydman, V.M., Gaskin, P., MacMillan, J.: Planta *118*, 123–132 (1974)
12. Sponsel (nee Frydman) V.M., MacMillan, J.: Planta *135*, 129–136 (1977)
13. Frydman, V.M., MacMillan, J.: Planta *125*, 181–195 (1975)
14. Ropers, H.-J., Graebe, J.E., Gaskin, P., MacMillan, J.: Biochem. Biophys. Res. Commun. *80*, 690–697 (1978)
15. Sponsel, V.M., MacMillan, J.: Planta *144*, 69–78 (1978)
16. MacMillan, J., Wels, C.M.: Phytochemistry *13*, 1413–1417 (1974)
17. Bowen, D.H., MacMillan, J., Graebe, J.E.: Phytochemistry *11*, 2253–2257 (1972)
18. Gaskin, P., Kirkwood, P.S., MacMillan, J.: Unpublished results
19. Durley, R.C., Sassa, T., Pharis, R.P.: Plant Physiol. *64*, 214–219 (1979)
20. Hiraga, K., Yokota, T., Murofushi, N., Takahashi, N.: Agric. Biol. Chem. *38*, 2511–2520 (1974)
21. Sponsel, V.M., Gaskin, P., MacMillan, J., Planta *146*, 101–105 (1979)

GA-Biosynthesis: The Development and Application of Cell-Free Systems for Biosynthetic Studies

J.E. GRAEBE [1]

Gibberellin biosynthesis has been studied in cultures of *Gibberella fujikuroi* and in cell-free systems from higher plants. The fungal cultures have been most useful for studying the overall pathway and for the identification of such intermediates as accumulate in the intact organism. Cell-free systems are needed for the study of individual enzymatic steps and for obtaining intermediates that do not accumulate when the entire pathway is operating. Cell-free systems of *G. fujikuroi* have been prepared, but they had variable and low activity and catalyzed only a limited part of the pathway. Most work on enzymic catalysis of GA biosynthesis has therefore been done with cell-free systems from higher plants. Incorporation of precursors into GA's by intact plants or parts of plants has not been successful.

The main systems that have been used for investigating GA biosynthesis are derived from immature seeds of *Marah macrocarpus, Cucurbita maxima,* and *Pisum sativum.* A cell-free system that makes both kaurene and other diterpenes has been derived from germinating *Ricinus communis.*

Figure 1 shows the GA pathway in the *Cucurbita* system, which catalyzes the most complete sequence. The pathway is conveniently regarded in three stages: (1) the conversion of mevalonate to kaurene (*ent*-kaur-16-ene), (2) the conversion of kaurene to GA_{12}-aldehyde, and (3) the conversion of GA_{12}-aldehyde to C_{20}- and C_{19}-GA s.

These and other aspects of GA biosynthesis have been recently described in two comprehensive reviews (1, 2), where the corresponding formulas are shown. This paper will mention recent progress obtained by others with the *Marah* system and describe in some detail recent work done by ourselves with the *Cucurbita* and pea systems.

The enzymes of the first stage of GA biosynthesis are soluble. Recent work on this part of the pathway has shown that the formation of kaurene from mevalonate is regulated by adenylate energy charge at the step in which isopentenyl pyrophosphate is formed from pyrophosphomevalonate (3). Since this step lies early in the sequence, its regulation affects the entire terpenoid biosynthesis. A characterization of the two-step cyclization of geranylgeranyl pyrophosphate to kaurene, including the effect of several growth retardants on these reactions, has also been published (4).

The second stage of GA biosynthesis, the conversion of kaurene to GA_{12}-aldehyde via several intermediates, is particle-bound. These steps are catalyzed by mixed-function oxidases (monooxygenases) requiring NADPH and O_2 for their activity. A detailed study of the cofactor requirement in the *Marah* system has also suggested a role for flavin nucleotides in the electron transfer chain associated with these enzymes (5).

1 University of Göttingen, 3400 – Göttingen, FRG

Furthermore, the presence of cytochromes P-450 and b_5 has been detected in microsomal preparations with which the mixed-function oxidases are associated (6).

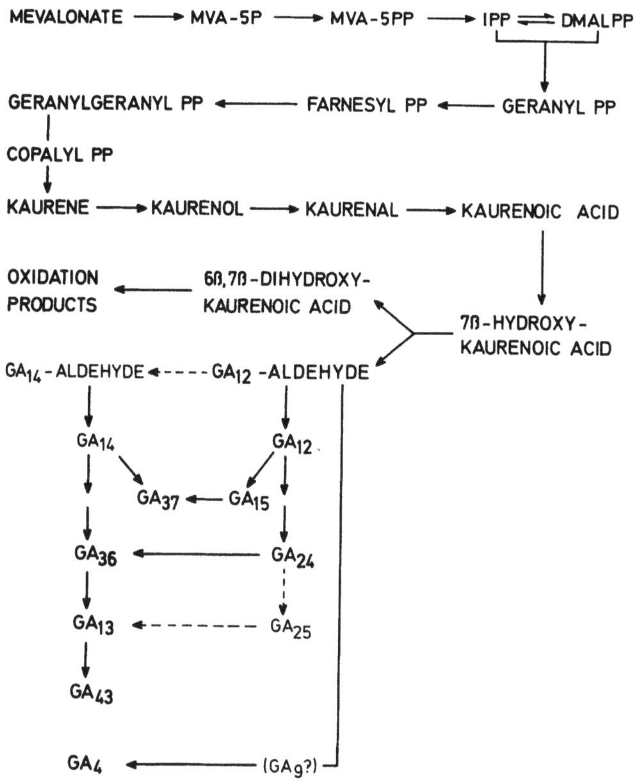

Fig. 1. Gibberellin biosynthesis in the cell-free system from immature seeds of *Cucurbita maxima* (the steps before kaurene have been inferred from other systems)

The second stage of GA biosynthesis is also the one which is sensitive to inhibition by ancymidol, a substituted pyrimidine and potent growth regulator (7). In the *Marah* systems this compound inhibits the oxidation of kaurene very efficiently (K_I = $2 \times 10^{-9} M$). It also inhibits the oxidation of kaurenol and kaurenal, but not that of kaurenoic acid. The inhibition seems to be specific for GA biosynthesis, since other mixed-function oxidases are not affected.

In cooperation with the laboratory of J. MacMillan we have studied the conversion of 7β-hydroxykaurenoic acid (*ent*-7α-hydroxykaurenoic acid) in the *Cucurbita* system. This reaction is interesting because it includes a ring contraction and because it constitutes a true branching point of the pathway, 6β,7β-dihydroxykaurenoic acid and GA_{12}-aldehyde both being formed from the same precursor and at about equal rates. Only GA_{12}-aldehyde is a precursor of GAs. Incubation of 7β-hydroxykaurenoic acid specifically labeled with tritium in the 6β-, 6α,6β-, and 7α-positions showed unequivocally that the 6β-H, and only it, is lost in the reaction (8, 9). Since this is true for the formation of both products, they are probably formed via a common carbonium ion, e.g., by the mechanism shown in Fig. 2. These incubations were done in the presence of Mn^{2+}, which blocks the pathway after GA_{12}, thus simplifying product analysis.

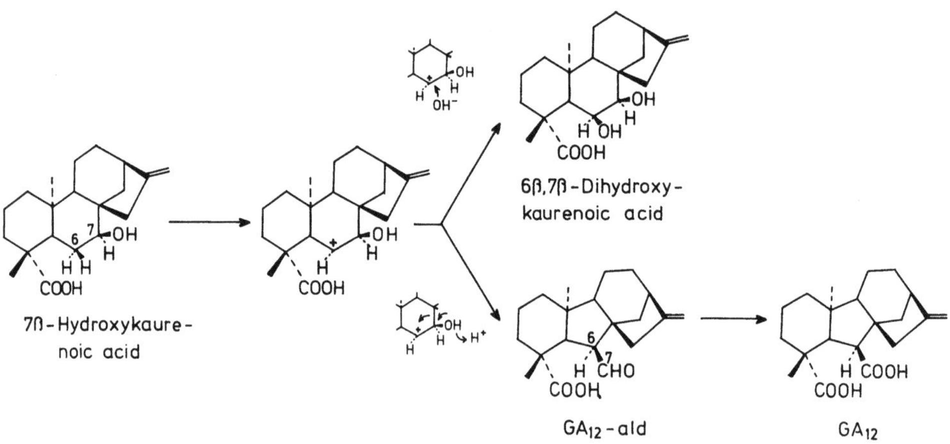

Fig. 2. Possible mechanism for the formation of 6β,7β-dihydroxykaurenoic acid and GA_{12}-aldehyde in the *Cucurbita* system

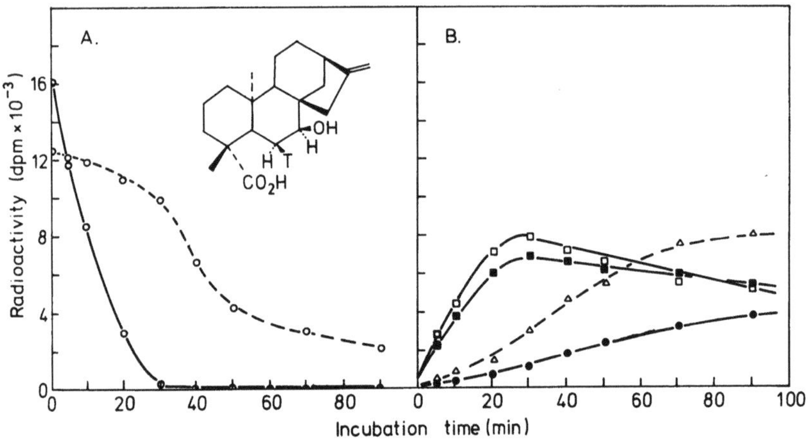

Fig. 3 A and B. Loss of the 6β-H and kinetic isotope effect in the conversion of 7β-hydroxy-[6β-^{3}H]kaurenoic acid by the *Cucurbita* system. **A** Substrate: a mixture of 7β-hydroxy-[6β-^{3}H]kaurenoic acid (- - -o- - -) and 7β-hydroxy-[^{14}C]kaurenoic acid (—o—). **B** Products: 6β,7β-dihydroxykaurenoic acid (▪), GA_{12}-aldehyde (□), GA_{12} (•) and H_2O (△). ^{14}C *(whole lines)*, ^{3}H *(broken lines)*

As expected in the removal of a tritium atom, there is a large isotope effect when the 6β-labeled substrate is converted (Fig. 3A). The ^{14}C-label, representing the bulk of substrate containing protium in the 6β-position, reacts at a normal rate. The tritium substrate reacts slower because of the higher energy required to break the carbon-tritium bond. Because of competition, the conversion of the tritium substrate only attains full rate when the protium substrate has been almost used up. The carbon compounds resulting from this reaction are not labeled with ^3H, but the label lost from the 6β-position is quantitatively recovered as water (Fig. 3B). Because of the isotope effect, the rate of formation of ^3H_2O cannot be used as a measure of the normal reaction rate in reactions where carbon-tritium bonds become broken and the ratio of $^3H/^{14}C$ increases severalfold as the reaction proceeds.

Calculation of the kinetic isotope effect from the data in Fig. 3A gives values from 10 to 12, indicating that removal of the 6β-H is a rate-limiting step in the reaction. No isotope effect was obtained with 7α-labeled substrate, and the ^3H was retained in both 6β,7β-dihydroxykaurenoic acid and GA_{12}-aldehyde. But it was lost in the conversion of the latter to GA_{12}.

Hafeman in our laboratory has begun an identification of the microsomal particles catalyzing the branching reaction (10). Figure 4 shows an isopycnic sucrose gradient centrifugation of a homogenate from *Cucurbita* seeds. The formation of GA_{12}-alde-

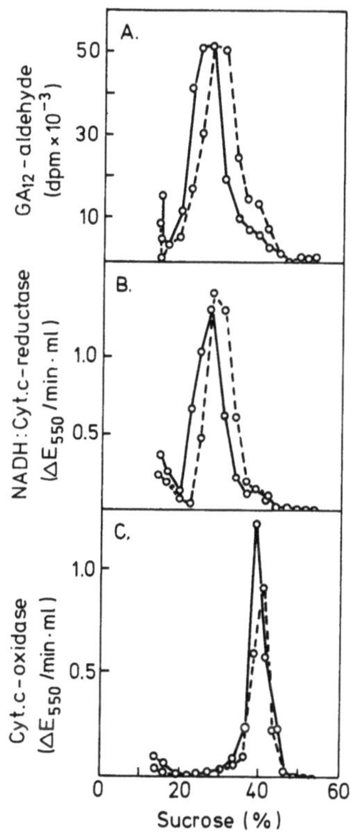

Fig. 4A–C. Density gradient centrifugation of a homogenate catalyzing the ring contraction reaction in *C. maxima*. Gradient with 0.1 m*M* $MgCl_2$ (—o—), gradient with 3.0 m*M* $MgCl_2$ (- - -o- - -). **A** Biosynthesis of GA_{12}-aldehyde from 7-hydroxykaurenoic acid; **B** marker enzyme activity for endoplasmic reticulum; **C** marker enzyme activity for mitochondria

hyde from 7β-hydroxykaurenoic acid (Fig. 4A) is catalyzed by a fraction with the same density as a fraction with NADH-cytochrome c-reductase activity (Fig. 4B), a marker for endoplasmic reticulum (ER). Cyt-c-oxidase activity, a marker for mitochondria, is associated with a fraction of higher density (Fig. 4C). The Mg^{2+}-dependent displacement of the curves in A and B to some degree supports the assignment of GA_{12}-aldehyde synthetase activity to ER. Low Mg^{2+} concentration furthers the dissociation of ribosomes from ER, thus producing smooth ER with a lower density (11). However, the displacement of the curves is too subtle to be conclusive and the data do not rule out the possibility that the activity is associated with plasmalemma fragments rather than with ER. We are working on the resolution of this question.

Using the experience obtained with the *Cucurbita* system, it has been possible to obtain an active cell-free system from immature pea seeds (12). Figure 5 shows the products identified by GC-MS in this system and their sequence. The 13-hydroxylation pattern, evident by the presence of GA_{53} and GA_{44}, indicates a direct relationship between the cell-free system and biosynthesis in vivo, since 13-hydroxylated GA s are typical for pea seeds (13). Several products still remain to be identified in the pea seed system.

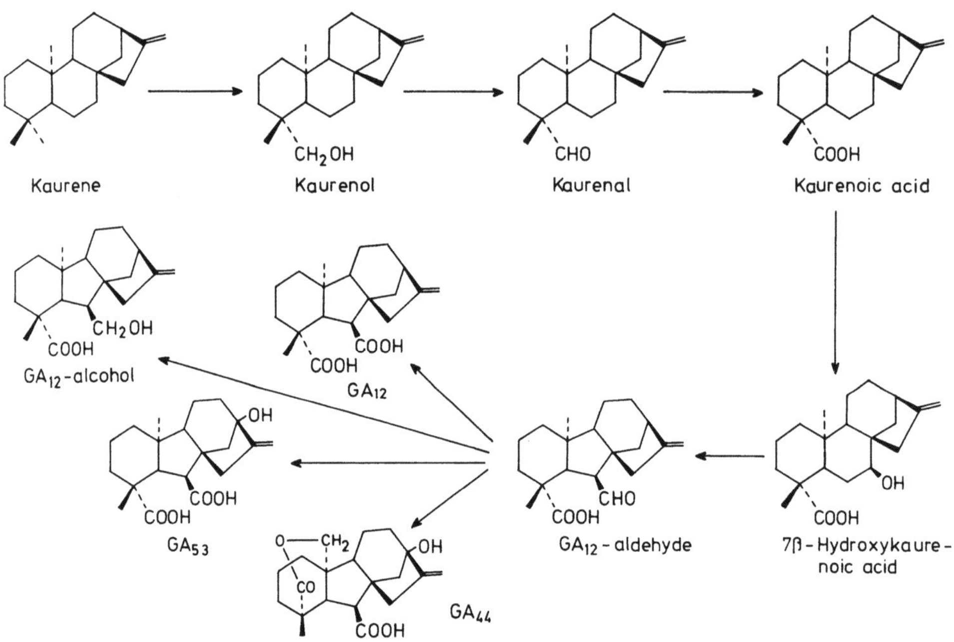

Fig. 5. GA biosynthesis in a cell-free system from immature pea seeds

Measurement of enzyme activities in cell-free extracts prepared from peas at different developmental stages are shown in Figs. 6 and 7. Kaurene synthesis reaches a maximum shortly before the seeds are fully developed and then disappears completely before the seeds become desiccated (Fig. 6). This curve is reminiscent of the one published by Coolbaugh and Moore (14) and of the variations in GA s in pea seeds in vivo (15). The conversions of 7β-hydroxykaurenoic acid and GA_{12}-aldehyde during development show maxima similar to the formation of kaurene, but the conversion of kaurene (mainly to kaurenol and kaurenal) was highest in the youngest seeds (Fig. 7). The

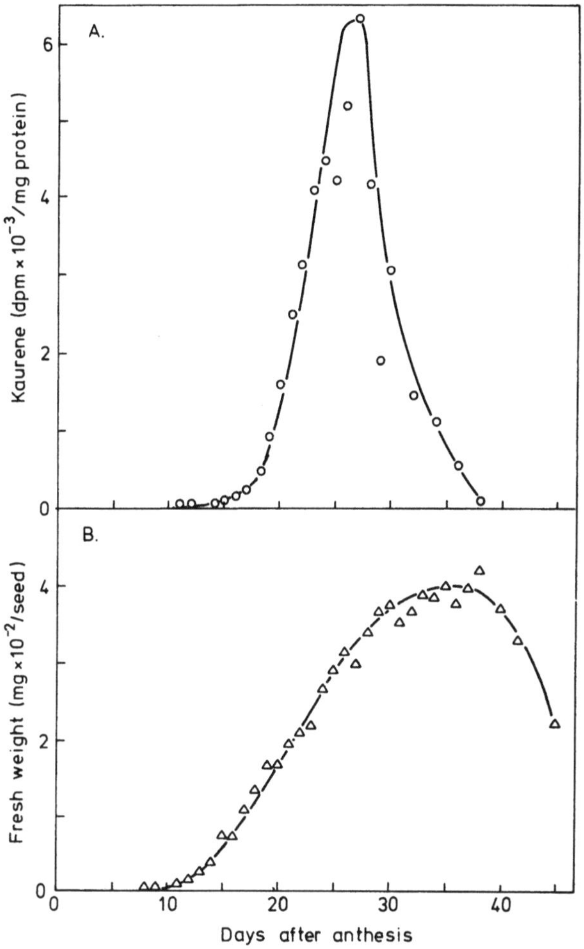

Fig. 6A and **B.** Kaurene biosynthesis in cell-free extracts of pea seeds prepared at different developmental stages. **A** Kaurene biosynthesis. **B** Development of the pea seeds

results in Fig. 7 support the view that the formation of kaurene is a limiting factor in GA biosynthesis. This would only be logical, since the formation of kaurene is the first committed step. However, the possibility remains that the preparation methods disfavor retention of full activity of the kaurene synthetase. The apparent regulation of kaurene synthesis here is superimposed on the regulation by energy charge mentioned earlier in this article, since it was measured in the presence of optimal amounts of ATP and an ATP-regenerating system.

For a detailed discussion of the third stage in GA biosynthesis the reader is referred to (1, 2).

Experimental

Experiment in Fig. 3: [6β-$^3H_1$]kaurenoic acid (synthesized in the laboratory of J. MacMillan) was mixed with 7β-hydroxy-[^{14}C]kaurenoic acid to final specific activity

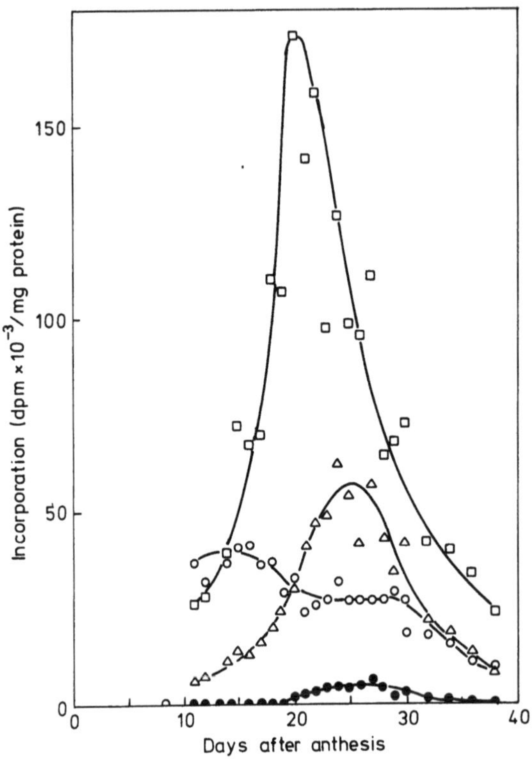

Fig. 7. Limiting character of kaurene biosynthesis in the pea seed system. Biosynthesis of kaurene from mevalonate (●), formation of products from kaurene (○), formation of products from 7β-hydroxykaurenoic acid (△), formation of products from GA_{12}-aldehyde (□)

3.5 mCi/mmol for ^{3}H and 4.3 mCi/mmol for ^{14}C and incubated (final concentration 45 μ*M*) with *C. maxima* endosperm (5000 × *g* supernatant dialyzed for 3 h against 0.05 *M* K-PO_4 buffer, pH 8.0). The incubation mixture contained endosperm preparation (0.4 ml), $MnCl_2$ (1 m*M*), and NADPH (0.5 m*M*). The reaction was started by mixing prewarmed substrate and endosperm preparation. At the times indicated, aliquots of 50 μl were removed and the reaction was stopped with aceton HCl. The products were extracted, separated by TLC, and counted by liquid scintillation.

Experiment in Fig. 4: Endosperm was removed from immature seeds of *Cucurbita maxima* and homogenized lightly in a medium (vol/vol) containing (in final concentrations): sucrose (0.55 *M*), K-PO_4 buffer (0.1 *M*, pH 8.0), EDTA (1 m*M*), and $MgCl_2$ (0.1 or 3.0 m*M*). The homogenate was filtered through cloth and centrifuged 6000 × *g* for 5 min. The supernatant was layered on a 30-ml 15%–55% continuous sucrose gradient containing K-PO_4 buffer (0.05 *M*, pH 7.6), EDTA (1.0 m*M*), and $MgCl_2$ (0.1 or 3.0 m*M*) and centrifuged in a Beckmann SW 27 rotor at 27,000 rpm for 2 h. The synthesis of GA_{12}-aldehyde was measured by incubation of 50,000 cpm of 7β-hydroxy-[^{14}C]kaurenoic acid (43.6 mCi/mmol), $MgCl_2$ (5 m*M*), $MnCl_2$ (1 m*M*), and NADPH (0.5 m*M*) at 30°C for 2 h. The products were separated by TLC, GA_{12}-aldehyde was scraped off and counted. NADH/cyt c-reductase was determined according to (16) and cyt c-oxidase according to (17, 18).

Experiments in Figs. 6 and 7: The pea seed system was prepared as described in (12) from daily collections of seeds grown in the field. Incubation mixtures contained $MgCl_2$ (10 m*M*), ATP (10 m*M*), PEP (20 m*M*), NADPH (1 m*M*), ^{14}C-labeled precursor (50,000 dpm, 43.6 mCi/mmol) and 75 μl of pea homogenate (2,000 × *g* supernatant) in a total volume of 0.1 ml. Incubation was at 30°C for 1 h. The products were extracted, separated by TLC, scraped off, and counted. Nitrogen was determined by a micro-Kjeldahl method.

Acknowledgments. We thank Mrs. G. Bodtke for skilled technical assistance. The work was supported by the Deutsche Forschungsgemeinschaft.

References

1. Graebe, J.E., Ropers, H.J.: In: Phytohormones and Related Compounds. Letham, D.S. Goodwin, P.B., Higgins, T.J.V. (eds.), Vol. I, pp. 107–204. Amsterdam: Elsevier-North Holland 1978
2. Hedden, P.; MacMillan, J., Phinney, B.O.: Annu. Rev. Plant Physiol. *29*, 149–192 (1978)
3. Knotz, J., Coolbaugh, R.C., West, C.A.: Plant Physiol. *60*, 81–85 (1977)
4. Frost, R.G., West, C.A.: Plant Physiol. *59*, 22–29 (1977)
5. Hasson, E.P., West, C.A.: Plant Physiol. *58*, 473–478 (1976)
6. Hasson, E.P., West, C.A.: Plant Physiol. *58*, 479–484 (1976)
7. Coolbaugh, R.C., Hirano, S.S., West, C.A.: Plant Physiol. *62*, 571–576 (1978)
8. Graebe, J.E., Hedden, P., MacMillan, J.: J. Chem. Soc. Chem. Commun. 161–162 (1975)
9. Graebe, J.E., Gaskin, P., MacMillan, J.: Unpublished
10. Hafemann, C.: Diplomarbeit, University of Göttingen, Germany (1978)
11. Lord, J.M., Kagawa, T., Moore, T.S., Beevers, H.: J. Cell. Biol. *57*, 659–667 (1973)
12. Ropers, H.J., Graebe, J.E., Gaskin, P., MacMillan, J.: Biochem. Biophys. Res. Commun. *80*, 690–697 (1978)
13. Sponsel, V.M., Gaskin, P., MacMillan, J.: Planta *146*, 101–105 (1979)
14. Coolbaugh, R.C., Moore, T.C.: Plant Physiol. *44*, 1364–1367 (1969)
15. Frydman, V.M., Gaskin, P., MacMillan, J.: Planta *118*, 123–132 (1974)
16. Ray, P.M.: Plant Physiol. *59*, 594–599 (1977)
17. Smith, L.: Methods Biochem. Anal. *2*, 427–434 (1955)
18. Simon, E.W.: Biochem. J. *69*, 67–74 (1958)

The Physiology of Gibberellin-Induced Elongation

R.L. JONES [1]

Despite the fact that the discovery of the gibberellins (GA s) resulted from the dramatic effect these compounds exert on stem elongation, our understanding of the physiology of this process has progressed slowly. Progress has been hampered by both methodological and conceptual limitations, particularly by the lack of an appropriate test system. Experiments have been confined largely to studies with whole plants (1, 2) and to studies with excised sections which show a very limited growth response to GA (3, 4) or some dependence on, or response to, auxins (3, 5, 6).

Among the conceptual limitations to progress in elucidating the mechanism of GA action, attempts to resolve the roles of cell division and cell elongation in the process of GA-induced elongation (7, 8) have commanded considerable attention. It is now clear, however, that only cell elongation, and not cell division, can be involved in the process of surface growth (11). Thus, GA may affect the process of cell division in intact plants or isolated plant parts (7–10), but elongation growth results only from the process of cell extension.

The primacy of the role of GA in the stimulation of growth in GA-sensitive plants has also been disputed (12, 13). Several workers have argued that the role of GA is to stimulate the synthesis of auxin (14, 15), and that it is auxin which functions to stimulate elongation growth. Although GA may indeed stimulate the synthesis of indoleacetic acid in intact plants, the evidence that it is auxin which stimulates elongation in GA-treated plants is weak.

The precise mechanism of GA-induced elongation has been the object of considerable experimentation. Although altered cell-wall plasticity was recognized to play a central role in auxin-stimulated growth as early as the 1930's (16), the relative contribution of pressure and osmotic potentials to the water potential of cells of GA-treated tissue have only recently been resolved. Indeed circumstantial evidence has led most workers to believe that unlike auxins, the GA s stimulated cell extension by influencing the osmotic potential of the cell (17–19).

The development of suitable excised test systems for studying the physiology of GA-induced growth has provided answers to many of the questions posed above. Kaufman and his colleagues (20–25) have explited the response of the excised internode of 45-day-old oat plants, and they have described many aspects of the physiology of GA-induced elongation in this tissue. The author's laboratory (26–30) has concentrated on aspects of the response of the excised lettuce hypocotyl to GA. This review will emphasize progress in elucidating the physiology of GA action in this system (26), although reference will be made to the work of others where appropriate.

1 Department of Botany, University of California, Berkeley, California 94720, USA

Auxins and Growth in GA-Responsive Test Systems

Excised tissue sections which are responsive to GA provide useful models to test the hypothesis that GA stimulates elongation via an increase in auxin biosynthesis. With an excised system, the hypothesis can be tested both directly by the addition of auxin and indirectly by the use of antiauxin (Fig. 1). Silk and Jones (26) reported that IAA

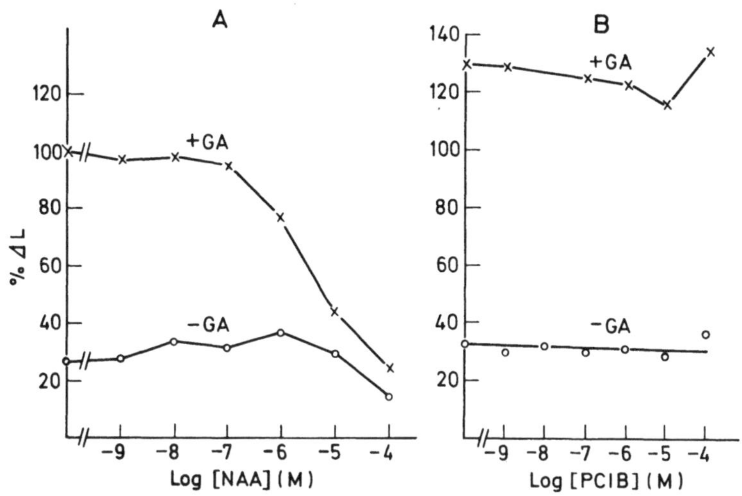

Fig. 1A and B. The effect of the synthetic auxin naphthaleneacetic acid (NAA) **(A)** and the antiauxin *p*-chlorophenoxyisobutyric acid (PCIB) **(B)** on the elongation of lettuce hypocotyl sections incubated in 5 μg ml^{-1} GA or H_2O

did not stimulate growth of excised lettuce hypocotyls. Neither did the antiauxin triiodobenzoic acid (TIBA) affect the response of the tissue to GA. We have now extended this study to include the application of synthetic auxins, e.g., 2,4-dichlorophenoxyacetic acid and naphthaleneacetic acid (NAA), neither of which stimulates elongation of the lettuce hypocotyl section (Fig. 1A). Also the powerful antiauxin *p*-chlorophenoxyisobutyric acid (PCIB) does not affect GA-induced growth at concentrations up to 10^{-4} M (Fig. 1B), and ethylene at concentrations up to 5,000 ppm does not affect the increase in length or weight of hypocotyls treated with GA (Fig. 2). This latter observation is particularly significant since auxin-responsive tissues, e.g., pea internodes, soybean hypocotyls, and cereal coleoptiles, exhibit a marked inhibition of elongation in response to ethylene, and this inhibition of elongation is generally accompanied by a pronounced swelling of the tissue.

Effects of GA on Cell Division and Cell Elongation

The GA s have been implicated as regulators of both cell division and cell elongation. Sachs (1) among others (35) has suggested that GA s can promote meristematic activity in whole plants, while Kaufman (20) has presented evidence that in *Avena* GA can

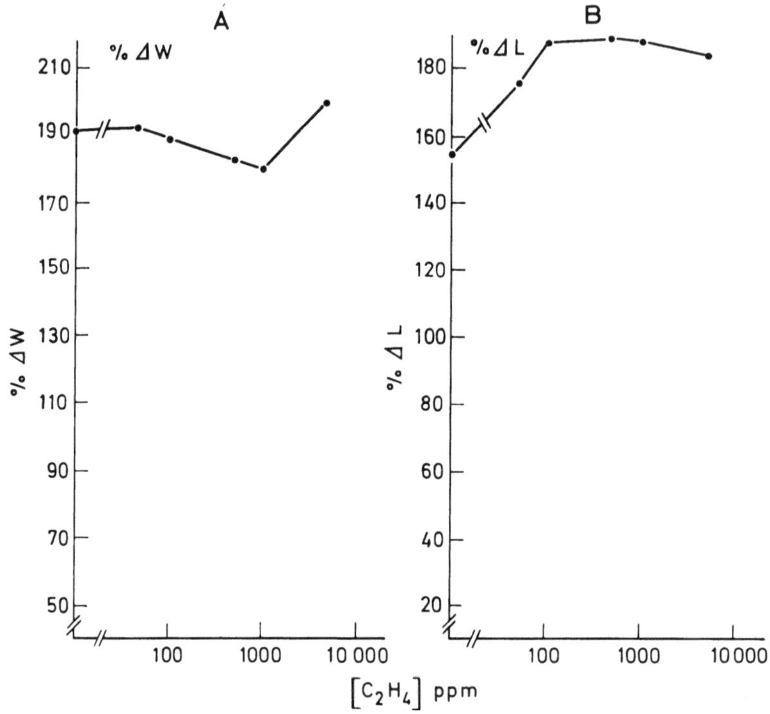

Fig. 2A and **B**. The effect of ethylene on fresh weight **(A)** and elongation **(B)** of lettuce hypocotyl sections incubated in GA or H_2O

inhibit cell division. In the excised hypocotyl of lettuce we have demonstrated that GA does not affect the process of cell division and that cell division has little if any influence on elongation (27). During the first 12 h of incubation of excised lettuce hypocotyls in either H_2O or GA, there is a 50% increase in the number of cells in the hypocotyl, but after 12 h few if any mitoses are observed. If hypocotyl sections are excised from seedlings of γ-irradiated seeds, cell division in the section is completely inhibited, but their GA-induced elongation is not significantly altered (27). Despite the fact that hypocotyl sections from seedlings of nonirradiated seeds have 50% more cells than those from irradiated seeds, their growth is essentially the same, indicating that factors other than cell number influence section length. These results emphasize the conclusions of Green (11) concerning the distinctions between the processes of cell elongation and cell division.

Although cell division does not contribute to the process of extension growth, it is often important to establish the extent to which cross-wall formation occurs after hormone treatment. If cell wall metabolism is to be studied in elongating tissue, for example, the contribution which new cross walls might make to the overall metabolism of polysaccharide could be significant.

There are other qualitative aspects of the response of lettuce hypocotyls to GA which distinguish it from the typical growth response to auxin. The effect of GA in this tissue is to overcome light-induced inhibition of elongation growth (26); in darkness elongation is rapid and unaffected by GA, while in blue and far-red light elongation is inhibited and the inhibition can be overcome by GA. In this respect the response of lettuce to GA is similar to that demonstrated in intact plants, e.g., pea (17), bean (13), and corn (31). The response of tissues to auxin, however, does not exhibit the same absolute light dependence; indeed, most auxin responses are unaffected by the presence or absence of light.

A further point of distinction between the GA and auxin response concerns the effect of a short pulse of hormone. Typically in auxin-responsive tissues, growth rates begin to decline to control rates soon after termination of the hormone pulse (32, 33), while in lettuce hypocotyl (34) and *Avena* internode sections (23) elongation growth is sustained for many hours after the hormone supply is withdrawn.

We conclude that in excised lettuce hypocotyl sections, elongation in response to GA is neither mediated through nor influenced by auxin, and that other physiological characteristics of the GA response of lettuce distinguish it from the auxin response.

Physiology of Cell Elongation in GA-Treated Tissue

From the foregoing it is axiomatic that a study of GA-induced elongation must emphasize changes in the rate of cell elongation. Does GA influence the rate of cell elongation by altering the tensile properties of the cell wall or by affecting the osmotic concentration of the cell sap? Although until recently, circumstantial evidence favored the latter, experiments with both lettuce hypocotyls (28, 29, 35) and *Avena* internodes (25) show that changes in cell wall plasticity accompany GA-induced growth. In lettuce, for example, both direct measurements of cell wall plasticity using Instron techniques (35) and indirect measurements of the properties of the cell walls of living tissue (28) indicate that elongation growth and increased cell wall plasticity are parallel events. During rapid growth, however, the osmotic concentration of the cell sap of GA-treated tissues decreases (29).

We have resolved the roles of altered pressure potential (ψ_p) and osmotic potential (ψ_π) in elongating lettuce hypocotyl tissue by incubating sections in dilute KCl or NaCl. When sections are incubated in 10 mM KCl in the absence of GA, there is no change in elongation rate, although the ψ_π of the tissue decreases (becomes more negative) because of KCl uptake (Fig. 3). Sections incubated in GA alone exhibit an elevated elongation rate with a concomitant increase in ψ_π. When GA-treated sections are incubated in KCl, however, the rate of elongation is further enhanced, and the ψ_π is maintained at a constant level (Fig. 3). From these data we conclude that elongation of lettuce hypocotyls is normally regulated by changes in cell wall extensibility, i.e., changed ψ_p, and that the elongation rate can be further modified by changes in ψ_π. In the absence of increased cell wall extensibility, an increase in the osmotic concentration of the cell sap has little direct influence on elongation.

Of the numerous hypotheses which have been advanced to account for changes in cell wall extensibility, the acid growth hypothesis has gained wide acceptance, particularly among those investigating the response of tissue systems to auxin (36, 37). We have examined both the elongation of lettuce hypocotyl sections at various pHs and the capacity of the tissue to secrete protons in response to GA treatment (30). Although

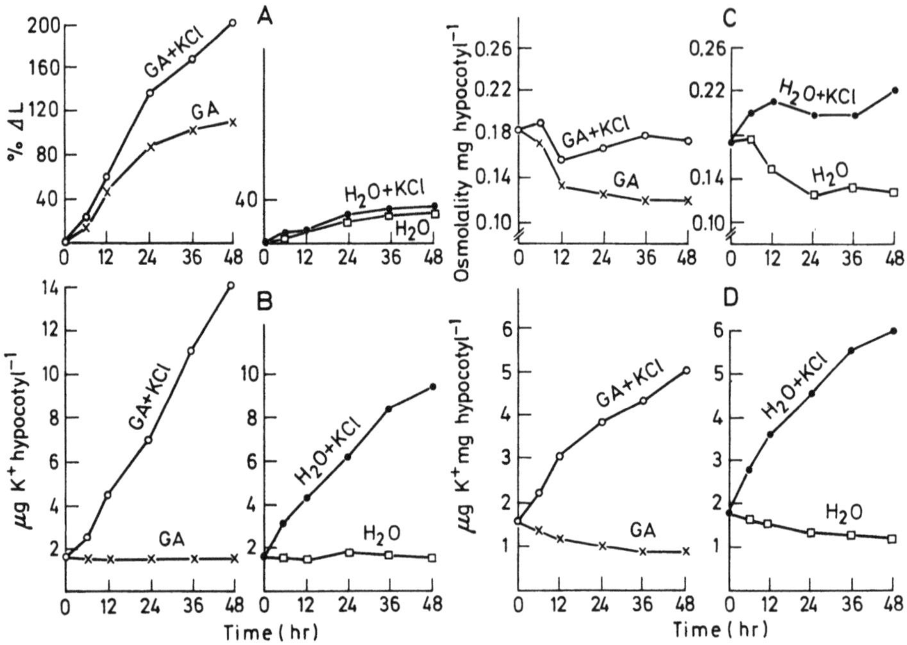

Fig. 3. Analysis of growth (A), K^+ uptake (B, D) and osmolarity of expressed scap (C) of lettuce hypocotyl sections incubated in GA or H_2O. From Stuart and Jones (28)

hypocotyl sections elongate in response to media of low pH, growth is only 25% over that of control sections incubated at pH 6.0, and furthermore, preincubation at low pH does not affect the magnitude of the GA response when sections are subsequently incubated in GA. Lettuce hypocotyl sections incubated in GA do not secrete protons, although under certain conditions of fusicoccin treatment sections will acidify the incubation medium. These data provide convincing evidence that for lettuce, elongation growth in response to GA treatment is not accompanied by proton secretion. These data contrast with the observations of Hebard et al. (25), who have shown that acid production accompanies GA treatment of *Avena* stem segments.

If protons are not responsible for changing the tensile properties of the cell wall, we must consider other possible agents. Enzymes which cleave specific bonds in the cell wall have been postulated by many to play an important role in regulating wall extensibility, but the evidence has been equivocal. Most of the experimental evidence is based on the extraction of enzymes from whole plants or tissue sections, and few attempts have been made to localize the extracted enzymes in the cell wall. The fact that extracted enzymes may use cell wall polysaccharides as substrates does not establish the enzymes as modifiers of the cell wall in intact tissue.

We have used a tissue centrifugation technique in an attempt to localize and characterize changes which might occur in cell wall polymers (40). This centrifugation technique has been used by Terry and Bonner (41) and Terry et al. (42) to examine changes in the polysaccharides of pea internode sections. Sections are infiltrated with water and then centrifuged at 1000 x *g*. This procedure removes the solution from the free space of the tissue with little detectable cytoplasmic contamination. Using this technique Terry (40) has shown that following auxin treatment of pea internode sections, there is an increase in the xyloglucan concentration of the free space solution which can be centrifuged from sections. This observation is in agreement with the experiments of Labavitch and Ray (43), who examined cell wall metabolism using conventional extraction techniques, and it provides evidence that the centrifugation technique does indeed measure changes in the cell wall of this tissue. We have begun to examine the proteins of the extra-protoplastic fluid from auxin-treated pea stem sections by means of electrophoresis on polyacrylamide, and our preliminary data indicate qualitative differences between control and auxin-treated tissue.

We propose to continue the application of this centrifugation technique to the problem of cell wall metabolism with emphasis on the role of GA in this process. Thus far we have examined changes in polysaccharides of GA-treated pea internode sections. In contrast to auxin, which increases the level of xylose and glucose in the free space of pea internodes, GA does not significantly affect xylose, glucose, arabinose, galactose, or rhamnose in the solution centrifuged from sections (Table 1). These data also

Table 1. GA and the neutral sugar composition of centrifugate from pea internodes [a]

μg/gm initial weight				
	Ara	Xyl	Gal	Glu
+ GA	8.6	10.1	21.2	19.2
− GA	9.5	10.6	22.2	19.9
+ GA/− GA	0.91	0.95	0.95	0.95
μg/gm final weight				
	Ara	Xyl	Gal	Glu
+ GA	7.2	8.4	17.7	16
− GA	8.7	9.7	20.3	18.2
+ GA/− GA	0.83	0.87	0.87	0.88

[a] GA growth = 148% of control

lend support to our observations on changes in water-soluble polysaccharides extracted by homogenization from GA-treated lettuce hypocotyl sections (Table 2). GA treatment does not enhance the metabolism of arabinose, xylose, galactose, or glucose in lettuce hypocotyl sections. We conclude that GA treatment of pea internode or lettuce hypocotyl sections does not result in an enhancement of the metabolism of xyloglucan or a similar alcohol-insoluble, water-soluble polysaccharide.

Table 2. The neutral sugar composition of the homogenate + external medium of lettuce hypocotyls incubated ± GA

μg/gm initial weight

	Ara	Xyl	Gal	Glu
+ GA	128.2	73.5	257.4	73.7
– GA	98.2	58.7	225.1	73.8
+ GA/– GA	1.31	1.25	1.14	1.00

μg/gm final weight

	Ara	Xyl	Gal	Glu
+ GA	56.6	32.3	114.2	35.2
– GA	69.9	34.4	133.5	42.3
+ GA/– GA	0.81	0.94	0.85	0.83

Conclusions

The evidence that the GA s serve to stimulate cell elongation in plant stems is convincing. The rate of elongation is enhanced by changes in the extensibility of the cell wall, but acidification of the cell wall does not seem to be involved in the process of cell wall softening in lettuce as it may be in auxin-responsive tissues. Also by contrast with auxins, there is no evidence for the turnover or synthesis of a xyloglucan or similar water-soluble polysaccharide in the tissues of pea or lettuce after GA treatment. Using a centrifugation technique to isolate the cell wall free-space solution, we are pursuing changes in both polysaccharides and proteins in the cell walls of GA-responsive tissues with the aim of correlating changes in these polymers with altered rates of elongation of the tissue.

Acknowledgment. I wish to thank Wendy K. Silk, David A. Stuart, Deborah J. Durnam, and Maurice E. Terry who completed much of the work cited above. This work was supported by grants from the National Science Foundation (BMS 75-18870 and PCM 78-13286).

References

1. Sachs R.M.: Annu. Rev. Plant Physiol. *16*, 73–96 (1965)
2. Sachs, R.M., Bretz, C., Lang, A.: J. Exp. Cell Res. *18*, 230–244 (1959)
3. Brian, P.W., Hemming, H.G.: Ann. Bot. *22*, 1–17 (1958)
4. Penny, D., Penny, P.: Can. J. Bot. *52*, 959–969 (1974)
5. Shibaoka, H.: Plant Cell Physiol. *13*, 461–469 (1972)
6. Galston, A.W., Warburg, H.: Plant Physiol. *34*, 16–22 (1959)
7. Nitsan, J., Lang, A.: Dev. Biol. *12*, 358–376 (1965)
8. Arney, S.E., Mancinelli, P.: New Phytol. *65*, 161–175 (1966)
9. Liu, P.B., Loy, J.B.: Am. J. Bot. *63*, 325–336 (1976)

10. Greulach, V.A., Haesloop, J.G.: Am. J. Bot. *45*, 566–570 (1958)
11. Green, P.B.: Bot. Gaz. *137*, 187–202 (1976)
12. Cleland, R.E.: In: The Physiology of Plant Growth and Development. Wilkins, M.B. (ed.), pp. 49–81. London: McGraw-Hill 1969
13. Brian, P.W.: Int. Rev. Cytol. *19*, 229–266 (1966)
14. Kuraishi, S., Muir, R.M.: Naturwissenschaften *50*, 337–338 (1963)
15. Skytt-Andersen, A., Muir, R.M.: Physiol. Plant. *22*, 354–363 (1969)
16. Heyn, A.N.J.: Proc. K. Akad. Wet. *33*, 1045–1058 (1930)
17. Lockhart J.A.: Plant Physiol. *35*, 129–135 (1960)
18. Paleg, L.G.: Annu. Rev. Plant Physiol. *16*, 319–322 (1965)
19. Cleland, R.E., Thompson, M.L., Rayle, D.L., Purves, W.K.: Nature (Lond.) *214*, 510–511 (1968)
20. Kaufman, P.B.: Physiol. Plant. *18*, 703–724 (1965)
21. Jones, R.A., Kaufman, P.B.: Plant Physiol. *24*, 491–497 (1971)
22. Kaufman, P.B., Petering, L.B., Adams, P.B.: Am. J. Bot. *56*, 918–927 (1969)
23. Montagne, M.J., Ikuma, H., Kaufman, P.B.: Plant Physiol. *51*, 1026–1032 (1973)
24. Adams, P.A., Motagne, M.J., Tepfer, M., Rayle, D.L., Ikuma, H.: Plant Physiol. *56*, 757–760 (1975)
25. Hebard, F.V., Amatangelo, S.J., Dayanandan, S.J., Kaufman, P.B.: Plant Physiol. *58*, 670–674 (1976)
26. Silk, W.K., Jones, R.L.: Plant Physiol. *56*, 267–272 (1975)
27. Stuart, D.A., Durnam, D.L., Jones, R.L.: Planta *135*, 249–255 (1977)
28. Stuart, D.A., Jones, R.L.: Plant Physiol. *59*, 61–68 (1977)
29. Stuart, D.A., Jones, R.L.: Plant Physiol. *61*, 180–183 (1978)
30. Stuart, D.A., Jones, R.L.: Planta *142*, 135–145 (1978)
31. Phinney, B.O.: Proc. Natl. Acad. Sci. USA *42*, 185–189 (1957)
32. De La Fuente, R.K., Leopold, A.C.: Plant Physiol. *45*, 19–24 (1969)
33. De La Fuente, R.K., Leopold, A.C.: Plant Physiol. *46*, 186–189 (1970)
34. Silk, W.K., Jones, R.L., Stoddart, J.L.: Plant Physiol. *59*, 211–216 (1977)
35. Silk, W.K.: Ph.D. Thesis, University of California, Berkeley (1977)
36. Cleland, R.E.: Proc. Natl. Acad. Sci. USA *70*, 3092–3093 (1973)
37. Cleland, R.E.: Plant Physiol. *58*, 210–213 (1976)
38. Masuda, Y., Yamamoto, Y.: Dev. Growth Differ. *11*, 287–294 (1970)
39. Fan, D.F., MacLachlan, G.A.: Can. J. Bot. *44*, 1025 (1966)
40. Terry, M.: Ph.D. Thesis, University of California, Davis (1978)
41. Terry, M., Bonner, B.A.: Plant Physiol. (in press, 1980)
42. Terry, M., Rubinstein, B., Bonner, B.A., Jones, R.L.: Abstr. 10th Int. Conf. Plant Growth Substances, p. 40 (1979)
43. Labavitch, J.M., Ray, P.M.: Plant Physiol. *53*, 669–673 (1974)

Ethylene

Chairman: H. KENDE

Ethylene and Seeds

M.A. HALL, M.A. ACASTER, T. BENGOCHEA, J.H. DODDS, D.E. EVANS, J.F. JONES, P.H. JERIE, G.C. MUTUMBA, B. NIEPEL, and A.R. SHAARI [1]

Some of the earliest reports on effects of ethylene on plant growth and development relate to studies on the breaking of seed dormancy [e.g., (1)]. In more recent times a number of workers have returned to this subject, and it appears that ethylene may break dormancy in seeds of a large number of species (2–4). Such effects have been taken to mean that ethylene is involved in the natural control of seed dormancy. Further interest in this possibility has been engendered by the discovery that ethylene is a natural component of the soil atmosphere (5).

Our own interest in this subject arose from work on *Spergula arvensis* which is typical of a large number of weed species in having seed the dormancy of which may be broken by ethylene at physiological concentrations ($1-10\ \mu l\ l^{-1}$). Such seeds almost always also have a light requirement (4). The duration of the dormant period varies greatly from species to species, ranging from a few days to many months. In general, the light and ethylene requirements diminish in parellel (4, 6).

Our efforts in the last few years have thus been directed towards attempting to elucidate the mode of action of ethylene in this system and also to seek an explanation for the loss of the ethylene requirement. In common with other workers on hormones and seed dormancy, our efforts on the first objective have not been attended by conspicuous success. Thus, a number of approaches have failed to pinpoint any process which can be deemed to constitute the principal vehicle for ethylene action as distinct from the many processes attendant on the breaking of dormancy such as mobilization of food reserves etc. We have, however, excluded the possibility suggested by Roberts (7) that the breaking of seed dormancy, by whatever means, involves a shift from the Embden-Meyerhof-Parnas pathway of respiration to the pentose-phosphate pathway. In *Spergula*, if anything, the converse is true (8). Likewise, electron microscopic studies, while yielding valuable information on the ultrastructure of the seed, did not provide a key to the problem.

One aspect of the work on respiration did, however, lead to an interesting observation, namely that the presence of CO_2 at atmospheric concentration was necessary for dormancy to be broken. The CO_2 and ethylene effects appear to be interdependent to some extent, but raising the CO_2 concentration above atmospheric had no further effect (9). Similar results have been obtained by other workers [e.g., (10)]. This observation will be further discussed below.

It was observed in cocklebur and peanut seeds, the dormancy of both of which may be broken by ethylene, that loss of dormancy is accompanied by an increase in

1 Department of Botany and Microbiology, The University College of Wales, Aberystwyth, Dyfed, United Kingdom

the capacity of the seed to emanate ethylene (11, 12). It was proposed therefore that the observed loss of dormancy was due to the increased ability of the seeds to produce ethylene. We have observed similar behavior in *Spergula*. Since rates of emanation of ethylene from plant organs are deemed to reflect rates of biosynthesis, it was assumed that the changing pattern of ethylene emanation from seeds represented a change in their capacity to synthesize ethylene. While this may well be so, wholly or in part, our most recent work casts some doubt on this conclusion.

It has always been assumed that the concentration of ethylene in the intercellular spaces of plant tissue bears a fairly constant relationship to that in the cell. Since water constitutes the major cellular component, it might be supposed that the air space and cell concentrations of ethylene would be related through the partition coefficient (C_p) of the gas in water. This is certainly true of the C_ps of ethylene in animal tissues (13, 14), which are of the same order as that in water, namely 0.1. On the other hand, in work with seeds and organs of a wide range of plants we obtained C_ps of up to 12 (15, 16) (Table 1). Although some of the lower figures we obtained might to some extent

Table 1. Apparent partition coefficients from $^{14}C_2H_4$ obtained with seed tissues from a range of leguminous species (15)

Species	Cp	Species	Cp
Ulex europaeus L.	0.25	*Vicia cracca* L.	2.15
Laburnum anagyroides Medic	0.29	*Phaseolus vulgaris* L.	2.50
Glycine max Merr. cv. Acme	0.32	*Medicago arabica* (L). All	4.60
Lupinus arboreus L.	0.89	*Vicia sativa* L.	4.80
Phaseolus multiflorus Lam.	0.90	*Vicia faba* L.	12.0

be explained by partial solution of ethylene in, for example, cell lipids, this is clearly not the case for the higher values. In one species – *Vicia faba* – it was found that rapid metabolism of ethylene to ethylene oxide – a substance with a C_p in water of about 80 – was responsible for the high apparent C_p in the tissue (17, 18).

In *Phaseolus vulgaris*, on the other hand, isolated developing cotyledons accumulate ^{14}C ethylene to give C_ps of between 1 and 2.5, which greatly exceed the value with steam killed tissue (0.05). The labelled ethylene is released only slowly (1%–10% h^{-1}) either into an air stream or into toluene. Heating to 60°C causes the immediate release of almost all the labelled ethylene. No significant metabolism is observed. Propylene and vinyl chloride competitively inhibit the accumulation of $^{14}C_2H_4$. The C_p of the cotyledons changes quite markedly during development, and it is notable that the rate of emanation of ethylene from the tissue shows an inverse relationship to the C_p.

Heating untreated cotyledons to 60°C in sealed vials results in the release of ethylene. This treatment causes the release of ethylene from all living tissues, but cotyledons release about 8 times more ethylene than other tissues with low C_ps (e.g. *Phaseolus* cotyledons, 9.47 nl g^{-1} f.m., *Phaseolus* pericarp 1.20 nl g^{-1} f.m., tomato pericarp 1.46 nl g^{-1} f.m.) We have suggested that this phenomenon represents the release of "compartmented" endogenously produced ethylene since as pointed out above, exogenously fed "compartmented" $^{14}C_2H_4$ may be released in this way. This suggestion is further supported by our observation that the amount of ethylene thus released from cotyledons of different ages follows closely the C_p of the tissue at any given time. Measurements of the quantity of ethylene released by heating and of rates of emanation suggest that *Phaseolus* cotyledons may hold ethylene in a compartmented form in sufficient amounts to account for at least 200 h of emanation (19).

If seeds are capable of "storing" amounts of ethylene of this magnitude, then clearly the results with *Spergula* and other seeds referred to above are susceptible to an alternative explanation, as indeed are similar data with other plant parts, since the effect in *Phaseolus* is not restricted to the developing cotyledons, although it is most pronounced there. Thus, it is not unreasonable to propose that changes in the rates of ethylene emanation from plant parts may reflect, at least in part, changes in the capacity of the tissue to compartment endogenously produced ethylene and/or the release of ethylene previously compartmented.

Further studies on the *Phaseolus* system have yielded other interesting data. Thus, we have been able to separate a cell-free fraction from *Phaseolus* cotyledons on continuous and discontinuous sucrose gradients which has the ability to bind ethylene with high affinity; indeed, it is possible to separate the ethylene:binding fraction association by centrifugation of cotyledon preparations derived from tissue pretreated with $^{14}C_2H_4$. Figure 1 shows an analysis of the equilibrium between ethylene and the binding sites obtained by a modified form of equilibrium dialysis. Clearly, the binding is saturable and calculations from Scatchard plots indicate a dissociation constant (K_D) for ethylene of between 4 and 9×10^{-10} M – this is equivalent to a concentration of ethylene in the gaseous phase of between about 0.08–0.18 $\mu l\ l^{-1}$. It has been pointed

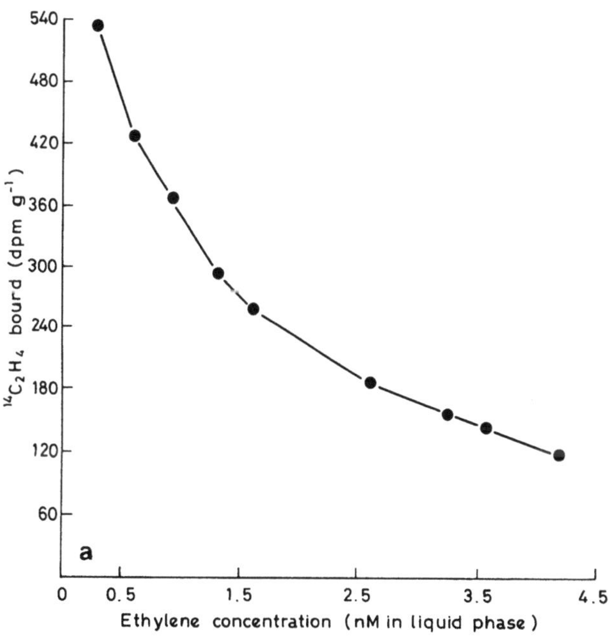

Fig. 1a and b. Binding of $^{14}C_2H_4$ to particulate preparations from *Phaseolus vulgaris.* a Bound radioactivity as a function of ethylene concentration. 1000 dpm $^{14}C_2H_4$ ml^{-1} gas (0.0842 nl ml^{-1} gas, 0.0062 nl ml^{-1} liquid) in all incubations

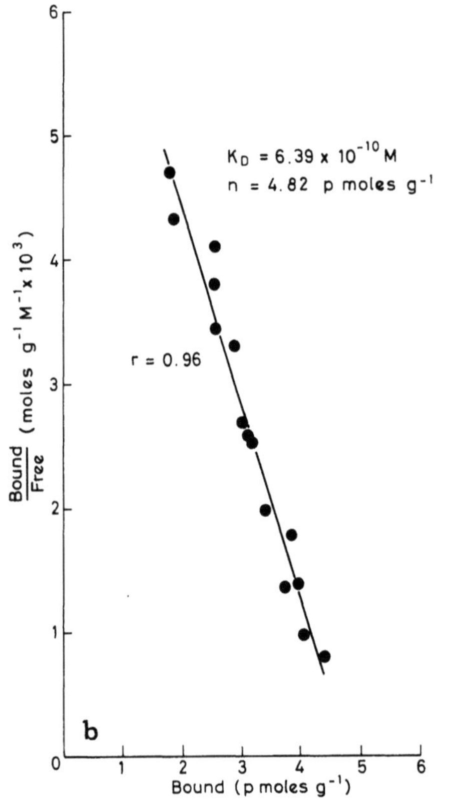

Fig. 1b. Scatchard analysis of data obtained by incubation of preparations with a range of concentrations of carrier free $^{14}C_2H_4$ (11,870 dpm nl^{-1}; 1 pmol $^{14}C_2H_4 \equiv 266$ dpm). Data derived by a modification of equilibrium dialysis. *r* coefficient of determination

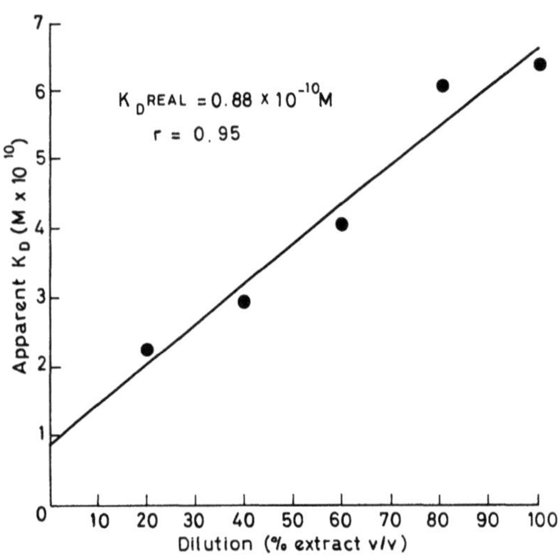

Fig. 2. Variation of K_D with concentration of binding site. Derivation of true K_D by method of Chang et al. (20). *r* coefficient of determination

out by Chang et al. (20) that where the concentration of receptor sites is greater than the true dissociation constant of the ligand, the apparent dissociation constant obtained from direct concentration binding curves varies as a linear function of the receptor concentration. Determination of apparent K_D at different concentrations of the receptor produced the data shown in Fig. 2 and this treatment yields a figure of 0.88×10^{-10}M for the K_D ($\equiv 0.018\ \mu l\ l^{-1}$ in the gaseous phase). The number of binding sites varies depending on the age of the cotyledons, being greater the greater the C_p of the tissue. However, since the dissociation of the ligand from the binding site is very slow, many of the sites are still occupied by endogenous ethylene during assay, thus making difficult the assessment of total binding site concentration. The binding activity is destroyed by heat or by incubation with proteolytic enzymes. Electron microscopy reveals that the binding fraction consists almost exclusively of membrane-bounded vesicles with diameters of between 0.5 and 2.0 μm (Fig. 3). Preliminary experiments with

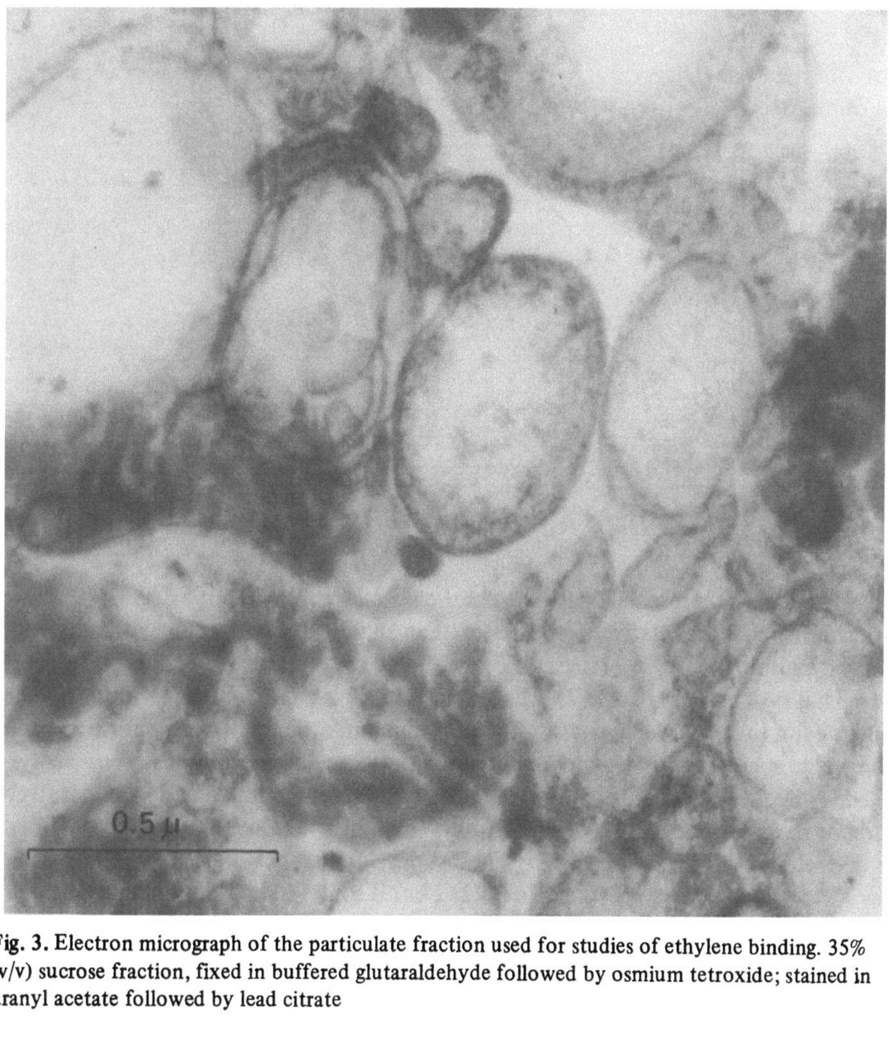

Fig. 3. Electron micrograph of the particulate fraction used for studies of ethylene binding. 35% (w/v) sucrose fraction, fixed in buffered glutaraldehyde followed by osmium tetroxide; stained in uranyl acetate followed by lead citrate

marker enzymes suggest that the binding activity is located on elements of the endoplasmic reticulum. We have now succeeded in solubilizing the binding activity with Triton X-100, which observation also makes unlikely the possibility that the system represents sequestration of ethylene into the vesicles.

^{14}C-ethylene may be released quantitatively from the association by heat treatment in the same way as in the intact cotyledon system. Freezing and subsequent thawing of extracts does not release the bound ethylene, which is similar to the situation in auxin binding systems (21). In addition to its saturability and high affinity for ethylene, the system shows a high specificity for the growth regulator. Structural analogues able to mimick the effect of ethylene in developmental systems competitively inhibit the binding of the growth regulator in the cell free system (Fig. 4, Table 2);

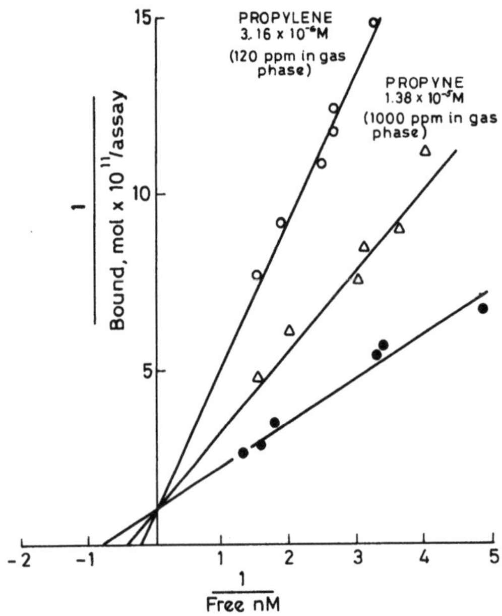

Fig. 4. Binding of $^{14}C_2H_4$ to particulate preparations from *Phaseolus vulgaris* in the presence and absence of propylene (○) and propyne (△). Control (●)

physiologically inactive analogues such as the alkanes were without effect. It is also noteworthy that the relative effectiveness of an analogue in competing with ethylene for the binding site (K_i/K_D) is very closely related to its relative effectiveness in developmental systems (22) (Table 2). The alkyne series has a rather higher affinity for the binding site than its effectiveness in the pea stem inhibition system used for comparison would suggest, but it should be noted that in some developmental systems acetylene may be more active than propylene [e.g., (23)]. Carbon dioxide also competitively inhibits the association of the ligand with the binding site. This would be expected from the observed behaviour of CO_2 in most developmental systems [the K_i for the gas is very close to the expected value (22)], but is somewhat curious in light of the work with *Spergula* and other seeds referred to above, where raising CO_2 concentrations to as much as 50% in the gaseous phase can neither completely replace ethylene nor can it inhibit the ethylene effect.

Table 2. Comparison of inhibitor constants for structural analogues of ethylene in the *Phaseolus vulgaris* binding site system

		K_i gas (M)	K_i liquid (M)	K_i gas ($\mu l\ l^{-1}$)	K_i gas (relative)	Relative ppm for half maximal activity on pea growth [a]	K_i liquid (relative)
Ethylene	$CH_2=CH_2$	8.15×10^{-9}	9.0×10^{-10}	0.18	1	1	1
Propylene	$CH_3CH=CH_2$	1.04×10^{-6}	5.6×10^{-7}	23.3	128	100	600
Vinyl chloride	$CH_2=CHCl$	3.8×10^{-6}	1.8×10^{-6}	85.1	466	1,400	2,022
Carbon monoxide	CO	8.7×10^{-6}	2×10^{-7}	195	1,068	2,700	222
Acetylene	$CH\equiv CH$	8.25×10^{-6}	1.03×10^{-5}	185	1,013	2,800	11.444
Vinyl fluoride	$CH_2=CHF$	9.3×10^{-6}	4.37×10^{-6}	208	1,139	4,300	4,800
Propyne	$CH_3C\equiv CH$	2.16×10^{-5}	6.7×10^{-6}	484	2,651	8,000	7,444
Vinyl methyl ether	$CH_2=CH\text{-}O\text{-}CH_3$	1.11×10^{-3}	1.11×10^{-4}	24,864	136,196	100,000	123,333
1-Butyne	$CH_3CH_2C\equiv CH$	1.81×10^{-4}	6.97×10^{-5}	4,054	22,206	110,000	77,427
1-Butene	$CH_3CH_2CH=CH$	4.9×10^{-3}	4.9×10^{-4}	109,760	601,227	270,000	544,444
Vinyl ethyl ether	$CH_2=CH\text{-}O\text{-}CH_2CH_3$		9.94×10^{-3}			300,000	11,100,000
Carbon dioxide	CO_2	1.15×10^{-3}		25,760	141,104	300,000	

[a] From Burg and Burg (22). Reproduced with permission. Inactive at concentrations ($\mu l\ l^{-1}$) in parentheses: Methane, ethane, propane cyclopropane (10^5) cis-2-butene, trans-2-butene (10^4)

It seems unlikely that a system such as this with its very high affinity and specificity for ethylene would evolve and persist without serving some function. There appear to be two main alternatives, namely:

1) that the system serves as a mechanism for regulating the ethylene status of the seed, either during the ripening phase or alternatively, during the early phases of germination, as already suggested;

2) that the system constitutes a receptor in the usual sense.

In support of the first alternative the data on "compartmentation", emanation, and total ethylene show that the system does indeed affect the ethylene status of the seed. It could be argued that since, unlike the situation obtaining with other growth regulators, in most plants, metabolism of ethylene occurs only at a very low level (24, 25) and since ethylene emanation from a dense tissue enclosed in a pod, such as a seed of *Phaseolus,* is likely to be restricted, that some means of regulating ethylene status may have evolved. The corollary, that the system represents a means of "storing" ethylene for release at an appropriate stage of ontogeny is rendered more plausible by the observation that similar phenomena are observable in *Spergula* (26) and may be related to the changing capacity of the seed to release ethylene as it ages. The capacity of *Phaseolus* cotyledons for holding ethylene in this way is substantial (up to 10 nl g^{-1} f.m. as calculated from heat treatment experiments or measurement of the total number of binding sites in intact tissue). Whether or not this ethylene is released at a sufficient rate to affect dormancy or processes occurring during germination in some seeds remains a matter for conjecture. It is also germane to reiterate the point raised above, that the possession by a plant of such mechanisms as these renders invalid the assumption that measurement of rates of emanation of ethylene or of air space concentrations of the gas reflect either rates of biosynthesis or cellular concentrations respectively.

The evidence in support of the second alternative is inconclusive. The binding activity is greatest in the developing cotyledon but is not restricted to it, and it is notable that abscission zones of *Phaseolus* have C_ps different from those of adjacent portions of the petiole.

The two alternatives are not mutually exclusive, however, and there is no reason why a receptor should not also function to regulate the status within the plant of the growth regulator for which it is specific. This possibility is not dissimilar in kind to the role suggested by Beyer (25, 27, 29) for the system bringing about the low levels of ethylene metabolism in a number of species; indeed the *Phaseolus* system may represent the same mechanism, the only difference being one of magnitude.

Alternatively, the system in the seed may perform a different function in other parts of the plant since the demonstration of binding sites for ethylene with the same K_D in different parts of a plant does not constitute proof that these binding sites are of a single type either within the same or in different tissues. Indeed, the range of developmental effects controlled by ethylene makes it unlikely that a single type of ethylene receptor mediates all such processes. In that all ethylene receptors would be likely to have dissociation constants in the range observed here, (which it should be noted represents an ethylene concentration in the gaseous phase close to that which produces a half-maximal effect in most developmental systems), until we can relate changes in the properties of rigorously isolated binding fractions as a result of ethylene treatment to particular biochemical events involved in a developmental process, the possibility remains that the system described in fact represents several binding sites with similar affinities for ethylene.

References

1. Vacha, G.A., Harvey, R.B.: Plant Physiol. *2,* 187–192 (1927)
2. Toole, V.K., Bailey, W.K., Toole, E.H.: Plant Physiol. *39,* 822–832 (1964)
3. Esashi, Y., Leopold, A.C.: Plant Physiol. *44,* 1470–1472 (1969)
4. Olatoye, S.T., Hall, M.A.: In: Seed Ecology. Heydecker, W. (ed.), pp. 233–249. London: Butterworth 1972
5. Smith, K.A., Russell, R.S.: Nature (Lond.) *222,* 769–771 (1969)
6. Hall, M.A., Wareing, P.F.: Proc. 11th Weed Control Conf., pp. 1173–1182 (1972)
7. Roberts, E.H.: Dormancy and Survival Symp. Soc. Exp. Biol. *23,* 161–192 (1969)
8. Jones, J.F., Hall, M.A.: Unpublished results
9. Jones, J.F., Hall, M.A.: Plant Sci. Lett. *16,* 87–93 (1979)
10. Keys, R.D., Smith, O.E., Kumamoto, J., Lyons, J.L.: Plant Physiol. *56,* 826–829 (1975)
11. Ketring, D.C., Morgan, P.W.: Plant Physiol. *45,* 268–273 (1970)
12. Katoh, H., Esashi, Y.: Plant Cell Physiol. *16,* 687–696 (1975)
13. Grollman, A.: J. Biol. Chem. *82*(2), 317–325 (1929)
14. Jerie, P.H., Zeroni, M., Hall, M.A.: Pestic. Sci. *9,* 162–168 (1978)
15. Jerie, P.H., Shaari, A.R., Zeroni, M., Hall, M.A.: New Phytol. *81,* 499–504 (1978)
16. Zeroni, M., Jerie, P.H., Hall, M.A.: Planta *134,* 119–125 (1977)
17. Jerie, P.H., Hall, M.A.: Proc. R. Soc. London Ser. B *200,* 87–94 (1978)
18. Dodds, J.H., Musa, S.K., Jerie, P.H., Hall, M.A.: Plant Sci. Lett. *17,* 109–114 (1979)
19. Jerie, P.H., Shaari, A.R., Hall, M.A.: Planta *144,* 503–507 (1979)
20. Chang, K.J., Jacobs, S., Cuatrecasas, P.: Biochim. Biophys. Acta *406,* 294–303 (1975)
21. Ray, P.M., Dohrman, U., Hertel, R.: Plant Physiol. *59,* 357–364 (1977)
22. Burg, S.P., Burg, E.A.: Plant Physiol. *42,* 144–152 (1967)
23. Crocker, W., Zimmerman, P.W., Hitchcock, A.E.: Contrib. Boyce Thompson Inst. *4,* 177 (1932)
24. Beyer, E.M., Jr.: Plant Physiol. *56,* 273–278 (1975)
25. Beyer, E.M., Jr.: This volume
26. Acaster, M.A., Hall, M.A.: Unpublished observations
27. Beyer, E., Jr.: Plant Physiol. *56,* 273–278 (1975)
28. Beyer, E., Jr.: Plant Physiol. *64,* 971–974 (1979)

Ethylene Metabolism and Its Possible Physiological Role in Plants

E.M. BEYER, Jr. and D.C. BLOMSTROM [1]

In contrast to the metabolism of the other plant hormones, ethylene metabolism has received surprisingly little attention until recently. The apparent lack of interest in this area occurred despite several early reports (1–7) indicating that a small but significant amount of labeled ethylene was metabolized by both fruit and vegetative tissues. These studies, however, failed to be convincing because of inadequate precautions with regard to the radiochemical purity of ethylene or to microbial contamination. Additional skepticism arose from the contradictory results of others (8, 9) and the subsequent finding by Jansen (5) that the common practice of regenerating labeled ethylene from mercuric perchlorate resulted in the formation of readily metabolized impurities.

These objections ultimately caused Abeles (10) and others (11, 12) to conclude that ethylene was probably not metabolized by plant tissues and that the small amount of incorporation observed by some workers was the result of experimental artifacts. This was a tenable position inasmuch as the volatility of ethylene permitted its dissipation, thus making metabolic inactivation seem unnecessary, and was consistent with the then current belief that ethylene maintained structural integrity during its action.

Because of the uncertainty that surrounded this area and the availability of $^{14}C_2H_4$ of high specific activity, a study was undertaken in 1973 to examine critically the suspected metabolic inertness of ethylene in plants. Surprisingly, this study (13) revealed a previously unrecognized ethylene metabolic system which oxidized the hormone to CO_2 and metabolized it to other less volatile water-soluble compounds. In this report, some of the characteristics of this system are examined, and it is suggested that this metabolism represents the initial biochemical steps in the ethylene action sequence.

Experimental Procedure

Since all samples of radioactive ethylene so far purchased have contained easily detectable labeled impurities, it is essential that the ethylene be purified prior to its use. For this purpose, a two-step preparative gas chromatographic procedure has been developed (14). Ethylene thus purified has been shown by high-sensitivity gas chromatography to be of ultrapure quality. In addition, specific trapping procedures have demonstrated

1 Central Research and Development Department, Experimental Station, E.I. duPont de Nemours & Co., Inc., Wilmington, Delaware 19898, USA

(13) that when ^{14}C-labeled ethylene is selectively removed from the incubation atmosphere, metabolism disappears.

Etiolated pea seedlings have been used extensively in this work because of their general sensitivity to ethylene and the relative ease with which they can be grown and treated aseptically with labeled ethylene. Figure 1 illustrates the general experimental protocol now routinely used for growing these seedlings and analyzing the tissue

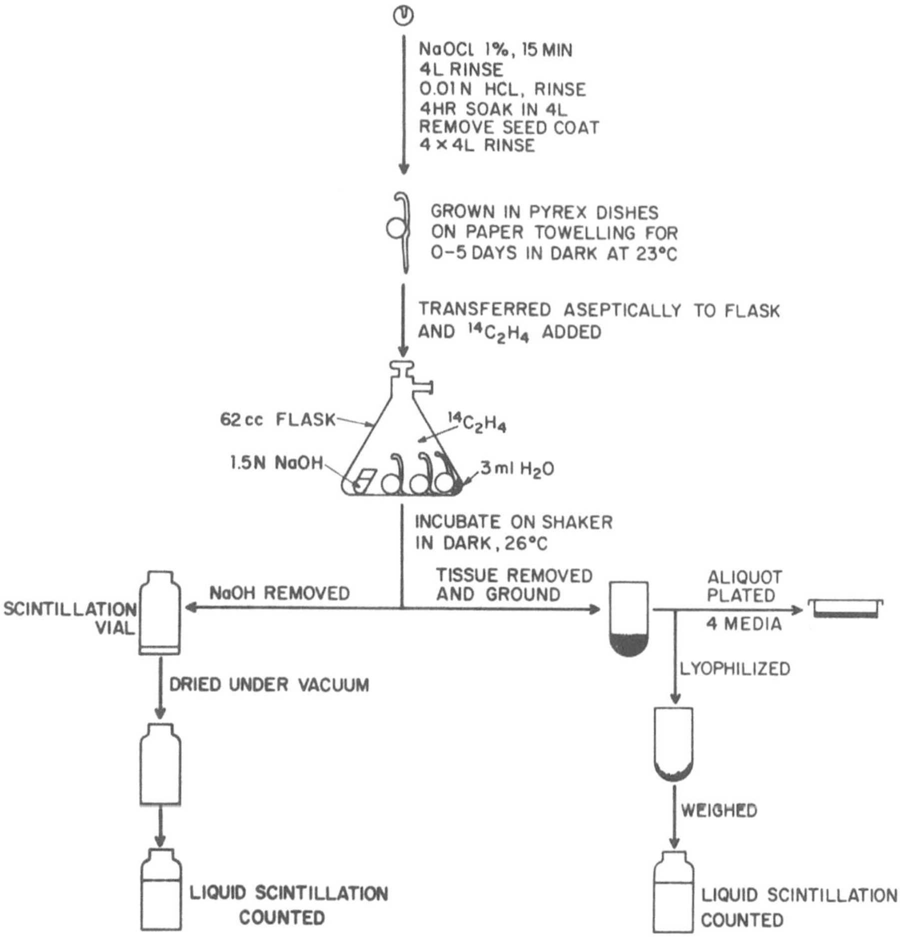

Fig. 1. Experimental procedure for growing and treating etiolated pea seedlings with purified $^{14}C_2H_4$ (14) and for analyzing the tissue and NaOH for radioactivity

and NaOH for radioactivity. This procedure produces two types of data. The radioactivity remaining in the left-hand liquid scintillation vial (Fig. 1) after the dissolved $^{14}C_2H_4$ has been removed by lyophilization, is referred to as ethylene oxidation and is expressed in specific activity units as DPM/mg dry weight/treatment period. The radioactivity in the right-hand vial contains the tissue radioactivity freed of dissolved

$^{14}C_2H_4$ and is referred to as tissue-^{14}C or tissue incorporation. This activity similarly is expressed in specific activity units.

General Characteristics of Ethylene Metabolism

Changes During Growth and Development

Like ethylene sensitivity and biosynthesis, the ability of plant tissues to metabolize ethylene changes markedly during development. This important feature of the ethylene metabolic system was first recognized in studies with intact etiolated pea seedlings (13). It was observed that during the first 20 h following imbibition, barely detectable rates of ethylene metabolism occurred in terms of either ethylene oxidation or tissue incorporation. However, as the root-shoot axis began rapidly to elongate, both activities increased markedly until 3 days after imbibition incorporation was up 11-fold and oxidation up 55-fold. Thereafter, both activities gradually declined to intermediate levels.

One curious phenomenon associated with ethylene metabolism in peas is the precipitous drop in ethylene oxidation and, to a lesser extent, in tissue incorporation when the cotyledons are detached from the root-shoot axis just prior to exposing both parts in the same flask to labeled ethylene (15). It appears that some factors must be provided by the root-shoot axis or vice versa for maintaining ethylene metabolism. Interestingly, detachment does not affect other metabolic activities such as ethylene biosynthesis or respiration and the addition of the plant hormones or cAMP does not overcome this effect.

During development, other tissues have also been found to vary considerably in their ability to metabolize ethylene. A predominant feature has been the apparent independence of ethylene oxidation from tissue incorporation. For example, during rapid petal expansion in cut carnations (16), the rate of tissue incorporation into the reproductive and receptacle tissues increased dramatically over a five-day period, while there was only a slight increase in the rate of oxidation. This independency was also apparent in a recent study with cherry tomato fruits (Fig. 2). When the fruits were still green but within a few days of starting to change color, the rate of ethylene oxidation greatly exceeded tissue incorporation. However, as the fruit matured, the rate of oxidation appeared to drop for a while whereas tissue incorporation increased steadily until at full maturity (red stage) both rates were about the same.

Another important feature is that metabolism and tissue responsiveness to ethylene often increase in parallel. In morning glory flowers (17), ethylene oxidation and tissue incorporation occurred at very low to nondetectable levels during the period of rapid bud elongation when the petals reportedly (18) do not respond to ethylene. However, about one day prior to opening, when the flowers do become responsive, the rate of ethylene oxidation and, to a lesser extent, tissue incorporation rapidly increased. Similarly in cotton (19), when the second true leaf blade of young plants was left intact, fairly constant low rates of ethylene oxidation and tissue incorporation were observed in the abscission zone. However, when the leaf blade was removed to induce abscission, the rate of ethylene oxidation increased fivefold. This increase preceded the first signs

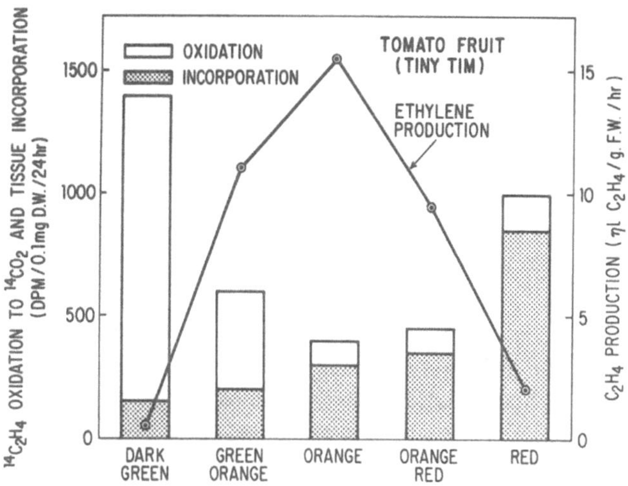

Fig. 2. $^{14}C_2H_4$ oxidation, tissue incorporation, and ethylene production in tomato fruit (*Lycopersicon esculentum* Mill. cv. Tiny Tim) during maturation and ripening. Fruit at various stages exposed for 24 h at 22°C to 10 μl/l of purified (14) $^{14}C_2H_4$ (120 mCi/mmol). Number of days after flowering were as follows: dark green 30–35, green-orange, 35–36; orange, 36–38; orange-red, 38–41; red, 41–44. The inverse relationship between ethylene oxidation and natural ethylene production suggests that the build-up in internal ethylene (20–50 μl/l) may have significantly diluted the specific activity of the applied $^{14}C_2H_4$

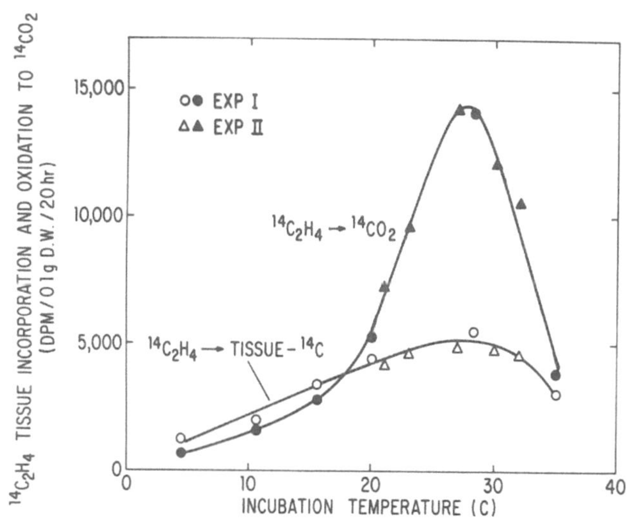

Fig. 3. Effect of temperature on $^{14}C_2H_4$ oxidation and on tissue incorporation. Excised 10-mm pea tips from 4-day-old etiolated seedlings were equilibrated for 1 h at various temperatures and then exposed to 3.5 μl/l of purified $^{14}C_2H_4$ (120 mCi/mmol) for 20 h (14)

of abscission and occurred simultaneously with a known increase in ethylene responsiveness.

Finally, a relatively close association exists between peaks in natural ethylene production and ethylene metabolism, but in no case have the two peaks coincided. In morning glory flowers (17) a peak of natural ethylene production lagged slightly behind the first large peak in ethylene oxidation, in pea seedlings (13) the opposite was true, while in cut carnations (16) peaks in ethylene metabolism occurred both before and after the large surge in ethylene production.

Temperature

Ethylene metabolism, especially the oxidation component, is very temperature-dependent. For example, in pea epicotyl tips oxidation increased 4.6-fold as the temperature was raised from 15°C to 25°C (Fig. 3). Over a 10°C lower range, the increase was 3.0-fold. Above 28°C, both activities declined. As reported earlier (15), brief high temperature treatments of 80°C or above destroy activity in two-day-old seedlings. The differences in the effects of temperature on oxidation and tissue incorporation of ethylene are further evidence that the two processes are independent.

Metabolic Inhibitors

Many types of data suggest that the ethylene metabolic system is delicate and tightly coupled to normal cellular metabolism. For example, as shown in Fig. 4, metabolic inhibitors have been found strongly to inhibit ethylene metabolism.

Oxygen

Anaerobic conditions also strongly inhibit ethylene metabolism (15). In a recent study with pea epicotyl tips, O_2 concentrations below 10% were found to rapidly limit both tissue incorporation and ethylene oxidation, with the latter being the more sensitive. At 10% O_2, tissue incorporation in pea epicotyl tips is almost O_2-saturated whereas ethylene oxidation requires 20% before maximum rates are attained.

Metal Chelators, CS_2, and COS

Chelating agents inhibit ethylene metabolism, suggesting the possible involvement of metal ions. Of the three tested as shown in Fig. 5, diethyldithiocarbamic acid (DTC) was the most inhibitory, follwed by 8-hydroxyquinoline (8-HQ). In contrast, ethylenediaminetetraacetic acid (EDTA) showed only a modest inhibition of oxidation as compared with the other two chelators (38% compared with $>$ 78% at 1 mM).

Light

Fluorescent or incandescent light has very little effect on ethylene metabolism in pea epicotyls when provided either during incubation or as a 12-h pretreatment to intact

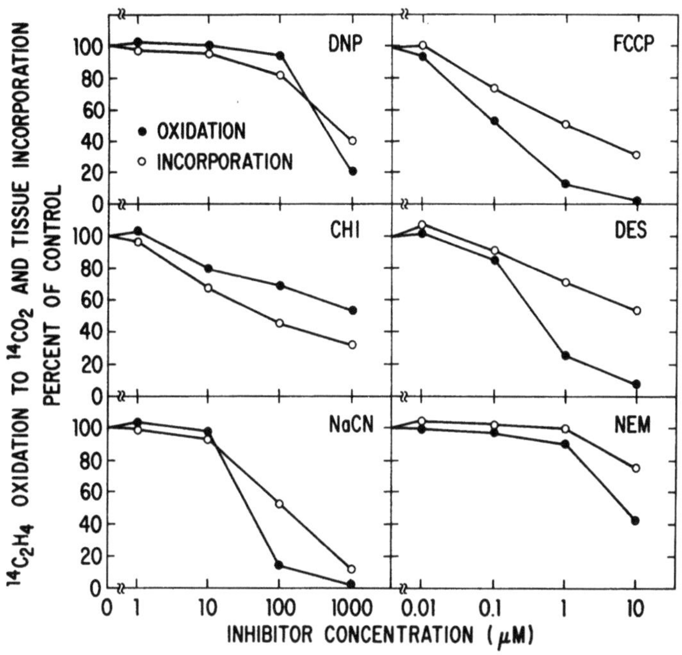

Fig. 4. Effect of metabolic inhibitors on $^{14}C_2H_4$ oxidation and on tissue incorporation. *DNP* 2,4-dinitrophenol; *CHI* cycloheximide; *NaCN* Na cyanide; *FCCP* carbonyl cyanide p-trifluoromethoxyphenylhydrazone; *DES* diethylstilbestrol; *NEM* N-ethylmaleimide. Excised pea tips from 5-day-old etiolated seedlings were treated with the various inhibitors and exposed to 2.5 μl/l of purified $^{14}C_2H_4$ (120 mCi/mmol) for 20 h at 24°C (14)

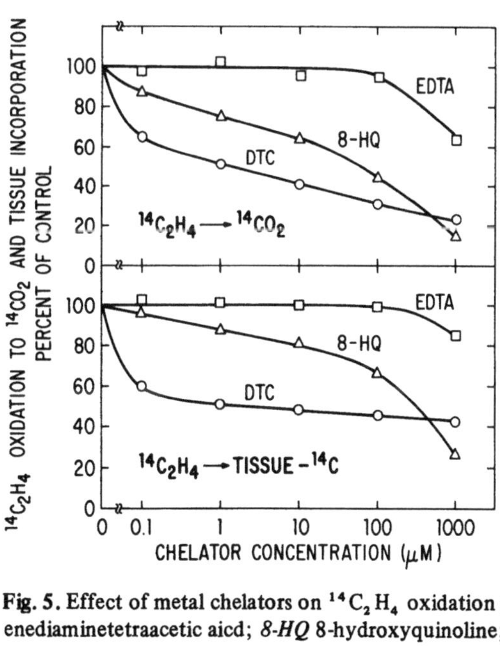

Fig. 5. Effect of metal chelators on $^{14}C_2H_4$ oxidation and on tissue incorporation. *EDTA* ethylenediaminetetraacetic aicd; *8-HQ* 8-hydroxyquinoline; *DTC* diethyldithiocarbamic acid. Excised pea hooks from 4-day-old seedlings exposed to 5 μl/l of purified $^{14}C_2H_4$ (120 mCi/mmol) for 20 h at 23°C (14)

seedlings which causes considerable greening of the tissue. Little is known at this time about the longer-term effects of light on ethylene metabolism.

Hormones

Like light, the plant hormones do not appear to have any direct effect on ethylene catabolism. Initially, IAA was thought to inhibit ethylene degradation but since this effect disappeared when auxin-induced ethylene production was inhibited with cobalt, it now appears the effect was simply due to a dilution of the specific activity of the applied $^{14}C_2H_4$. Longer-term, more indirect effects of hormone treatments have been examined for auxin and abscisic acid on cotton abscission (19). Auxin delayed both ethylene oxidation and abscission, while abscisic acid had the opposite effect.

Ethylene Concentration

Increasing the $^{14}C_2H_4$ concentration from 0.2 to 100 µl/l progressively increases the rate of both oxidation and tissue incorporation (15, 20). Below about 0.4 µl/l, tissue incorporation occurs at a greater rate than oxidation, whereas at higher concentrations oxidation gradually exceeds tissue incorporation until at 10 µl/l it is about 3 times greater. These results provide still further evidence for the independence of the two processes.

Ethylene Tissue Metabolites

Ethylene

The majority of the $^{14}C_2H_4$ metabolites in pea seedlings are readily extractable in 50%–80% ethanol, and upon ion exchange fractionation nearly all the label can be recovered in either the neutral or basic fractions. Paper or thin layer chromatography can be used to separate these fractions further into several basic and two neutral compounds (21). Since short-term experiments have demonstrated that the neutral metabolites are formed first, and in some cases (e.g., carnation ovary tissue) are the only significant metabolites formed, most of the metabolite identification work has centered on this fraction.

Good separation of the two neutral compounds, designated X ($R_f = 0.52$) and Y ($R_f = 0.65$), can be obtained (21) with descending paper chromatography and a butanol/acetic acid/water (3/3/2 v/v) solvent system. The chemical identification of Y was attempted first because pulse-chase experiments had indicated it was formed before X. Y was chromatographically compared to a large number of neutral standards including oxidation products of ethylene. In liquid, gas, paper, and thin layer chromatography with various solvent systems, it cochromatographed precisely with authentic ethylene glycol. Extraction of nearly 3 kg of ethylene-treated pea epicotyl tips followed by preparative high-performance liquid chromatography provided enough pure Y for its identification as ethylene glycol by gas chromatography-mass spectrometry (30).

In subsequent work X was identified as the glucose conjugate of ethylene glycol (Fig. 6).

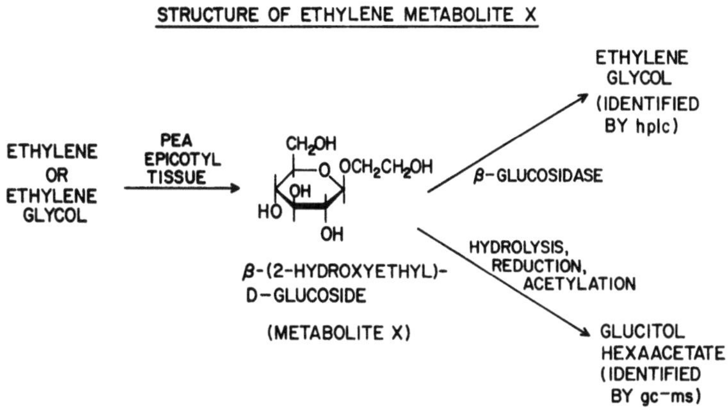

Fig. 6. Structure of ethylene metabolite X as determined by gas chromatography-mass spectrometry and hydrolysis by β-glucosidase treatment

Propylene

Similar studies (22) with propylene have demonstrated that it too is metabolized to water-soluble neutral tissue metabolites. Oxidation also occurs but, in contrast to ethylene, at a slower rate than tissue incorporation. With the solvent system used to separate ethylene glycol (Y) from its glucose conjugate (X), the neutral propylene metabolites were separated into three compounds designated A ($R_f = 0.76$), B ($R_f = 0.68$), and C ($R_f = 0.58$). Gas chromatography combined with mass spectrometry has been used to rigorously characterize A as 1,2-propanediol (30); C is its glucose conjugate, while B is at present unknown. From its overall metabolism and the fact that propylene effectively competes in ethylene metabolism, it seems certain that the same metabolic system is involved in the formation of ethylene- and propylene metabolites.

Possible Physiological Role

Ethylene Removal

By analogy with the physiological chemistry of other plant hormones, ethylene metabolism might be expected to serve as a mechanism for ethylene inactivation, but this does not appear to be the case. First, only a small percentage of the ethylene normally synthesized by the tissue is metabolically degraded. With morning glory flowers, for example (17), less than 0.2% of the ethylene produced is metabolized even during peak periods of metabolism. Secondly, when ethylene metabolism is severely and selectively inhibited, there is not a concomitant increase in the amount of ethylene produced by the tissue (unpublished data). If metabolism were removing a significant portion of the ethylene being synthesized, such an increase would be expected. Thirdly, the ability of the tissue to metabolize ethylene seldom corresponds with maximum rates of ethylene production (13, 16, 17) (Fig. 2) when ethylene removal or detoxification would seem most needed; and finally, such an inactivation system would seem to be superfluous since ethylene rapidly diffuses out of the tissue.

Mode of Action

What then might be the purpose of this metabolic system? The possibility that it is an essential part of the ethylene action mechanism is currently being explored.

It is worthwile to consider the close interrelationships that have been observed between ethylene metabolism, action, and biosynthesis. As the ethylene biosynthetic pathway is activated to initiate or modify development, ethylene metabolism and tissue responsiveness appear as closely associated phenomena. This was certainly the case with flower senescence in morning glory (17), where ethylene oxidation appeared and increased rapidly just as the petals became responsive to the rapidly increasing amounts of ethylene being synthesized by the tissue. Similarily, ethylene oxidation increased several-fold prior to cotton leaf abscission (19), and it was inhibited when abscission was delayed with auxin or stimulated when abscission was accelerated with ABA.

More convincing studies have been conducted with antiethylene agents like $Ag+$ and CO_2, both of which significantly inhibit ethylene metabolism. In peas for example (20), a striking quantitative relationship was found between the ability of $Ag+$ to counteract ethylene-induced growth retardation and to decrease tissue incorporation. A treatment with 100 mg/liter of $AgNO_3$ overcame nearly 50% of the growth inhibition caused by 0.2 μl/l of ethylene and inhibited tissue incorporation by the same amount. As the concentration of ethylene was increased, the effectiveness of the $AgNO_3$ treatment diminished in a remarkably parallel fashion in the two processes. Interestingly, $Ag+$ was highly selective in its effect since tissue incorporation was inhibited, but ethylene oxidation was unaffected. It seems highly unlikely that ethylene metabolism would respond to $Ag+$ in this unique manner if metabolism were not in some way linked to ethylene action.

Like $Ag+$, CO_2 (70%) was also found significantly to inhibit ethylene metabolism (20), but unlike $Ag+$ only ethylene oxidation was inhibited. These selective inhibitory effects of $Ag+$ and CO_2 clearly substantiate separate pathways for ethylene oxidation and tissue incorporation. When either pathway was inhibited, ethylene action was also inhibited, hence both pathways appear necessary for the full expression of an ethylene response. The functional relationship between the two pathways, however, is not yet clear.

It is interesting to note that many features of the ethylene metabolism model for ethylene action are consistent with the earlier proposal by Burg and Burg (23, 24) that ethylene acts by binding to a metalloenzyme, causing enzyme activation and a conversion of some unknown substrate to labile product. In their original scheme, they proposed that O_2 must interact directly with the metal site. Although this scheme was conceptually different in that ethylene underwent no permanent metabolic change, it nevertheless had several striking similarities with the ethylene metabolism idea for ethylene action. If in their scheme ethylene were oxidized during activation, rather than being released unchanged from the metalloenzyme-ethylene-O_2 complex after activation, the two schemes would be largely reconciled into one.

The metal involved in this oxidative process is thought to be copper (25). This deduction is supported by the known tendency of ethylene to form complexes with copper (26), the sensitivity of this system to Cu chelators (Fig. 5), COS and CS_2 (16), and the fact that in nonbiological aqueous systems Cu-C_2H_4 complexes can undergo oxidation to yield ethylene oxide, a likely precursor of ethylene glycol (27). Moreover, ethylene oxide has been shown to be a major product of ethylene metabolism in *Vicia faba* L. cotyledons (28), and we have very recently found that the radioactivity trapped in the NaOH in our work was derived from labeled ethylene converted by the tissue to ethylene oxide and CO_2.

Surprisingly, ethylene oxide has been found to act synergistically with ethylene in the "triple response" of peas (unpublished data) rather than antagonistically as previously reported (29). These studies raise the intriguing possibility that some aspect of ethylene action involve the controlled alkylation of key cellular components via ethylene oxide. This idea is currently being investigated. A general scheme illustration possible mechanisms for ethylene oxide, CO_2 and glycol formation, and the interaction of both pathways with ethylene action is shown in Fig. 7.

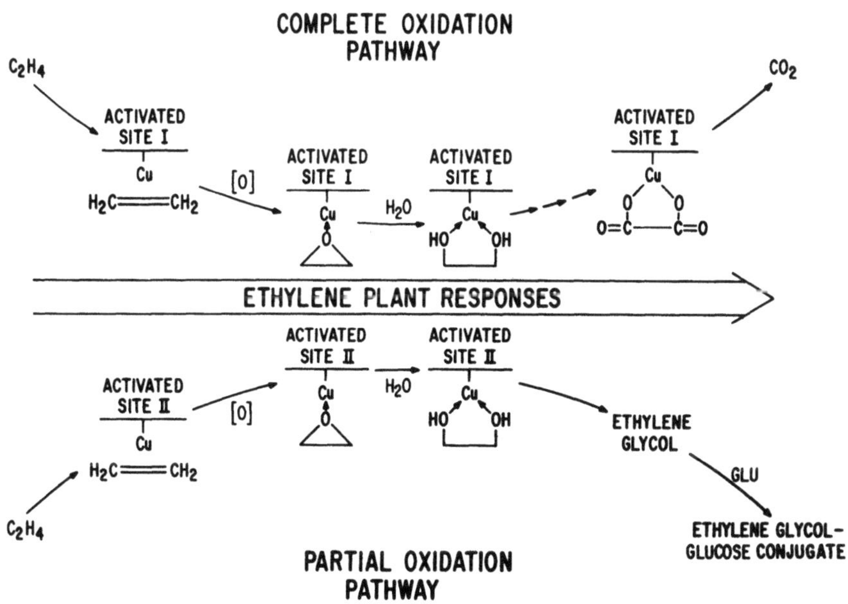

Fig. 7. Scheme illustrating possible mechanisms of ethylene oxide, CO_2, and ethylene glycol formation in plant tissues

References

1. Buhler, D.R., Hansen, E., Wang, C.H.: Nature (Lond.) *179*, 48–49 (1957)
2. Hall, W.C., Miller, C.S., Herrero, F.A.: 4th Plant Growth Regul. Proc. Int. Conf. 1959. *4*, 751–773 (1961)
3. Jansen, E.F.: J. Biol. Chem. *238*, 1552–1555 (1963)
4. Jansen, E.F.: J. Biol. Chem. *239*, 1664–1667 (1964)
5. Jansen, E.F.: 1st Food Sci. Technol. Proc. Int. Congr. 1962. *1*, 475–481 (1969)
6. Shimokawa, K., Kasai, Z.: Agric. Biol. Chem. *32*, 680–682 (1968)
7. Shimokawa, K., Yokoyama, K., Kasai, Z.: Mem. Res. Inst. Food Sci. Kyoto *30*, 1–7 (1969)
8. Behmer, M.: Klosterneuderg, Aust. Höhere Bundeslehr- und Versuchsanst. f. Wein-, Obst- und Gartenbau Ser. B. Obst Garten Mitt. *8*, 257–273 (1958)
9. Carlton, B.C., Peterson, C.E., Tolbert, N.E.: Plant Physiol. *36*, 550–552 (1961)
10. Abeles, F.B.: In: Ethylene in Plant Biology, p. 302. London, New York: Academic Press 1973
11. McGlasson, W.B.: In: Biochemistry of Fruits and Their Products. Food Sci. Tech. Monogr. Ser. London, New York: Academic Press 1970
12. Varner, J.E., Ho, D.T.: In: Plant Biochemistry. Bonner, J., Varner, J.E. (eds.), p. 925. London, New York: Academic Press 1976
13. Beyer, E.M., Jr.: Nature (Lond.) *255*, 144–147 (1975)
14. Beyer, E.M., Jr.: Plant Physiol. *55*, 845–848 (1975)
15. Beyer, E.M., Jr.: Plant Physiol. *56*, 273–278 (1975)
16. Beyer, E.M., Jr.: Plant Physiol. *60*, 203–206 (1977)
17. Beyer, E.M., Jr., Sudin, O.: Plant Physiol. *61*, 896–899 (1978)
18. Kende, H., Hanson, A.D.: Plant Physiol. *57*, 523–527 (1976)
19. Beyer, E.M., Jr.: Plant Physiol. *63* (in press, 1980)
20. Beyer E.M., Jr.: Plant Physiol. *63*, 169–173 (1979)
21. Giaquinta, R., Beyer, E.M., Jr.: Plant Cell Physiol. *18*, 141–148 (1977)
22. Beyer, E.M., Jr.: Plant Physiol. *61*, 893–895 (1978)
23. Burg, S.P., Burg, E.A.: Science *148*, 1190–1196 (1965)
24. Burg, S.P., Burg, E.A.: Plant Physiol. *42*, 144–152 (1967)
25. Beyer, E.M., Jr.: Plant Physiol. *58*, 268–271 (1976)
26. Coates, G.E., Green, M.L.H., Wade, K.: In: Organometallic Compounds, Vol. II, p. 376. London: Methuen
27. Buxton, G.V., Green, J.C., Sellers, R.M.J.: J. Chem. Soc. Dalton Trans. 2160–2165 (1976)
28. Jerie, P.H., Hall, M.A.: Proc. R. Soc. London Ser. B *200*, 87–94 (1978)
29. Asen, S., Lieberman, M.: Florists Rev. *131*, 27–30 (1963)
30. Blomstrom, D.C., Beyer, E.M. Jr.: Nature (Lond.) *283*, 66–68 (1980)

Mechanism and Regulation of Ethylene Biosynthesis [1]

S.F. YANG, D.O. ADAMS, C. LIZADA, Y. YU, K.J. BRADFORD, A.C. CAMERON, and N.E. HOFFMAN [2]

Introduction

Ethylene is a plant hormone which initiates fruit ripening and regulates many aspects of plant growth and development (1). The role of methionine as a biological precursor of ethylene was first reported by Lieberman et al. (2), and now it is well established that methionine is the common precursor of ethylene throughout the diversity of higher plant tissues (3). In this conversion, C-1 of methionine is converted to CO_2, C-2 to formic acid and C-3,4 to ethylene. The sulfur atom, however, is retained in the tissue (3, 4). Since the conversion of methionine to ethylene is greatly inhibited by uncouplers of oxidative phosphorylation, Burg (4), and Murr and Yang (5) have proposed that SAM , formed from methionine and ATP, is an intermediate between methionine and ethylene. Adams and Yang (6) have presented evidence showing that MTA and its hydrolysis product, MTR, are derived from the CH_3S group of methionine during its conversion to ethylene and have substantiated the role of SAM as an intermediate in the biosynthesis of ethylene from methionine. More recently, Adams and Yang (7) have identified ACC as an intermediate between SAM and ethylene. In this paper we summarize our recent results pertinent to the role of ACC in ethylene biosynthesis and its regulation.

ACC, a Metabolic Intermediate in the Conversion of Methionine to Ethylene

It has been established for some time that endogenous ethylene production, as well as the conversion of methionine to ethylene, ceases in plant tissues placed in an anaerobic atmosphere, and that a surge of ethylene production occurs upon returning the tissue to air. These observations are interpreted to indicate that an intermediate accumulates during anaerobic incubation and is subsequently converted to ethylene upon exposure to oxygen (8). Adams and Yang (7) have recently examined the metabolism of methionine in apple tissue under these conditions. In air, L-[U-^{14}C]methionine was efficiently converted to ethylene, as reported previously; in nitrogen, however, it was not me-

---

1 Abbreviations: ACC, 1-aminocyclopropane-1-carboxylic acid; AVG, aminoethoxyvinylglycine, or 2-amino-4-(2'-aminoethoxy)-trans-3-butenoic acid; CCCP, carbonylcyanide *m*-chlorophenylhydrazone; DNP, 2,4-dinitrophenol; MTA, 5'-methylthioadenosine; MTR, 5-methylthioribose; SAM, S-adenosylmethionine

2 Department of Vegetable Crops, University of California, Davis, California 95616, USA

tabolized to ethylene but was instead converted to ACC. When apple tissues were fed with L-[methyl-^{14}C]methionine or L-[^{35}S]methionine and incubated in nitrogen, radioactivity was found largely in MTR with trace amounts in MTA. This suggests that methionine is first converted to SAM which is in turn fragmented to ACC and MTA. MTA is then hydrolyzed to MTR. The conclusion that ACC is an intermediate in the conversion of methionine to ethylene is supported by the following observations: (a) Labeled ACC was efficiently converted to ethylene by apple tissue incubated in air,

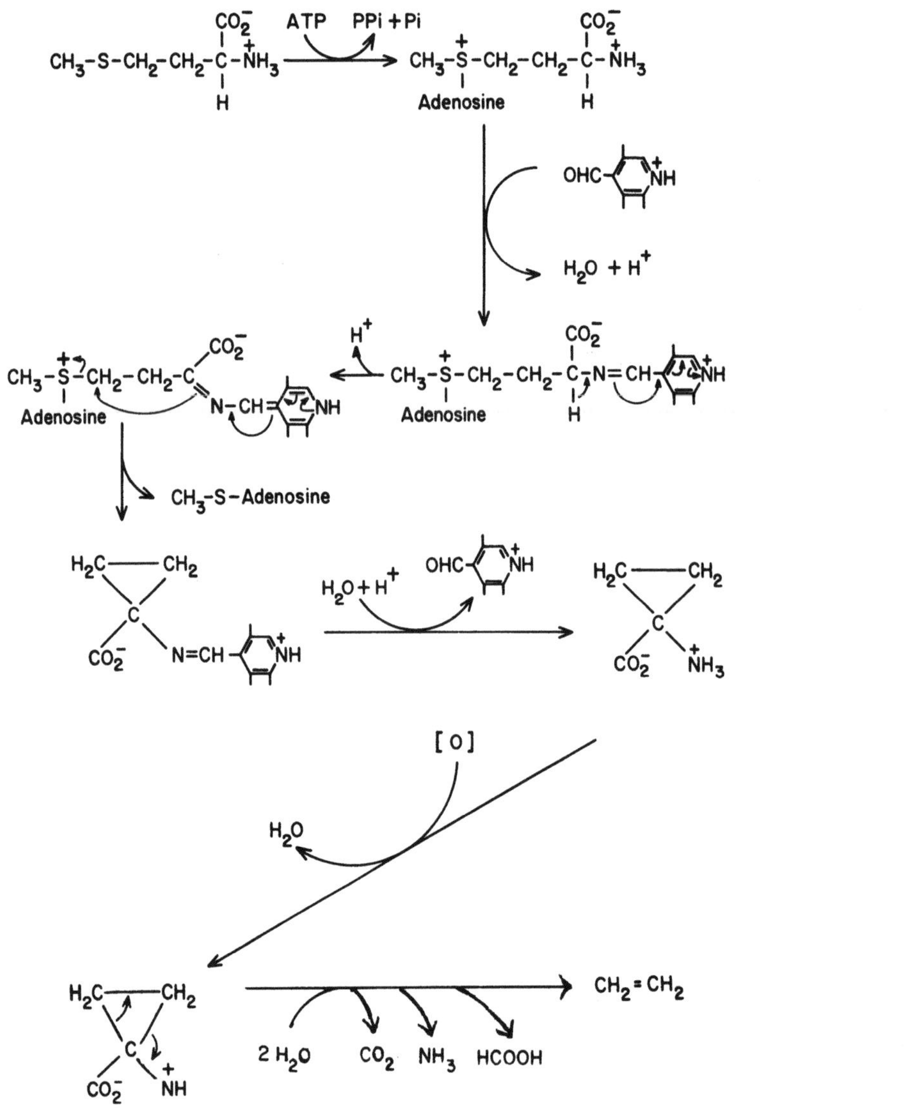

Fig. 1. A postulated mechanism for the biosynthesis of ethylene from methionine. The substituted pyridinecarboxaldehyde stands for pyridoxal phosphate. This scheme is modified from that of Adams and Yang (7)

(b) the conversion of labeled methionine to ethylene was greatly reduced in the presence of unlabeled ACC, but the conversion of labeled ACC to ethylene was little affected by the presence of unlabeled methionine, and (c) AVG, a potent inhibitor of pyridoxal-phosphate mediated enzyme reactions (9, 10), greatly inhibited the conversion of methionine to ethylene but did not inhibit conversion of ACC to ethylene. These data indicate the following sequence for the pathway of ethylene biosynthesis in apple tissue: methionine → SAM → ACC → ethylene. A postulated mechanism which accounts for these reactions is shown in Fig. 1. The first step is the activation of methionine by ATP to produce SAM, which then reacts with pyridoxal phosphate to form a Schiff base. This is in keeping with the observation that pyridoxal phosphate inhibitors interfered with the in vivo conversion of SAM to ACC (7). It is well known that pyridoxal enzymes are capable of catalyzing the elimination of the γ-substituent group along with β-H of amino acids (1,2-elimination) resulting in the formation of olefins (11). However, the conversion of SAM to ACC and MTA represents a new reaction in which the α-H is eliminated along with the γ-substituent (1,3-elimination) resulting in the formation of a cyclopropane ring. It is assumed that SAM and pyridoxal phosphate form a Schiff base. The elimination of a proton from the α-carbon of an amino acid by a pyridoxal enzyme is well documented (11). Once a carbanion is formed, an internal nucleophilic displacement reaction occurs resulting in the elimination of the MTA moiety. ACC synthase, which catalyzes the conversion of SAM to ACC, has been recently demonstrated in extracts of tomato by Boller et al. (12). As predicted by the in vivo study of Adams and Yang (7), this enzyme specifically utilized SAM as substrate and was inhibited by AVG. Although S-adenosylethionine also serves as a substrate for this enzyme, its efficiency is much lower than that of SAM; S-adenosylethionine, however, inhibits the conversion of SAM to ACC, when they are present together (13). The requirement for pyridoxal phosphate for ACC synthase activity was also confirmed (13). It is interesting to note an analogous organic reaction in which ACC was synthesized from a derivative of S-methylmethionine sulfonium salt (14). In the enzymatic reaction, the elimination of the α-H as a proton is accomplished by the pyridoxal phosphate coenzyme, while in the organic reaction it is carried out by a base. Both SAM and S-methylmethionine possess a sulfonium function in their γ-substituent group, which greatly facilitates the 1,3-elimination.

In vivo studies indicate that the conversion of ACC to ethylene requires oxygen and is insensitive to AVG inhibition. Although the enzymic conversion of ACC to ethylene catalyzed by a pea seedling extract has been reported by Konze and Kende (15), there is much doubt that the system described by them is indeed a physiological one. The chemical oxidation of substituted cyclopropylamines to ethylene via nitrenium ion intermediates is known (16). Analogous to the chemical oxidation, it is proposed that in vivo ACC is oxidized to the corresponding nitrenium ion intermediate by hydroxylation on the amino nitrogen followed by elimination of the hydroxide ion. The nitrenium ion intermediate is then fragmented into ethylene and other products by a concerted elimination process that is facilitated by the electrophilic nitrenium ion and the nucleophilic attack of water on C-1, as shown in Fig. 1. Decarboxylation of ACC prior to the formation of ethylene in vivo is ruled out, because cyclopropylamine does not serve as a precursor of ethylene in plant tissues. The formation of CO_2, formic acid, and ethylene from the carboxyl-C-1, and C-2,3, respectively, of ACC is in keeping with the results obtained with methionine as substrate in which they are derived from C-1, C-2, C-3,4 of methionine, respectively.

ACC was first isolated from perry pears and cider apples by Burroughs in 1957 (17). It is now clear that ACC plays a most significant role in fruit ripening by serving as the precursor of ethylene.

A Simple and Sensitive Assay for ACC

Because of the important role of ACC in ethylene biosynthesis, a sensitive and simple assay for ACC in plant tissues is essential. It has been reported that 1-phenylcyclopropylamine can be oxidized to ethylene and benzonitrile by NaOCl (16). While unaware of such a report, Lizada and Yang (18) observed that ACC was degraded to ethylene in good yield by NaOCl (commercial bleach solution) in the presence of Hg^{2+} and that ACC can be quantitated by assaying ethylene evolved with gas chromatography. The method is quite specific and can detect as low as 5 pmol of ACC. Routinely, the plant extracts were purified by a cation exchange resin prior to NaOCl degradation. The efficiency of the conversion of ACC to ethylene was determined by adding a known amount of authentic ACC as internal standard to another sample before degradation with NaOCl. Calculation of the amount of ACC present in samples was based on the determined conversion efficiency of the internal standards (18).

Mechanism of Auxin-Induced Ethylene Production

Auxins are known to stimulate ethylene production in a wide variety of plant tissues (1). Many of the effects of auxin on growth and on other processes, such as epinasty, hook opening, inhibition of growth, root induction and geotropism, are now attributed to its ability to induce ethylene production (1). In vegetative tissues, the rate of ethylene production is thought to be regulated by the internal level of free auxin (1). Although the regulation of ethylene production in vegetative tissues is different from that in fruit tissues, the characteristics of ethylene production in these two different types of tissues are quite similar. Both utilize methionine as the precursor (3), with ACC being the intermediate. Since the pathway of ethylene biosynthesis from methionine is now known (7), it is possible to determine at which step in the sequence auxin exerts its hormonal effect. We have therefore examined the mechanism of auxin-induced ethylene synthesis by the hypocotyls of mung bean seedlings. Although the conversion of methionine to ethylene requires the presence of auxin (3), the tissues are capable of actively converting ACC to ethylene in the absence of auxin (19). Even though IAA stimulated ethylene production about 500 times over that of the control and AVG at 10 μM completely abolished this IAA-induced ethylene production, the incorporation of label from methionine into SAM was little affected by either IAA or IAA plus AVG treatment. It is obvious that the conversion of methionine to SAM is not subject to either IAA induction or AVG inhibition. When ethylene production was induced by IAA, active conversion of methionine to ethylene, active conversion of methionine to ACC, and accumulation of endogenous ACC were observed (20). In the presence of AVG, IAA-induced ethylene production, conversion of methionine to ethylene and conversion of methionine to ACC were all abolished (20). Since auxin did not influence the conversion of methionine to SAM but caused active conversion of methionine to ACC, auxin must exert its regulatory function on a step involved in the conversion of SAM to ACC. It is known that auxin-induced ethylene production is characterized by a substantial lag period and is vulnerable to inhibitors of RNA and protein synthesis (1). These characteristics are interpreted to indicate that auxin-induced ethylene production requires synthesis of a new enzyme (1). Such a view leads to the proposal that auxin stimulates ethylene production by inducing the synthesis of ACC synthases. Indeed, ACC synthase activity was found only in the tissue treated with IAA (20). The marked increase (100 times) in the ACC level over that of the control is an expected consequence of the induced synthesis of ACC synthase in the IAA-treated tissues. Obviously, the limited quantity of ACC in the control tissue restricted ethylene production, since once ACC was provided, it was readily converted without dependence upon IAA (19). It is thus concluded that subsequent to the synthesis of methionine, the only auxin-regulated step is the conversion of SAM to ACC. Similar results were obtained with the ethylene production system in mung bean hypocotyls induced by Ca^{2+} plus cytokinin.

Regulation of Ethylene Production

Ethylene production in vivo is known to be regulated by various physiological and environmental factors. For example, ethylene production is induced during certain stages of growth, such as germination, ripening of fruits, and abscission, by wounding, disease, radiation and other physical and chemical stresses, and by treatment with IAA or other growth regulators (11).

A surge in ethylene production at the onset of ripening in climacteric-type fruits is well documented (1, 21). To determine the regulation of ethylene production during the ripening of fruits, we have examined the change in internal ACC levels during ripening, as well as the effect of exogenous application of ACC on ethylene production in preclimacteric fruits. In contrast to vegetative tissues in which ethylene production greatly increases following ACC application, preclimacteric apple and cantaloup tissues exhibit only slight increase in ethylene production following the application of ACC (D.O. Adams and M.L. Bliss, unpublished observations). These observations coupled with the previous observation from tracer studies (22) that preclimacteric fruits are unable to convert SAM to ACC, suggest that both the conversion of SAM to ACC and the conversion of ACC to ethylene are restricted in preclimacteric fruit. This is in contrast to vegetative tissues in which only the conversion of SAM to ACC is rate-limiting. Unlike the vegetative tissues, the onset of ethylene production in climacteric fruit tissues is not regulated by auxin. Thus, fruit tissue and vegetative tissue share a common pathway for ethylene biosynthesis, but the mechanism of regulation appears to be different. Inability of preclimacteric fruit tissues to synthesize ACC was further supported by our recent observations that the endogenous ACC level is near nil in preclimacteric avocado fruit, increases to 45 nmol/g at the later stage of the climacteric rise, and declines thereafter; the accumulation of ACC parallels the onset of ethylene production (23). In banana, ACC increases from near 0 to 8 nmol/g and declines several days after the ethylene has declined, while in tomato, ACC levels increase from near 0 to 9 nmol/g. It is worth noting that the ACC level in these overripe fruit tissues is high, although their ethylene production rates were quite low. Obviously, the conversion of ACC to ethylene is restricted in these tissues.

In vegetative tissue, application of ACC caused marked increases in ethylene production in various leaf, stem and root tissues, which normally produce little ethylene (19, 24). These results are in agreement with the view that the rate-limiting reaction in ethylene biosynthesis occurs at a step prior to the formation of ACC, and that the unavailability of ACC in these tissues restricts ethylene production. In vegetative tissues the rate of endogenous ethylene production is thought to be regulated by the internal levels of auxin, and application of exogenous auxin often stimulates ethylene production. However, the ethylene production of some vegetative tissues is not induced by IAA. For example, ethylene production in the leaf tissues of sugar beet and the petioles of the diageotropica mutant of tomato does not increase following the application of IAA, but marked increases in ethylene production were observed following the application of ACC [N. Aharoni, unpublished observation; (25)]. These data suggest that there must be some other stimuli which can also cause the synthesis of ACC. Increased ethylene production under various types of stresses is well documented. This trauma-induced ethylene can originate from physical, chemical, or biological stress and does not appear to be associated with the normal auxin-mediated ethylene production exhibited by young developing plant parts (1, 26). The mechanism of stress ethylene production in waterlogged plants will be discussed in the following section.

The biochemistry of ethylene formation in wounded tissues has been studied in a number of plant materials, and in all of them methionine seems to be the precursor [for recent review, see (26)]. If the pathway for wound ethylene is similar to that of ripening fruit or auxin-induced vegetative tissue, one would expect ACC to accumulate as a result of wounding. To test this hypothesis we have induced stress ethylene production in mung bean hypocotyls by chemical treatment with Cu^{2+} (27) or in citrus albedo tissue (28) and in preclimacteric avocado fruit by slicing. Both ethylene production and the level of ACC were very low immediately following the treatments, but the level of ACC increased along with the rise in ethylene production during subsequent incubation (Y. Yu, unpublished observation). It is not known how the synthesis of stress ethylene is induced, but it is clear that the pathway of stress ethylene synthesis is similar, if not identical, to other ethylene-producing systems.

Ethylene production and the conversion of methionine to ethylene are greatly inhibited by rhizobitoxine and its ethoxy analog, AVG, in a number of plant tissues (9). We have previously shown that the in vivo conversion of ACC to ethylene was insensitive to AVG inhibition (7, 19). Since AVG greatly inhibits conversion of methionine to ethylene, methionine to ACC, and auxin-dependent or wounding-dependent ACC accumulation, but does not affect the conversion of methionine to SAM, or the conversion of ACC to ethylene, it is apparent that AVG inhibits the conversion of SAM to ACC, the same step upon which auxin or wounding exerts its inductive effect. This was further confirmed by the demonstration that AVG effectively inhibited ACC synthase activity (12, 20).

Lau and Yang (29) observed that Co^{2+} inhibited ethylene production in all systems tested, including those of fruit tissues and vegetative tissues treated with IAA, kinetin, IAA plus kinetin, Ca^{2+}, kinetin plus Ca^{2+}, or Cu^{2+}. They have therefore suggested that Co^{2+} inhibited ethylene production by inhibiting an essential reaction common to all of the systems. Since Co^{2+} also inhibited the conversion of methionine to ethylene, it is concluded that Co^{2+} inhibits ethylene production at a step subsequent to methionine formation (29). To identify the step at which Co^{2+} exerts its inhibition, we have compared the effect of various concentrations of Co^{2+} on the IAA-dependent and ACC-dependent ethylene production in hypocotyls of mung bean seedlings. Co^{2+} effectively inhibited both IAA-dependent and ACC-dependent ethylene production. Since both systems are inhibited by Co^{2+} in similar fashion, it is concluded that Co^{2+} exerted its inhibitory effect at a common step, the conversion of ACC to ethylene (20). If this is the case, one may predict the endogenous ACC levels in IAA-treated tissues should increase as a result of the inhibition of ethylene production by Co^{2+} application. Indeed, in the presence of 50 μM Co^{2+}, ethylene production was inhibited 75%, but the ACC level was 40% higher than in tissues in the absence of Co^{2+}. At a higher concentration of Co^{2+} (0.5 mM), ethylene production was further diminished while ACC accumulated (20). These results are in contrast to AVG-inhibited ethylene production, in which the ACC level decreased dramatically when IAA-induced ethylene production was inhibited by AVG. These data strongly support the notion that Co^{2+} inhibits ethylene production by inhibiting the conversion of ACC to ethylene, although the mode of action of Co^{2+} in this process is unknown.

DNP, an uncoupler of oxidative phosphorylation, has long been known to inhibit ethylene production and was shown by Murr and Yang (5) to inhibit conversion of methionine to ethylene. It has been suggested that DNP acts by depriving the tissue of ATP thereby preventing incorporation of methionine into SAM. However, further studies revealed that DNP and CCCP at 50 μM effectively inhibited ethylene production and caused accumulation of endogenous ACC, but did not block the conversion of methionine to SAM; inhibition of such conversion occurred only at a much higher concentration [(22), Y. Yu, unpublished]. These results suggest that these uncouplers exert their inhibitory effect by interfering with the conversion of ACC to ethylene. Such a view was further supported by the observations that DNP and CCCP also inhibited ACC-dependent ethylene production by mung bean hypocotyls. However, it has not been determined whether the inhibition is due to their action in uncoupling oxidative phosphorylation or in disrupting membrane function. Ethylene evolution by plant parts is well known to be greatly reduced when the temperature is raised above 35°C to 40°C. By studying the dependence of ethylene production by mung bean hypocotyls, it was similarly concluded that the conversion of ACC to ethylene is vulnerable to high temperature.

Xylem Transport of ACC in Waterlogged Plants

It is well known that waterlogging of tomato plants results in epinasty of the petioles, adventitious root formation, and leaf senescence. Recent studies have established that waterlogging creates anaerobic conditions in the root zone, which trigger elevated synthesis of ethylene in the shoot. The elevated internal ethylene levels in turn stimulate epinastic growth, adventitious root formation and senescence (30, 31). It has been suggested that a "signal" is synthesized in the anaerobic roots and transported to the shoot where it stimulates ethylene synthesis (30, 31). Since ACC accumulates under anaerobic conditions and is readily converted to ethylene in aerobic tissues, ACC is a logical candidate to serve as the "signal". To test this hypothesis, xylem sap was collected from detopped tomato plants. ACC in the sap or in the root was quantitated by the NaOCl reagent as described in a previous section, and its chemical identity was verified by paper chromatography. ACC levels in both roots and xylem sap were very low in the control plants, but increased markedly in response to waterlogging or root anaerobiosis (Fig. 2). These data indicate that ACC is synthesized in the anaerobic root

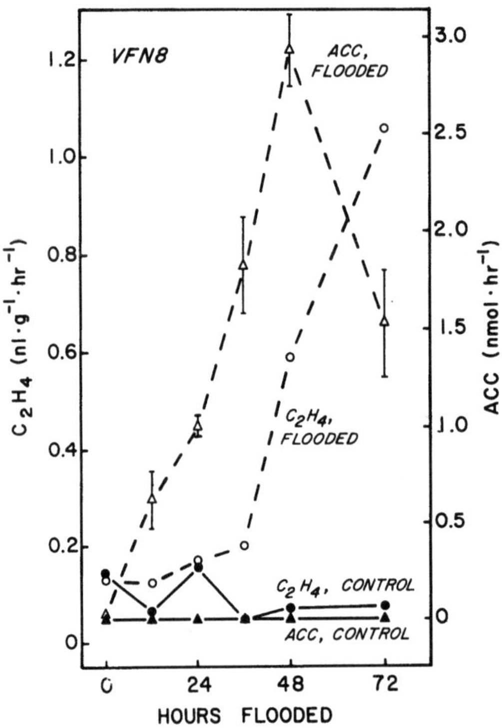

Fig. 2. Time course of the appearance of ACC in the xylem sap and changes in petiolar ethylene production following flooding of tomato plant (32)

and exported to the shoot through the xylem. Furthermore, the appearance of ACC in the xylem sap of flooded plants preceded both the increase in ethylene production and epinastic growth. Plants flooded and then drained showed a rapid, simultaneous drop in ACC flux and ethylene synthesis rate. To confirm that ACC could indeed be responsible for the increased synthesis of ethylene and epinasty in the shoot, we supplied ACC through the cut stem of tomato shoots at concentrations comparable to those found in xylem sap. As predicted, these treatments caused increased ethylene production and epinasty (32).

There are two mechanisms which could account for the accumulation of ACC in anaerobic roots. First, lack of O_2 prevents conversion of ACC to ethylene, so that the ethylene normally synthesized in the root would instead accumulate as ACC. Alternatively, anaerobiosis may acutally stimulate the ethylene biosynthetic pathway, resulting in ACC accumulation. The normal rate of ethylene synthesis in the root is less than the rate of shoot ethylene synthesis in response to root anaerobiosis (31). Yet, the ACC exported from anaerobic roots is adequate to sustain the elevated shoot ethylene synthesis in response to anaerobic root treatment. Thus, a mere blockage of ethylene synthesis in the root by anaerobiosis is insufficient to account for the accelerated ACC transport. Anaerobic stress must therefore not only block conversion of ACC to ethylene, but also stimulate ACC synthesis. However, the regulatory mechanism of ACC synthesis in such anaerobically stressed tissues remains to be elucidated.

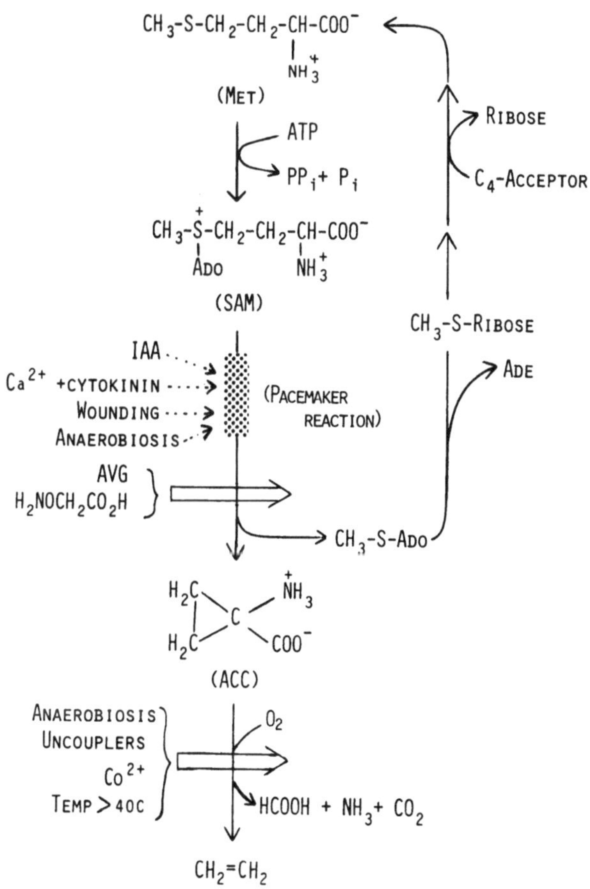

Fig. 3. Regulation of ethylene biosynthesis. - - - - →: induction of synthesis of the enzyme; ⟹: inhibition of the conversion. *Met, Ado,* and *Ade* stand for methionine, adenosine, and adenine, respectively

The classical definition of a hormone is an endogenous compound which is synthesized at one site and is transported to another site where it exerts a physiological effect. Due to its gaseous nature, ethylene exerts a physiological effect only at or near a site where it is synthesized. Thus, this classical definition does not apply to ethylene. The present work, however, has shown that the ethylene precursor, ACC, is synthesized in the anaerobic root, is readily transported, and exerts its effect through conversion to ethylene in the aerobic shoot.

The current status of knowledge with respect to pathway and regulation of ethylene biosynthesis is summarized in Fig. 3. Finally, we wish to suggest "ethenine" [ethene (ethylene) + ine (to denote an amino aicd)] as the common name for ACC. The term gives an indication of its chemical nature as well as its intimate relation to ethylene.

Acknowledgment. Supported by a research grant from the National Science Foundation (PCM 78-09278).

References

1. Abeles F.B.: Ethylene in Plant Biology. London, New York: Academic Press 1973
2. Lieberman, M., Kunishi, A.T., Mapson, L.W., Wardale, D.A.: Plant Physiol. *41*, 376–382 (1966)
3. Yang, S.F.: The Chemistry and Biochemistry of Plant Hormones. Runeckles, V.C., Sondheimer, E., Walton, D.C. (eds.), pp. 131–164. London, New York: Academic Press 1974
4. Burg, S.P.: Proc. Natl. Acad. Sci. USA *70*, 591–597 (1973)
5. Murr, D.P., Yang, S.F.: Plant Physiol. *55*, 79–82 (1975)
6. Adams, D.O., Yang, S.F.: Plant Physiol. *60*, 892–896 (1977)
7. Adams, D.O., Yang, S.F.: Proc. Natl. Acad. Sci. USA *76*, 170–174 (1979)
8. Burg, S.P., Thimann, K.V.: Proc. Natl. Acad. Sci. USA *45*, 335–344 (1959)
9. Lieberman, M., Kunishi, A.T., Owens, L.D.: In: Facteurs et Regulation de la Maturation des Fruits, pp. 161–170. Paris: C.N.R.S. 1975
10. Rando, R.R.: Science *185*, 320–324 (1974)
11. Davis, L., Metzler, D.E.: In: The Enzymes. Boyer, P.D. (ed.), Vol. 7, pp. 33–74. London, New York: Academic Press 1972
12. Boller, T., Herner, R.C., Kende, H.: Planta *145*, 293–303 (1979)
13. Yu, Y., Adams, D.O., Yang, S.F.: Arch. Biochem. Biophys. *198*, 280–286 (1979)
14. Rich, D.H., Tam, J.P.: Synthesis *1978*, 46
15. Konz, J.R., Kende, H.: Planta *146*, 293–301 (1979)
16. Hiyama, T., Koide, H., Nozaki, H.: Bull. Chem. Soc. Jpn. *48*, 2918–2921 (1975)
17. Burroughs, L.F.: Nature (Lond.) *179*, 360–361 (1957)
18. Lizada, C., Yang, S.F.: Anal. Biochem. *100*, 140–145 (1979)
19. Yu, Y., Adams, D.O., Yang, S.F.: Plant Physiol. *63*, 589–590 (1979)
20. Yu, Y., Yang, S.F.: Plant Physiol. *64*, 1074–1077 (1979)
21. Pratt, H.K., Goeschl, J.D.: Annu. Rev. Plant Physiol. *20*, 541–584 (1969)
22. Adams, D.O.: Ph.D. Thesis, University of California, Davis (1979)
23. Hoffman, N.E., Yang, S.F.: J. Amer. Soc. Hort. Sci. *100* (in press, 1980)
24. Cameron, A.C., Fenton, C.A.L., Yu, Y., Adams, D.O., Yang, S.F.: HortScience *14*, 178–180 (1979)
25. Bradford, K.J., Yang, S.F.: Plant Physiol. *65*, 322–326 (1980)

26. Yang, S.F., Pratt, H.K.: In: Biochemistry of Wounded Plant Tissues. Kahl, G. (ed.), pp. 595–622. Berlin: de Gruyter 1978
27. Lau, O., Yang, S.F.: Plant Physiol. *57*, 88–92 (1976)
28. Hyodo, H.: Plant Physiol. *59*, 111–113 (1977)
29. Lau, O., Yang, S.F.: Plant Physiol. *58*, 114–117 (1976)
30. Jackson, M.B., Campbell, D.J.: New Phytol. *76*, 21–29 (1976)
31. Bradford, K.J., Dilley, D.R.: Plant Physiol. *61*, 506–509 (1978)
32. Bradford, K.J., Yang, S.F.: Plant Physiol. *65*, 322–326 (1980)

Enzymes of Ethylene Biosynthesis [1]

H. KENDE [2], J.R. KONZE [3], and T. BOLLER [4]

In 1964, Lieberman and Mapson (1) discovered that free radicals, generated through the reaction of copper and ascorbate in aqueous solution, oxidized methionine to yield ethylene. Model systems of ethylene synthesis based on similar principles were subsequently described by others (2, 3). In 1966, Lieberman et al. (4) reported that methionine was also the precursor of ethylene in apples. Although this finding was confirmed for many other plant tissues [see (5)], the mechanism by which methionine was converted to ethylene remained unknown until very recently. It was not clear whether ethylene synthesis in vivo was a result of a free-radical-mediated reaction with methionine as the direct precursor of ethylene or whether methionine was converted to ethylene in one or several enzyme-mediated steps. In vivo conversion of methionine to ethylene through the action of free radicals was mainly advocated by Lieberman (5), largely on the grounds that in vivo ethylene synthesis was inhibited by free-radical scavengers such as n-propyl gallate and benzoate. Involvement of at least one enzyme in the generation of ethylene from methionine was first suggested by Burg (6), who hypothesized that S-adenosylmethionine (SAM) might be an intermediate in the pathway of ethylene biosynthesis.

Results of recent work strongly indicate that ethylene is not formed in plants by a direct reaction of methionine with free radicals (7) but rather is derived from methionine via two intermediate products. First, Adams and Yang (8) and Konze and Kende (7) presented evidence supporting Burg's hypothesis that ethylene synthesis proceeded via SAM. Second, Adams and Yang (9) and, independently, Lürssen et al. (10, 11) discovered that 1-aminocyclopropane-1-carboxylic acid (ACC) was a precursor of ethylene, and both groups postulated that ACC was derived from SAM. Therefore, the following pathway of ethylene synthesis was suggested: methionine → SAM → ACC → ethylene.

Until recently, no ethylene-synthesizing system had been isolated from plant tissues that exhibited characteristics similar to those of the in vivo ethylene-generating system. In the following, we shall summarize recent results on the isolation and char-

1 Abbreviations. ACC, 1-aminocyclopropane-1-carboxylic acid; AVG, amino-ethoxyvinylglycine, the aminoethoxy analog of rhizobitoxine, L-2-amino-4-(2-aminoethoxy)-trans-3-butenoic acid; SAM, S-adenosylmethionine

2 MSU-DOE Plant Research Laboratory, Michigan State University, East Lansing, MI 48824, USA

3 Present address: Biochemisches Labor, Institut für Botanik, Technische Universität, Arcisstr. 21, D-8000 München, FRG

4 Present address: Botanisches Institut der Universität Basel, Schönbeinstr. 6, CH-4056 Basel, Switzerland

acterization of three enzymes or enzyme systems that account for the three proposed steps of ethylene biosynthesis from methionine.

Methionine Adenosyltransferase

During our work on the pathway of ethylene synthesis in morning glory flower tissue and etiolated pea stem sections, we found that selenomethionine and selenoethionine enhanced ethylene production several-fold while methionine applied at the same concentration had very little if any effect (12). Both selenoamino acids were converted to ethylene by the same route as methionine and appeared to be better substrates than methionine. We made use of these properties of the seleno analogs of methionine to explore which of the proposed mechanisms of ethylene synthesis operated better with selenomethionine as substrate rather than with methionine.

In three model systems involving oxidation of the amino acid by free radicals, selenomethionine was much less effective as an ethylene precursor than methionine (7). Based on these results, we concluded that in vivo ethylene synthesis was not the result of a direct reaction between free radicals and methionine. Next, we investigated whether methionine adenosyltransferase (ATP: L-methionine S-adenosyltransferase, EC 2.5.1.6.), the enzyme that forms SAM from methionine, operated more efficiently with selenomethionine as substrate than with methionine (7). We found that the rate of conversion of selenomethionine to Se-adenosylselenomethionine was higher than that of methionine to SAM. An investigation of the kinetic parameters of this enzyme showed that the V_{max} for selenomethionine was twice that for methionine and that the K_m for methionine was lower than the K_m for selenomethionine. In a typical experiment, the V_{max} for selenomethionine was 0.31 μmol h^{-1} mg $protein^{-1}$ and the K_m 0.61 mM, while the V_{max} for methionine was 0.16 μmol h^{-1} mg $protein^{-1}$ and the K_m 0.37 mM. As a consequence of these properties of methionine adenosyltransferase, we expected that methionine added to the reaction mixture containing high levels of selenomethionine would slow the rate of product formation to a rate approximating the V_{max} for methionine. Under these conditions methionine would act as a competitor of selenomethionine. It would be preferentially bound to the enzyme while its conversion to the reaction product would be relatively slower. Results of testing this hypothesis are given in Fig. 1. This figure shows the V_{max} for methionine and selenomethionine and the effect of methionine at different concentrations on the reaction rate at a constant level of selenomethionine. As predicted, addition of methionine decreased the rate of product formation.

In order to test whether the stimulatory effect of selenomethionine on the activity of methionine adenosyltransferase and on ethylene synthesis exhibited similar characterstics, we examined the effect of added methionine on selenomethionine-enhanced ethylene formation in senescing flower tissue of the morning glory *(Ipomoea tricolor)* (7). The high levels of ethylene formed in the presence of selenomethionine were reduced to the level of ethylene formed by methionine-treated tissue when methionine was added to the incubation medium containing selenomethionine (Fig. 2). Thus, methionine overrides the enhancement of ethylene synthesis by selenomethionine, just as it overrides the selenomethionine-enhanced activity of methionine adenosyl-

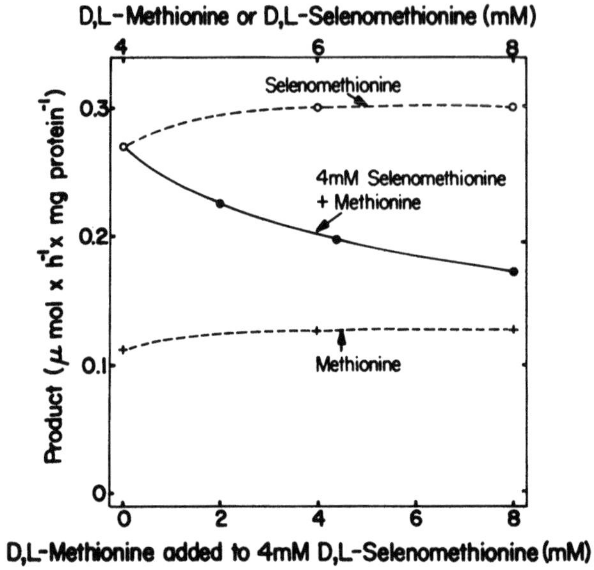

Fig. 1. Reduction of selenomethionine-dependent methionine adenosyltransferase activity by methionine in homogenates of *Ipomoea tricolor* flower buds. SAM formed with D,L-methionine as substrate (+- - - -+); SeAM formed with selenomethionine as substrate (○- - - -○); SAM + SeAM formed with 4 mM D,L-selenomethionine and increasing concentrations D,L-methionine as substrates (●- - - -●). From (7)

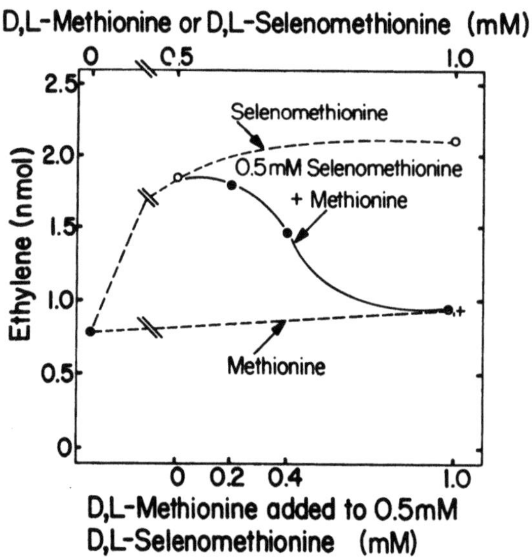

Fig. 2. Reduction of selenomethionine-enhanced ethylene production by methionine in rib segments of *Ipomoea tricolor* flowers during senescence. Rib segments were excised between 4:00 and 5:00 P.M. on the day before flowering and were incubated overnight and throughout the day of flowering in batches of 15 in 25-ml Erlenmeyer flasks on distilled H_2O (●); 0.5 mM and 1.0 mM D,L-selenomethionine (○——○); 1 mM D,L-methionine (+); 0.5 mM D,L-selenomethionine plus 0.2–1.0 mM D,L-methionine (●— — —●). Figure shows the total amount of ethylene produced by rib segments between 10:30 A.M. and 11:15 P.M. From (7)

transferase. The similar kinetic characteristics of both methionine adenosyltransferase and of the ethylene-forming system are evidence that methionine adenosyltransferase is involved in ethylene synthesis and that SAM is an intermediate in the pathway of ethylene formation.

Enzymatic Formation of 1-Aminocyclopropane-1-Carboxylic Acid (ACC)

A prerequisite for finding the ACC-forming enzyme was the availability of a sensitive and rapid assay for this amino acid. We developed such an assay based on the liberation of ethylene from ACC at pH 11.5 in the presence of pyridoxal phosphate, $MnCl_2$ and H_2O_2 (13). Using a partially purified homogenate from tomato fruit, we found an

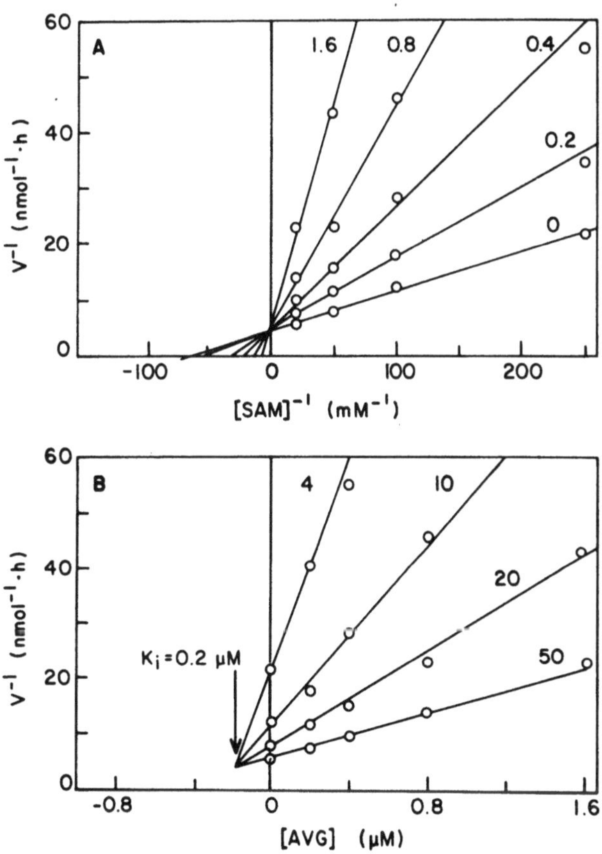

Fig. 3A and B. Competitive inhibition of the ACC-forming enzyme by AVG. Equal samples of an enzyme preparation (400 μl) from the pericarp of pink tomato were incubated for 2 h with various concentrations of AVG and SAM, and the rate of ACC formation was determined. **A** Double-reciprocal plot of the data. Straight lines connecting data points obtained in the presence of the indicated AVG concentration (μM) intersect on the ordinate. **B** Dixon plot of the same data. Straight lines connecting data points obtained in the presence of the indicated SAM concentration (μM) yield a K_i value of 0.2 μM for AVG. From (13)

enzymatic activity that specifically converted SAM to ACC (13). The ACC-forming enzyme had a pH optimum of 8.0–8.5, it was localized in the soluble fraction of the homogenate, and it was stable in the cold, but its activity decayed rapidly above 40°C. The enzyme had a K_m value of 13 μM for SAM and required pyridoxal phosphate as cofactor.

Aminoethoxyvinylglycine (AVG), an analog of rhizobitoxine, is a potent inhibitor of ethylene synthesis in many plant tissues (5). Since AVG inhibited formation of ethylene from [^{14}C]methionine but not from [^{14}C]ACC, Adams and Yang (9) postulated that the AVG-sensitive step in ethylene biosynthesis was between methionine and ACC. Using the enzyme preparation from tomato fruits, we showed that AVG was a very potent inhibitor of the ACC-forming enzyme (Fig. 3A, B). The inhibition was competitive, reversible, and the K_i value for AVG was 0.2 μM (Fig. 3B).

In vivo ethylene formation is also inhibited by metal chelators like EDTA, by free-radical quenchers like n-propyl gallate, by L-canaline and under anaerobic conditions (5). Of the above substances, n-propyl gallate, $CoCl_2$, and L-canaline inhibited the ACC-forming enzyme, though much less effectively than AVG did. The activity of the ACC-forming enzyme was not reduced under nitrogen (13).

Ethylene synthesis seems to be regulated at the level of ACC formation. When tomato fruits were sampled at different physiological stages, we found a positive correlation between the activity of the ACC-forming enzyme, the level of ACC in the tissue, and the capacity of the tissue to form ethylene (13). This observation was confirmed with mature, wild-type tomatoes which produce copious amounts of ethylene and tomatoes of the *rin* mutant which do not ripen normally and which form very little ethylene. Similarly, indoleacetic acid (IAA)-induced ethylene synthesis in etiolated pea stem sections appears to depend on the induction of ACC synthesis by IAA (14). Without IAA, the activity of the ACC-forming enzyme was below a detectable level, and the ACC content of the tissue was very low. With IAA, the activity of the ACC-forming enzyme was enhanced, and the level of ACC increased in parallel with ethylene production.

From these experiments, we conclude that the ACC-forming enzyme isolated from tomato fruits and peas is involved in ethylene biosynthesis and that SAM is indeed the precursor of ACC as has been postulated (9, 11).

In Vitro Ethylene Formation from ACC

Homogenates of etiolated pea shoots formed ethylene upon incubation with ACC (15). Induction of ethylene synthesis by IAA prior to homogenization of the tissue was not necessary to obtain this activity. The time course of ethylene formation in the presence of various concentrations of ACC is shown in Fig. 4A. A 10- to 20-min lag in ethylene evolution was observed at all ACC concentrations. Figure 4B shows the Lineweaver-Burk plot from the rates of ethylene production in Fig. 4A and from two additional ACC concentrations (2 and 2.5 mM). Since the regression line almost went through the origin, it was concluded that ethylene formation in homogenates of etiolated pea seedlings could only be saturated at very high ACC concentrations, if at all.

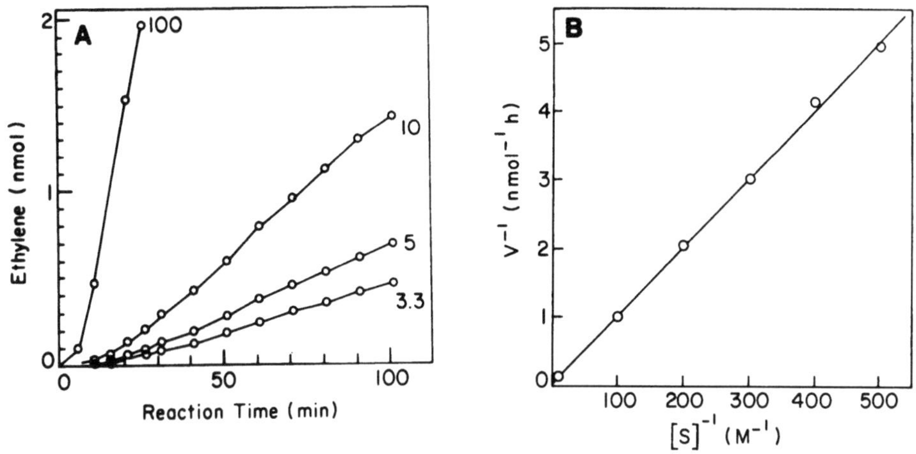

Fig. 4A and B. Time courses and Lineweaver-Burk plot of ethylene formation in homogenates of etiolated pea shoots in the presence of different concentrations of ACC. Homogenates of etiolated pea shoots were centrifuged at 11,000 *g* for 15 min. A 100 μl of supernatant was incubated with 3.3, 5, 10, or 100 mM ACC and 150 mM Tris-HCl, pH 7.9, at 28°C in a total volume of 0.4 ml. The time course of ethylene production was determined for each concentration. **B** Time courses of ethylene production in the presence of 2.0, 2.5, 3.3, 5.0, 10.0, or 100 mM ACC were monitored, and the rates of ethylene production were calculated from the linear range of the time courses. The rates of ethylene production were plotted against the concentrations of ACC in the Lineweaver-Burk plot. From (15)

When supernatants obtained after a short centrifugation of pea homogenates were passed through a Sephadex G-25 column, only 3.7% of the applied ethylene-forming activity was recovered in the excluded fraction. The retarded fractions likewise had very low ethylene-forming activity. The ethylene-synthesizing capacity of the excluded fraction increased substantially after aliquots of the retarded fractions had been added back to it. Figure 5 shows the elution pattern of components that stimulate ethylene production in the excluded fraction by later-eluting fractions. Two peaks containing substances that increased the ethylene-forming activity of the excluded fraction were evident in the eluate of the Sephadex column. The first peak moved slightly faster than the second maximum of UV absorbance, the other followed the third maximum of UV absorbance. These results confirm that one or more low-molecular-weight factors are necessary for the excluded fraction to display full ethylene-synthesizing activity.

When an aliquot of the pooled excluded fraction was heated at 80°C for 2 min, it lost its ability to catalyze the production of ethylene from ACC. The stimulating activity of the retarded fractions was not destroyed by the same heat treatment. ACC-dependent ethylene synthesis in pea homogenates proceeded optimally at 30°C. At 40°C and above, it began to decay within 30 min or less. It had a pH optimum of 7.9 and appeared to be associated, at least in part, with the particulate fraction of the homogenate (15).

Ethylene synthesis from ACC in pea homogenates was severely inhibited under anaerobic conditions and by catalase. It was also inhibited by n-propyl gallate, cyanide, azide and, most effectively, by EDTA. $CoCl_2$ at 0.1 and 0.3 mM inhibited ethylene formation from ACC, but higher and lower concentrations of $CoCl_2$ stimulated rather than inhibited ethylene formation. AVG did not inhibit ethylene formation from ACC in pea homogenates (15).

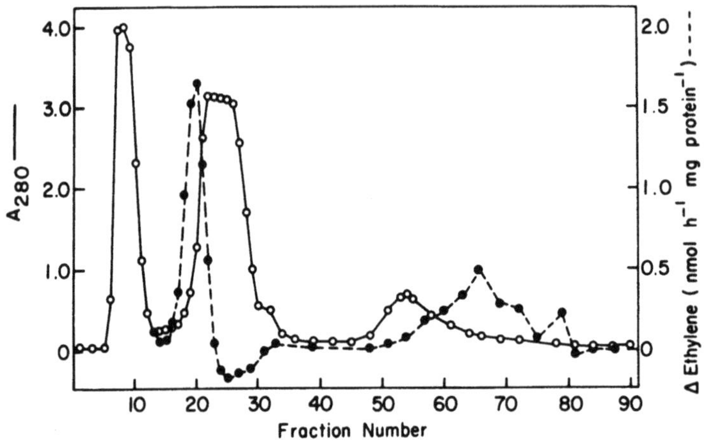

Fig. 5. Elution from a Sephadex G-25 column of the components required for ethylene synthesis in homogenates of etiolated pea shoots. Homogenates of etiolated pea shoots were centrifuged at 11,000 *g* for 15 min. 5 ml of supernatant was applied to a Sephadex G-25 column (20 cm × 2.5 cm). The column was eluted with 20 mM Tris-HCl, pH 7.9, and fractions of 2.7 ml were collected. The absorption of the fractions at 280 nm was recorded. Fractions 7–11 represented the excluded, early-eluting fraction and were pooled. For determination of ethylene formation, 100 μl of each later-eluting fraction and a combination of both were incubated with 10 mM ACC at 28°C in a total volume of 0.4 ml for 1 h. The ethylene production resulting from the combination of the excluded and the later-eluting fractions (= Δ ethylene) was calculated by subtracting from the total amount of ethylene produced in each combined fraction the separate contributions of the excluded and the respective later-eluting fraction. From (15)

In addition to ACC, 2-keto-4-mercaptomethyl butyrate (KMB) also proved to be a substrate for ethylene formation in pea homogenates, but methional, methionine, and S-methylmethionine did not yield significant amounts of ethylene under the same conditions. The reaction utilizing KMB as substrate differed significantly from that utilizing ACC in its kinetic properties, inhibitor sensitivity, and cofactor requirement. Therefore, we concluded that KMB was converted to ethylene by a different mechanism than was ACC (15).

The fact that ethylene formation in homogenates of etiolated pea seedlings can only be saturated at high ACC concentrations or cannot be saturated at all can be interpreted in two ways. Ethylene production in pea homogenates may be mediated by an enzyme that needs a heat-stable cofactor and that has a very low affinity to its substrate, ACC. Alternatively, conversion of ACC to ethylene may not be directly catalyzed by an enzyme but may be the result of a chemical reaction between ACC and the product of a reaction between an enzyme and a heat-stable cosubstrate:

$$\text{Heat-labile protein + heat-stable cosubstrate} + O_2 \longrightarrow \text{enzyme reaction product} \xrightarrow{+\text{ACC}} \text{ethylene}$$

The inhibition of ethylene production in pea homogenates by catalase indicates that H_2O_2 participates in the conversion of ACC to ethylene. However, our experiments also indicate that H_2O_2 is not directly responsible for the oxidation of ACC in pea homogenates (15).

We do not know whether in vivo conversion of endogenous ACC to ethylene is mediated by the same mechanism as that found in pea homogenates. The sensitivity of the cell-free system to inhibitors is in close agreement with the inhibitor sensitivity of the ACC-dependent in vivo system. Most notably, both are oxygen-dependent and neither is inhibited by AVG (9, 11, 15). However, there is one major discrepancy between in vivo and in vitro ethylene formation from ACC which we have not been able to explain. As discussed above, conversion of ACC to ethylene in pea homogenates is either not saturable or only saturable at very high ACC concentrations (above 400 mM), while conversion of ACC to ethylene in pea stem sections is saturated at 1 mM ACC (14). This discrepancy may reflect real differences in the reaction mechanisms or may have arisen because of the presence of a saturable uptake system in pea stem tissue. The identification of the active principles in both the heat-labile and the heat-stable fractions of pea homogenates will be an important step in elucidating the mechanism by which ACC is converted to ethylene.

Conclusions

The three in vitro enzymes or enzyme systems from homogenates of higher plant tissues that have been described above and in detail elsewhere (7, 13, 15) account for all proposed steps of ethylene biosynthesis from methionine. The first enzyme, methionine adenosyltransferase, utilizes selenomethionine more efficiently than methionine as a substrate, and this may be the reason for the observed stimulation of ethylene synthesis by selenoanalogs of methionine. The second enzyme, which catalyzes the conversion of SAM to ACC, appears to be under regulatory control. It seems to be the only enzyme of the ethylene synthesis pathway which needs to be formed or activated for ethylene biosynthesis to occur. Inhibition of its activity by AVG is most likely responsible for the inhibition of in vivo ethylene formation by this compound. The third step in ethylene synthesis, the conversion of ACC to ethylene, is thought to be mediated either directly or indirectly by an enzyme system consisting of a heat-labile protein and a heat-stable cofactor or cosubstrate. It is that part of the ethylene-synthesizing pathway which is most sensitive to EDTA, n-propyl gallate, cyanide, azide, and anaerobic conditions. The demonstration and characterization of the above three enzyme systems lend direct support to the hypothesis (9, 11) that ethylene synthesis in plant tissues proceed from methionine via SAM and ACC.

Acknowledgment. This research was supported by the U.S. Department of Energy under Contract No. EY-76-C-02-1338, by the National Science Foundation through Grant No. PCM 77-08522 to H.K., by a fellowship of the Deutsche Forschungsgemeinschaft to J.R.K., and a fellowship of the Swiss Natonal Science Foundation to T.B.

References

1. Lieberman, M., Mapson, L.W.: Nature (Lond.) *204*, 343–345 (1964)
2. Konze, J.R., Elstner, E.F.: FEBS Lett. *66*, 8–11 (1976)
3. Yang, S.F., Ku, H.S., Pratt, H.K.: J. Biol. Chem. *242*, 5274–5280 (1967)
4. Lieberman, M., Kunishi, A., Mapson, L.W., Wardale, D.A.: Plant Physiol. *41*, 376–382 (1966)
5. Lieberman, M.: Annu. Rev. Plant Physiol. *30*, 533–591 (1979)
6. Burg, S.P.: Proc. Natl. Acad. Sci. USA *70*, 591–597 (1973)
7. Konze, J.R., Kende, H.: Plant Physiol. *63*, 507–510 (1979)
8. Adams, D.O., Yang, S.F.: Plant Physiol. *60*, 892–896 (1977)
9. Adams, D.O., Yang, S.F.: Proc. Natl. Acad. Sci. USA *76*, 170–174 (1979)
10. Lürssen, K., Naumann, K., Schröder, R.: Naturwissenschaften *66*, 264–265 (1979)
11. Lürssen, K., Naumann, K., Schröder, R.: Z. Pflanzenphysiol. *92*, 285–294 (1979)
12. Konze, J.R., Schilling, N., Kende, H.: Plant Physiol. *62*, 397–401 (1978)
13. Boller, T., Herner, R.C., Kende, H.: Planta *145*, 293–303 (1979)
14. Jones, J.F., Kende, H.: Planta *146*, 649–656 (1979)
15. Konze, J.R., Kende, H.: Planta *146*, 293–301 (1979)

Abscisic Acid

Chairman: F. T. ADDICOTT

Introductory Comments: Abscisic Acid in the Physiology of Plants

F.T. ADDICOTT [1]

Research on abscisic acid had its beginnings in widely separated investigations that detected an acidic growth inhibitor in extracts and diffusates from such diverse sources as potato skins, dormant buds, autumn leaves, and aborting fruit. From those modest origins, an extensive body of knowledge has emerged, and abscisic acid is now recognized as one of the central endogenous regulators of higher plants with important hormonal functions.

The chemistry and biochemistry of abscisic acid have been the subject of highly competent investigations. We now have substantial knowledge of analogs and derivatives, of biosynthesis and degradation.

In relation to abscission, the function of abscisic acid as a hormonal accelerator has been verified repeatedly, and applied abscisic acid has promoted abscission of leaves, flowers, and fruits of numerous species.

In seed dormancy, abscisic acid is possibly the most widespread naturally occurring inhibitor (hormonal) of germination. Its action in the biochemistry of germination has been studied intensively. In bud dormancy, abscisic acid has strong inhibitory effects, but its role in the coordination of dormancy remains an enigma.

The striking acceleration of abscisic acid synthesis in response to water stress and similar stresses continues to receive much attention, particularly with respect to the mechanism of its synthesis as well as to its action in the responses of guard cells and in membrane transport generally.

The close correlation of abscisic acid with maturation and senescence in several fruits supports the recent suggestion that abscisic acid may function as a trigger in fruit ripening.

Another important area of current investigation is that of root growth and geotropism, where ABA and IAA appear to interact in the functions of control.

In other physiological processes, the involvement of abscisic acid is less clear, but at times it strongly influences both growth and flowering. Research on abscisic acid's original role as a growth inhibitor has been almost completely neglected. Such research is deserving of further attention.

The reports in this session present a selection of some of the most recent developments in research on abscisic acid.

1 Department of Botany, University of California, Davis CA 95616, USA

A Role for Abscisic Acid in Drought Endurance and Drought Avoidance

W.J. DAVIES, T.A. MANSFIELD, and A.R. WELLBURN [1]

Abscisic Acid and the Responses of Plants to Water Stress

Abscisic Acid (ABA) and Stomata

The delayed reopening of stomata by plants that have experienced a period of wilting provided early evidence for the existence of a stress-induced factor which could influence the water balance of plants (1). In 1969, Wright and his co-workers found that wheat leaves which had been allowed to wilt contained greatly increased amounts of the growth regulator ABA (2, 3). Subsequently, Wright and others established that in a number of species, ABA levels started to increase at a critical water deficit and from this point the amount of ABA produced was related to the level of water stress (4). Following Wright's early work, Jones and Mansfield demonstrated that an external application of ABA to leaves resulted in closure of the stomata (5, 6). Stomatal behaviour in plants treated with ABA resembled that exhibited by plants which experienced a period of water stress. This observation, coupled with the work of Milborrow and co-workers which showed that the chloroplasts are the main site of ABA formation in green leaves (7, 8), suggests an attractive hypothesis for the control of stomata in water-stressed plants. Namely, when a critical level of leaf water stress is reached, ABA can act as a "distress signal" moving from the chloroplasts to the guard cells to promote the maintenance of leaf turgor. Additional evidence for this hypothesis was provided by Loveys, who showed that the bulk of the ABA produced in the leaf is not produced in the guard cells but is transported there from the mesophyll (9).

Additional Turgor-Maintaining Processes

It is now apparent that there are several processes other than stomatal closure which might contribute to the maintenance of plant turgor during times of water shortage. Of significance in this regard is the observation that the application of ABA to root tissues can result in a reduction in a radial resistance to water flow across the root (10, 11), but it is not clear whether this effect is due to a change in membrane permeability to water or to an effect of ABA on active solute transport into the xylem (12). It is tempting to suggest that ABA might act to maintain turgor by decreasing transpiration while simultaneously promoting an increased flux of water into the roots. There is some doubt that water stress can induce an increased production of ABA in roots (13),

1 Department of Biological Sciences, University of Lancaster, Bailrigg, Lancs., United Kingdom

but it is known that ABA can move throughout the phloem to roots from water-stressed leaves (14, 15). It remains to be shown whether this ABA influences water uptake by the stressed plant.

Malek and Baker have found that ABA inhibits the active proton flux considered to be involved in the co-transport of sucrose during phloem loading in *Ricinus* (16). Such action might result in an accumulation of photosynthates in leaves with a consequent reduction in solute potential. As a result of solute accumulation, leaves can generate low water potentials while maintaining high levels of turgor (17), an obvious advantage to a plant growing in drying soil. The phenomenon of osmoregulation might be particularly effective if ABA reduced the rate of cell expansion and the ultimate size of leaf cells (18).

Stress-Stimulated Stomatal Closure Before ABA Levels Increase

Beardsell and Cohen have shown that there is a good correlation between stomatal behaviour and changes in abscisic acid levels in several cultivars of both maize and *Sorghum* plants subjected to a soil drying treatment (19). Careful study of their data suggests, however, that the stress-induced closure of stomata in maize takes place *before* the levels of ABA in the leaf increase significantly. This observation and others of a similar nature prompted several groups to search for compounds other than abscisic acid which build up in the plant in response to water stress. Several such compounds have been identified. One of these, all-trans farnesol, has been found in water-stressed *Sorghum* plants. The stress-induced build-up of this compound shows a close correlation with stress-induced stomatal closure (20). The destructive action of farnesol when applied directly to guard cells (21) suggests, however, that while the build-up of these compounds may be implicated in the stomatal response to water stress, they may not function as separate endogenous antitranspirants.

Some important measurements of ABA distribution in stressed and unstressed leaves have provided an alternative explanation of the observation that stress-induced stomatal closure can precede the build-up of ABA in the plant (9). Loveys has shown that nearly all the ABA in the leaves of well-watered plants is contained in the chloroplasts, and the result of a period of only mild stress is a release and redistribution of ABA throughout the cell (9). An investigation by Wellburn and Hampp of the effect of greening on the penetrability of plastid envelopes to ABA suggests that the changes in membrane permeability could be used in a regulatory manner (22). If an increase in membrane permeability to ABA occurred during the early stages of water stress, then a rapid closure of stomata and an extremely fine control of leaf turgor could result, without the necessity for any new synthesis of ABA. One would postulate, however, that synthesis of new ABA in the chloroplasts would be promoted by its release into the cytoplasm. In the light of these suggestions, an understanding of the control and effects of ABA redistribution within the cell should lead us to a new appreciation of the role of ABA as a stress hormone.

The Effects of a Soil Drying Treatment on Plant Water Relations, Stomatal Behaviour and Leaf Extension

If plants are subjected to a water stress treatment of gradually increasing severity, then a time course of the plant's responses to water stress can be constructed and the possible significance of a redistribution of ABA within the cell can be seen. Using a viscous flow porometer (23) and a linear displacement transducer (24), continuous measurements of stomatal behaviour (Fig. 1) and leaf extension rate (Fig. 2) of maize plants can be made. Partial stomatal closure and a limitation on leaf extension are recorded on days 3 and 4 of a drying treatment, at a soil water potential above 0.01 MPa

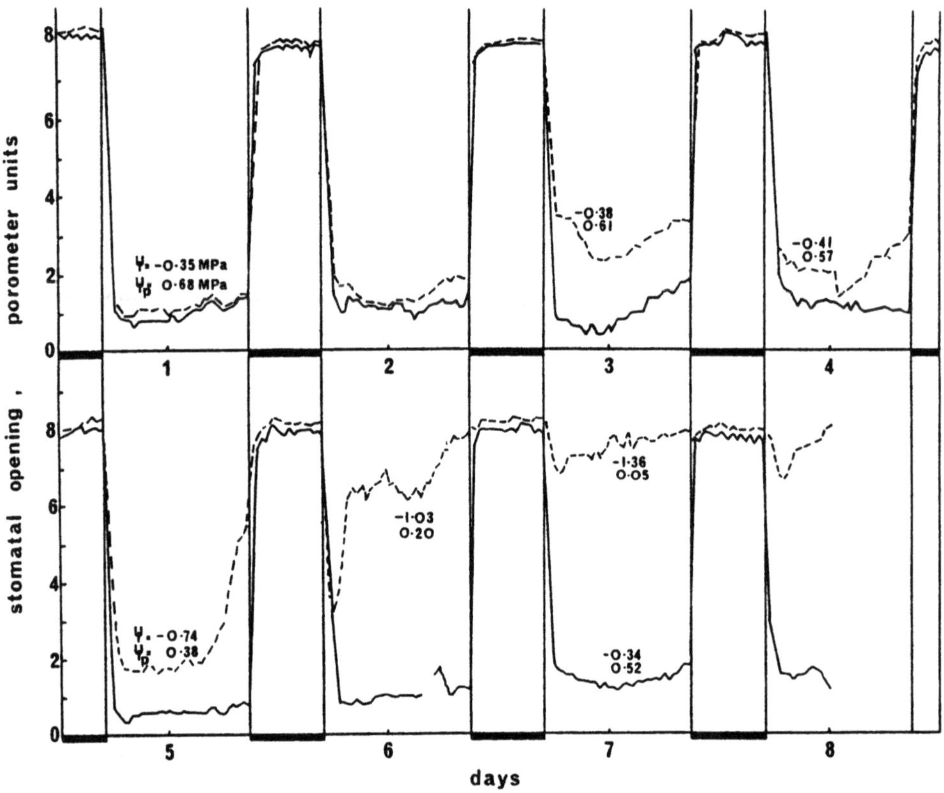

Fig. 1. Stomatal behaviour of well-watered plants (———) and plants subjected to a drying treatment (- - - -). Midday leaf water potentials and turgor levels are indicated. *Vertical lines* indicate lights on and off (25)

and *before* a reduction in leaf water potential and leaf turgor can be measured (Figs. 1 and 2) (25). In similar experiments, a net increase in the level of ABA in the leaf is not recorded until leaf water potential has dropped to around – 0.8 to 1.0 MPa, or until day 6 in this experiment, the stage in the drying cycle where substantial stomatal closure is observed. The implication of these observations is that partial stomatal closure and a limitation of leaf extension are imposed by redistribution of ABA from the chloroplasts which occurs as a result of only a very small reduction in water potential or turgor.

It is well known that in many systems ABA has an inhibitory effect on growth (26). In the results shown above, the increase in the rate of leaf extension (Fig. 2) when the lights were switched off on days 3, 4 and 5 suggests that cell expansion during the light period was limited by lack of turgor. It is significant, however, that during the dark period when there is more than adequate leaf turgor for maximal rates of leaf extension, the rate of leaf extension of the water-stressed plant is still significantly reduced. Correlative evidence that the limitation on leaf extension in the water-stressed plant

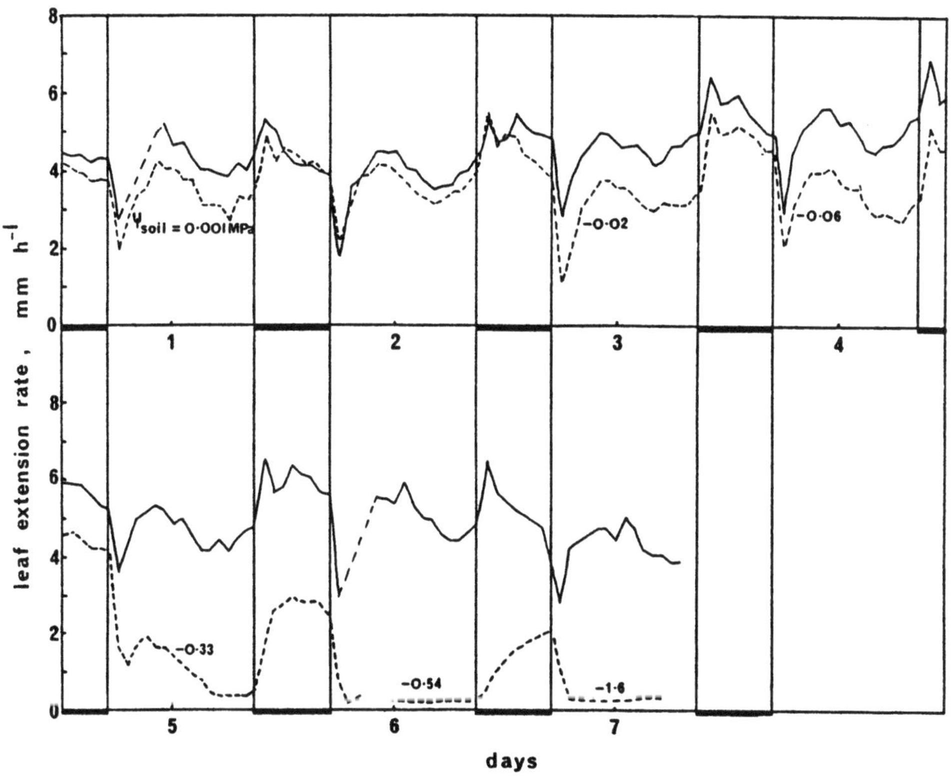

Fig. 2. Leaf extension rates of well-watered plants (——) and plants subjected to a drying treatment (- - -). Soil water potentials at midday of the plants subjected to a drying treatment are shown. *Vertical lines* indicate lights on and off (25)

is caused by ABA may be found in an experiment where a surface application of a 10^{-4} M ABA solution to well-watered and water-stressed leaves of maize plants reduces the rate of leaf extension in both cases (Davies, unpublished data).

A reduction in the rate of leaf area development is in itself an advantage to a plant which is suffering from reduced water availability. In addition to this, the increased availability of assimilates as a result of reduced leaf growth can result in an accumulation of osmotically active substances in leaves and roots, and thus in turgor maintenance in plants growing in drying soil. Sharp and Davies have shown that solutes can accumulate in roots of mildly water-stressed maize plants while the stomata are still open to a significant extent (27). These authors have suggested that the redistribution of solutes to roots and the resulting turgor maintenance may explain the net increase in root growth that occurs in mildly stressed plants. An increase in the root/shoot ratio is a common response when plants are subjected to a water stress treatment (17) and is of some advantage in increasing the amount of soil water available to the plant.

The combination of turgor-maintaining events which are described above, rather than the alternative of complete stomatal closure, will mean that photosynthesis can continue at a relatively high rate for an extended period. A continuous supply of photosynthates is important for continued solute regulation and continued growth, especially of the roots. It seems likely that turgor-maintaining processes might be promoted by the redistribution of ABA in the leaf at a comparatively high water potential. It might be argued, therefore, that in this situation ABA provides the plant with some drought-enduring capacity. While turgor is maintained in the leaf, ABA production is maintained at a low level and only later in the drying cycle, when water potential and leaf turgor have dropped to a low and potentially damaging level, will abnormal ABA synthesis occur. This synthesis will promote almost complete stomatal closure and therefore a drought-avoiding state.

If solute regulation is an important means of turgor maintenance in leaves, which will delay the onset of abnormal ABA synthesis, it seems likely that abnormal ABA production should be linked more closely to reductions in leaf turgor than to reductions in leaf water potential. Recently, Pierce and Raschke have shown that this may be the case (28).

The Release of ABA from Mesophyll Chloroplasts

In an earlier paper, Mansfield et al. proposed a hypothesis to explain the involvement of ABA in promoting stress-induced stomatal closure (29). Recent findings [(9), and Milborrow, this volume] would suggest that changes in the permeability of the mesophyll plastid envelopes induced by water stress are of considerable significance. An alternative explanation, namely the conversion of "bound ABA" to free ABA by water stress, seems less likely since the ratio of these two forms is apparently not responsive to water stress treatment (30).

Fenton et al. have shown that all-trans farnesol and also long-chain unsaturated fatty acids are capable of inhibiting the metabolic function of isolated chloroplasts and that these effects appear to be the outcome of the influence of the compounds upon envelope membranes (31). In view of the relationship between water stress and the level of farnesol in the leaf and the membrane activity of the compound, Mansfield et al. have proposed that stress-stimulated farnesol production could be involved in the release of ABA from chloroplasts (30). They have suggested that when the water potential (or the turgor) of the chloroplast falls, there is a block in the conversion of farnesyl pyrophosphate (FPP) to geranyl-geranyl pyrophosphate (Fig. 3), and some production of farnesol may occur. This could alter the nature of the chloroplast envelope membranes and as a consequence release existing ABA into the cytoplasm. Fresh biosynthesis of ABA in the chloroplast will follow because of the raised level of the precursor FPP and because the continued release of ABA into the cytoplasm prevents end-product inhibiton.

mevalonate ← ← CO_2

↓

isopentenyl pyrophosphate

↓

geranyl pyrophosphate

↓

farnesol ← farnesyl pyrophosphate → abscisic acid

↓

geranyl-geranyl pyrophosphate

Fig. 3. Part of the pathway of terpenoid biosynthesis to show the point of origin of farnesol and abscisic acid

In our most recent studies (32) several *Sorghum* and maize cultivars have been assayed for "antitranspirant activity" using the *Commelina* epidermis bioassay (33). A substantial proportion of the activity of all cultivars which correlates well with stomatal behaviour in the intact plant has been found to be due to the presence of unsaturated fatty acids. Willmer et al. have shown that levels of short-chain fatty acids increase with water stress and that they too are capable of inducing stomata to close (34). Interestingly, the concentrations of fatty acids required to close stomata have an effect on isolated epidermis which is closely similar to that of farnesol, namely lysis of the guard cell chloroplasts and cell death (21). In view of these effects it is difficult to envisage an antitranspirant role for endogenous fatty acids which involves a direct effect on the guard cells.

In considering a likely role for fatty acids during the development of water stress, Mansfield and Wilson noted that the involvement of fatty acids in the processes contributing to the ageing of chloroplasts and to chilling injury has been firmly established (35). It is well known that ultrastructural changes comparable to those which are seen in guard cells when epidermal strips are treated with fatty acids will occur as a result of chilling injury and chloroplast ageing (36, 37). These changes are associated with an inhibition of electron flow and energy-linked reactions (38, 39), an alteration of the physicochemical properties of membranes (40, 41) and conformational changes of proteins. Many of these changes can be induced if some of the fatty acids normally found in chloroplast membranes are applied to freshly isolated chloroplasts (38).

Interestingly, water stress can cause changes in chloroplast ultrastructure which are comparable to those described above (42–44). Vierra Da Silva et al. have attributed the changes which they observed to a water stress-stimulated increase of enzymatic activity on the chloroplast lipids, causing a release of free fatty acids (42). Water stress will directly influence electron transport, photophosphorylation and carbon fixation by isolated chloroplasts (45).

The ultrastructural and biochemical changes described above are nearly always the result of severe water-stress treatment. Information is needed on the effects of moderate levels of water stress on the biosynthesis and metabolism of lipids and on the various stages of the terpenoid biosynthetic pathway. It seems likely that low levels of the compounds which we have found to be associated with water stress in maize and *Sorghum* could act in a more subtle manner than the high levels found in severely stressed tissue. The increase in linolenic acid, which we have found to be associated with water stress in maize and *Sorghum,* might point to an increase in the unsaturation of membranes leading to changes which could affect the release of ABA.

The Transport of ABA Through the Plant and Its Mode of Action in the Leaf

Movement of ABA

We have argued that ABA may play a multifaceted role as a stress hormone, influencing several processes throughout the plant. The bulk of the stress-induced ABA production apparently occurs in the mesophyll chloroplasts and moves from thence to the sites of action. Since ABA is apparently not produced in the guard cells, the movement of the hormone to them from the place of biosynthesis fits the classical concept of hormone action. Not surprisingly, ABA is found in the xylem and in the phloem. It seems unlikely, however, that these pathways need to be involved in the transport of ABA to the guard cells, since the direct route to them from the mesophyll need only involve 4 cells (29). Whether the movement from the chloroplast occurs via an apoplastic or symplastic route is a matter for speculation. The nature of this route is of some significance, however, since it must inevitably have some control over the speed of response to hormone redistribution.

Weyers and Hillman have shown that radioactive ABA accumulates quickly in the stomatal complexes when applied to isolated epidermal tissue (46). ABA is apparently accumulated by guard cells when the stomata are both open and closed and irrespective of whether its control function has been achieved. This suggests that no regulatory mechanism over uptake exists in the guard cells themselves. In most plants there are no plasmodesmata between subsidiary and guard cells (47), which means that movement of the hormone at this point cannot be symplastic. Up to this point, symplastic or apoplastic transport may occur but apoplastic transport would tend to be rather undirected since movement to sites of evaporation will occur and the guard cells are only one of several such sites (48). Movement along the shortest linear route, i.e. via the symplasm to the subsidiary cells cannot be assumed, but on the present available evidence it seems unnecessary to suggest more complicated pathways.

Since guard cells apparently accumulate ABA indiscriminately, it remains a possibility that they have a way of storing the compound by compartmentalization or sequestration. Mansfield and Wilson have suggested that this would allow controlled access of the hormone to the receptor site (35). Such a mechanism would explain the prolonged effect of externally applied ABA, which can suppress stomatal opening for many days.

The Mechanism of Action of ABA

The regulation of stomata by ABA could be achieved if the hormone were to interfere with the ionic fluxes that are the essential basis for changes in guard cell turgor (49). Certainly, the outcome of the action of ABA on stomata is to inhibit both the uptake of K^+ ions and the disappearance of guard cell starch, which normally accompany stomatal opening (50).

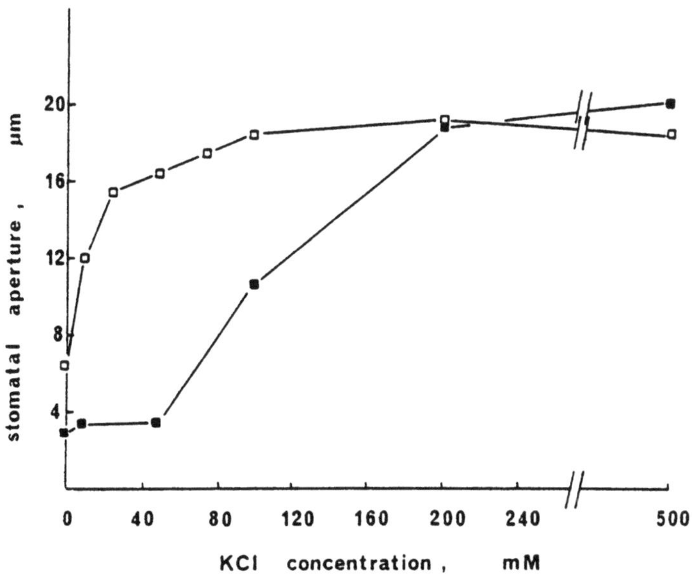

Fig. 4. Apertures of isolated stomata of *Commelina communis* incubated for 2.5 h in a range of KCl concentrations, with (▪) or without (▫) abscisic acid. Each point is the mean of 30 measurements. Standard errors were within the limits of the insignia. Before being transferred to the final incubation media, stomata were wide open (17.0 ± 0.3 μm). (53)

Raschke has concluded that ABA acts by inhibiting an H^+ expulsion mechanism in the plasmalemma of the guard cells (51). Stomata respond within 5 min to an application of a 10^{-7} M solution (52), which suggests that effects upon the plasmalemma are involved. Wilson et al. have found that the response of isolated stomata to ABA may be counteracted by high levels of K^+ ions in the incubation medium (Fig. 4). ABA had its greatest effect at a potassium concentration around 50 mM (53). These authors suggest that their findings could be interpreted in terms of an effect of ABA blocking "channels" in the plasmalemma for K^+ entering the guard cells. Thus, although ABA does ultimately affect H^+ expulsion from the guard cells, this could result from a linkage between this process and the numbers of K^+ ions entering. It is not possible, therefore, to say at which point the inhibitory effect of ABA operates, but a close connection with ion transport through the plasmalemma seems to be indicated.

Recently, Dubbe et al. have argued that the plant's responses to ABA and CO_2 may combine to provide highly sophisticated control of gas exchange related to the plant's environment and immediate physiological state (54). Mansfield and Wilson have pointed out that data are to be found in the literature which suggest that ABA may act in a manner which amounts to a crude and arbitrary control in relation to external conditions (35). A good example of such a condition is the often-reported after-effect of water stress on stomata (55, 56) which can be explained by the action of ABA (57, 58). The advantages of such a prolonged action of ABA can perhaps be appreciated when considered in ecological terms (29) if not in an agricultural context (59). There is little eivdence, however, that the imposition of a "ceiling" on stomatal opening is in any way related to the plant's immediate physiological state.

It seems possible that several of the effects of water stress in photosynthesis might be brought about by high levels of ABA within the cell. So far, studies of the effects of ABA upon isolated organelles have failed to demonstrate similar responses to those observed in vivo during water stress. Keck and Boyer demonstrated that most photosynthetic functions, and in particular those of photophosphorylation, were inhibited during water stress (60). This diminished ability to form ATP has been shown to be due to conformational changes of the chloroplast coupling factor particles (61). There is, however, no significant effect of ABA (10^{-4} M) on photosystem I activity of isolated chloroplasts, and only a slight decrease in photosystem II activities (62), whilst ABA (10^{-5} M and lower) has no effect upon cyclic photophosphorylation linked to membrane-dependent electron transport using diaminoduerene as electron donor and methyl viologen as acceptor (Robinson and Wellburn, unpublished observations). These results suggest that the redistribution of ABA in the leaf at mild levels of water stress may promote some degree of turgor maintenance but will have little direct effect on photosynthesis. At concentrations around 10^{-4} M, ABA acts as an uncoupler of oxidative phosphorylation (63), but at lower concentration this effect disappears (Robinson and Wellburn, unpublished observations). Nevertheless both isolated mitochondria and etioplasts may show increases in the rate of malate uptake amounting to 30%, after pre-incubation for 15 min with ABA over the concentration range 10^{-4} to 10^{-7} M (Hampp, personal communication). The implication is that changes in the permeability of envelope or plasma membranes are a fundamental part of the mechanisms involved.

Exogenous Application of ABA and Breeding for Drought Resistance

The use of ABA as an artifical controller of transpiration has been discussed in detail elsewhere (59). It is widely recognised that while the immediate consequences of an application of an antitranspirant will be a saving of water, another inevitable consequence will be the inhibition of CO_2 exchange. There seems some possibility that ABA applications to plants might result in an increase in water use efficiency, but the main aims in such an application must be a delay in the onset of water stress, desiccation injury and possibly death. Some reduction in photosynthesis and growth can be tolerated as an inevitable sacrifice for the protection of the cellular equipment required for that photosynthesis which will be carried out in the future. Several workers have suggested that antitranspirant application may result in an increase in growth relative to that of water-stressed, untreated plants, presumably due to a more judicious use of available water [(65) and Watts and Davies, unpublished observations]. Interestingly, ABA application to a well-watered plant seems to mimic the effect of water stress and may result in an increase in root growth relative to shoot growth (Watts and Davies, unpublished observations), a result that might be predicted from the hypothesis proposed above.

While antitranspirants provide a short-term solution to water stress problems, in the long term, problems may be alleviated by breeding for drought resistance. Taking into account many of the arguments advanced in the present paper, Mansfield and Wilson have suggested that the following patterns of behaviour should be sought in crop plants (35):

1. A quick release of any ABA from mesophyll chloroplasts of mildly stressed plants, followed by a rapid synthesis of new ABA as soon as more severe water stress is experienced.

2. An early recovery to a normal pattern of stomatal behaviour after the turgor of the plant has been restored.

These authors have concluded that attention has perhaps been misdirected in the past to the amount of ABA produced under water stress, based upon the supposition that a greater quantity should be indicative of drought resistance. We would suggest that the rapid release of a quantity of ABA just sufficient to cause the maintenance of plant turgor would be a more desirable character in crops than an accumulation of sufficient ABA to close the stomata for a prolonged period of time.

References

1. Glover, J.: J. Agric. Sci. *53*, 412–416 (1959)
2. Wright, S.T.C.: Planta *36*, 10–20 (1969)
3. Wright, S.T.C., Hiron, R.W.P.: Nature (Lond.) *224*, 719–720 (1969)
4. Wright, S.T.C.: Planta *134*, 183–189 (1977)
5. Jones, R.J., Mansfield, T.A.: J. Exp. Bot. *21*, 714–719 (1970)
6. Jones, R.J., Mansfield, T.A.: Physiol. Plant. *26*, 321–327 (1972)
7. Milborrow, B.V., Robinson, D.R.: J. Exp. Bot. *24*, 537–548 (1973)
8. Milborrow, B.V.: Phytochemistry *13*, 131–136 (1974)
9. Loveys B.R.: Physiol. Plant.*40*, 6–10 (1977)
10. Glinka, Z., Reinhold, L.: Plant Physiol. *48*, 103–105 (1971)
11. Collins, J.C., Kerrigan, A.P.: In: Ion Transport in Plants. Anderson, W.P. (ed.), pp. 589–594. London, New York: Academic Press 1973
12. Clarkson, D.: Ion Transport and Cell Structure in Plants. London: McGraw-Hill 1974
13. Torrey, J.G.: Annu. Rev. Plant Physiol. *27*, 435–459 (1976)
14. Hocking, T.J., Hillman, J.R., Wilkins, M.B.: Nature (Lond.) New Biol. *235*, 124–125 (1972)

15. Hoad, G.V.: Planta *113*, 367–372 (1973)
16. Malek, T., Baker, D.A.: Plant Sci. Lett. *11*, 233–239 (1978)
17. Hsiao, T.C., Acevedo, E., Fereres, E., Henderson, D.W.: Philos. Trans. R. Soc. London Ser. B *273*, 479–500 (1976)
18. Quarrie, S.A., Jones, H.G.: J. Exp. Bot. *28*, 192–203 (1977)
19. Beardsell, M.F., Cohen, D.: Plant Physiol. *56*, 207–212 (1975)
20. Ogunkanmi, A.B., Wellburn, A.R., Mansfield, T.A.: Planta *117*, 293–302 (1974)
21. Fenton, R., Davies, W.J., Mansfield, T.A.: J. Exp. Bot. *28*, 1043–1053 (1977)
22. Wellburn, A.R., Hampp, R.: Planta *131*, 95–96 (1976)
23. Allaway, W.G., Mansfield, T.A.: Can. J. Bot. *47*, 1995–1998 (1969)
24. Sharp, R.F., Osonubi, O., Wood, W.A., Davies, W.J.: Ann. Bot. *44*, 35–45 (1979)
25. Sharp, R.F., Davies, W.J.: Manuscript in preparation (1980)
26. Milborrow, B.V.: Annu. Rev. Plant Physiol. *25*, 259–307 (1974)
27. Sharp, R.F., Davies, W.J.: Planta *147*, 43–49 (1979)
28. Pierce, M., Raschke, K.: Plant Physiol. Suppl. *61*, 131 (1978)
29. Mansfield, T.A., Wellburn, A.R., Moreira, T.J.S.: Philos. Trans. R. Soc. London, Ser. B *284*, 471–482 (1978)
30. Milborrow, B.V.: J. Exp. Bot. *29*, 1059–1066 (1978)
31. Fenton, R., Mansfield, T.A., Wellburn, A.R.: J. Exp. Bot. *27*, 1206–1214 (1976)
32. Wilson, J.A., Davies, W.J.: Plant Cell Environ. *2*, 49–57 (1979)
33. Ogunkanmi, A.B., Tucker, D.J., Mansfield, T.A.: New Phytol. *72*, 277–282 (1974)
34. Willmer, C.M., Don, R., Parker, W.: Planta *139*, 281–287 (1978)
35. Mansfield, T.A., Wilson, J.A.: Proc. Univ. Nottingham Easter School. (in press, 1980)
36. Siegenthaler, P.A.: Plant Cell Physiol. *10*, 801–810 (1969)
37. Siegenthaler, P.A.: Plant Cell Physiol. *10*, 810–820 (1969)
38. Siegenthaler, P.A.: Biochim. Biophys. Acta *275*, 182–191 (1972)
39. Perdersen, T.A., Kirk, M., Bassham, J.A.: Biochim. Biophys. Acta *112*, 189–203 (1966)
40. Wintermans, J.F.G.M.: In: Biochemistry of Chloroplasts. Goodwin, T.W. (ed.), vol. 1, p. 115. London, New York, Academic Press 1966
41. Helmsing, P.J.: Biochim. Biophys. Acta *178*, 519–533 (1969)
42. Vierra Da Silva, J., Naylor, A.W., Kramer, P.J.: Proc. Natl. Acad. Sci. USA *71*, 3243–3247 (1974)
43. Freeman, T.P., Duysen, M.E.: Protoplasma *83*, 131–145 (1975)
44. Giles, K.L., Beardsell, M.F., Cohen, D.: Plant Physiol. *54*, 208–212 (1974)
45. Boyer, J.S.: Philos. Trans. R. Soc. London Ser. B *273*, 501–512 (1976)
46. Weyers, J.D.B., Hillman, J.R : Planta *144*, 167–172 (1979)
47. Carr, D.J.: In: Intercellular Communication in Plants: Studies on Plasmodesmata. Gunning, B.E.S., Robards, A.W. (eds.), pp. 243–289. Berlin, Heidelberg, New York: Springer 1976
48. Meidner, H.: J. Exp. Bot. *26*, 666–673 (1975)
49. Mansfield, T.A.: In: Plant Physiology. Sutcliffe, J.F., Steward, F.C. (eds.), vol. VII. London, New York: Academic Press 1981
50. Mansfield, T.A., Jones, R.J.: Planta *101*, 147–158 (1971)
51. Raschke, K.: Annu. Rev. Plant Physiol. *26*, 309–340 (1975)
52. Cummins, W.R., Kende, H., Raschke, K.: Planta *99*, 347–351 (1971)
53. Wilson, J.A., Ogunkanmi, A.B., Mansfield, T.A.: Plant Cell Environ. *1*, 199–201 (1978)
54. Dubbe, D.R., Farquhar G.D., Raschke, K.: Plant Physiol. *62*, 413–417 (1978)
55. Allaway, W.G., Mansfield, T.A.: Can. J. Bot. *48*, 513–521 (1970)
56. Fischer, R.A.: J. Exp. Bot. *21*, 386–404 (1970)
57. Boussiba, S., Richmond, A.E.: Planta *129*, 217–219 (1976)
58. Dörfling, K., Streich, J., Kruse, W., Muxfeldt, B.: Z. Pflanzenphysiol. *31*, 43–56 (1977)
59. Davies, W.J., Mansfield, T.A., Orton, P.J.: Proc. Joint BCPC and BPGRG Symp. – Opportunities for chemical plant growth regulation (1978)
60. Keck, R.W., Boyer, J.S.: Plant Physiol. *53*, 474–479 (1974)
61. Younis, H.M., Boyer, J.S., Govindjee Plant Physiol. Suppl. *61* (1978)

62. Zima, J., Sestak, Z.: Biol. Plant. *20,* 232–233 (1978)
63. Hemberg, T.: Physiol. Plant. *43,* 65–67 (1978)
64. Mansfield, T.A.: Philos. Trans. R. Soc. London Ser. B (1976)
65. Poljakoff-Mayber, A., Gale, J.: In: Water Deficits and Plant Growth. Kozlowski, T.T. (ed.), vol. III. London, New York: Academic Press 1972

Abscisic Acid and Other Naturally Occurring Plant Growth Inhibitors

G. SEMBDNER [1], W. DATHE [1], V.I. KEFELI [2], and M. KUTÁČEK [3]

Introduction

Some aspects of the physiological role of naturally occurring plant growth inhibitors have been studied in collaboration between the Institute of Plant Biochemistry, Halle/Saale, the Institute of Plant Physiology, Moscow, and the Institute of Experimental Botany, Prague. These include both the physiological activity and natural occurrence of phenolics in plants [cf. (1–7)] as well as the biochemistry and physiology of ABA metabolism. As an example, the occurrence of ABA and ABA conjugate in bleeding sap, wood, and bark of birch trees will be reported and discussed with respect to their role in breaking dormancy.

Occurrence of ABA and ABA Conjugate in Branches and Bleeding Sap of Birch

ABA has been shown to occur naturally in the branches of woody plants, i.e., *Pinus radiata* (8,9), *Pinus sylvestris, Picea abies* (10), *Abies balsamea* (11), *Prunus domestica,* and *Acer pseudoplatanus* (12). Within the branches it was found mainly in the bark of *Persica vulgaris* (13) and in the cambial region of *Picea sitchensis* (14). ABA has been identified also in bleeding sap from other sources: *Acer saccharum, Actinidia chinensis,* and *Tecomaria capensis* (15), *Betula* species (16), *Malus* species (15, 17), *Persica vulgaris* (15, 18), and *Salix* species (15, 17, 19, 20). ABA is assumed to control bud dormancy [cf. (21–23)]. With respect to this function the physiological significance of ABA translocated in spring-bleeding sap cannot be explained.

In order to elucidate the role of ABA in dormancy of *Betula pubescens,* translocation of the hormone in bleeding sap as well as its distribution in bark and wood of branches have been investigated. Branches 3–6 mm in diameter from birch trees about 30 years old were collected both in Moscow and Halle at 4 different stages (see Table 1). In addition, dormant branches (stage I) of varying size were studied in order to get further information on the inhibitor gradient. The branches were separated into wood (xylem) and bark (containing phloem), fixed in liquid nitrogen and lyophilized.

1 Institute of Plant Biochemistry, Halle, Academy of Sciences of the German Democratic Republic

2 K.A. Timiryazev Institute of Plant Physiology, Moscow, USSR Academy of Sciences

3 Institute of Experimental Botany, Prague, CSSR Academy of Sciences

Bleeding stap (stage II) was collected by cutting branches 15–20 mm in diameter. The sap dripping from the surface was frozen and lyophilized. Freeze-dried wood and bark tissue was extracted with 80% methanol, and the aqueous concentrates as well as the bleeding sap – redissolved in small quantities of H_2O – were partitioned at pH 2.5 with ethyl acetate and subsequently with water-saturated *n*-butanol. The ethyl acetate and *n*-butanol extracts were purified using both DEAE-Sephadex A-25 and silica gel columns. Detection of ABA-like inhibitor components was achieved by bioassaying aliquots of each fraction using wheat seedlings [for further details see (24, 25)].

Table 1. Changes in the levels of ABA and "neutral ABA conjugate". The inhibitor contents were determined as ng ABA-equiv./g fr. wt in wheat seedling bioassay of fractions after chromatography on DEAE-Sephadex A-25 and silica gel Woelm

Stage		Bark		Wood	
		ABA	ABA conj.	ABA	ABA conj.
I	Winter dormancy	2160	10400	1460	2450
II	Buds swelling, bleeding sap flowing	4300	(–)	450	(–)
III	Bud burst, first leaf visible	3100	(–)	25	(–)
IV	Bud burst first leaves unfolded	75	7900	0	1700

(–) not investigated

In both wood and bark tissue two inhibitors were detected: one corresponds chromatographically to ABA and the other is a more polar compound. The former has been identified as (+)-abscisic acid by means of ORD measurement and combined GC-MS (24). The latter is eluted in the neutral fraction by anion exchange chromatography and proved to be a new ABA conjugate (see below). By partition of aqueous solutions at pH 2.5 with ethyl acetate, it is extracted only to a slight extent (ca. 30%); the remainder is extractable with *n*-butanol.

Bleeding sap, studied during several years, contained fairly constant amounts of ABA. For example, in 1973 the ABA content was 50–60 ng/ml as measured by bioassay. In 1974 it was about 30 ng ABA/ml sap as determined by bioassay and 40 ng/ml by ORD measurement. The different samples of bleeding sap also contained the neutral "ABA conjugate". However, the amounts varied greatly, e.g., between 10 and 800 ng ABA equiv./ml. Quantification of the conjugate was only performed by bioassay (24). In the wheat seedling bioassay, its biological activity was of the same order as that of free ABA.

Both in bark and wood tissue the ABA content varied considerably from dormancy to bud break (Table 1). In the bark the ABA level was high during dormancy (stage I) and increased further during sap flow. After bud burst (III–IV) the ABA content in the bark decreased greatly. In the wood the ABA level was highest in winter (I) and decreased rapidly from the period of sap flow to bud break. At the last stage (IV), in the wood of thin branches (3–6 mm ϕ) no ABA was detectable. In contrast, very high levels of ABA conjugate persisted during transition from dormancy to bud break, both in the bark and in the wood (24).

The distribution of inhibitors was studied in dormant branches collected in the middle of January 1977, which were cut into sections and sorted into 5 groups based on diameter size (see Figs. 1 and 2).

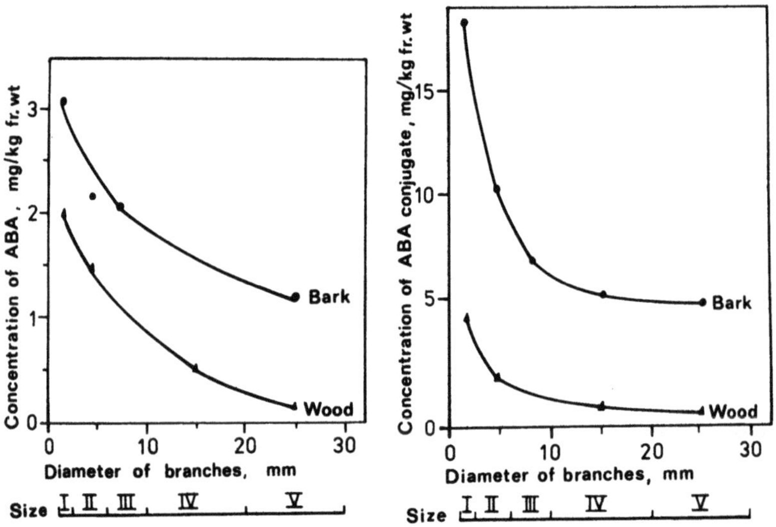

Fig. 1. Concentration of ABA and ABA conjugate (mg/kg fr. wt) in bark and wood of birch branches differing in diameter

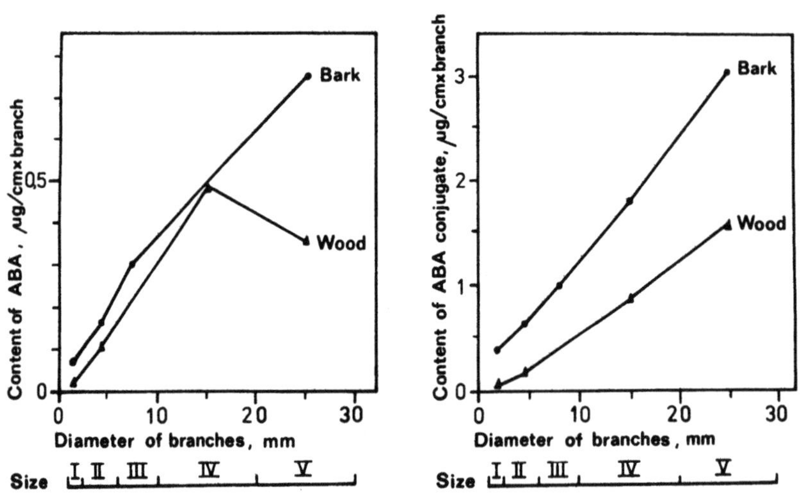

Fig. 2. Content of ABA and ABA conjugate (µg/cm × branch) in branch parts of different diameter

Regardless of branch diameter, the concentration of ABA as well as of ABA conjugate was higher in the bark than in the wood (Fig. 1). ABA conjugate concentrations in both bark and wood were higher than those of free ABA. Differences were greatest in the bark of young branches. The concentration of ABA conjugate was highest in the apical region and decreased with increasing branch thickness. Branches thicker than about 15 mm did not differ in their ABA conjugate concentration. In the dormant branches the concentration of free ABA also was highest in the apical region in both bark and wood tissue and decreased with increasing branch diameter.

Quite opposite relations are obtained when ABA and ABA conjugate contents are based not on unit fresh wt (g) but on unit length (cm) of branches (Fig. 2). The total amounts of both free and conjugated ABA were higher in the thicker parts of the branches than in the thin part near the tips. The very high amounts of ABA conjugate increased both in wood and bark from branch size I to IV. The free ABA content per cm also increased with thickness of the branch. In the wood, maximal values were reached in branch parts 10–20 mm in diameter (size IV).

Characterization of the New ABA Conjugate

The β-d-glucopyranosyl ester of (+)-ABA is well known to be a naturally occurring ABA conjugate. It was identified unequivocally in fruits of *Lupinus luteus* (26) and in the pseudocarp of *Rosa arvensis* (27). Moreover, the occurrence in higher plants of "conjugated ABA forms" has been described repeatedly; in most cases they presumably represent the glycosyl ester of ABA, though the polar compounds have been characterized only partially [cf. (28)]. Recently, β-hydroxy-β-methylglutarylhydroxyabscisic acid has been detected as a new type of conjugate in seeds of *Robinia pseudacacia* (29). It derives not directly from ABA but from its 6'-hydroxy-metabolite. Furthermore, phaseic acid and dihydrophaseic acid also seem to form conjugates, the structures of which are not yet completely elucidated (30, 31).

The ABA conjugate detected both in wood and bark of birch branches showed chromatographic (TLC, CC) properties similar to authentic ABA glucosyl ester (24). For further characterization and identification, a 560-g sample of thin branches (1–4 mm) was collected at bud break, lyophilized, extracted, and partitioned at pH 2.5 with water-satured *n*-butanol. The residue of the *n*-butanol extract was purified on a silica gel column (Fig. 3) and subsequently on a DEAE-Sephadex A-25 column (Fig. 4). Aliquots of each fraction were bioassayed using wheat seedlings. Repeated TLC using two different systems [(1) silical gel PF_{254}, chloroform-methanol-acetic acid 70/12/1, R_F 0.10–0.20 and (2) polyamide, chloroform-methanol-methylethylketone 80/5/14, R_F 0.33–0.42] yielded 4.2 mg of the purified inhibitor. This compound – after acetylation with acetanhydride/pyridine – was examined by GC on SE 30. The only detectable peak (R_t = 9.7) differed significantly from that of acetylated ABA-β-D-glucopyranosyl ester (R_t = 7.8).

Acid hydrolysis of the inhibitor with 1 N HCl (70°C, 30 min) yielded ABA, the identity of which was confirmed after purification by GC and rearrangement to trans-ABA by irradiation. The hydrolysate was purified on DEAE-Sephadex A-25. Besides ABA it contained a sugar component which after acetylation behaved similarly to

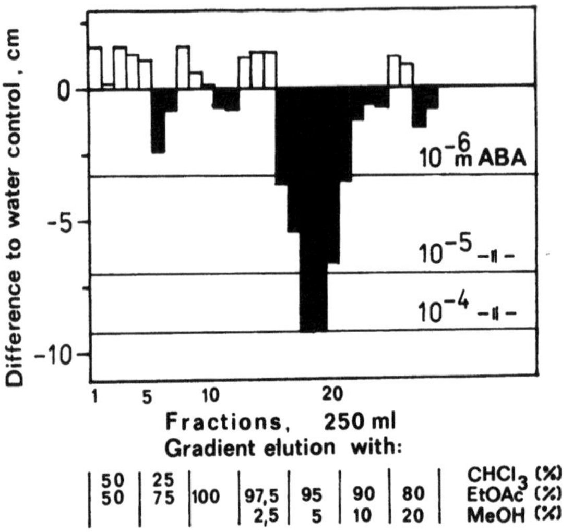

Fig. 3. Purification of ABA conjugate from dormant birch branches (560 g). Partition chromatography of the *n*-butanol extract on silica gel and subsequent bioassay (wheat seedlings) of aliquots of each fraction

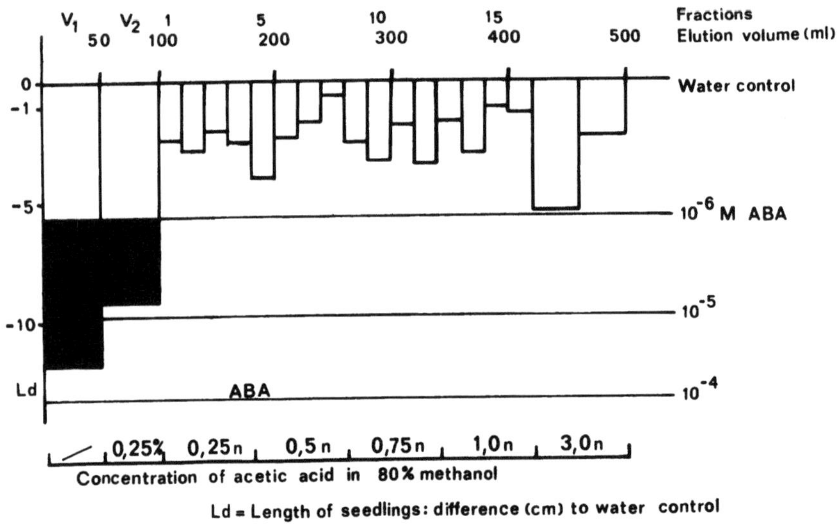

Fig. 4. Anion exchange chromatography on DEAE-Sephadex A-25 of purified ABA conjugate fraction (see Fig. 3). Detection by wheat seedling bioassay

glucose in GC on SE 30. Further characterization was achieved by reduction of the sugar component to the corresponding alcohol, subsequent acetylation and GC on OV 225. Two peaks were found, one of which corresponds to the respective glucose derivative ($R_t = 10.0$) and a second one with a retention time corresponding to a disaccharide derivative ($R_t = 11.8$). The neutral character of the inhibitor found by ion exchange chromatography indicates that the conjugating sugar moiety is linked to the carboxy group of ABA. This was confirmed by the fact that the conjugate can be hydrolyzed by alkali (1 N NaOH, room temperature). According to chromatographic results the conjugate seems to be an ABA disaccharide, containing at least one glucose unit. This assumption is supported by preliminary NMR data. Final structural elucidation is still under investigation using both NMR and MS as well as enzymatic studies (32).

Discussion

A number of studies have dealt with changes in levels of free ABA and its conjugate in plant organs as related to physiological processes. For example, in buds of *Ribes nigrum* and *Fagus silvatica* the highest level of free ABA was shown to occur in the autumn at the onset of winter dormancy (33). The ABA content declined throughout the winter and reached its lowest value before bud burst. Inversely, the level of the conjugate increased throughout the autumn and winter. Comparable investigations were made with nodes of vine carrying dormant buds, and with leaves as well (34). Progress in breaking dormancy of *Prunus amygdalus* buds was shown to be related to a decrease in total free ABA and a correlated increase in conjugated ABA (32). Some further physiological results on the role of ABA have been published regarding stratification of seeds (36), maturation of fruit (37, 38), and wilting (39, 40). From these data it may be concluded that conjugation of ABA is apparently involved in regulation of the physiologically active hormone level. The ABA conjugate possibly functions as a storage and/or translocation form. However, its physiological role cannot be considered to be fully established. Thus in shoots of tomato and *Beta vulgaris* it has been shown that the ABA glucosyl ester is not reconverted to free ABA during wilting and, therefore, conjugation seems to be irreversible (41).

In bleeding sap of *Salix* the ABA content was found to decrease during bud break (20). In *Betula pubescens*, bleeding sap flowed only for a period before bud burst. At that time the sap contained about 50 ng ABA/ml. This is in the same range as found in other species. In wood of birch branches ABA content was high during winter dormancy and decreased markedly during bud break, indicating that ABA may be stored in the wood and leached out by flowing sap. In contrast, ABA in the bark increased during sap flow and decreased subsequently. This may have been due to ABA accumulation in the bark during sap flow and basipetal translocation after bud break (24). This conclusion was confirmed by studies using radioactively labeled ABA (39). The ABA applied to the cut surface of branches at the stage of bud swelling and bursting was translocated acropetally in xylem sap and carried to the bark where it accumulated. No transport to the buds was observed. These data agree with recent results (43) showing that in apricot trees basipetal translocation of exogenous ABA is restricted before bud break.

Interpreting the results reported and discussed, it may be concluded that ABA in spring sap – at least of birch trees – apparently has no physiological significance with respect to the transition from dormancy to bud break. The ABA conjugate, preliminarily identified as a disaccharide ester, is present in very high amounts in both the bark and wood. Its content remains high up to the stage when the buds burst and the first leaves unfold. Apparently there is no leaching of ABA conjugate from the wood via spring sap. Regarding the gradient in branches, it may be concluded that the ABA conjugate continues to accumulate over the years, without remobilization. Therefore, it can be assumed that the ABA disaccharide ester represents an immobilized hormone form, irreversibly conjugated. As the ABA conjugate is stored in the bark in rather high amounts also during cambial activity, it is expected to be either biologically inactive and/or localized apart from the cambium cells. The biological activity observed in the wheat seedling bioassay may be due to hydrolysis in the course of the bioassay, as is well known to be the case for conjugated gibberellins [cf. (44)].

References

1. Kefeli, V.I., Kutaček, M.: In: Plant Growth Regulation. Pilet, P.E. (ed.), pp. 181–188. Berlin, Heidelberg, New York: Springer 1977, 305 pp.
2. Kefeli, V.I.: In: Natural Plant Growth Inhibitors and Phytohormones, 277 pp. The Hague, Boston: Junk 1978
3. Turetskaya, R.Kh., Kefeli, V.I., Kutaček, M., Vackova, K., Chumakovskii, N., Krupnikova, T.: Biol. Plant. *10,* 205–221 (1968)
4. Kutaček, M., Kefeli, V.I., Vackova, K., de Almeida, R.M.: Phytochemistry (in press, 1980)
5. Kefeli, V.I., Loznikova, V., Khlopenkova, L.P., Kof, E.M., Sidorova, K.K., Khvostova, V.V., Turetskaya, R.Kh., Chailakhyan, M.Kh.: Izv. Akad. Nauk SSSR Ser. Biol. *5,* 681–688 (1973)
6. Kefeli, V.I., Chailakhyan, M.Kh., Turetskaya, R.Kh., Kof, E.M., Kochankov, V.G., Loznikova, V.N., Khlopenkova, L.P., Janina, L.I., Komizerko, E.I., Mazin, V.V., Frolova, I.A., Jakovleva, L.V., Vlasov, P.V., Podolnyi, V.Z.: Fiziol. Rast. *22,* 1291–1298 (1975)
7. Kefeli, V.I., Turetskaya, R.Kh.: Dokl.. Akad. Nauk SSSR *170,* 472–475 (1966)
8. Jenkins, P.A., Shepherd, K.R.: New Phytol. *71,* 501–511 (1972)
9. Jenkins, P.A., Shepherd, K.R.: N. Z. J. For. Sci. *4,* 511–519 (1974)
10. Andersson, B., Haggstrom, N., Andersson, K.: J. Chromatogr. *157,* 303–310 (1978)
11. Little, C.H.A., Strunz, G.M., La France, R., Bonga, J.M.: Phytochemistry *11,* 3535–3536 (1972)
12. Wareing, P.F., Ryback, G.: Endeavour *29,* 84–88 (1970)
13. Bottini, R., Goleniowski, M., Correa, N.: Phyton (Buenos Aires) *36,* 19–24 (1978)
14. Little, C.H.A., Heald, J.K., Browning, G.: Planta *139,* 133–138 (1978)
15. Davison, R.M., Young, H.: Planta *109,* 95–98 (1973)
16. Harrison, M.A., Saunders, P.F.: Planta *123,* 291–298 (1975)
17. Hoad, G.V.: Proc. Res. Inst. Pomol. Skierniewice Ser. E *3,* 17–30 (1973)
18. Davison, R.M., Young, H.: Plant Sci. Lett. *2,* 79–82 (1974)
19. Lenton, J.R., Bowen, M.R., Saunders, P.F.: Nature (Lond.) *220,* 86–87 (1968)
20. Alvim, R., Hewett, E.W., Saunders, P.F.: Plant Physiol. *57,* 474–476 (1976)
21. Eagles, C.F., Wareing, P.F.: Physiol. Plant. *17,* 697–709 (1964)
22. Düring, H., Bachmann, O.: Physiol. Plant. *34,* 201–203 (1975)
23. Wareing, P.F.: Philos. Trans. R. Soc. London, Ser. B *284,* 483–498 (1978)
24. Dathe, W., Sembdner, G., Kefeli, V.I., Vlasov, P.V.: Biochem. Physiol. Pflanz. *173,* 238–248 (1978)
25. Dathe, W., Schneider, G., Sembdner, G.: Phytochemistry *17,* 963–966 (1978)
26. Koshimizu, K., Inui, M., Fukui, M., Mitsui, T.: Agric. Biol. Chem. *32,* 789–791 (1968)
27. Milborrow, B.V.: J. Exp. Bot. *21,* 17 -29 (1970)
28. Sembdner, G., Gross, D., Liebisch, H.W., Schneider, G.: In: Biosynthesis and Metabolism of Plant Hormones. Encyclopedia of Plant Physiology, New Series. Berlin, Heidelberg, New York: Springer (in press, 1980)
29. Hirai, N., Fukui, H., Koshimizu, K.: Phytochemistry *17,* 1625–1627 (1978)

30. Milborrow, B.V.: Phytochemistry *14,* 123–128 (1975)
31. Zeevaart, J.A.D., Milborrow, B.V.: Phytochemistry *15,* 493–500 (1976)
32. Dathe, W., Schneider, G., Sembdner, G., Schreiber, K.: in preparation
33. Wright, S.T.C.: J. Exp. Bot. *26,* 161–174 (1975)
34. Düring, H., Alleweldt, G.: Vitis *12,* 26–32 (1973)
35. Leshem, Y., Philosoph, S., Wurzburger, J.: Biochem. Biophys. Res. Commun. *57,* 526–531 (1974)
36. Diaz, D.H., Martin, G.C.: J. Am. Soc. Hortic. Sci. *97,* 651–654 (1972)
37. Goldschmidt, E.E., Goren, R., Even-Chen, Z., Bittner, S.: Plant Physiol. *51,* 879–882 (1973)
38. Rudnicki, R., Pieniazek, J.: Bull. Acad. Pol. Sci. Ser. Sci. Biol. *19,* 421–423 (1971)
39. Hiron, R.W., Wright, S.T.C.: J. Exp. Bot. *24,* 769–781 (1973)
40. Dörffling, K., Sonka, B., Tietz, D.: Planta *121,* 57–66 (1974)
41. Milborrow, B.V.: J. Exp. Bot. *29,* 1059–1066 (1978)
42. Vlasov, P.V., Kefeli, V.I., Lehmann, H.: unpublished
43. Kaska, N.: Acta Hortic. *80,* 219–224 (1978)
44. Sembdner, G., Borgmann, E., Schneider, G., Liebisch, H.W., Miersch, O., Adam, G., Lischewski, M., Schreiber, K.: Planta *132,* 249–257 (1976)

Regulation of Abscisic Acid Metabolism

B.V. MILBORROW [1]

Introduction

The dramatic changes which occur in the abscisic acid content of leaves during wilting and recovery have been the subject of considerable research interest since the phenomena were discovered by Wright and Hiron (1). Other stress conditions, such as heat, waterlogging, and salinity, cause similar increases in the amounts of abscisic acid (ABA) present in leaves but there is some uncertainty as to whether these treatments themselves caused the rise or whether a secondary, incipient wilting effect they induced was responsible (2).

Enough measurements of the distribution, concentration, and metabolism of ABA are available to enable a hypothesis to be put forward which can account for the changes in amounts of ABA that occur under conditions of stress.

In addition, a balance sheet of the products of metabolism of (±)-[2-^{14}C]ABA by apple seeds has been drawn up and a new class of metabolite has been characterized.

Materials and Methods

The growth of the plants, extraction procedures, and other manipulations have been described elsewhere (3–5). The only unreported methods are HPLC conditions for the glucosides, enzyme assays and chemical ionization and field desorption mass spectrometry.

HPLC

A Waters Associates, Model M6000A pump, U6K injector, and a reverse-phase C_{18} column (7.9 mm i.d. × 300 mm Bondpapak, 10-μm particle size) was used to separate the samples of glucoside in an ethanol/water/acetic acid (95/899/1 v/v for PAGS, retention volume 44 ml; 190/810/1 v/v for ABAGS, retention volume α = 80 ml, β = 48 ml) solvent mixture after the acetone extracts had been preliminarily separated on silica-gel tlc plates in chloroform/ethanol/water (75/22/3, v/v). A Phillips s.p.6. UV spectrometer, fitted with a 9-μl Z-cell, was used to monitor the effluent solution, and metabolites of (±)-[2-^{14}C]ABA were detected by their ^{14}C content.

1 School of Biochemistry, University of New South Wales, P.O. Box 1, Kensington, N.S.W. 2033, Australia

Glucose Oxidase Test

"Tes-Tape" is a urine glucose analysis paper (Eli Lilly, Australia) entirely specific for glucose (27, 28). Glucose (1 μg) in water (1 μl) gave an area (7 mm) of green colour within a few seconds of application. One-μl standards produced a set of spots with which glucose derived from the ABA metabolites could be compared.

α- and β-Glucosidase Assays

α-Glucosidase from yeast (43 units/mg; in solution at 5.0 mg/ml) and β-glucosidase from almonds *(Prunus amygdalus)* (5.1 units/mg) were obtained from Sigma Chemical Co. Both enzymes were made up in Na_2HPO_4/citrate buffer (0.05 M) to give solutions calculated to have 100 times the capacity to hydrolyze all of the ABA conjugate (100 ng in 25 μl) in 10 min at 20°C (based on conventional substances, the α at pH 6.8, the β at pH 5.2). The enzyme solutions (25 μl) were added to the dried substrates in a small tube and stirred. Samples (10 μl) were withdrawn after 2 and 10 min, mixed with 0.1 M citric acid solution (0.4 ml) and the ABA that had been released was extracted into ether, methylated with ethereal diazomethane, and determined by GLC.

Chemical Ionization Mass Spectrometry

The glycone from the equivalent of 2 μg MeABA of the conjugate (the sample had hydrolyzed on methylation) was trimethylsilylated at 90°C with "Tri-Sil Z" (Trimethylsilylimidazole in dry pyridine, 1.5 mEq/ml, Pierce Chemical Co., Rockford, Illinois, USA) in a "Reacti-vial". The silylated products were injected into a 1.5% OV 1 on Gas Chrom Q (100–200 mesh) column which was held at 150°C for 1 min then programed to rise at 10°/min to 300°C. Mass spectra were obtained by Dr. A. Duffield of the UNSW Biomedical MS Unit in a Finnigan Model 3200 Chemical Ionisation GC/MS system with methane as carrier gas (20 ml/min) and ionization gas (source pressure 0.8/0.9 Torr). The GC was interfaced to a Finnigan Model 6115 Data System. This method of ionization characteristically protonates molecular ions. An added complication is that the cracking patterns of MeABA and TMS ABA are very different from the familiar electron impact pattern (6).

Field Desorption Mass Spectrometry

Field desorption mass spectrometry of α-ABAGS was carried out by Dr. K. Murray of C.S.I.R.O., N. Ryde, Sydney, using a Varian 311A mass spectrometer with a combined electron impact, chemical ionization and field desorption source and a high-temperature emitter prepared by Dr. Murray. Fast, repetitive scanning (3 sec over the molecular ion region during rapid, linear programed heating of the emitter (22 mA, 0.2 mA/sec) showed the presence of a cluster of ions corresponding to the intact, cationised molecule (potassium) at $M + K^+ + 1/_z$, $M + K^+ - H/_z$ and $M + K^+ - H_2/_z$. A total of 2 μg α-ABAGS was used for 3 runs.

Results and Discussion

A Proposed Mechanism for the Regulation of ABA Contents

The following observations are to be considered. Chloroplasts are the site of ABA biosynthesis (7), and they are normally quite impermeable to ABA (8). Loveys (9) found that chloroplasts isolated from turgid leaves contained 96% of the leaf's ABA while those from wilted leaves contained some 15%. In the wilted leaves a large proportion of the ABA must be in the cytosol. Gillard and Walton (10) have found that ABA is oxidised by a microsomal fraction from wild cucumber fruit, so the majority of the oxidative metabolism can be expected to occur in the cytosol. This is in accord with Harrison and Walton's (11) results where the turnover of (+)-[2-^{14}C]ABA in wilted bean *(Phaseolus vulgaris)* leaves was surprisingly rapid, half turning over about every 3 h. These authors proposed that there was no discrete metabolic signal to initiate breakdown of the "extra" wilt-induced ABA once the leaves had regained turgor, there was merely a cessation of rapid biosynthesis, and the "extra" ABA would be rapidly destroyed by the excess degradative capacity of the cells.

If, as Loveys' (9) data suggest, virtually all the ABA in a turgid leaf is confined to the chloroplasts, then the high concentration within the organelles would effectively "switch off" biosynthesis. The synthesis of the "extra" wilt-induced endogenous (+)-ABA has been shown to be prevented by the supply of large amounts of (±)-ABA to tomato leaves before they were wilted (12). If, as is now proposed, stress makes the plastid membranes permeable to ABA, then this could act as the mechanism modulating the overall ABA content of leaves (5). A similar proposal has been suggested by Mansfield et al. (13). Stress would cause the ABA to leak out of plastids, as found by Loveys. Four consequences can be expected to follow from this:

1. The ABA would be available for transport to other cells and so stress could begin to close stomata even before an overall rise in leaf ABA is detectable, as reported by Beardsell and Cohen (4). Indeed it is possible that the leakage of ABA from mesophyll chloroplasts could cause stomatal closure at the same time as a reduction occurred in the overall leaf content of ABA if degradation were to exceed biosynthesis transiently. Higher concentrations than normal would be available for transport out of stressed leaves (15, 16).
2. The availability of ABA for oxidation by cytosolic oxygenases will be detectable as a rapid turnover of ABA, as observed by Harrison and Walton (11), and a rapid rise in the amounts of phaseic acid (PA) and dihydrophaseic acids (DPA; epi-DPA) present.
3. A fall in the concentration of ABA within the leaf chloroplasts would allow rapid biosynthesis of ABA to begin initially to raise the concentration of ABA in the cell volume open to the entry of ABA, and also to compensate for losses by degradation and translocation.
4. On regaining turgor most of the extrachloroplastic ABA will be destroyed and, once the chloroplasts are no longer permeable to ABA, the intrachloroplastic concentration will return to a level which inhibits further biosynthesis.

The high concentrations of ABA in ripening avocado fruits and their apparently unregulated biosynthesis of ABA (17) are now considered to be the result of the increased permeability of membranes which occurs in the climacteric phase.

Any treatments that increase the permeability of chloroplast membranes to ABA should cause a rise in the overall concentration in leaves. However, it may be that it is not so much loss through the outer membrane as leakage of bound material from membranes. Farnesol has been found to close stomata (18), but it has also been observed to cause visible damage to chloroplasts (19). Phenylmercuric acetate has been found to close stomata (20), and it is a powerful inhibitor of photosynthetic reactions in chloroplast fragments which depend on membrane integrity. It was considered, therefore, that the action of these two anti-transpirants could be interpreted in terms of their effects of chloroplast membrane permeability. Spray application of both compounds to spinach leaves caused the amounts of ABA they contained to rise almost as much as did wilting (Table 1). The hypothesis can provide an interpretation of why stress phenomena, other than wilting, cause a change in the amount of ABA in leaves

Table 1. Increase in the content of ABA of spinach leaves by mercuric acetate, farnesol and wilting

Treatment	Wt. of spinach leaves	Percent loss on wilting	ABA μg/kg original fresh wt.
Control, turgid	17.2	–	17.4
Control, wilted	16	18	212
Farnesol, turgid	17.3	–	227
Farnesol, wilted	17.1	21	262
Phenylmercuric acetate, turgid	16.8	–	185
Phenylmercuric acetate, wilted	16.6	19	187

The plants were sprayed with a 2×10^{-3} M solution of farnesol (mixed isomers), 1×10^{-3} M phenylmercuric acetate, and analysed 23 h later. The wilt treatment was given for the last 6 h

(2, 21). Any treatment that increases chloroplast permeability would be expected to increase the overall content of ABA. The observations of Zabadal (22), Beardsell and Cohen (14), and Wright (23) show that there is a sharp demarcation at about – 9 bars water potential beyond which wilting and the rise in content of ABA occur. Furthermore, the amounts of ABA present is linearly related, at least up to – 11.5 bars, to the negative water potential. This suggests that the modulation of the membrane's permeability is progressive. If this is so, then the concentrations measured after 6 h by Wright (23) represent equilibrium values between the rate of leakage and the rate of degradation.

The proportion of the cellular volume open to the entrance of ABA, the volume occupied by the chloroplasts, and the rate of turnover of ABA in the cytosol should define the magnitude of the rise in the overall content of ABA. An attempt was made to discover whether or not ABA could enter the vacuoles by collecting vacuolar sap from vesicles of citrus fruit. ABA and ABA released by alkaline hydrolysis were present in much higher overall concentrations in ruptured cell debris, cell wall material, and unbroken cells than in the vacuolar sap (Table 2). This suggests that low concentrations of ABA and its conjugates are present in vacuolar sap.

Two sets of results appear to be in disagreement with the hypothesis. The first is that the amount of [^{14}C] mevalonate incorporated into ABA by a lysed chloroplast preparation (7) was doubled by the addition of (±)-ABA (28 μg/ml). However, this does not necessarily indicate that net biosynthesis was increased; dilution of the newly synthesised [^{14}C]ABA could have doubled the efficiency of recovery of the labelled ABA during the isolation to radiochemical purity. The concentration of the (±)-ABA added would have been less than the concentration of (+)-ABA in the protoplasm of intact avocado fruit cells (8 μg/g total fruit) if vacuoles, fat droplets, and cell walls do not contain ABA and occupy four-fifths or more of the total fruit's volume.

Table 2. Presence of free and alkali-hydrolysable abscisic acid in citrus vesicles

	Wt. of fraction extracted (g)	Wt. free ABA (mg/kg)	Wt. ABA released by alkaline hydrolysis (mg/kg)
Mandarin			
Protoplasm and cell walls	0.65	0.75	1.79
Vacuolar sap	0.52	0.14	0.48
Grapefruit			
Protoplasm and cell walls	1.35	3.98	6.91
Vacuolar sap	0.64	0.77	2.68

The second anomaly is that chloroplasts from wilted bean leaves (9) contained 70% more ABA than those from turgid ones. This could have been caused by slight damage to the chloroplast envelopes during isolation (there is no guarantee that chloroplasts from the two leaf samples encountered identical mechanical stresses during isolation from crisp, turgid, or flaccid wilted leaves). If virtually all the ABA in normal, unwilted leaves were in the chloroplasts, then any damage during isolation could cause leakage of a considerable part of the free ABA before the membrane could reseal. In the wilted leaves the ABA would be expected to be present at high concentrations inside and outside the chloroplasts and so damage might not cause such severe losses. Loveys also noted that hypotonic lysis of chloroplasts from turgid leaves caused leakages of a small quantity of ABA into the supernatant, while a considerable proportion remained bound to the pellet from which it could be extracted with acidic methanol. This may indicate that release from a bound (i.e., noncovalently linked) form of ABA could also be involved in the leakage of ABA from chloroplasts.

Quantitative Aspects of the Metabolism of (±)-ABA

A report by Rudnicki and Czapski (24) that (±)-[1-^{14}C]ABA was decarboxylated by apple seeds prompted a repetition of the experiment, but using (±)-[2-^{14}C]ABA, in an attempt to find a neutral C_{14} or other product formed by decarboxylation. The experiment was repeated twice, with various elaborations, but no neutral, ether-soluble, labelled products were found (4). [^{14}C]CO_2 was not produced by surface-sterilised seeds, but a very little was collected from seeds that had been collected by hand and not sterilised (Fig. 1). Phaseic acid, DPA, and epi-DPA were formed by the isolated embryos and seed coats but not by cultures of microorganisms isolated from decomposing fruit and soil. The solid residues were combusted by a modified Van Slyke method (25) and a balance sheet of radioactivity drawn up. Over 90% of the (±)-[2-^{14}C]ABA supplied at the beginning of the experiment was accounted for after 49 days incubation (Table 3).

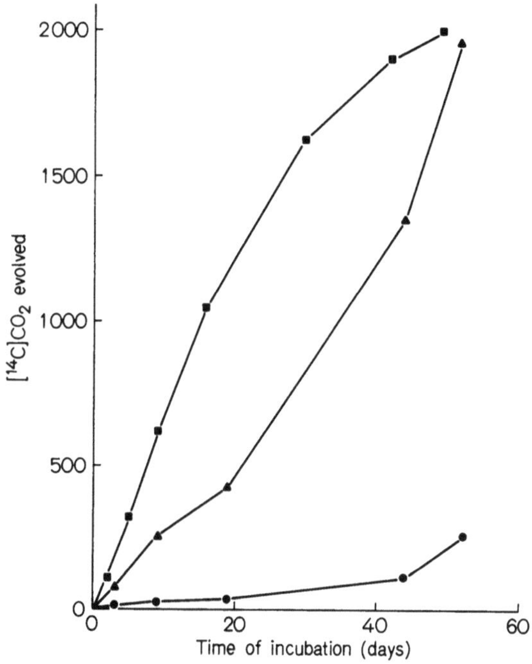

Fig. 1. Evolution of [^{14}C]CO_2 from sterile (●) and nonsterile (▲) apple seeds. In a preliminary experiment the seeds were removed from the fruit by hand (■). 200 seeds absorbed (±)-[2-^{14}C]ABA (2 ml, 26.5 μg/ml, 11.5 μCi/μmol) and were held at 4°C during the experiment except for periods of a few hours when the CO_2 was collected in an ethanolamine/methoxyethanol (1/3 v/v) trap

One batch of seeds was separated into husks and total, undamaged embryos; these were soaked separately in a solution of (±)-[2-^{14}C]ABA. When extracted after 71 days, 26% of the ^{14}C taken up by the embryos remained in the aqueous fraction after mild alkaline hydrolysis and ethyl acetate extraction. At the same time Dr. B.R. Loveys of C.S.I.R.O., Adelaide, found polar conjugates of (±)-[2-^{14}C]ABA in extracts of tomato plants, and the isolation and identification of the compounds were undertaken jointly. We have isolated and identified two compounds from tomato shoots (6), the first is 1′-O-abscisic acid-β-D-glucopyranoside (β-ABAGS).

The occurrence of β-ABAGS *(1)* had not been previously suspected because the 1′-tertiary hydroxyl group of ABA cannot be acetylated, and such forcing conditions have to be used to add TMS to the 1′-O position that a large proportion of the ABA is destroyed (26). It was probably tacitly assumed that the failure to react was caused by steric hindrance, but Dreiding and space-filling models indicate that the glycosyl moiety can be readily accomodated at C-1′ of ABA. There are, however, indications of strong interactions between the glucose and the diene system of the ABA side chain. These manifest themselves as an overall broadening and upfield shift of some of the

Table 3. Distribution and recovery of ABA fed to apple seeds. (±)-[2-^{14}C]ABA fed to 200 intact apple seeds (2.0 ml, 26.5 μg/ml, 11.5 μCi/μmol, i.e., 4.97 × 10^6 dpm/53 μg per 8.90 g seed fresh weight). For the 71-day treatment, husks and seeds were used separately (2.49 × 10^6 dpm 27 μg, 1 ml). All data are for dpm/200 seeds

Duration of stratification (days)	ABA not taken up (dpm)	dpm remaining on filter paper (wet combustion)	Wet cumbustion of solid residues		Total acetone extract		CO_2 trapped dpm	Total dpm accounted for	(±)-[^{14}C]ABA accounted for
			Husk (dpm	Embryo (dpm)	Husk (dpm)	Embryo (dpm)			
7	1239100	1308200	690273	27232	1493626	13754	1359	4773544	96.0%
49	1425200	440718	192436	6317	2750440	65690	3968	4884769	98.3%
100 seeds separated into husk and embryo after taking up the (±)-[2-^{14}C]ABA solution. 2.49 × 10^6 dpm, 27 μg ABA in 1.0 ml									
Husk 71	591046	342297	47938	–	1072052	–	4495	2057828	82.6%
Embryo 71	1174319	161130	–	84552	–	930337	8885	2359223	94.7%

signals in the NMR spectrum of the carbon-borne hydrogen atoms of the glucose residue. There are also differences between the mass spectra (chemical ionisation) of ABATMS and the ABATMS moiety lost from trimethylsilylated ABAGS in the mass spectrometer. Most importantly ABAGS is unstable to methylation giving MeABA and glucose.

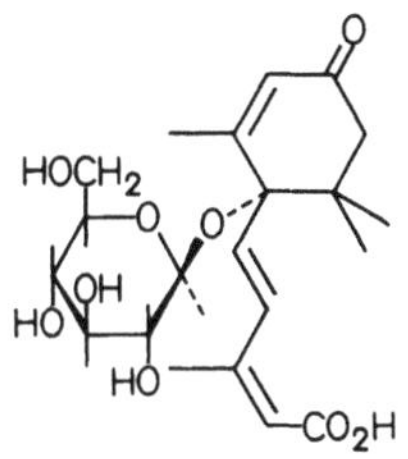

(1)

1'-O-(+)-Abscisic acid-β-D-glucopyranoside
(β-ABAGS)
(ABA β-glucoside)

The glucose released by methylation of ABAGS was identified by GLC/MS after trimethylsilylation and by applying approximately 1 μg in 1 μl water to "Tes-Tape". Trimethylsilylated β-ABAGS was subjected to chemical ionisation MS with probe insertion by Dr. Alan Duffield (UNSW Biomedical Mass Spectrometry Unit); no parent ion could be detected. The cracking pattern showed peaks at m/z 319 and 337 attributable to the C-1 trimethylsilylated ABA residue's losing the 1'-O-glucosyl residue with, and without, the C-1' oxygen atom. There was also a peak at m/z 117 which is attributed to COOTMS+H, an insignificant fragment from C-1-TMS ABA, but the major peak at m/z 183 seen with this latter derivative (side chain + TMS) was small with TMS ABAGS. These differences are attributed to the interaction of the side chain with the glucosyl residue.

A closely related compound, separated with difficulty from β-ABAGS and present in similar amounts in tomato shoots, has been identified as α-ABAGS *(2)*. Like the

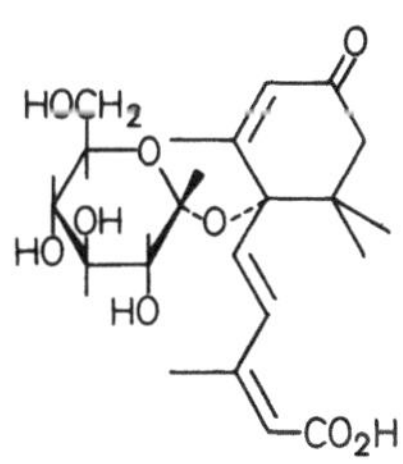

(2)

1'-O-(+)-Abscisic acid-α-D-glucopyranoside
(α-ABAGS)
(ABA α-glucoside)

β-epimer it usually, but not invariably, hydrolyses spontaneously on methylation. Its NMR spectrum is almost identical (Fig. 2) to that of β-ABAGS except that the signal of the anomeric C-1 proton of the α-glucosyl residue is upfield at 6.06 δ and shows coupling of about 0.04 δ, consistent with an α-configuration. The other difference is that the signal of the C-2′ methyl of ABA in the α-glucoside is 0.026 δ further upfield than that of the analogous signal of the β (1.759 δ). This difference is also attributed to deshielding of the methyl group by the glucosyl residue. Conversely the deshielding

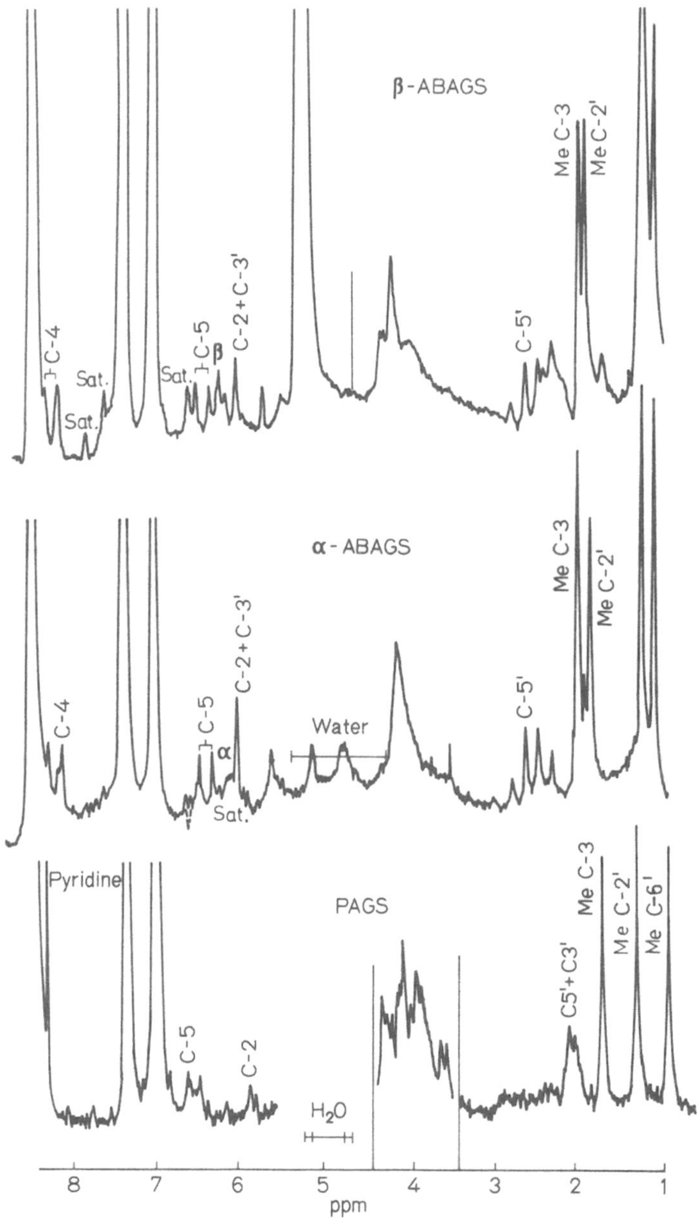

Fig. 2. Proton NRM spectra of 1′-O-abscisic acid β-D-glucopyranoside (β-ABAGS), α-ABAGS and 1′-O-phaseic acid β-D-glucopyranoside (β-PAGS) in D_5 pyridine. The spectra were obtained in a Jeol 100 MHz Fourier-transform machine on 2–5 μg after 35,000 scans

effect of the diene system displaces the signals of the glucosyl protons upfield to varying degrees. Furthermore there is a broadening of the signals of all the carbon-borne protons of glucose so that they appear as an unresolved envelope. The α- and the β-anomers are hydrolysed by both α- and β-glucosidase, the α-ABAGS more rapidly than β-ABAGS by the α-enzyme and β-ABAGS more rapidly by β-glucosidase than by α (Table 4). Both enzymes were specific for their substrates when assayed on maltose (α) and phloridzin (β).

Table 4. Action of α- and β-glucosidases on α-ABAGS and β-ABAGS

Substrate	Enzyme	Time	ABA released (ng)
α-ABAGS	α-glucosidase	2 min	28
	α-glucosidase	10 min	39
α-ABAGS	β-glucosidase	2 min	16
	β-glucosidase	10 min	25
β-ABAGS	α-glucosidase	2 min	20
	α-glucosidase	10 min	20
β-ABAGS	β-glucosidase	2 min	47
	β-glucosidase	10 min	48

The specific activity of the β-ABAGS sample was lower than that of the (±)-[2-^{14}C] ABA fed, consequently it had been diluted with endogenous (+) and so it occurs naturally. The O.R.D. spectrum of the ABA, released from both samples on methylation, showed an excess of (–) so, presumably, the formation of the glucosides is not stereospecific. Dr. G. Ryback of Shell Research, England, kindly measured these spectra on the instrument used previously.

Another glucoside, which crystallised from pyridine, has been isolated from apple seeds and from tomato leaves, it has been identified as 1′-O-phaseic acid-β-D-glucopyranoside (PAGS). This compound *(3)* is much more stable than ABAGS, it does not

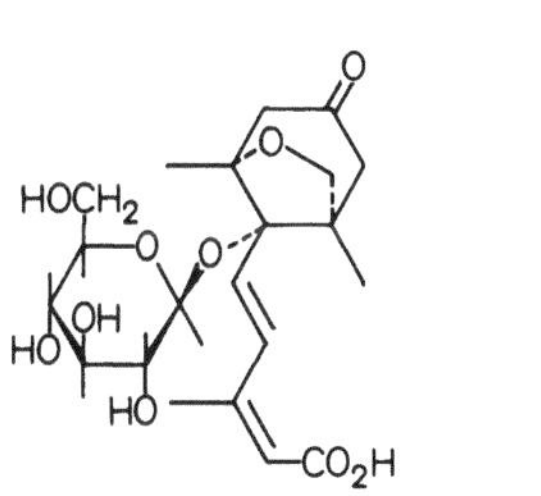

(3)

1′-O-Phaseic acid-β-D-glucopyranoside

cleave on methylation, and is hydrolysed by mild alkali with difficulty. In accord with these characteristics the glucosyl NMR signals are unbroadened. However, that of the

C-2′ methyl is downfield by 0.3 δ in comparison with that in the free phaseic acid. So far only the β-glucoside of PA has been detected; like ABAGS, PAGS occurs naturally.

These compounds comprise a new class of metabolites of ABA, but their separate roles in the ABA economy have yet to be established. ABAGS is unstable to mild alkaline hydrolysis and so it contributes, with the glucose ester (ABAGE), to the "ABA released by alkaline hydrolysis" fraction. Evidence has recently been presented *(3)* to show that ABA is not released from the conjugated fraction of tomato shoots in vivo, even when the shoots are wilted. The demand for the "extra" ABA is apparently met by fresh biosynthesis.

"Fast" and "Slow" Reactions of Phytohormones

The measurements of response time of growth and stomatal closure after applications of IAA and ABA have established that there are "fast" responses which can be detected within a few minutes of the start of treatment. On the other hand there are "slow" responses which require one or more hours to be detectable and involve changes in the pattern of protein synthesised.

A question to be answered is whether a fast reaction is a necessary preliminary to the change in metabolism seen in the slow reaction. This question can now be answered in the negative for ABA.

There are two reports of the action of the (+) and (–) enantiomers of ABA on stomatal closure. Cummins and Sondheimer (29) found that (+)-ABA was highly effective in barley leaves while (–)-ABA was considerably less so, and its weak action may have been caused by contaminating (+). Kriedemann et al. (30) compared (+)-with (±)-ABA on *Xanthium* stomata; they reported that the activity of the racemate was half that of the (+), a result consistent with (–) being inactive.

On the other hand (–)-ABA has been shown to be equally, or almost equally, active in a number of tests which involve slow growth reactions [seed germination (31), growth of bean barley (32), root and shoot axes]. It appears, therefore, that the structural requirements of the active sites for fast and slow responses to ABA are different, and it seems reasonable to conclude that the reactions are quite separate.

Acknowledgment. I thank Mr. G. Vaughan and Dr. A.G. Netting for help with many of the experiments, Drs. A. Duffield and K. Murray for the MS, and Mr. G. Grossman for the NMR. The work was supported by the Australian Research Grants Committee.

References

1. Wright, S.T.C., Hiron, R.W.P.: Nature (Lond.) *224*, 719–720 (1969)
2. Hiron, R.W.P., Wright, S.T.C.: J. Exp. Bot. *24*, 769–781 (1973)
3. Milborrow, B.V.: J. Exp. Bot. *29*, 1059–1066 (1978)
4. Milborrow, B.V., Vaughan, G.: J. Exp. Bot. *30*, 983–995 (1980)
5. Milborrow, B.V.: Aust. J. Plant Physiol. (in press, 1980)
6. Loveys, B.R., Milborrow, B.V.: (in press, 1980)
7. Milborrow, B.V.: Phytochemistry *13*, 131–136 (1974)

8. Wellburn, A.R., Hampp, R.: Planta *131*, 95–96 (1976)
9. Loveys, B.R.: Physiol. Plant. *40*, 6–10 (1977)
10. Gillard, D.F., Walton, D.C.: Plant Physiol. *58*, 790–795 (1976)
11. Harrison, M.A., Walton, D.C.: Plant Physiol. *56*, 250–254 (1975)
12. Milborrow, B.V.: In: Plant Growth Substances 1970. Carr, D.J. (ed.), pp. 281–290. Berlin, Heidelberg, New York: Springer 1972
13. Mansfield, T.A., Wellburn, A.R., Moreira, T.J.S.: Philos. Trans. R. Soc. London Ser. B (1978)
14. Beardsell, M.F., Cohen, D.: Plant Physiol. *56*, 207–212 (1975)
15. Hoad, G.V.: Planta *113*, 367–372
16. Zeevaart, J.A.D.: Plant Physiol. *59*, 788–791 (1977)
17. Milborrow, B.V., Robinson, D.R.: J. Exp. Bot. *24*, 537–548 (1973)
18. Ogunkanmi, A.B., Wellburn, A.R., Mansfield, T.A.: Planta *117*, 293–302 (1974)
19. Fenton, R., Davies, W.J., Mansfield, T.A.: J. Exp. Bot. *28*, 1043–1053 (1977)
20. Squire, G.R., Jones, M.B.: J. Exp. Bot. *22*, 890–891 (1971)
21. Mizrahi, Y., Richmond, A.E.: Plant Physiol. *50*, 667–670 (1972)
22. Zabadal, T.J.: Plant Physiol. *53*, 125–127 (1974)
23. Wright, S.T.C.: Planta *134*, 183–189 (1977)
24. Rudnicki, R., Czapski, J.: Ann. Bot. *38*, 189–192 (1974)
25. Watson, G.R., Williams, J.P.: Anal. Biochem. *33*, 356–365 (1970)
26. Most, B.H., Gaskin, P., MacMillan, J.: Planta *92*, 41–49 (1970)
27. Comer, J.P.: Anal. Chem. *28*, 1738–1750 (1956)
28. Seltzer, H.S., Loveall, M.J.: JAMA *167*, 1826–1830 (1958)
29. Cummins, W.R., Sondheimer, E.: Planta *111*, 365–369 (1973)
30. Kriedemann, P.E., Loveys, B.R., Fuller, G.L., Leopold, A.C.: Plant Physiol. *49*, 842–847 (1972)
31. Milborrow, B.V.: Abscisic acid. In: Phytohormones and Related Compounds – A Comprehensive Treatise. Letham, D.S., Goodwin, P.B., Higgins, T.J.V. (eds.), vol. I, pp. 295–347. Elsevier/North Holland: Biomedical Press 1978
32. Sondheimer, E., Galson, E.C., Chang, Y.P., Walton, D.C.: Science *174*, 829–831 (1971)

Studies on the Role of Abscisic Acid in Stomatal Movements

K. DÖRFFLING, D. TIETZ, J. STREICH, and M. LUDEWIG [1]

Introduction

It is well known that abscisic acid (ABA) was discovered in investigations on bud dormancy and abscission. Numerous attempts have been undertaken to establish a regulatory role of ABA in both of these processes. However, so far we have only little evidence that ABA functions as a dormancy hormone or as a regulator of abscission (1).

In recent years another possible function of ABA in higher plants has interested many workers, namely the control of stomatal action. The hypothesis that ABA is involved in the regulation of guard cell movement under stress conditions is based on several observations, especially the following:

1. Open stomata close rapidly, and closed stomata do not open in the presence of ABA (2).

2. Water stress not only induces stomatal closure, but also causes a considerable rise in the level of ABA in leaves (3).

3. Under certain conditions a close correlation exists between the amount of ABA within leaves and the degree of stomatal opening (4).

There is evidence that ABA is produced in the chloroplasts of the mesophyll cells, from where it is assumed to be released into the cytosol and to migrate to the guard cells (5), which are presumably unable to synthesize ABA (6). That ABA is transported to guard cells where it accumulates has been shown by Itai et al. (7).

Despite the bulk of evidence that ABA is à regulator of stomatal reactions under stress conditions, some problems remain. Among them are the following:

1. Does ABA accumulate rapidly enough and in sufficient amounts to account for the fast reaction of stomata in response to water stress? Several authors (8–10) have observed that stomata of different plant species close in response to stress before an increase of the ABA level is observable. This discrepancy may be explained by the assumption that only a very small amount of ABA, below the limit of detection, is transported from the mesophyll to the guard cells.
2. Up to now the assumption that ABA is produced in the mesophyll and transported to the guard cells is based on results obtained with only one plant species, *Vicia faba* (6). Therefore further data are necessary to confirm this result.
3. In many plant species the opening movement of the stomata after a period of water stress is delayed for several hours or even days. This reaction occurs although the

---

1 Institut für Allgemeine Botanik und Botanischer Garten, Universität Hamburg, 2000 Hamburg, FRG

leaf water deficit is eliminated within a much shorter time. Stålfelt (11), who first described this phenomenon, supposed it to be a "safety mechanism" against serious water loss. Direct evidence that ABA might be involved was first presented by Wright (12). In subsequent papers, however, other workers (8, 10) concluded that impaired stomatal function after water stress was not correlated with high residual amounts of ABA.

This report deals with experiments on the problems outlined above, especially on the relationship between ABA and after-effect of water stress on stomatal opening in some plants of different ecological adaptation. Moreover, some results on the metabolism of ABA after a wilting period will be reported.

Materials and Methods

Materials and methods are described in detail elsewhere (13–15).

Plant Material

Most studies on the relationship between leaf water status and ABA were performed with leaves of *Vicia faba, Tradescantia x andersoniana* and, more recently, with *Commelina communis. Vicia faba* belongs to the mesophytes, whereas the ecological relationship of *Tradescantia* and *Commelina* is not fully clear; the morphological character of these plants resembles that of hygrophytes, but their ability to withstand long periods of water stress characterizes them as xerophytes. Other mesophytes used were: *Pisum sativum, Helianthus annuus,* and *Sorghum bicolor.* Typical hygrophytes used were: *Mentha aquatica* and *Menyanthes trifoliata.*

Treatment of Leaves

Isolated Leaves. To induce a water deficit, leaves cut from well-irrigated plants were placed on dry filter paper under fluorescent lamps. For recovery, the stressed leaves were transferred to waterfilled boxes. Changes in the leaf water status were followed by measuring either the fresh weight or the leaf water potential. The variation in the stomatal pore width was measured microscopically or with a diffusion porometer (Lambda Instruments, Model Li 60). The content of ABA was analyzed by gas-liquid chromatography. In *Commelina,* ABA contents were measured in whole leaves as well as in the isolated lower epidermis and in the mesophyll.

Application of ABA. To investigate the possibility of modifying the after-effect duration, 10^{-4} M racemic ABA (Fluka) was added to the water to which the wilted leaves were transferred.

Attached Leaves. Several experiments were performed with intact *Vicia faba* plants cultivated in garden soil. They were stressed by withholding irrigation. Recovery from stress was induced by rewatering.

Metabolism of ABA

Pea seedlings cultivated in nutrient solution were wilted for three days. During recovery, they were pulse-labeled with ABA-1-^{14}C (Bionuclear Corporation). After 24 h the seedlings were extracted with ethanol and the radioactive products analyzed by thin-layer chromatography, autoradiography, and combined gas chromatography–mass spectrometry (GC-MS).

Results and Discussion

Stomatal Reaction and ABA Levels after the Onset of Stress

To determine whether the increased production of ABA in a water-stressed leaf precedes stomatal closure, the time course of the stomatal reaction and the change of the ABA level were followed in both isolated and attached leaves.

Isolated leaves subjected to a wilting period closed their stomata rapidly, *Commelina* leaves within 5 min, *Vicia* and *Sorghum* leaves within 15 min. The ABA level, however, measured as total leaf ABA, increased only after about 30 min in *Vicia* (Fig. 1) and in *Sorghum*, and after about 1 h in *Commelina*. Thus, stomatal closure precedes the increase of the ABA level.

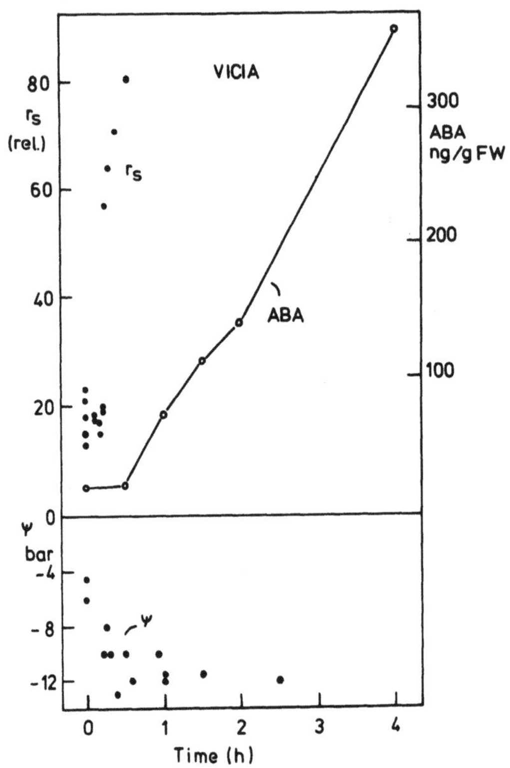

Fig. 1. Relationship between stomatal resistance (r_S), water potential (ψ), and abscisic acid level in water-stressed isolated leaves of *Vicia faba*

The objection can be raised, however, that measurements of total leaf ABA content may not give conclusive information on whether the guard cells receive the stress signal from the mesophyll within the time between the onset of stress and the start of stomatal closure. Therefore, we have studied the distribution of ABA between the mesophyll and the epidermis in the *Commelina* leaves. When stressed leaves were separated into the lower epidermis and the mesophyll, a significant rise of the ABA level was observed in both tissues one hour after the onset of stress. However, when the epidermis was stressed separately after isolation from the unwilted leaf, no change in ABA content was observed (Fig. 2). These results confirm the observation of Loveys

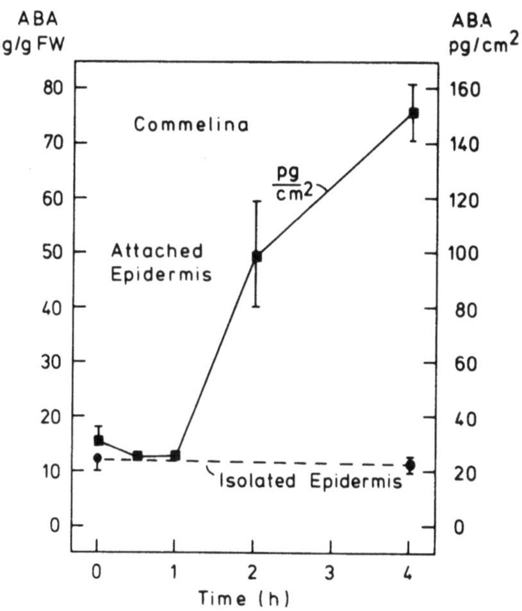

Fig. 2. Variation in the level of abscisic acid in the lower epidermis of *Commelina* leaves, stressed either before (attached epidermis) or after (isolated epidermis) isolation from the leaf

(6) that the epidermis is not able to produce ABA in response to stress. However, they do not support the hypothesis that stomatal closure in response to water stress is affected by ABA, since the stomata closed within 5 min, long before the ABA content in the epidermis increased. Similar results have been obtained recently by Pierce and Raschke (24) with *Commelina.*

A better correlation between stomatal closure and change in the ABA level was found in attached leaves of *Vicia faba.* Water stress developed much slower in attached leaves (third nodes) of four-week-old bean plants than in isolated leaves, although the environmental conditions were the same. The stomata closed after about one day, and this reaction was accompanied by a simultaneous increase in the ABA level (Fig. 3). It is possible that the slow stomatal reaction in attached leaves is induced by the increased ABA level, whereas the fast reaction in detached leaves may be caused in another way. However, in both cases stomatal closure has to be regarded as hydroactive, since the reaction of the guard cells is accompanied by an efflux of potassium ions (24). More information is necessary to resolve the problem of whether ABA induces stomatal closure under stress conditions.

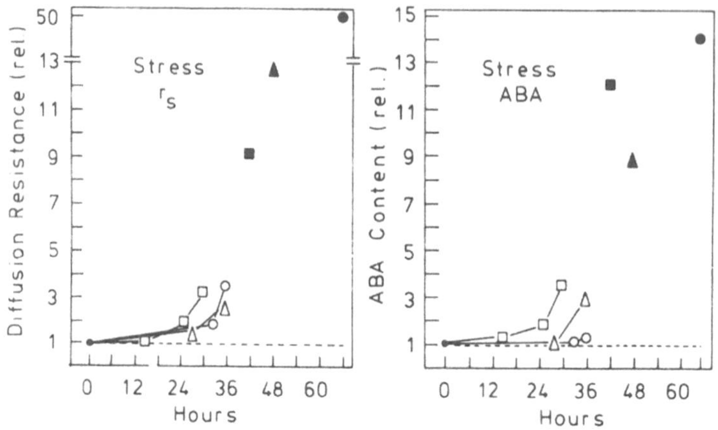

Fig. 3. Relationship between stomatal resistance and abscisic acid levels in water stressed attached leaves of *Vicia faba*. The symbols represent different experiments

After-Effects of Stress and Their Relation to ABA

Among the several attempts to test the hypothesis that ABA is involved in the regulation of stomatal movements, an investigation on the relationship between after-effects of stress and ABA levels in isolated leaves of different plant species provided evidence for such a role.

Experiments with Isolated Leaves. When isolated leaves of *Vicia faba* were wilted for a few hours at moderate temperature under fluorescent lamps and then rewatered, the opening of the stomata was clearly delayed in relation to the disappearance of the water deficit. The duration of this after-effect was related to the duration of the wilting period. In leaves wilted for 2.5 h it lasted 2–3 h, whereas in leaves wilted for 4 h, 3.5–4.5 h passed before the stomata started to open again (Fig. 4 and Table 1). The

Table 1. Relationship between duration of stress, delay of stomatal opening, and levels of "stress-ABA" in isolated leaves of *Vicia faba*

Duration of stress (h)	Delay of stomatal opening (h)	Stress-ABA a the time of rewatering (ng per leaf)	Stress-ABA at the time of stomatal opening (ng per leaf)
1	1–2	126	136
2.5	2–3	290	336
4	3.5–4.5	404	410

recovery of the water status, measured as the increase of fresh weight or leaf water potential, began immediately upon transfer of the wilted leaves to water. Full turgidity was regained sooner than full stomatal opening. This is evidence that the after-effect of stress on stomatal opening is not due to the persistence of leaf water deficit. The duration of the after-effect was not only related to the length of the stress period, but also to the change of the ABA level. The amount of ABA increased during the wilting

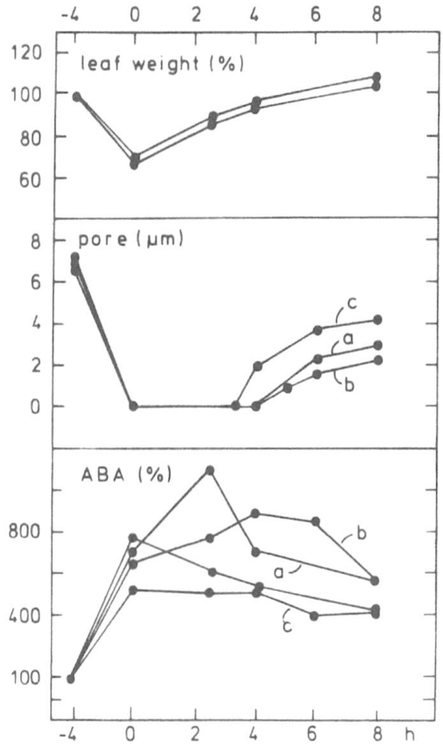

Fig. 4. Changes in leaf weight, abscisic acid content, and stomatal opening during and after a wilting period lasting 4 h in isolated leaves of *Vicia faba. a, b, c* represent different experiments

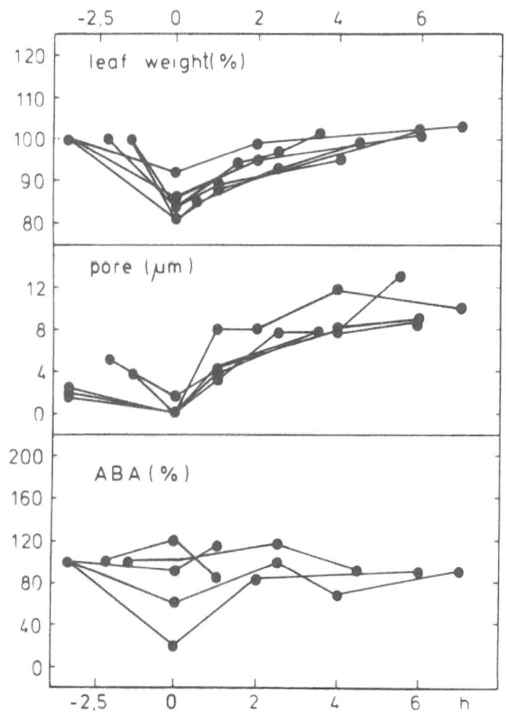

Fig. 5. Changes in leaf weight, abscisic acid level, and stomatal opening of isolated *Tradescantia* leaves during and after a 2.5-h stress period

period and even during the first phase of recovery. Afterwards it decreased slowly, reaching control levels in about 24 h. It seems noteworthy that the stomata began to open when the ABA level had just reached its maximum. Moreover, stomatal opening began at quite different ABA levels. The total amount of ABA and the amount of additionally produced ABA ("stress-ABA") at the time of stomatal opening was about three times higher in leaves stressed for 4 h than in leaves wilted for only 1 h (Table 1). At higher levels of stress-ABA, however, the rate of increase of the pore width seemed to be slower than at lower ABA levels. Similar results were obtained with the other mesophytic species investigated, *Pisum* and *Helianthus.*

Leaves of *Tradescantia* (Fig. 5), *Menyanthes,* and *Mentha* showed quite different behavior. In all three species no delay in the opening reaction was observed, although the water stress (measured as fresh weight decrease) was of the same magnitude as in the three species mentioned before. The stomata opened immediately after the stress period and this paralleled the recovery of leaf turgor. Moreover, the ABA level did not change significantly in *Menyanthes,* whereas in *Tradescantia* (Fig. 5) and *Mentha* it actually decreased during the wilting period. Thus the immediate opening reaction of the stomata in these plants is correlated with the fact that no increase in the ABA level occurred.

If a causal relationship exists between the delay in stomatal opening and the increased ABA level, it should be possible to modify the duration of the after-effect by the addition of exogenous ABA. This possibility has been investigated in *Vicia faba* and in *Tradescantia. Vicia faba* leaves were wilted for 1 h and then rewatered with an

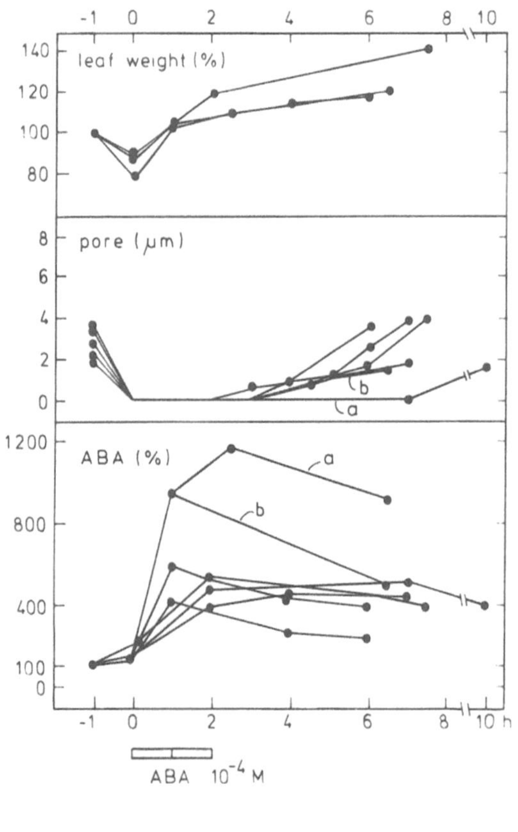

Fig. 6. Effect of ABA applied to leaves of *Vicia faba* (wilted for 1 h) on the internal ABA level and on the delay of stomatal opening

aqueous solution containing ABA (10^{-4} M). The ABA solution was replaced with water after 1 or 2 h. This procedure resulted in increased levels of ABA (up to 12-fold) and in an increased time delay of stomatal opening (up to 7 h – Fig. 6) in comparison with control leaves not supplied with ABA. In those control leaves a stress period of 1 h resulted in an increase of the ABA level of about 160% and in a time delay of stomatal opening of only 1 or 2 h (Table 1). In leaves of *Tradescantia* a time delay of stomatal opening of 3–7 h could be induced by exogenously applied ABA (Fig. 7). The

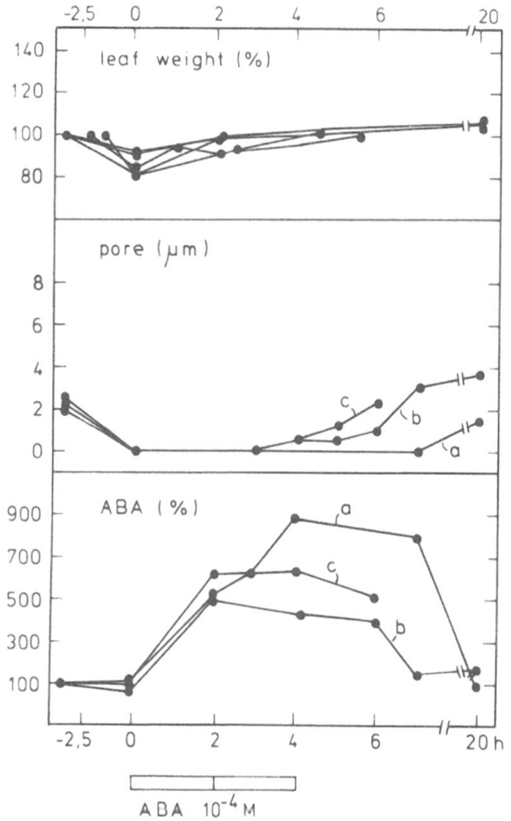

Fig. 7. Effect of ABA applied to wilted leaves of *Tradescantia* on the internal level of ABA and on the opening movement of the stomata. *a, b, c* represent different experiments

ABA level increased five- to ninefold. This experiment provides clear evidence that a delay in stomatal opening after a stress period in detached leaves is causally related to the time course of the ABA increase and decrease during and after the stress period. It is difficult to account for the stomata opening immediately or soon after the attainment of the maximal ABA level in the leaf. Two explanations are possible:

1. The sensitivity of the guard cells is reduced by prolonged stress or enhanced ABA level, consequently rendering a possible threshold ABA level relative to concentration and duration (16, 17).
2. Reduction of the level of ABA at its active site in the guard cells may be accomplished by processes other than degradation, for example, removal into storage sites where it cannot act on the stomata (18).

Experiments with Attached Leaves. In contrast to the experiments with isolated leaves, no clear relationship was found between after-effects of stress and ABA levels in attached leaves of *Vicia faba.* When *Vicia faba* plants were rewatered after a stress period of 30 h, water potential increased quickly, reaching normal levels within 6 h. The diffusion resistance also decreased quickly, but remained at enhanced values for about 30–50 h, indicating a long-lasting after-effect of stress on stomatal opening. This became especially evident when the stress period was repeated several times (Fig. 8). The

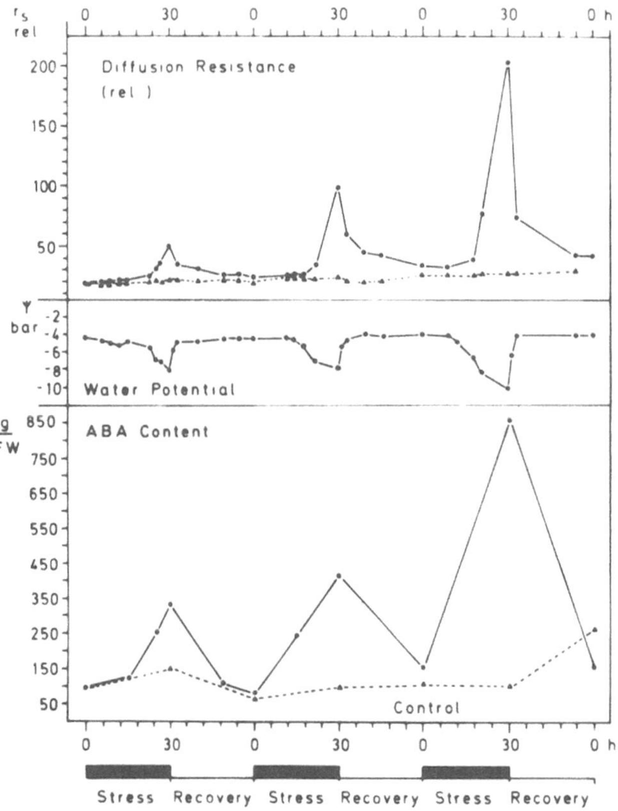

Fig. 8. Diffusion resistance, water potential, and ABA levels in leaves of *Vicia faba* plants during and after a repeated stress period

amount of ABA, however, returned to the control level within 24 h. Thus the after-effect in attached leaves cannot be related exclusively to the kinetics of ABA. Other "stress factors" such as farnesol (5) may be involved in addition to ABA, which may be produced only after prolonged or repeated stress, but no experimental evidence is yet available.

In summary, it seems evident that ABA is responsible in some way for the delay in stomatal opening after short stress periods in isolated leaves. The question remains to what extent ABA is also responsible for after-effects of longer duration. Evidence for and against a role of ABA in stress-induced stomatal closure was obtained, and this process needs further investigation.

Degradation of ABA After a Stress Period

After a wilting period, the level of free ABA decreases. This can occur either by conjugation processes, especially by the formation of the glucoside, or by degradation to phaseic acid, dihydrophaseic acid and other, still unknown metabolites. Evidence for both pathways is available (3, 19). In pea seedlings recovering from stress, we have shown (20) that the level of free ABA is not reduced by the formation of "bound" ABA, but by the formation of at least two acidic metabolites. Evidence is now available that phaseic acid (PA), dihydrophaseic acid (DPA), and a new metabolite, with the proposed structure of 4′-desoxy-ABA, are among them.

Thin-layer chromatograms of ethanol extracts obtained from wilted pea seedlings, which had been supplied for 1 h with radioactive ABA (1-^{14}C-ABA), revealed the presence of several radioactive metabolites (Fig. 9). The less polar fractions A, B, C,

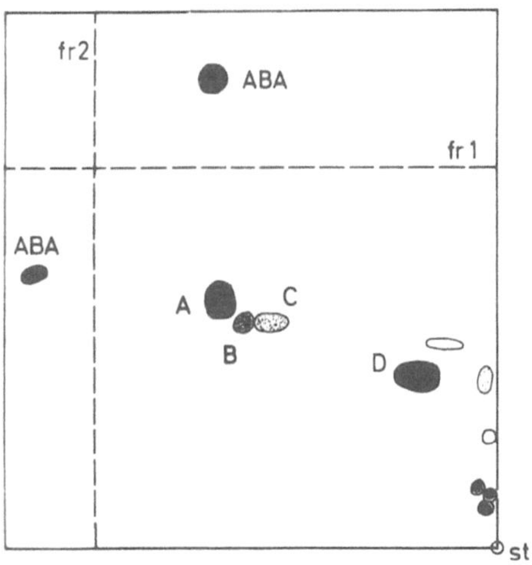

Fig. 9. Autoradiography of a two-dimensional thin-layer chromatogram from water-stressed pea seedlings supplied with ABA-1-^{14}C at the time of rewatering and extracted 24 h later. Chromatography in (1) chloroform/methanol/water 75/22/3 and in (2) toluene/ethyl-acetate/acetic acid 50/30/4. *st* start; *fr 1* and *fr 2* front in solvent systems 1 or 2. Spot *A* corresponds to marker-ABA, *B* to 4′-desoxy-ABA, *C* to phaseic acid, *D* to dihydrophaseic acid

and D were analyzed by GLC after methylation. The retention time of A exactly corresponded to that of ABA. B, C, and D also gave pronounced peaks. The mass spectra of A, C, and D clearly corresponded to the mass spectra of ABA, PA, and DPA published by Zeevaart and Milborrow (21). The mass spectrum of B (Fig. 10) is interpreted as being derived from methylated 4′-desoxy-ABA (Fig. 11). This compound has not been previously described as a naturally occurring metabolite of ABA. The mol peak of 4′-desoxy-ABA-Me is at m/e 262. The difference of 16 in the molecular weights between methylated ABA and substance B can be explained by the loss of an oxygen atom. Theoretically, the oxygen of ABA could also be lost at C-1. The resulting compound, 1′-desoxy-ABA, has been synthesized (22, 23). Contradicting this possibility, however, there is the intense loss of water (M-18) from 4′-desoxy-ABA indicated

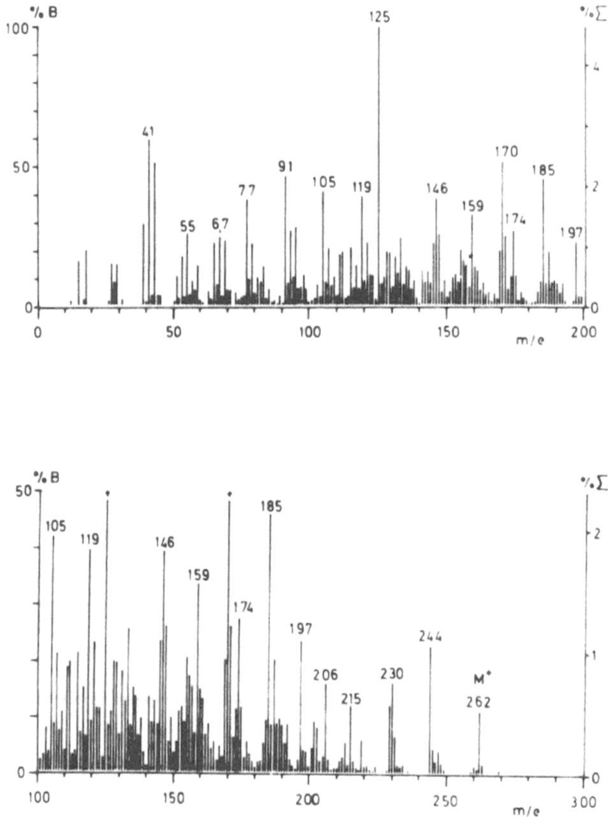

Fig. 10. Mass spectrum of the fraction B (= 4′-desoxy-ABA) in Fig. 9

especially at m/e 244, which is difficult to explain without the presence of an alcoholic group. In fact, the mass spectrum of 1′-desoxy-ABA does not contain the respective fragment at m/e 244 (personal communication of Dr. Oritani). Further details of the fragmentation are presented elsewhere (15).

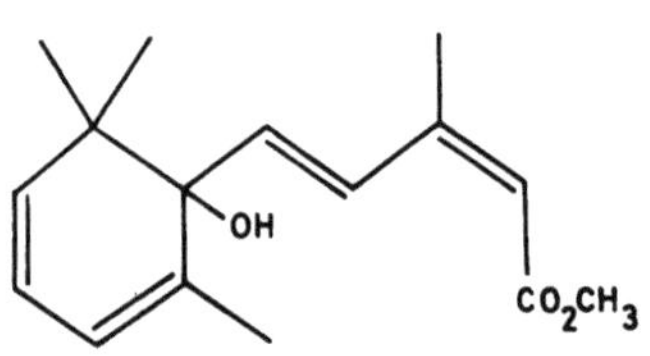

Fig. 11. Structure of 4′-desoxy-ABA

The role of 4′-desoxy-ABA in the pathway of ABA degradation and its biological role are not clear at present. Moreover, the high number of unknown polar metabolites (Fig. 9), possibly conjugates, needs further investigation.

Acknowledgment. The expert technical assistance of Mrs. Dorothea Martin and Mrs. Vera Handke is gratefully acknowledged. Thanks are due also to Mrs. Inge Erxleben and Prof. Dr. D. Wöhrle (University of Bremen) for the performance and interpretation of the mass spectra. This work was supported by the Deutsche Forschungsgmeinschaft.

References

1. Wareing, P.F.: Philos. Trans. R. Soc. London Ser. B *284*, 483–498 (1978)
2. Jones, R.J., Mansfield, T.A.: J. Exp. Bot. *21*, 714–719 (1970)
3. Hiron, R.W.P., Wright, S.T.C.: J. Exp. Bot. *24*, 769–781 (1973)
4. Raschke, K., Zeevaart, J.A.D.: Plant Physiol. *58*, 169–174 (1976)
5. Mansfield, T.A., Wellburn, A.R., Moreira, T.J.S.: Philos. Trans. R. Soc. London Ser. B *284*, 471–482 (1978)
6. Loveys, B.R.: Physiol. Plant. *40*, 6–10 (1977)
7. Itai, C., Weyers, J.D.B., Hillman, J.R., Meidner, H., Willmer, C.M.: Nature (Lond.) *271*, 652–653 (1978)
8. Beardsell, M.F., Cohen, D.: Plant Physiol. *56*, 207–212 (1975)
9. Walton, D.C., Galson, E., Harrison, M.A.: Planta *133*, 145–148 (1977)
10. Bengtson, C., Falk, S.O., Larsson, S.: Physiol. Plant. *41*, 149–154 (1977)
11. Stålfelt, M.G.: Physiol. Plant. *8*, 572–593 (1955)
12. Wright, S.T.C.: Planta *36*, 10–20 (1969)
13. Dörffling, K., Streich, J., Kruse, W., Muxfeldt, B.: Z. Pflanzenphysiol. *81*, 43–56 (1977)
14. Tietz, D., Dörffling, K.: Mitt. Inst. Allg. Bot. Hamburg *16*, 101–114 (1978)
15. Tietz, D., Dörffling, K., Wöhrle, D., Erxleben, I., Liemann, F.: Planta *147*, 168–173 (1979)
16. Kriedemann, P.E., Loveys, B.R., Fuller, G.L., Leopold, A.C.: Plant. Physiol. *49*, 842–847 (1972)
17. Raschke, K.: Annu. Rev. Plant Physiol. *26*, 309–340 (1975)
18. Cummins, W.R.: Planta *114*, 159–167 (1973)
19. Harrison, M.A., Walton, D.C.: Plant Physiol. *56*, 250–254 (1975)
20. Dörffling, K., Šonka, B., Tietz, D.: Planta *121*, 57–66 (1974)
21. Zeevaart, J.A.D., Milborrow, B.V.: Phytochemistry *15*, 493–500 (1976)
22. Oritani, T., Yamashita, K.: Agric. Biol. Chem. *34*, 108–114 (1970)
23. Milborrow, B.V.: In: Biochemistry and Physiology of Plant Growth Substances. Wightman, F., Setterfield, G. (eds.), pp. 1531–1545. Ottawa: Runge Press 1968
24. Pierce, M., Raschke, K.: 10th Int. Conf. Plant Growth Substances, Madison, Wisc. 1979, List of Abstracts

New Growth Factors

Chairman: C. A. WEST

New Growth Factors – Summary of Session

C.A. WEST [1]

One session of twelve contributed papers and a group of five posters were concerned with substances other than well-established plant growth factors which nonetheless influence plant growth and development. A major emphasis was on one group of steroidal substances termed brassins which appear to invoke a unique set of growth responses and thus are suggested to be a new class of plant hormones.

The discovery of brassins is based on the observation of Mitchell et al. (1) that extracts of *Brassica napus* (rape) pollen cause elongation, swelling, and splitting of the bean second-internode. On the basis of this bioassay, 4.5 mg of an active substance named brassinolide have been isolated in pure form from 40 kg of rape pollen (2). The structure was determined by mass spectrometry and X-ray crystallography to be that of the steroidal lactone (22*R*, 23*R*, 24*S*)-2α,3α,22,23-tetrahydroxy-24-methyl-6,7-seco-5α-cholestano-6,7-lactone. Brassinolide-like growth promoters with similar biological, chromatographic, and structural properties have also been demonstrated in pollen from alder and orange (3). Brassinolide possesses some structural features reminiscent of the ecdysone family of insect hormones. A number of ecdysones and synthetic structural analogs of brassinolide have been tested for brassin biological activity (4). The ecdysones are uniformly inactive. The α-orientation of hydroxyl groups in the 2 and 3 positions appears to be an important requisite for activity. The presence of the 22- and 23-hydroxyl and 24-methyl (or ethyl) substituents was necessary for activity, but some stereochemical variation in the positioning of these substituents is tolerated. Other aspects of the relation of structure to biological activity were also probed with these synthetic brassinosteroids.

The biological activity of brassinolide and purified brassin preparations has been assessed in a large number of bioassay systems (5–8). A wide variety of plants have been shown to respond under appropriate conditions. Brassinolide enhances the response to IAA and NAA in some test systems (but not the response to chlorinated phenyls). The augmentation of auxin responses in the bean internode and the mung bean epicotyl has been made the basis for quantitative bioassays. Brassin or brassinolide produces responses characteristic of auxins, of gibberellins, and of cytokinins in some of the assay systems developed for each of these groups of hormones, but not all test systems for a given hormone respond. Suppression of rooting is a characteristic response. Although the responses are characteristic of gibberellins in many respects, brassin does not promote the growth of dwarf maize. Light appears to be necessary for a response to brassins. Taken together, the results indicate a unique role for brassinolide in its interactions with the other known classes of plant hormones.

---

1 Department of Chemistry, University of California, Los Angeles, California 90024, USA

The variable capacity of avocado cuttings to root has been correlated with the level of a rooting promoter extractable from the leaves (9). A preliminary characterization of the rooting promoter suggests that it is not an auxin because of (i) its greater activity than IBA at optimal concentrations in promoting rooting and the absence of inhibition at supraoptimal concentrations and because (ii) there is no evidence of a chromogen resembling IAA in the active extract. The active material is widely distributed in avocado and can also be detected in other plants.

Two groups of substances with well-established roles in animals have been implicated as possible factors in plant growth. Catecholamines have been shown to act as synergists for gibberellin-stimulated elongation of lettuce hypocotyls in the same manner as the natural cotyledon factor, dihydroconiferyl alcohol (10). Evidence has also been obtained for the natural occurrence of dopamine in extracts of lettuce seedlings. Polyamines have been proposed as possible second chemical messengers in the regulation of macromolecular metabolism, mitosis, and senescence from studies with oat leaves and protoplasts (11).

Finally, the actions of several groups of naturally occurring inhibitors were described. The structure of a new cucurbitacin isolated from seed of *Trewia nudiflora* L. was reported (12). This substance inhibits both the endogenous and gibberellin-stimulated growth of the rice second-leaf sheath. Thirty-seven nor- and *bis*-norditerpenoid lactones of the podolactone type have been identified from various *Podocarpus* species (13). Many of these substances have strong plant growth inhibitory properties. Two low-molecular-weight glycopeptides have been isolated from culture filtrates of the fungus *Phialophora cinerescens*, a pathogen of carnation. These glycopeptides inhibited root and hypocotyl growth of carnation plants in a manner which indicates that they are responsible for some of the symptoms of infection resulting from the fungus-plant interaction (14).

References

1. Mitchell, J.W., Mandava, N., Worley, J.F., Plimmer, J.R., Smith, M.V.: Nature (Lond.) *225*, 1065–1066 (1970)
2. Grove, M.D., Spencer, G.F., Mandava, N., Worley, J.F., Warthen, Jr., J.D., Steffens, G.L., Flippen-Anderson, J.L.: Abstract 501 [1], p. 25 (1979)
3. Mandava, N., Worley, J.F.: Abstract 553 [1], p. 49 (1979)
4. Thompson, M.J., Mandava, N.B., Flippen-Anderson, J.L., Worley, J.F., Sutky, S.R., Robbins, W.E.: Abstract 503 [1], p. 25 (1979)
5. Meudt, W.J., Worley, J.F., Gregory, L.E., Mandava, N., Buta, J.G., Steffens, G.L., Thompson, M.J.: Abstract 502 [1], p. 25 (1979)
6. Yopp, J.H., Ladd, D., Jacques, D., Mandava, N.B.: Abstract 504 [1], p. 25 (1979)
7. Buta, J.G., Meudt, W.J.: Abstract 551 [1], p. 49 (1979)
8. Gregory, L.E., Mandava, N.B., Cina, D.K.: Abstract 552 [1], p. 49 (1979)
9. Raviv, M., Reuveni, O., Goldschmidt, E.E.: Abstract 508 [1], p. 26 (1979)
10. Kamisaka, S.: Abstract 507 [1], p. 26 (1979)
11. Galston, A.W., Sawhey, R., Altman, H., Flores, H.: Abstract 554 [1], p. 50 (1979)
12. Ganguly, S.N., Saha, P.K., Ganguly, T.: Abstract 506 [1], p. 25 (1979)
13. Sasse, J.M., Wardrop, J.J., Rowan, K.S.: Abstract 513 [1], p. 26 (1979)
14. Pugin, A., Rouillon, R., Dubouchet, J.: Abstract 512 [1], p. 26 (1979)

1 Abstract numbers and corresponding page numbers refer to the Abstracts of the Tenth International Conference on Plant Growth Substances, Madison, Wiconsin, July 1979

Hormonal Regulation in Plant Reproductive Development

Chairman: A. LANG

The Hormonal Control of Tuberisation in Potato

P.F. WAREING and A.M.V. JENNINGS [1]

Introduction

Just as the photoperiodic control of flowering appears to involve a flowering stimulus produced in the leaves under inductive conditions, so there is analogous evidence for the existence of a tuber-inducing stimulus in potato and other tuber-forming species.

That tuberisation may involve a stimulus of a hormonal nature was suggested by several earlier workers (1), but it was the work of Gregory (2) which first provided strong experimental evidence for the existence of such a stimulus. He found that cuttings from induced parent plants (which had been maintained under short days) rapidly formed tubers from axillary buds on the stem, whereas similar cuttings from non-induced (long day) plants did not do so. When induced stem pieces were grafted on to non-induced ones, the latter formed tubers, indicating the existence of a graft-transmissible stimulus. Gregory concluded that tuber-forming stimulus arises in the plant under specific conditions of photoperiod and temperature and that the stimulus occurs throughout the plant. These results were effectively confirmed by Chapman (3) and by Kumar and Wareing (4).

From the results of experiments in which he exposed either (1) only young developing leaves at the shoot tip, or (2) only the mature leaves, Chapman (3) concluded that the tuberisation stimulus is formed mainly at the apices of aerial shoots. Similar conclusions had been reached by Zimmerman and Hitchcock (1) for *Helianthus tuberosus* , but these results were not confirmed by Hamner and Long (5) who found that the locus of photoperiodic perception in this species is not the shoot tip but the mature leaves. Fabian (6) also concluded that in *Ullucus tuberosus* primarily the leaves are the organs of photoperiodic perception. More recently, Hamnes and Beyers (7) have shown that both young expanding leaves and mature leaves promote tuberisation in potato plants maintained under short days, but that young leaves have an *inhibitory* effect when maintained under long days.

Assuming that the tuberisation stimulus is hormonal in nature, the crucial question arises, as with flowering, whether there is a special tuberisation "hormone" or whether the response is regulated by endogenous growth substances, possibly by an interaction between growth-promoting and growth-inhibiting substances, as has been suggested by a number of workers (8–10).

1 Department of Botany and Microbiology, University College of Wales, Penglais, Aberystwyth, Dyfed SY23 3DA, United Kingdom

Most of the evidence bearing on this question relates to the effects of exogenous growth substances on tuberisation and on variations in levels of endogenous growth substances in relation to tuber-inducing and non-inducing conditions.

Gibberellins (GA s)

There appears to be general agreement that exogenous GA s inhibit tuberisation in potato whether applied to whole plants or to stem cuttings (8, 9, 11, 12). Similar results have been demonstrated for *Helianthus tuberosus* (13), and for the formation of root tubers in *Dahlia variabilis* (14).

Studies on the effects of photoperiod on endogenous GA levels have shown that there is a rapid decline in extractable GA s in young, mature leaves of potato plants when they are transferred from long-day to short-day conditions (15). Thus it would appear that, in potato, tuberisation is inhibited under long days by the high endogenous GA levels prevailing under these conditions. On the other hand, developing potato tubers contain significant amounts of endogenous GA's (16).

Cytokinins

Palmer and Smith (17, 18) were the first to demonstrate the promotion of tuberisation in isolated stolons by exogenous kinetin. Isolated stolons of *S. tuberosum* were first cultured on the medium of Murashige and Skoog (19) but in the absence of kinetin, and were then transferred to the same medium to which kinetin had been added. After 7 days these stolons produced terminal swellings which later developed into morphologically defined tubers. This kinetin-induced tuberisation response was dependent upon the supply of a concentration of at least 6% sucrose in the medium.

Studies have been carried out by Sattelmacher and Marschner (20, 21) on the relations between nitrogen nutrition, cytokinin activity, and tuberisation in *S. tuberosum.* When plants of cv. Ostara are grown in water culture and supplied with high nitrogen, tuberisation is inhibited even under 12-hour photoperiods, but withdrawal of N rapidly leads to initiation of tuberisation and the following changes in endogenous cytokinin levels: (1) there is a temporary increase in the cytokinin activity in the roots, but a decrease in root exudate; (2) the cytokinin activity increases markedly in the stolon tips; (3) in the shoots, after an initial decrease, the cytokinin levels rise markedly. The authors conclude that despite the close correlation between tuberisation and cytokinin activity, cytokinins are not directly responsible for the onset of tuberisation, although they play an important role in tuber growth.

Mauk and Langille (22) have also shown that there is a twofold increase in the levels of endogenous zeatin riboside in the below-ground parts of potato plants (including roots and stolons) on transfer to short days. They also showed that tuberisation can be induced in isolated stolons of *S. tuberosum* by application of zeatin riboside.

Abscisic Acid (ABA)

El-Antably et al. (23) reported that tuberisation was promoted in *S. andigena* grown under long days when ABA was applied to the aerial shoots. Subsequently other authors reported no effects of exogenous ABA in various tests (17, 24). Palmer and Smith (17) reported that kinetin-induced tuberisation in isolated potato stolons was inhibited by 10^{-5} M ABA.

On the other hand, it has been reported that exogenous ABA promotes tuberisation in *Dahlia* (14) and *H. tuberosus* (25), and we have recently obtained the following further evidence which indicates that ABA may, indeed, be involved in the tuberisation response in *S. andigena.*

If single-node cuttings, bearing a leaf, are taken from induced plants of *S. andigena* (which have received 20 short-day cycles) and maintained in a moist medium under mist, the axillary buds develop into a small tuber within 5–7 days. However, if the leaves are removed from such single-node cuttings, the buds rapidly grow out into orthotropic, elongated shoots before any roots have been formed on the stem (Fig. 1),

Fig. 1. Single-node stem cuttings from *induced* plants of *S. andigena. Left* defoliated

whereas little growth of such buds occurs on leafless stem cuttings taken from non-induced plants over the same period (26). However, if leafless nodal cuttings from induced plants are placed with their bases in 10^{-5} M ABA and maintained in the dark, in 4–5 days the buds develop into small tubers (Fig. 2), whereas with nodal cuttings from noninduced plants there is little or no growth and no tuberisation (Wareing, unpublished observations). ABA can therefore replace the effect of the leaf in promoting tuberisation in *induced* cuttings.

The question therefore arises as to whether the effect of the leaf is mediated by the export of endogenous ABA from the leaf to the bud, and whether there is any difference in this respect between induced and non-induced leaves. Extracts were made of induced and non-induced leaves, and the levels of free and "bound" ABA were determined by gas-liquid chromatography (GLC) using ^{14}C-ABA as an internal standard. As has been found previously (27) the endogenous ABA levels were slightly higher in non-induced (LD) than in induced (SD) leaves, although in the present work there was some indication of a higher proportion of "bound" ABA in the induced leaves.

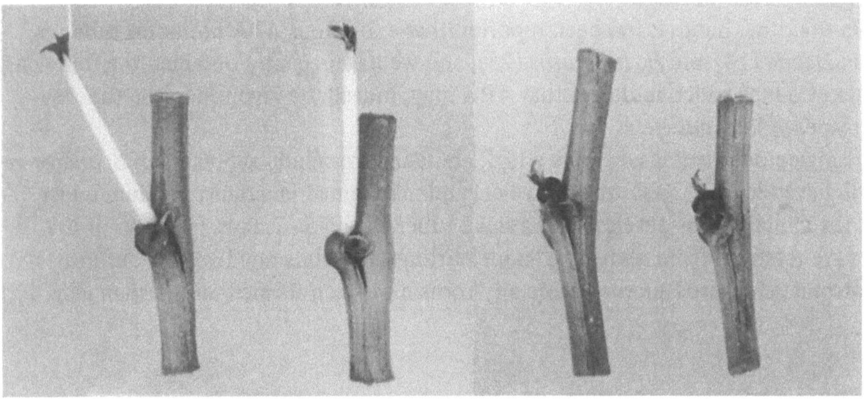

Fig. 2. Single-node stem cuttings from *induced* plants of *S. andigena. Left side* supplied with 10^{-5} M ABA. *Right side* supplied with water

Experiments were also carried out to determine whether there were differences in the rate of export of ABA from induced and non-induced leaves by standing the bases of detached leaves in a solution of 20 mM EDTA for 18 h, using the technique of King and Zeevaart (28). The solution was then "spiked" with ^{14}C-ABA as an internal standard and the levels of free and bound ABA which had diffused into the EDTA solution were determined by GLC. Slightly higher amounts of free and bound ABA were present in the diffusates from induced leaves, but whether these differences are significant was not determined.

In further experiments ^{14}C-labelled ABA was applied to the lamina tip or to a cut in the petiole of single-node cuttings, with leaves attached, from induced and non-induced plants, and the levels of radioactivity in the various parts of the cuttings were determined after 36 and 72 hours of feeding. It was found that in both induced and non-induced cuttings substantial levels of radioactivity were present in the buds, indicating that a proportion of the ABA exported from the leaves is accumulated in the buds.

The foregoing observations suggest that one of the roles of the leaf in promoting tuberisation in cuttings of *S. andigena* is mediated via the endogenous ABA which is exported from the leaf to the bud. However, significant amounts of ABA could be diffused from both induced and non-induced leaves, so that if ABA supplied from the leaves to the buds is essential for tuberisation, then non-induced (LD) leaves should be as effective in this respect as induced ones. Grafting experiments were therefore carried out, in which induced and non-induced leaves were grafted on to stem cuttings from induced and non-induced plants, in all combinations. Stem cuttings from induced

Fig. 3. Single-node stem cutting from an induced plant of *S. andigena* with a leaf from a non-induced plant (long-day grown) grafted onto it

plants produced tubers when either induced or noninduced leaves (Fig. 3) were grafted onto them. On the other hand, stem cuttings from non-induced plants produced tubers when induced leaves, but not when non-induced leaves, were grafted onto them. In further experiments it was shown that induced stems of *S. andigena* will also form tubers when tomato leaves are grafted on them, presumably as a result of the endogenous ABA exported from these leaves (Fig. 4).

The results of these grafting experiments indicate that:

1. Both induced and noninduced leaves can promote tuberisation in induced stem cuttings, and the leaves can be replaced by exogenous ABA
2. A "second factor", evidently supplied by induced, but not by non-induced leaves, accumulates in the stems of induced plants, so that a grafted non-induced leaf is then effective in promoting tuber formation. The possible nature of this second factor is discussed below.

Further experiments were carried out to determine whether ABA plays a specific role in tuberisation or merely inhibits elongation growth. The initiation of tuber development involves arrest of activity in the apical meristem of the stolon and stimulation of both cell division and cell expansion in the subapical region (29). Thus, it is possible that the effect of ABA is to inhibit the activity of the apical meristem. If so, other agents which inhibit apical activity might also be effective. Since sucrose has been found to promote tuberisation in *S. tuberosum* the effects of 0.1 M and 0.5 M sucrose were tested on single-node stem cuttings from fully induced plants of *S. andigena* and compared with those of mannitol, to determine whether growth inhibition due to osmotic effects might also promote tuberisation. From the results presented in

Fig. 4. Single-node cuttings from induced plants of *S. andigena* with leaves of tomato *(Lycopersicon esculentum)* grafted onto them

Table 1 it is seen that both sucrose and mannitol did, indeed, increase tuberisation, although they were not as effective as ABA. These results appear to be consistent with the hypothesis that ABA promotes tuberisation by the inhibition of growth of the apical meristem. However, there may also be a more specific effect of ABA, since the buds of the tubers formed in response to application of ABA remained dormant when the tubers were transferred to water but were stimulated into growth by the application of gibberellic acid.

Table 1. Tuber formation by stem sections from induced plants of *S. andigena* in response to applied abscisic acid, sucrose and mannitol (38 sections in each treatment)

Treatment	H_2O	ABA	Sucrose		Mannitol	
Molar Conc.	–	10^{-5}	0.1	0.5	0.1	0.5
Percent of sections forming tubers	24	69	38	21	57	40

The "Second Factor" in Tuberisation

As indicated, there is a striking difference between the behaviour of axillary buds of induced and non-induced leafless stem cuttings when supplied only with water, in that the former grow out strongly and rapidly into elongated shoots, whereas the latter grow out more slowly, if at all. Thus it would appear that the second factor is a growth promoter. The observations of Palmer and Smith (17) on the promotion of tuberisation by kinetin in isolated stolons of *S. tuberosum* indicate that the second factor may be cytokinin. We have explored this possibility in some detail, using stem sections of clones of *S. andigena* which have an obligate short-day requirement for tuberisation.

When leafless nodal stem sections (bearing a single axillary bud) from non-induced plants are supplied with Murashige and Skoog medium, including kinetin and IAA, the buds grow into etiolated shoots when kept in the dark, but such shoots do *not* form tubers, either without or with ABA. On the other hand, kinetin does induce more than 50% tuberisation in fully *induced* stem cuttings of *S. andigena.* Thus, cytokinin does not replace the short-day requirement in these clones of *S. andigena,* but it does promote tuber development in *induced* cuttings.

Evidently there is a difference between the response to exogenous cytokinin of the isolated stolons of *S. tuberosum* (17,18) and the non-induced stem cuttings of *S. andigena* in their response to exogenous cytokinin. This difference may stem from the fact that the clones of *S. andigena* we used have an obligate short-day requirement, whereas many cultivars of *S. tuberosum* have no absolute requirement for short days, but give a quantitative SD response. Thus, it is possible that the second factor is present in *S. tubersoum* independently of the photoperiodic pretreatment; we have no information regarding the photoperiodic responses of the cultivar (Norgold Russett) of *S. tuberosum* used by Palmer and Smith, but this suggestion is supported by Forsline and Langille's finding that stem segments from both induced and non-induced cuttings of *S. tuberosum* cv. "Katahdin" produced tubers without kinetin treatment, although kinetin increased the percentage of tubers formed by non-induced cuttings (30). Thus, in this latter cultivar the second factor would appear to be present in stem tissue even from non-induced plants, albeit in lower amounts than from induced plants. Thus, in the experiments of Palmer and Smith the exogenous kinetin need not have been replacing, but rather complementing the second factor, possibly by inhibiting the activity of the apical meristems of the stolons and so promoting cell division in the subapical region, as the authors themselves suggested (18).

We have attempted to extract material(s) capable of promoting tuber formation (in non-induced stem cuttings) from induced shoots of *S. andigena.* Tests of basic butanol-soluble fractions which would include the endogenous cytokinins of *S. andigena,* as well as acidic butanol-soluble fractions and fractions soluble in petroleum ether or ethyl acetate so far have been entirely negative; as have tests of diffusates from induced leaves of *S. andigena.* Substantial amounts of phloem exudate were obtained from potato leaves by pretreatment with EDTA, using the technique of King and Zeevaart (28), but the residual EDTA in the exudate was toxic to the buds of *S. andigena* used in our assay system.

Thus, the isolation of the second factor in tuberisation is proving as intractable as the isolation of florigen. The observation of Nitsch (31) that tuberisation can be induced in stocks of Jerusalem artichoke *(H. tuberosus)* by grafting on them leaves of *H. annuus,* which does not itself form tubers, indicates that in *H. tuberosus* the tuberisation stimulus is non-specific. However, our own findings that tuberisation cannot be stimulated in non-induced cuttings of *S. andigena* by treatment with any of the known types of growth substance suggest that the second factor in tuberisation may be of a new type, the effects of which are not specific to tuberisation.

References

1. Zimmerman, P.W., Hitchcock, A.E.: Am. J. Bot. *23*, 690 (1936)
2. Gregory, L.E.: Am. J. Bot. *43*, 281–288 (1956)
3. Chapman, H.W.: Physiol. Plant. *11*, 215–224 (1958)
4. Kumar, D., Wareing, P.F.: New Phytol. *72*, 283–287 (1973)
5. Hamner, K.C., Long, E.M.: Bot. Gaz. *101*, 81–90 (1939)
6. Fabian, I.: Z. Bot. *33*, 305–357 (1938)
7. Hamnes, P.S., Beyers, E.A.: Potato Res. *16*, 68–72 (1973)
8. Okazawa, Y.: Proc. Crop Sci. Soc. Jpn. *28*, 129–133 (1959)
9. Okazawa, Y.: Proc. Crop Sci. Soc. Jpn. *29*, 121–124 (1960)
10. Okazawa, Y., Chapman, H.W.: Physiol. Plant. *15*, 413–419 (1962)
11. Tizio, R.: C.R. Acad. Sci. *259*, 1187–1190 (1964)
12. Tizio, R.: Potato Res. *14*, 193–204 (1971)
13. Claver, F.K.: Phyton (Buenos Aires) *23*, 5 (1966)
14. Biran, I., Gur, I., Halevy, A.H.: Physiol. Plant.*27*, 226–230 (1972)
15. Railton, I.D., Wareing, P.F.: Physiol. Plant.*28*, 88–94 (1973)
16. Bialek, K.: Z. Pflanzen. Physiol. *71*, 370–372 (1973)
17. Palmer, C.E., Smith, O.E.: Plant Cell Physiol. *10*, 657–664 (1969)
18. Palmer, C.E., Smith,O.E.: Plant Cell Physiol. *11*, 303–314 (1970)
19. Murashige, T., Skoog, F.: Physiol. Plant. *15*, 473–496 (1962)
20. Sattelmacher, B., Marschner, H.: Physiol. Plant.*44*, 65–68 (1978)
21. Sattelmacher, B., Marschner, H.: Physiol. Plant.*44*, 69–72 (1978)
22. Mauk, C.S., Langille, A.R.: Plant Physiol. *62*, 438–442 (1978)
23. El-Antably, H.M.M., Wareing, P.F., Hillman, J.: Planta *73*, 74–90 (1967)
24. Claver, F.K.: Phyton (Buenos Aires) *27*, 25–29 (1970)
25. Charnay, D., Courduroux, J.-C.: C.R. Acad. Sci. Ser. D *275*, 2351–2354 (1972)
26. Ewing, E.E., Wareing, P.F.: Plant Physiol. *61*, 348–353 (1978)
27. Kumar, D., Wareing, P.F.: New Phytol. *73*, 833–840 (1974)
28. King, R.W., Zeevaart, J.A.D.: Plant Physiol. *53*, 96–103 (1974)
29. Cutter, E.G.: In: The Potato Crop. The Scientific Basis for Improvement. Harris, P.M. (ed.), pp. 70–152. London: Chapman & Hall 1978
30. Forsline, P.L., Langille, A.R.: Can. J. Bot. *54*, 2513–2516 (1976)
31. Nitsch, J.P.: Bull. Soc. Bot. Fr. *112*, 333–340 (1965)

Inhibition of Flowering in Short-Day Plants [1]

W.P. JACOBS [2]

Introduction

In 1925, Garner and Allard (5) reported that placing the top leaves of the SDP *Cosmos* in LD's inhibited the flower-stimulating effect from the lower leaves that were in SD's. The top leaves of the controls were in darkness. Since then a variety of experiments have confirmed that LD's, rather than being merely neutral, may inhibit flowering in SDP's (1, 8, 14). Lang (14) has reviewed the early literature. I shall give references only to papers not cited by him.

Maximal inhibition typically results from LD leaves that are inserted ("interposed") between the SD leaf and the responding apex, whether the LD leaf is on the same or another stem. On a plant with decussate leaf arrangement, such as *Perilla,* LD on one leaf will slow flowering resulting from SD on its sister leaf, although the inhibition is not so strong as if the LD leaf were above the SD leaf and thus interposed between the latter and the apex (14). In both *Kalanchoë* (8) and *Xanthium* (6) if the SD leaf is interposed between the apex and the LD leaf, the latter has little or no inhibitory effect.

Surprisingly, inhibition of flowering occurs even if LD is given only to the basal half of a leaf the apical half of which is given SD [shown for *Perilla* (14), *Kalanchoë* (8), *Salvia* (1), *Xanthium* (6)]. LD on the apical half does not inhibit flowering resulting from SD on the basal half (*Salvia* is an exception).

A second type of evidence for floral inhibition by LD comes from experiments in which the whole shoot is given LD's intercalated in a series of SD's (1, 3, 14). Schwabe (14) calculated that each intercalated day of CL cancelled the inductive effect of 1.4 SD's for *Kalanchoë*. For *Salvia,* a single day of CL, if given after 8 to 10 SD's (the time of maximal sensitivity to CL), delayed flowering by 11 days (1).

If LD leaves and SD leaves act in opposite directions, it is natural to wonder if SD leaves on an SDP are necessary except to counter the effect of the LD leaves. That is, SD plants will perhaps flower autonomously if the LD inhibition is removed. The first such evidence came from LDP's, Lang and Melchers' early paper on *Hyoscyamus* being one of the best (14). They found that *Hyoscyamus,* if defoliated, would flower in LD, SD, or CD. Similar results were reported subsequently for several SD plants (Table 1).

---

1 Abbreviations: CD, continuous dark; CL, continuous light; DN, day neutral; DNP, day-neutral plant; LB, light break (i.e., light given during the dark period); LD, long day; LDP, long-day plant; SD, short day; SDP, short-day plant

2 Biology Department, Princeton University, Princeton, N.J. 08544, USA

While grafting has been popular as a way to stimulate flowering in non-induced receptor plants – this response usually considered to result from transmission of a floral stimulus –, only a few authors have used grafting or similar experiments as evidence for floral inhibitors. Fratianne (4) used the angiosperm parasite *Cuscuta* to connect photoinduced Biloxi soybean plants with Biloxi plants kept in LD; there was no evidence for transmission of a floral stimulus to the plants in LD, but the plants in SD took longer to flower when connected via *Cuscuta* with plants in LD. Fratianne interpreted these results as evidence for transmission of a floral inhibitor from the LD-treated plants.

Table 1. Floral primordia formed by SDP's in noninductive lightening conditions

Genus and conditions	Author(s) and year [a]
Chenopodium, in LD if leaves excised and sucrose supplied	Lona, 1948
Perilla, if old and in CD	Lona, 1949
Glycine, in CL (decapitating the plants pushed primordia further into full flowering)	Haupt, 1954
Fragaria, in CL if leaves excised	Thompson and Guttridge, 1960
Perilla, in LD on aseptic cultures of apical bud	Raghavan and Jacobs, 1961

[a] Reference for each paper is in Lang's review (14)

Lang et al. (15) recently reported on extensive cross-grafting with tobacco of various photoperiodic responses (see following article). The LDP *Nicotiana silvestris* in SD inhibited flowering of the day-neutral Trapezond cultivar but the SDP Maryland Mammoth tobacco when kept in LD had no detectable inhibitory effect on flowering in Trapezond. (In SD Maryland Mammoth speeded flowering of Trapezond by 17 days – a statistically significant effect.)

Grafts Between the SDP Coleus fredericii and the DNP C. blumei

Despite the number of reports on transmission of floral stimulus by grafting, a relatively minute sample of the thousands of species has been used in interspecific grafting experiments. To extend this sampling, we tried various interspecific grafts using the SDP *Coleus fredericii* G. Tayl. and the DN Princeton clone of *C. blumei* Benth.

Some evidence that the Princeton clone of *C. blumei* is DN was provided in (10), where the effects of SD's were compared with those of LD's of 14 h or with those of SD's with 4 h LB's in the middle of the dark period. More detailed information is given in Fig. 1. *Coleus fredericii* was described as an SDP by Laibach (12), who generously mailed seeds to me. Halaban (7) determined more exactly that *C. fredericii* was an SDP, with a critical day-length of 12–13 h (her Fig. 1).

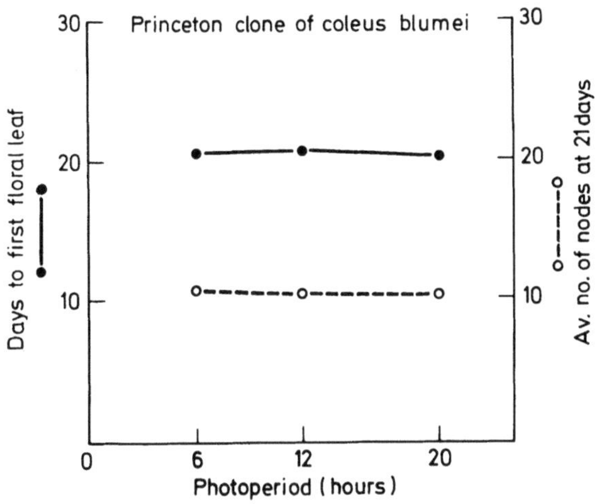

Fig. 1. Evidence from its response to various photoperiods that Princeton clone of *Coleus blumei* is a day-neutral plant

We used Zeevaart's (14) grafting technique, modified to the extent that we used the whole leaf blade rather than cutting it down to a standard area. Fig. 2 is a diagram of one of our graft combinations. The scion (a leaf, serving as potential donor of flowering promoter or inhibitor) was grafted into a slit in internode 3 (the third internode from the apex) of the *C. blumei* stock. Leaves at nodes 4, 5, and 7 were cut off, so that only leaf pair 6 remained on the stock. The axillaries at node 4 were the only axillary shoots allowed to remain on the stock: their development was checked regularly, newly unfolded leaves were excised, and the date of the first macroscopically visible inflorescence primordium was recorded. The *C. fredericii* stocks had more

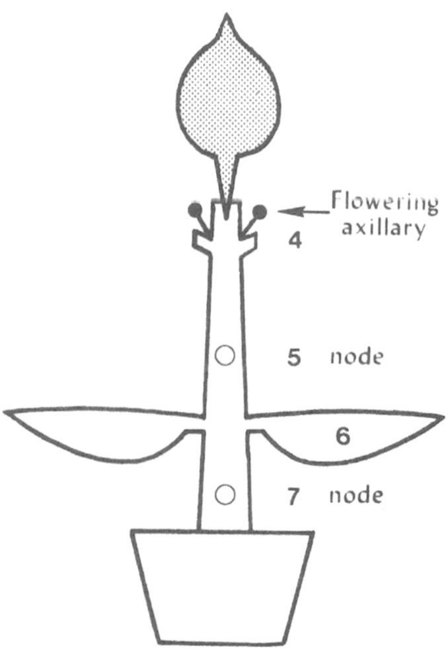

Fig. 2. The *Coleus* graft combinations with the scion leaf *(stippled)* as donor and the axillary branches at node 4 below as receptors

nodes than the *C. blumei* stocks because the plants had to be grown for a longer time before the third internode reached the 20–30 mm length required for successful grafting.

As the noninductive regime we used SD's with LB's in the middle of the dark periods to avoid the possible confounding of photoperiodic effects with those resulting from the extra hours of photosynthesis available if true LD's are used, as critically discussed in (9). The LB's were 2 h long, at a light intensity of 2900 μW cm^{-2}. The SD's had 8 h of light at 4900 μW cm^{-2}. These treatments were given in growth chambers after the new grafts had been left for 6 days in the cool work area under low-level CL to allow time for tissue union. Thereafter all grafts were kept in the LB chamber until the end of each experiment.

Donor leaf blades were between 60 and 100 mm long, and were closely matched within that size range. The assignments of donor leaves (scions) to treatments was by a mathematically random method. Donor leaves were either from vegetative plants (*C. blumei* too young to be in flower, or *C. fredericii* kept vegetative by LB's) or from flowering plants (older *C. blumei*, or *C. fredericii* photoinduced with 8h SD's).

A total of 11 experiments was run. The significance of differences was analyzed by standard statistical tests. Here I shall only summarize the main findings, illustrating them with the pooled data of Fig. 3.

Consider first the flowering response of *C. fredericii*, both as stock and as intact plant (row A of Fig. 3). Its SD character is clearly expressed in the intact plants, which developed a macroscopically visible inflorescence only 20 days after being placed in

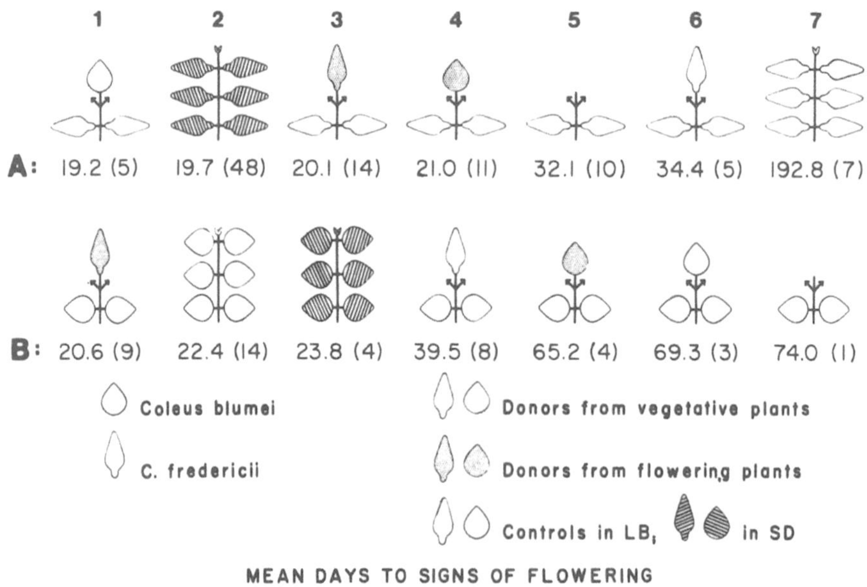

Fig. 3. Summary of pooled data from *Coleus* grafting experiments. The *numbers* under each diagram represent the mean days to the appearance of a macroscopic inflorescence primordium, based on the sample size shown in parentheses. Plants that had not flowered when each experiment was terminated are not included in the numbers, nor are any grafts in which the scion did not stay alive until the inflorescence primordium was visible

the SD chamber, compared to the 193+ days required in the LB chamber. (Although seven plants averaged the 193 days shown in Fig. 3, nine others in the LB regime still had not developed visible inflorescences when the experiments were terminated.)

C. blumei donor leaves stimulated floral initiation just as much whether they came from vegetative plants (the 19-day mean of A1) or from flowering ones (the 21-day mean of A4). These data fully agree with the classic view that DNP's produce floral stimulus in either SD's, LD's, or SD's with LB's. Donor leaves from SD-treated, flowering *C. fredericii* were just as stimulatory as the *C. blumei* donors (compare the 20-day mean of A3 to the 32-day mean of A5, the control group). Donor leaves from *C. fredericii* that had been kept vegetative with LB's inhibited flowering on *C. fredericii* hosts (A6). Inhibition was actually stronger than indicated by the mean values in Fig. 3, because the mean of 32 days for controls (A5) is based on all 10 of the plants so treated, whereas the mean of 34 days for the vegetative *C. fredericii* donors does not include the 5 receptor plants still vegetative when one of the experiments was ended at 47 days. Hence, the true mean value for plants with vegetative *C. fredericii* donors is "41+ days", rather than 34.

The flowering responses of *C. blumei* are shown in row B of Fig. 3. This clone of *C. blumei* shows its day-neutral character in the response of the intact plants to LB's and SD's: under either photoperiod the intact plants develop inflorescence primordia after 22–24 days (B2 and B3). Similarly, when grafts onto *C. blumei* stocks are considered, no difference in effectiveness is seen between donor leaves from vegetative or from flowering *C. blumei* (B5, B6). These values probably are not different from those for the controls (B7), although all three sample sizes were too small to engender much confidence in the means. (The single control showing an inflorescence primordium at 74 days should be considered with 5 other such controls still vegetative at 69 days and 5 others still vegetative at 50 days, at which times the respective experiments were terminated.)

A donor leaf from a flowering *C. fredericii* plant was strikingly more effective than a *C. blumei* donor leaf in stimulating floral initiation in the *C. blumei* receptor (the 21-day mean, B1). The *C. fredericii* leaf from plants that had been kept vegetative with LB's was markedly less effective than the leaf from a flowering *C. fredericii,* averaging "more than 40 days" to floral initiation (in addition to the 8 plants providing the mean of 40 days of B4, there were 4 more such graft combinations that were still vegetative at the 45 day termination of the experiment).

The conclusions we may draw from these *Coleus* experiments are:

1. Leaves of the DNP *C. blumei* provide floral stimulus in either SD's or SD's with LB's, with no discernible difference in effectiveness between the two photoperiodic treatments whether one considers whole plants or individual donor leaves of *C. blumei* grafted onto *C. blumei* or *C. fredericii* stocks. Surprisingly, a single leaf from *C. blumei* induced flowering just as quickly in the axillary branches of a noninduced individual of the SDP *C. fredericii* as placing *entire* plants of *C. fredericii* under SD's induced flowering in the apex of the main shoot. Furthermore, the donor leaves from *C. blumei* were just as effective in speeding flowering of the SDP as was a donor leaf from *C. fredericii* that had been photoinduced. These results quantitatively confirm the familiar hypothesis that DNP's produce florigen irrespective of photoperiod, that

florigen is not species-specific, and that it is the same in DNP's and SDP's [see Table 7 of Lang (14) for other examples].

2. Treatment of intact *C. fredericii* plants with SD's caused a striking acceleration of flowering compared to plants given SD's with LB's. A leaf from such a photoinduced, flowering plant continued to produce floral stimulus even after it was removed from SD's, grafted onto stocks, and the grafts subsequently kept under SD's with LB's. One such *C. fredericii* leaf that had been preinduced with SD's was just as effective in speeding flowering of its receptor as was placing the whole intact receptor in SD's and leaving it there; this was true whether the receptor was *C. fredericii* or *C. blumei.*

3. Leaves of SDP *C. fredericii* subjected to SD's with LB's inhibit flowering rather than being merely neutral. This is indicated by the great increase in days to floral initiation of the intact plants in SD's with LB's (193+ days) compared to the graft control plants with only two lower leaves in this regime (averaging 32 days). More specific evidence comes from the delayed flowering resulting from grafting an LB-leaf to the stocks. Such a leaf slows flowering to "41+ days" from the 32 days of the control. That is, the LB-leaf as donor slows flowering of *C. fredericii* as much as, or more than, the SD leaf speeds it up.

4. When *C. blumei* was the receptor, the difference in days to floral initiation resulting from SD- and LB-donor leaves of *C. fredericii* was approximately the same 20+ days as when *C. fredericii* was the receptor; however, from the *C. blumei* data alone one could not conclude that the LB donor leaves were inhibiting flowering because of the puzzling slowness of *C. blumei* stocks to initiate flowering either when their own leaves acted as donors or in the ungrafted controls (B5–7).

Possible Explanations of the LD Inhibition

One of the earliest hypotheses to explain flower inhibition by LD leaves in SDP's was that more photosynthesis occurred in leaves treated with LD and that the resulting stream of photosynthate moving down the stem somehow interfered with the upward movement of the hypothetical floral stimulus. The data on *Kalanchoë* from Harder's group seemed, at first, to fit this view (8), because LD inhibition was restricted to leaves or leaf-halves interposed between the SD leaf and the responding apex. Chailakhyan and Butenko (14) supplied an SD or LD leaf of *Perilla* with $^{14}CO_2$ and estimated the amount of radioactivity in various regions of the plant by means of autoradiographs. Their often-reproduced summary diagrams indicate that more radioactivity went into shoot apices from SD leaves that were in positions in which they also stimulated flowering. That is, florigen follows the translocation stream. (They also found that the ^{14}C distribution was no different from $^{14}CO_2$ supplied to an LD leaf than from an SD leaf, as long as the age and location of the leaves were the same.) Estimating the amount of radioactivity from autoradiographs is obviously at best semiquantitative. More precise information came from King and Zeevaart's work on *Perilla* (11). They also used $^{14}CO_2$, but actually counted extracted ^{14}C, then tried to correlate the amount of such "photosynthate" with the quantity of the floral stimulus (estimated by "average days to flowering") in various deblading patterns. Their conclusion was similar to that of Chailakhyan and Butenko: the movement of photosynthates and floral stimulus seemed to be coordinated. When an LD leaf inhibited flowering of a given axillary shoot, it also decreased the amount of ^{14}C moved into that shoot from the $^{14}CO_2$-treated, SD-donor leaf. [1]

Further considered evidence for the hypothesis that LD inhibited by an "opposing photosynthate stream" were several reports that decreasing the light intensity on an inhibiting LD leaf decreased the inhibition (11, 14). Harder et al. (14) found that an LD leaf exposed to 5000 lx of artificial light completely inhibited flowering of *Kalanchoë*, as did 35% of natural light in June at Göttingen. An intensity of 3000 1x was only partly effective as an LD inhibitor.

The only direct evidence that interposed LD's inhibit by means of photosynthate seems to be in Chailakhyan's 1947 paper on *Perilla* (14). A 3% sugar solution substituted for an inhibiting, interposed leaf slowed the apperance of floral buds by 25 days, compared to the 27 days' retardation from the LD leaf. Substituted for an LD leaf that was opposite the SD-treated leaf (and therefore slowed the appearance of floral buds by only 9 days), sugar solution slowed their appearance by 6 days. Considering the importance of this experiment it is regrettable that it apparently has not been repeated by others. (Even when it is cited, however, the fact is usually not mentioned that Chailakhyan also found that sugar could *not* substitute to any degree for the SD-inhibiting leaves of the LDP *Rudbeckia* – a fact indicating that noninductive photoperiods may inhibit by different mechanisms in SDP's and LDP's.)

Not fitting the hypothesis that LD inhibition is based on the pattern of photosynthate translocation are several types of evidence that such inhibition is photoperiodic. Van Senden (14) first showed that true LD's, providing extra hours of photosynthesis, were not required for inhibition in *Kalanchoë:* LB's of only 2 min given to an interposed leaf in the middle of the long nights of SD cycles completely inhibited the flowering response to a photoinduced leaf below. Light breaks cause inhibition of flower formation in *Xanthium* [(6); Lincoln et al., 1956 (14)], *Perilla* [Wellensiek, 1959 (14)], *Salvia* (1), and *Coleus* (this paper). The very low intensity of the light effective as an LB in both *Kalanchoë* and *Xanthium* (900 and ca. 500 1x, respectively) also indicates a photoperiodic rather than a photosynthetic action. Gibby and Salisbury (6) have, in addition, measured the inhibition from 5 min LB's given at different times during a 16h dark period to the basal half of an otherwise photoinduced leaf. The LB resulted in maximal inhibition when given in the middle of the dark period, again indicating a true photoperiodic effect. They concluded that "when the leaf is not producing florigen, it is actively producing some inhibitory effect."

1 The conclusions of King and Zeevaart should be checked against their data with care. In my opinion they tend to overstate their results as fitting their conclusions. For instance, a 33% increase in days to flower (20 days as compared to 15) is treated as no difference, and their reference to a "slight delay" providing a "minor inconsistency" actually refers to a more than 100% increase in days to flower (39 days as compared to 19). Unfortunately, as in so many of the papers on flowering, no statistical tests of significance are provided

Nature of the Inhibiting Chemicals Produced by LD Leaves in SDP's

In evaluating the reports of inhibiting chemicals, we should keep in mind the notorious lack of specificity common to inhibitory effects. The fact that a certain chemical "off the shelf" can stop or slow development is not strong evidence for its being the natural inhibitor unless one can at least substitute the endogenous amount of that chemical with consequent exact replacement of the natural effect. No such evidence is available for any chemical in the case of flower inhibition. Probably closest to meeting these requirements comes indole-3-acetic acid (IAA), known since 1937 (2) to inhibit flowering when exogenously applied to a plant [see (13)]. Flowering in DN *C. blumei,* which is inhibited by the axillary branches, was shown to be also inhibited to the same extent by IAA applied at the sites of the excised axillaries at concentrations shown in other experiments to replace exactly the normal IAA supply (10). For SDP's, Raghavan (18) showed that IAA at 1 mg/liter completely inhibited flowering of excised and cultured apical buds or *Perilla,* duplicating the inhibiting effect of unfolded leaves implanted in the medium along with the apices. However, he provided no evidence that such an IAA inhibition was either solely, or stronger, from LD- than from SD-treated *Perilla* leaves.

The inhibitor from LD leaves of *Kalanchoë* has been studied by Schwabe and coworkers. Juice pressed from such leaves inhibited flowering when injected into SD-treated leaves, whereas injections of juice from SD leaves gave results not significantly different from those from control injections (19). The chemical responsible has not yet been identified although abscisic acid or xanthoxin (injected at 50 and 100 mg/l, respectively) did cause significant inhibition of flowering. The essentiality of statistical testing, as well as an indication of the variability of the "bioassay", are demonstrated by the fact that injections of 5% sucrose gave average values that were not significantly different from, although they were 30% above, the controls, and that the effect of cobaltous chloride was not significantly different from the controls although only 57% of their average value. [A report that gallic acid was the endogenous inhibitor of *Kalanchoë* (17) could not be confirmed in other studies from the same laboratory (20).]

Conclusions

Long days have been shown to inhibit flowering of several SDP's, either when interposed leaves were treated with LD's or when LD's were intercalated in a series of SD's given to the whole shoot. A single leaf of the SDP *Coleus fredericii,* if taken from a plant that had been induced to flower with SD's, stimulated flowering when grafted onto a stock and subsequently kept in non-inductive conditions (SD's with LB's) just as efficiently as placing an intact plant under SD's. This was true whether the stock was *C. fredericii* itself of the DNP *C. blumei.* In contrast, a single leaf taken from *C. fredericii* plants kept vegetative with LB's inhibited flowering of the stock. With either *C. fredericii* or *C. blumei* as stocks, the difference in the time to flowering with a pre-induced and an LB-treated leaf of *C. fredericii* was ca. 20 days. The grafting results fit the hypothesis advanced by earlier workers that leaves of SDP's produce a floral stimulus in SD's and a floral inhibitor in LD's or SD's with LB's.

While physiologists generally prefer to have a single explanation for a phenomenon, there is at present no compelling evidence for believing that all the LD inhibitions found in SDP's have a common cause. However, from the efficacy of low-intensity LB's in both *Kalanchoë* and *Xanthium* it is unlikely that in these genera bulk photosynthesis is involved in the LD inhibition. Whether LD inhibition results from transmissible inhibitors or diversion of the floral stimulus (16) still needs to be demonstrated in most SDP's, and the possibility that LD's inhibit by interfering with circadian rhythms (7) has hardly been considered by investigators.

Acknowledgment. The skilled and extensive aid of Research Assistant Hannah B. Suthers, particularly in carrying out the *Coleus* grafting experiments, is gratefully acknowledged.

References

1. Bhargava, S.C.: Meded. Landbouwhogesch. Wageningen *64* (12), 1–70 (1964)
2. Dostál, R., Hosek, M.: Flora (Jena) *131*, 263–286 (1937)
3. Evans, L.T.: Aust. J. Biol. Sci. *15*, 291–303 (1962)
4. Fratianne, D.G.: Am. J. Bot. *52*, 556–562 (1965)
5. Garner, W.W., Allard, H.A.: J. Agric. Res. *31*, 555–566 (1925)
6. Gibby, D.D., Salisbury, F.B.: Plant Physiol. *47*, 784–789 (1971)
7. Halaban, R.: Plant Physiol. *43*, 1894–1898 (1968)
8. Harder, R.: Symp. Soc. Exp. Biol. *2*, 117–138 (1948)
9. Jacobs, W.P.: In: Plant Hormones and Plant Development. New York: Cambridge University Press 1979
10. Jacobs, W.P., Davis, R.V., Jr., Bullwinkel, B.: In: Photoperiodism and Related Phenomena in Plants and Animals. Withrow, R.B. (ed.), pp. 393–407. Washington, D.C.: AAAS 1959
11. King, R.W., Zeevaart, J.A.D.: Plant Physiol. *51*, 727–738 (1973)
12. Laibach, F.: Field Information Agency (Office of Military Government for Germany, U.S.) Tech. Rep. No. *1135*, 1–20 (1947)
13. Lang, A.: Encycl. Plant Physiol. *14*, 909–950 (1961)
14. Lang, A.: Encycl. Plant Physiol. *15* (Part 1), 1380–1536 (1965)
15. Lang, A., Chailakhyan, M.K., Frolova, I.A.: Proc. Natl. Acad. Sci. USA *74*, 2412–2416 (1977)
16. Ogawa, Y., King, R.W.: Plant Physiol. *63*, 643–649 (1979)
17. Pryce, R.J.: Phytochemistry *11*, 1911–1918 (1972)
18. Raghavan, V.: Am. J. Bot. *48*, 870–876 (1961)
19. Schwabe, W.W.: Planta *103*, 18–23 (1972)
20. Schwabe, W.W., Wimber, R.H.: Perspect. Exp. Biol. *2*, 41–57 (1976)

Inhibition of Flowering in Long-Day Plants [1]

A. LANG [2]

Literature Survey

Literature reports on inhibition of flower formation in LDP's by noninductive photoperiodic conditions are less numerous than analogous reports for SDP's which have been reviewed by Dr. Jacobs (see preceding article in this volume), but the evidence is derived from the same three types of experiments: subjecting plants to alternations of inductive and noninductive photoperiodic conditions; subjecting different parts of a plant – usually different leaves, but in some cases different parts of the same leaf – simultaneously to inductive and noninductive conditions; and defoliation experiments.

An example of the first type of experiment is given in Table 1. Plants of *Hyoscyamus niger* (an annual, non-cold-requiring strain) were given a constant number of LD's, known to be sufficient for flower induction, but after every one or two LD's one or more SD's were intercalated. Except for one regime, all plants formed flowers, and flower initiation was delayed at most by a number of days corresponding to the total number of intercalated SD's (this fact is not apparent in the simplified table); i.e., the intercalated SD's suspended but did not reduce the inductive action of the LD's. The phenomenon has been called summation of fractional inductions. The one exception was the regime of single LD's followed by periods of 12 SD's, when some of the plants failed to form flowers, and those that initiated them did so after a significant delay. However, in this treatment the LD's were spread over a period of 66 days, so that the first and last LD's most probably were not perceived by the same leaves and the reduced flowering response may have been the consequence of less induction of the individual leaves rather than of inhibition by the intercalated SD's.

Fractional-induction experiments have also been done with sugar beet (21, 29) and *Lolium temulentum* (12); in several other LDP's, Wellensiek (35) has studied the effect of an intercalated two-day period of darkness. The responses differed in different species and were dependent on the character of the noninductive photoperiods. Thus, in *Rudbeckia bicolor* intercalated SD's were more effective in delaying and even suppressing flowering if sunlight rather than low-intensity (ca. 20 lx) light from incandescent lamps was used in the light periods (30). In general, however, it appears that the inhibitory effect of intercalated noninductive photoperiods on flower formation, if at all

1 Abbreviations: DN, dayneutral; DNP, dayneutral plant; LD, long day; LDP, long-day plant; SD, short day; SDP, short-day plant; HYO, *Hyoscyamus niger* (annual strain); MM, *Nicotiana tabacum* cv. Maryland Mammoth; NS, *N. silvestris;* TR, *N. tabacum* cv. Trapezond.

2 MSU-DOE Plant Research Laboratory, Michigan State University, East Lansing, MI 48824, USA

present, is considerably less marked in LDP's than in SDP's, where one intercalated LD may completely suppress the inductive action of the following 1.5–2 SD's, and a single LD, intercalated between two sequences of six SD's, reduces the flowering response of *Kalanchoë blossfeldiana* by ca. 50% (32, 33). [Some investigators have applied fractional induction to LDP's until obtaining a flowering response, or to the end of the given experiment. With this procedure, however, plants in different treatments may not receive the same number of inductive photoperiods and comparisons of the responses may not be valid. Such reports (e.g. 10, 31) therefore have not been used in this survey, although most results do not seem to disagree with above conclusions.]

Table 1. Fractional induction in *Hyoscyamus niger*

Treatment	No. plants with flower buds/no. vegetative	Treatment	No. plants with flower buds/no. vegetative
A. From (27)		*B. From (1), two different experiments combined*	
2 LD	0/6	6 LD	14/0
6 LD	6/0	6 × (1 LD + 1 SD)	14/0
3 × (2 LD + 2 SD)	6/0	4 × (1 LD + 2 SD)	14/0
3 × (2 LD + 5 SD)	6/0	6 × (1 LD + 12 SD)	4[a]/10
3 × (2 LD + 10 SD)	6/0	6 × (2 LD + 1 SD)	14/0
		6 × (2 LD + 2 SD)	14/0
		6 × (2 LD + 12 SD)	14/0

[a] Flower formation retarded

In analogy to findings in numerous SDP's, leaves maintained on noninductive photoperiods can reduce the flower-promoting effect of leaves given inductive treatment if they are inserted between the induced leaves and the responding meristem, i.e., if noninduced tissue is interposed (Dr. Jacob's term) between the induced and the target tissue. Examples of this effects are given in Table 2; the experiments were carried out with plants defoliated except for one to three leaves or two leaf pairs. The response, like that to fractional induction, varies with the plant, from markedly delayed flower formation in *Sinapis alba* and *Silena armeria* to no effect in *Rudbeckia*, and may depend on experimental conditions. In *Urtica pilulifera*, for example, noninduced leaves inhibited flower formation if given "regular" SD's, but not when given SD's with only 1–3 h of light per day (28); unexpectedly they were inhibitory also in continuous darkness. On the average this effect of interposed noninductive leaves, similar to that of intercalated noninductive photoperiods, is less in LDP's than in SDP's. In *Sinapis*, flower formation was considerably delayed if the apical half of a leaf was treated with LD's and the basal half with SD's, as compared to the reverse treatment (4). In *Xanthium*, the results of analogous experiments have been considered as evidence for a flowering inhibitor (19), but reexamination (40) has shown that this interpretation is not obligatory.

In some cases, however, SD leaves can affect the flowering response in an LDP even when *not* interposed between the induced leaf or leaves and the target meristem. Evans

Table 2. Effect of SD leaves on the flowering response of several LDP's. Data for *Silene* (36), all others (5, 6)

Upper leaf/ leaves on	.......SD	LD	LD	–[a]	SD	SD
Lower leaf/ leaves on	.......SD	LD	–[a]	LD	LD	LD
Sinapis [b]	27	15			12	23
Raphanus [b, c]	∞	15			17	21
Spinacia [b]	∞	14			17	20
Rudbeckia [b]	∞	9			10	10
Silene [d]	0	6.6	6.1	4.9	6.5	2.6

a Leaf removed
b Days to flower buds in terminal inflorescence or to onset of stem elongation
c In SD/SD and LD/LD, 3 leaves; in LD/SD and SD/LD, 1 leaf in LD, 2 leaves in SD
d Leaf *pairs;* response in "stages": 0 = vegetative, 1, 2 . . . = increasing response

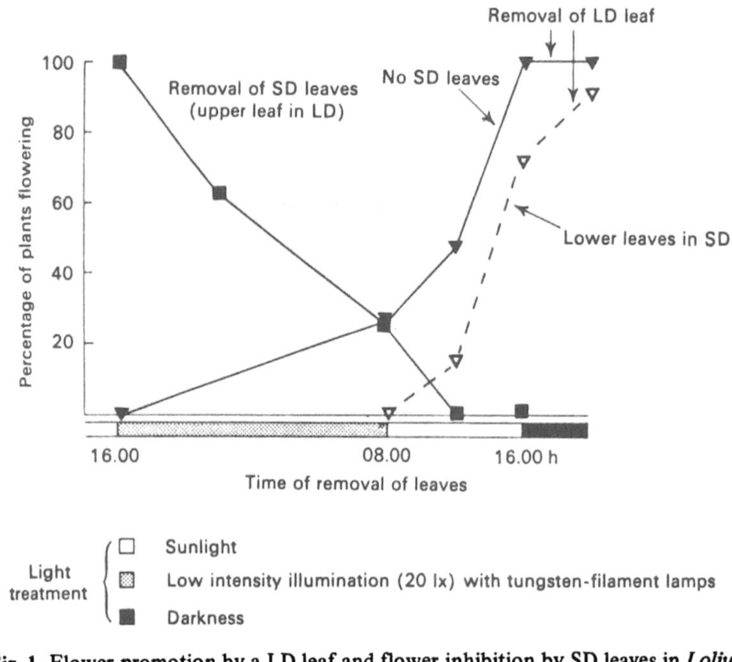

Fig. 1. Flower promotion by a LD leaf and flower inhibition by SD leaves in *Lolium temulentum* as dependent on the time the leaves are allowed to remain on the plant. After data of Evans (13) from Vince-Prue (34), by permission. The uppermost leaf of an SD-grown plant was given one 24-h treatment consisting of 8 h sunlight (starting at 08.00 hours) and 16 h light from electric lamps (16.00 hours to 08.00 hours the next day); the other leaves were either removed at the start of the inductive treatment, or were continued on SD's. At 16.00 hours, and at various intervals thereafter, either the LD leaf (in plants with the lower leaves either removed: △———△, or kept continuously in SD's: △- - - - -△) or the SD leaves were removed

(13) showed that the flowering response of *Lolium temulentum* plants whose uppermost leaves were given one LD was reduced if older leaves were left on the plants and given SD's; this effect increased with increasing number of noninduced leaves and with the length of time these leaves were allowed to remain on the plant (Fig. 1). Similar

experiments, on a smaller scale but with comparable results, have been done in *Sinapis* (23) and in *Blitum capitatum* and *B. virgatum* (22); in the latter two species mature leaves inhibited the flowering response even if maintained on LD's, but much less than if maintained on SD's. Experiments with spinach (37) also indicate an inhibitory effect of noninterposed SD leaves. However, leaves that developed during the treatment were apparently kept under SD's and may have contributed to this effect. In *Lolium*, the inhibitory effect of SD leaves was abolished if these leaves were exposed to a nitrogen atmosphere, whereas the flower-inducing effect of the LD leaf was not affected by this treatment (14).

In experiments with annual *Hyoscyamus*, defoliation had a striking effect. Continued removal of the leaves abolished the photoperiodic dependence of the plant, and flowers were initiated in LD, SD, and even in darkness [(27); see Table 3]. To obtain

Table 3. Effect of defoliation on the photoperiodic response of *Hyoscyamus niger*. [(27), and unpublished]

Treatment of plant	Light regime	No. plants with flower buds	Days [a]	Leaf increment [b]	No. plants vegetative
Intact	LD	10	10.5	11.0	
	SD				20
	CD [c]				8
Defoliated	LD	26	8.0; 18.3 [d]	22.5; 32.6 [d]	9
	SD	32	7.5; 16.5 [d]	24.3; 29.3 [d]	2
	CD	7	10.5	30.3	
Defoliated except for one leaf	LD	10	9.2	22.3	
	SD	2 [e]	(18.2)	(23.0)	8
Defoliated, one leaf regrafted	LD	6	14.8	24.5	
	SD	2 [e]	(17.5)	(27.5)	6

a In intact plants, to visible stem elongation; in defoliated plants, to first flower primordium
b No. of leaves formed from beginning of treatment to formation of first flower bud
c Continuous darkness
d Two different experiments
e Leaf or grafted leaf in poor condition

this result one has to use plants with a well-developed tuber and to remove all expanded or expanding leaves. Leaving only one mature leaf on the plant (Lang, unpublished results) or regrafting such a leaf on a defoliated plant (27) restores the photoperiodic sensitivity, at least for as long as the leaf remains healthy. Defoliation in other LDP's [*Nicotiana silvestris* (27); beet, radish, *Rudbeckia* (9); *Silene, Crepis tectorum* (Lang, unpublished results)] did not result in flower formation under either SD or LD conditions. Evidently, these defoliated plants are incapable of continued growth and development, so that their failure to form flowers does not invalidate the positive results obtained with *Hyoscyamus*.

As reviewed by Vince-Prue (34), three hypotheses have been advanced to explain inhibitory effects of noninduced leaves on flower formation in photoperiodic plants: changes of the pattern of translocation of flower-promoting substance(s), i.e., florigen; interference with florigen synthesis; and production of transmissible flower inhibitor(s). The inhibitory effect of interposed SD leaves in LDP's can be interpreted according to the first hypothesis, based on our general information about the distribution of photoassimilates in plants and assuming that florigen moves with the assimilate stream. As a rule, a shoot meristem is primarily supplied with photoassimilates from the nearest photosynthesizing leaf. If in an LDP this leaf is in LD's, the meristem will also receive florigen, but if the leaf is kept in SD's, florigen produced in a more distant LD leaf may reach the meristem only in smaller quantities. Differences in the flower-inhibiting effect of interposed SD leaves in different LDP's (Table 2) may reflect differences in how rigorously this translocation pattern has developed. This idea could be tested in studies of photoassimilate movement with radioactive tracers, as done in the SDP *Perilla* (7, 25), where they provided strong circumstantial support for the translocation pattern hypothesis. In the only such study so far on an LDP, *Lolium temulentum,* the conclusion was that the floral stimulus and photoassimilates are *not* transported together, the latter moving considerably faster than the former (18).

The results of the fractional-induction experiments may be interpreted, as they have been in SDP's, on the basis of the second hypothesis, i.e., inhibition of florigen formation in the leaf. However, in LDP's these results are too limited to support a firm conclusion; the very existence of inhibition of flower formation in LDP's by intercalated SD's may need further examination.

The inhibitory effects of noninterposed SD leaves on flower formation in LDP's have been interpreted, particularly by Evans (13, 16), as evidence for inhibitor(s) of flower formation being formed in SD leaves, translocated to the meristem, and there counteracting the effect of florigen coming from LD leaves. The flowering response curve obtained in *Lolium* plants with one LD leaf when SD leaves are removed at different times during induction can be considered as the time-course curve for the export of the inhibitor(s) from the leaf; comparison with the analogous curve obtained when the LD leaf is removed at various times shows that the export of the inhibitor(s) precedes that of florigen (Fig. 1).

The results of the defoliation experiments with *Hyoscyamus* can obviously also be interpreted in terms of removal of the source (SD leaves) of transmissible inhibitor(s) of flower formation. However, the evidence from both types of experiments, i.e., those with inhibitory but noninterposed SD leaves and the defoliation experiments, is indirect and is subject to alternative interpretation. Thus, the SD leaves could side-track florigen or materials necessary for florigen production in LD leaves (27, 38). Evans (16) has rejected this explanation because photoassimilates from SD-treated older leaves of *Lolium* were primarily transported not to the shoot apex but to other parts of the plant, mainly the root (17). However, since photoassimilates and flower stimulus in *Lolium* seem to be transported independently [(18); see above], this argument is not quite conclusive.

Evidence for Graft-Transmissible Flowering Inhibitors in Long-Day Plants

To answer the question of the existence of hormone-like flowering inhibitors Professor Chailakhyan and I have utilized grafting experiments. This work was started in 1975–1976, when I spent a sabbatical leave at the K.A. Timiryazev Institute of Plant Physiology in Moscow, and has been continued as a joint project in East Lansing and Moscow. As experimental material we chose the LDP's *Hyoscyamus niger* (annual strain) and *Nicotiana silvestris* (HYO and NS, respectively), the SDP *N. tabacum* Maryland Mammoth (MM), and the DNP *N. tabacum* Trapezond (TR), because in these plants the existence of graft-transmissible, flower-promoting material(s) (florigen) has been amply demonstrated; likewise by grafting experiments. However, whereas in studies that led to the florigen concept the influence of a plant capable of flower formation (an induced photoperiodic plant or a DNP) on a graft partner incapable of flowering (a photoperiodic plant in the noninductive daylength regime) was examined, we now examined effects of a nonflowering plant on a graft partner entirely competent to flower formation. Thus, we used TR, the DNP, as the "assay" material throughout and

Fig. 2. Grafts NS on TR. *Left* on LD; *right* on SD. From (26)

HYO, NS, and MM, kept on their respective noninductive photoperiods, as potential "donors" of inhibitor(s) of flower formation.

When NS/TR and HYO/TR grafts were maintained on LD's, flowering in the DN partner was significantly accelerated, indicating transmission of florigen from the induced LDP to the DNP, but when these grafts were maintained on SD's the DN partner failed to flower [(26), and unpublished data; Table 4; Figs. 2 and 3]. Moreover, in

Fig. 3. Grafts HYO on TR; as in Fig. 5. (Original)

grafts kept on SD's the growth of the TR shoots was markedly modified: the stem became thicker and the internodes shorter than normal, the entire habit somewhat approaching the rosette growth habit characteristic of most LDP's, including NS and HYO, grown under SD conditions. Noninduced LD partners also may suppress the development of initiated flower buds of TR, at least when these are small. In some grafts, the TR indicator shoot did form flower buds, presumably before tissue union had been fully established, but these buds did not develop further and a vegetative lateral shoot with the modified growth habit took over (Fig. 4). The entire syndrome is highly persistent. In a few grafts kept from August 1976 to August 1977 the TR shoots, while continuing to grow, remained strictly vegetative and maintained the modified growth habit. In contrast, if the grafts are transferred to LD the syndrome soon disappears; the TR partners have visible flower buds after ca. 4 weeks and open flowers 3–4 weeks later, and their stems attain normal thickness and internode length (Fig. 5).

Table 4. Flowering and growth response in grafts between the dayneutral tobacco cultivar Trapezond (TR), the short-day cultivar Maryland Mammoth (MM), and the long-day plants *Nicotiana silvestris* (NS) and *Hyoscyamus niger* (HYO). [(26), and unpublished]

Combination	Photo-period	No. grafts	No. with flowers	No. vegetative	Mean no. days to first visible flower bud	Mean no. nodes on indicator shoot
TR/TR	LD	15	15	0	50	18
(control)	SD	15	15	0	49	22
MM/TR	LD	16	15	1[a]	> 48[b]	20
	SD	16	16	0	32	15
NS/TR	LD	15	15	0	23	14
	SD	15	0	15		36
HYO/TR	LD	16	16	0	34	17
	SD	16	0	16		34

[a] Indicator shoot growing poorly
[b] Indicator shoots of 3 grafts produced no visible flower buds but contained microscopic ones; growth of these shoots was slow

Thus, NS and HYO, two LDP's known to produce flower-promoting graft-transmissible material(s) (florigen) under LD's, under SD's produce flower-inhibiting, graft-transmissible material(s) (antiflorigen), which are evidently instrumental in maintaining the plants in the vegetative state and in the rosette habit. Florigen is known to be produced in the leaves; grafting one mature leaf of NS onto TR was enough to inhibit flower formation in the latter, and to modify its growth habit, if the graft was kept on SD's, and as long as the grafted leaf remained healthy (unpublished data).

To gain some insight into the interaction of florigen and antiflorigen Dr. R. Kaiss Chapman made double grafts, consisting of a shoot of NS and one of MM both grafted onto an MM plant, and varied the amount of inductive and noninductive tissue by leaving two, five, or ten leaves on either partner, in all combinations. When the grafts were kept on SD's the MM partners formed flowers the sooner the more leaves were present on the MM scion, and the later the more leaves were left on the NS scion, although flower formation occurred ultimately even in grafts with two MM and ten NS leaves (Table 5). Thus, florigen and antiflorigen appear to interact in a rather simple, "stoichiometric" manner, most probably in the shoot meristem.

Do SDP's maintained on LD's produce antiflorigen? In grafts between MM and TR we did not find much evidence that this was the case (26). However in grafts of NS on MM, Chailakhyan et al. (9a) did obtain marked inhibition of flower formation in the

Fig. 4. Graft HYO on TR. TR indicator shoot *(right)* formed small flower buds but these did not develop further; instead, an axillary bud developed into a shoot that grew vegetatively. (Original)

NS partner when the grafts were kept on LD's and the MM partner had six or more leaves, while Jacobs and Suthers (see preceding article) observed a similar response in a DN *Coleus* under the influence of a SD graft partner.

Concluding Remarks

The results of the grafting experiments with NS, HYO, and TR, described above, fully agree with the results of Evans' experiments with *Lolium temulentum* in which a young leaf was given LD's, and the older leaves SD's. Our conclusions also agree with his that, depending on the photoperiod to which the leaves of an LDP are subjected, they produce materials which promote or inhibit flower formation; both types of material(s) are transported to and interact in the shoot meristem. In addition, our work has shown that "antiflorigen" is graft-transmissible, strengthening the idea that it is of a hormonal nature, and – most importantly – that it is not specific for a given photo-

Fig. 5. Grafts NS on TR, both started on SD. Graft on the right remained on SD, graft on the left was transferred after 12 weeks to LD. Photograph taken ca. 8 weeks after the transfer SD to LD. (Original)

periodic type: it is evidently active at least in "our" LDP's, i.e., NS and HYO, but can also inhibit flower formation and modify the growth pattern of a DNP and an SDP. It is in these respects similar to florigen, which is also interchangeable between, and thus most likely identical in, the various photoperiodic plant types.

The idea of inhibitory phenomena in flower formation is not new. F.G. Gregory (20) wrote in 1948: "We must suppose that the necessary genes (for flower formation) are already present in the fertilized ovum, and that if external factors are such that no flowers are formed there must be inhibitory factors at work; in a word, the problem may quite as properly be considered as one of 'failure to flower' as of promoting flowering". Similar ideas have been expressed before and since. Of course, the fact that all seed plants are genetically programed for flowering is a truism which in itself does not help us understand why a given plant does or does not flower. Gregory's statement also implies that promotion and inhibition of flowering are alternatives, and this viewpoint has been even more strongly espoused by others. Our results, like those

Table 5. Flowering of MM indicator shoots in double grafts with an MM and an NS scion, each with various numbers of leaves. (Chapman and Lang, unpublished)

		MM indicator shoot	
No. of leaves on MM scion	No. of leaves on NS scion	Days to first flower bud	No. of leaves
10	0	29	27
10	2	47	34
10	5	89	41
10	10	122	46
5	0	37	29
5	2	75	42
5	5	113	44
5	10	136	46
2	0	83	41
2	2	127	47
2	5	136	49
2	10	145	54

of Evans and of Jacobs and Suthers, show that such a view is too restrictive. At least "our" Solanaceae, Evans' *Lolium*, and Jacobs and Suthers' SD *Coleus* all posses a dual control mechanism for flower formation, involving the production of two hormone-like materials. In inductive photoperiods these plants produce exclusively or predominantly florigen; in noninductive ones, antiflorigen. Both materials arise in the leaves and act in the meristems, where they seem to interact in a simple manner. If the plant is transferred from one photoperiodic regime to the other, the production of one material is stopped or greatly reduced, and that of the other is turned on or greatly stepped up.

However, while evidence for "florigen" derived from grafting experiments is available in at least eight taxonomically widely divergent families of angiosperms (dicotyledons), similar evidence for "antiflorigen" is presently far more limited. One objective for further research on inhibitory effects in flower formation should therefore be to determine how widespread antiflorigen is in the plant kingdom, and also whether or not it is equally present in LDP's, SDP's, and other photoperiodic response types. And another, obvious objective is isolation and chemical identification of antiflorigen(s); in this regard, one can only hope that this work will be less frustating than the analogous work on florigen has so far proven.

When one is confronted with inhibitory phenomena in plant development, one's attention nowadays focuses on abscisic acid (ABA) and ethylene. Several investigators have reported inhibition of flowering in LDP's by ABA (2, 3, 8, 11, 15, 24), but in most cases the effect is to delay rather than completely suppress flower formation and has been observed either under conditions which do not result in a maximal flowering response or with unphysiological doses (up to 500–1000 mg/liter). Complete and persistent inhibition of flower formation, such as we have found in a DNP under the influence of antiflorigen from two LDP's, has not been obtained with ABA. It, therefore, is not a likely antiflorigen candidate. As to ethylene, I am not aware of reports of effects on flower formation in LDP's. In SDP's, both inhibition of flower formation under SD conditions and promotion under LD conditions have been observed, and the flower-inducing effect of ethylene in pineapple and other members of the Bromeliaceae is now well known [see (39)]. The effect of antiflorigen on the growth pattern of dayneutral tobacco (thickened stems, shortened internodes) is reminiscent of some ethylene effects on the growth of peas, etc., but other features of the ethylene syndrome (epinasty and loss of geotropic orientation) are absent; thus, it is uncertain whether those similarities have any common physiological basis.

Acknowledgments. The author's research at Michigan State University reported in this paper was supported by the US Department of Energy under Contract No. EY-76-C-02-1338; his stay at the Timiryazev Institute in Moscow was under the Scientific Exchange Program between the National Academy of Sciences of the USA and the Academy of Sciences of the USSR.

References

1. Carr, D.J.: Physiol. Plant. *8*, 512–526 (1951)
2. Cathey, H.M.: Proc. Am. Soc. Hortic. Sci. *93*, 560–568 (1968)
3. Cathey, H.M.: Proc. Am. Soc. Hortic. Sci. *93*, 693–698 (1968)
4. Chailakhyan, M.Kh.: C.R. Dokl. Acad. Sci. URSS *47*, 220–224 (1945)
5. Chailakhyan, M.Kh.: C.R. Dokl. Acad. Sci. URSS *54*, 735–738 (1946)
6. Chailakhyan, M.Kh.: C.R. Dokl. Acad. Sci. URSS *54*, 837–840 (1946)
7. Chailakhyan, M.Kh., Butenko, R.G.: Fiziol. Rast. *4*, 450–462 (1957)
8. Chailakhyan, M.Kh., Kochankov, V.G.: Dokl. Akad. Nauk SSSR *188*, 477–480 (1969)
9. Chailakhyan, M.Kh., Samygin, G.A.: Dokl. Akad. Nauk SSSR *62*, 549–552 (1948)
9a. Chailakhyan, M.Kh., Yanina, L.I., Lotova, G.N.: Dokl. Akad. Nauk SSSR (in press, 1980)
10. Curth, P.: Z. Pflanzenzücht. *47*, 254–276 (1962)
11. El-Antably, H.M., Wareing, P F., Hillman, J.: Planta *73*, 74–90 (1967)
12. Evans, L.T.: New Phytol. *59*, 163–174 (1960)
13. Evans, L.T.: Aust. J. Biol. Sci. *13*, 429–440 (1960)
14. Evans, L.T.: Aust. J. Biol. Sci. *15*, 281–290 (1962)
15. Evans, L.T.: Science *151*, 107–108 (1966)
16. Evans, L.T.: In: The Induction of Flowering – Some Case Histories. Evans, L.T. (ed.), pp. 457–480. Ithaca, N.Y.: Cornell University Press 1969
17. Evans, L.T., Wardlaw, I.F.: Aust. J. Biol. Sci. *17*, 1–9 (1964)
18. Evans, L.T., Wardlaw, I.F.: Planta *68*, 310–326 (1966)
19. Gibby, D.D., Salisbury, F.B.: Plant Physiol. *47*, 784–789 (1971)
20. Gregory, F.G.: Proc. Soc. Exp. Biol. *2*, 75–102 (1948)
21. Hamner, K.C.: Bot. Gaz. *101*, 658–687 (1940)
22. Jacques, M.: Physiol. Vég. *9*, 461–474 (1971)
23. Kinet, J.M., Bodson, M., Alvinia, A.M., Bernier, G.: Z. Pflanzenphysiol. *66*, 49–66 (1971)
24. Kinet, J.M., Bodson, M., Jacqmard, A., Bernier, G.: Z. Pflanzenphysiol. *77*, 70–74 (1975)
25. King, R.W., Zeevaart, J.A.D.: Plant Physiol. *51*, 727–738 (1973)
26. Lang, A., Chailakhyan, M.Kh., Frolova, I.A.: Proc. Natl. Acad. Sci. USA *74*, 2412–2416 (1977)
27. Lang, A. Melchers, G.: Planta *33*, 653–702 (1943)
28. Lona, F.: In: Lavori di Botanica, Vol. pubbl. in occasione del 70° genetliaco di Prof. G. Gola, pp. 285–311. Turin: Rosenberg & Sellier 1947
29. Naylor, A.W.: Bot. Gaz. *102*, 557–585 (1941)
30. Samygin, G.A.: Dokl. Akad. Nauk SSSR *60*, 1433–1436 (1948)

31. Schmalz, H.: Z. Pflanzenzücht. *39*, 97–112 (1958)
32. Schwabe, W.W.: Ann. Bot. *20*, 1–14 (1956)
33. Schwabe, W.W.: J. Exp. Bot *10*, 317–329 (1959)
34. Vince-Prue, D.: Photoperiodism in Plants, 444 pp. London: McGraw-Hill 1975
35. Wellensiek, S.J.: Proc. K. Ned. Akad. Wet. C *63*, 155–158 (1960)
36. Wellensiek, S.J., Elings, C.G.: Proc. K. Ned. Akad. Wet. C *70*, 187–191 (1967)
37. Withrow, A.P., Withrow, R.B., Biebel, J.P.: Plant Physiol. *18*, 294–298 (1943)
38. Zeevaart, J.A.D.: In: Environmental Control of Plant Growth. Evans, L.T. (ed.), pp. 289–310. London, New York: Academic Press 1963
39. Zeevaart, J.A.D.: In: Phytohormones and Related Compounds – A Comprehensive Treatise. Letham, D.S., Goodwin, P.B., Higgins, T.J.V. (eds.), Vol. II, pp. 291–327. Amsterdam, Oxford, New York: Elsevier/North Holland 1978
40. Zeevaart, J.A.D., Brede, J.M., Cetas, C.B.: Plant Physiol. *60*, 747–753 (1977)

Regulation of Flowering in the Grapevine (Vitis vinifera L.)

M.G. MULLINS [1]

Introduction

Detailed information on the control of flowering is available for numerous annual and biennial plants, including species of little or no economic importance, but such information is scarce for many of the world's major crop plants. Fruit trees, grapevines, and most other broad-leaved woody perennials are among the plants which have received relatively little attention in this respect.

There are several reasons for this paradox. First, trees and vines are inconvenient experimental material. They are large and difficult to manage in glasshouses or growth rooms. Seedlings of most woody perennials have a protracted juvenile or nonflowering phase and adult, sexually mature individuals produce flowers only once a year. With normal techniques the turnover of experiments is slow, and this has severely limited the acquisition of knowledge on the physiology of flowering in trees and vines.

Second, the control of flowering was first studied in plants which are highly responsive to changes in light or temperature. Subsequently, much research on the mechanisms by which apices are transformed from the vegetative to the reproductive mode of development has been with plants specially selected for sensitivity to photoperiod or to vernalization, such as *Xanthium, Pharbitis, Lolium, Hyoscyamus, Nicotiana, Kalanchoë,* and *Bryophyllum.* These are relatively small, fast-growing, and easily grown herbaceous plants.

Third, there is the tacit assumption that the information which emerges from research on "plants of convenience" has wide applicability. Textbooks in plant physiology, for example, seldom refer to the fact that flowering in many species, including economically important plants, is regulated by mechanisms which are either unknown or which seem to lack a close relationship with the familiar light- or temperature-controlled systems.

This paper describes the control of flowering in the grapevine, the most widely grown of fruit plants (10.3×10^6 ha in 1976). Special attention is given to effects of cytokinins, but the discussion will not be limited to a single class of growth substance because flowering in the grapevine is a sequence of events widely separated in time and now known to involve different growth substances.

Mention will be made of the structural peculiarities of grapevines and of the origin of inflorescences. It will be shown that whilst the route to flowering in *Vitis* is signifi-

1 Department of Agronomy and Horticultural Science, University of Sydney, N.S.W. 2006, Australia

cantly different from that of the more commonly studied herbaceous species there are, nevertheless, some parallels in the roles of growth substances in the regulation of flowering.

Origin of Inflorescences

The complex morphology of the grapevine shoot system and the origin of inflorescences (2, 5, 6, 20, 21) have been clarified recently by scanning electron microscopy (28). In brief, flowering in the mature grapevine is normally a three-step process. The first step is the formation of *Anlagen* or "uncommitted primordia" by the apices of specialized lateral buds (latent buds) on shoots of the current season. Next, the Anlagen develop either as inflorescence primordia or as tendril primordia, and shortly thereafter the latent buds enter into dormancy. In some circumstances Anlagen may produce shoot primordia instead of inflorescences or tendrils. Finally, the formation of flowers from the inflorescence primordia occurs at the time of bud burst in the next season.

Formation of Anlagen and Tendrils

The apex of the latent bud produces up to five leaf primordia (Fig. 1a) and then divides into two almost equal parts (Fig. 1b). The part opposite the youngest leaf primordium is the Anlage (28). The first step in the differentiation of the Anlage is the formation of a bract (Fig. 2). Next, the apex subtended by the bract divides to form two branches or arms. This is a critical stage in the reproductive development of the

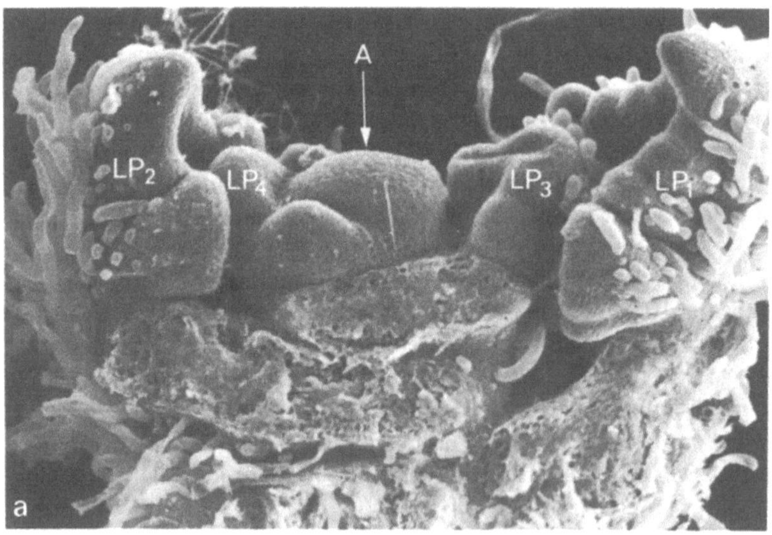

Fig. 1a. Structure of the latent bud before Anlagen formation. The apex *(A)* has produced four leaf primordia (LP 1–4). X 100

grapevine because two-branched Anlagen have the potential to produce inflorescence primordia, tendril primordia, or shoot primordia (Fig. 2).

The control of shoot formation from Anlagen (28, 31) will not be considered further here. In the case of flowering, however, it is evident that control can be exercised at two levels. The first is a coarse control and involves the formation of Anlagen. The second is a finer level of control and concerns the switching of the two-branched Anlage into either the inflorescence or the tendril pathway.

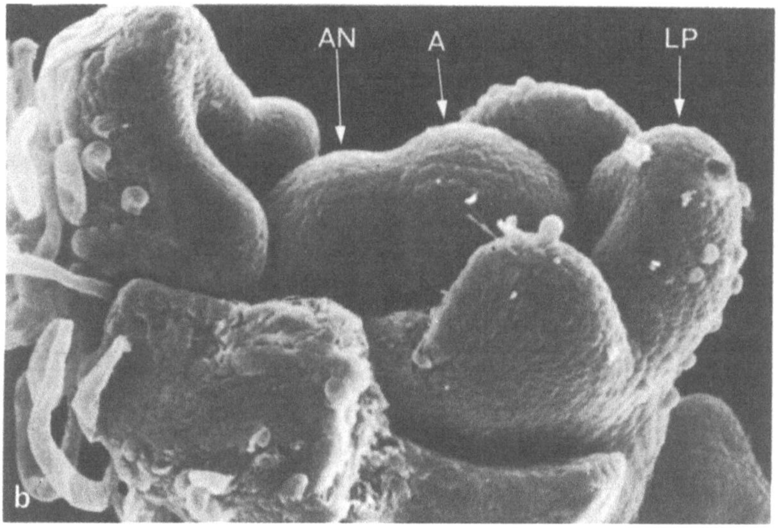

Fig. 1b. Anlagen *(AN)* formation. × 200

Gibberellin is involved in both Anlagen formation and in the determination of Anlagen development (31). When container-grown vines are treated with gibberellic acid (GA_3, 8–30 μM) there is premature sprouting and elongation of latent buds and precocious formation of Anlagen, i.e., the first Anlagen are formed at the 2nd or 3rd nodes from the base of the stem instead of at the 4th or 5th nodes as is normal. Anlagen formed in GA-treated plants grow only into tendrils and formation of inflorescence primordia is inhibited. Anlagen formation and tendril elongation are suppressed by chlormequat (2-chlorethyltrimethyl ammonium chloride), an inhibitor of GA synthesis (13).

Differentiation of Inflorescences

Effects of Exogenous Cytokinins

A characteristic of *Vitis* is that tendrils and inflorescences are homologous organs, both arising from the same Anlage. Tendrils are generally regarded as vegetative appendages which provide support for climbing plants, but grapevine tendrils can be regarded as potential reproductive organs.

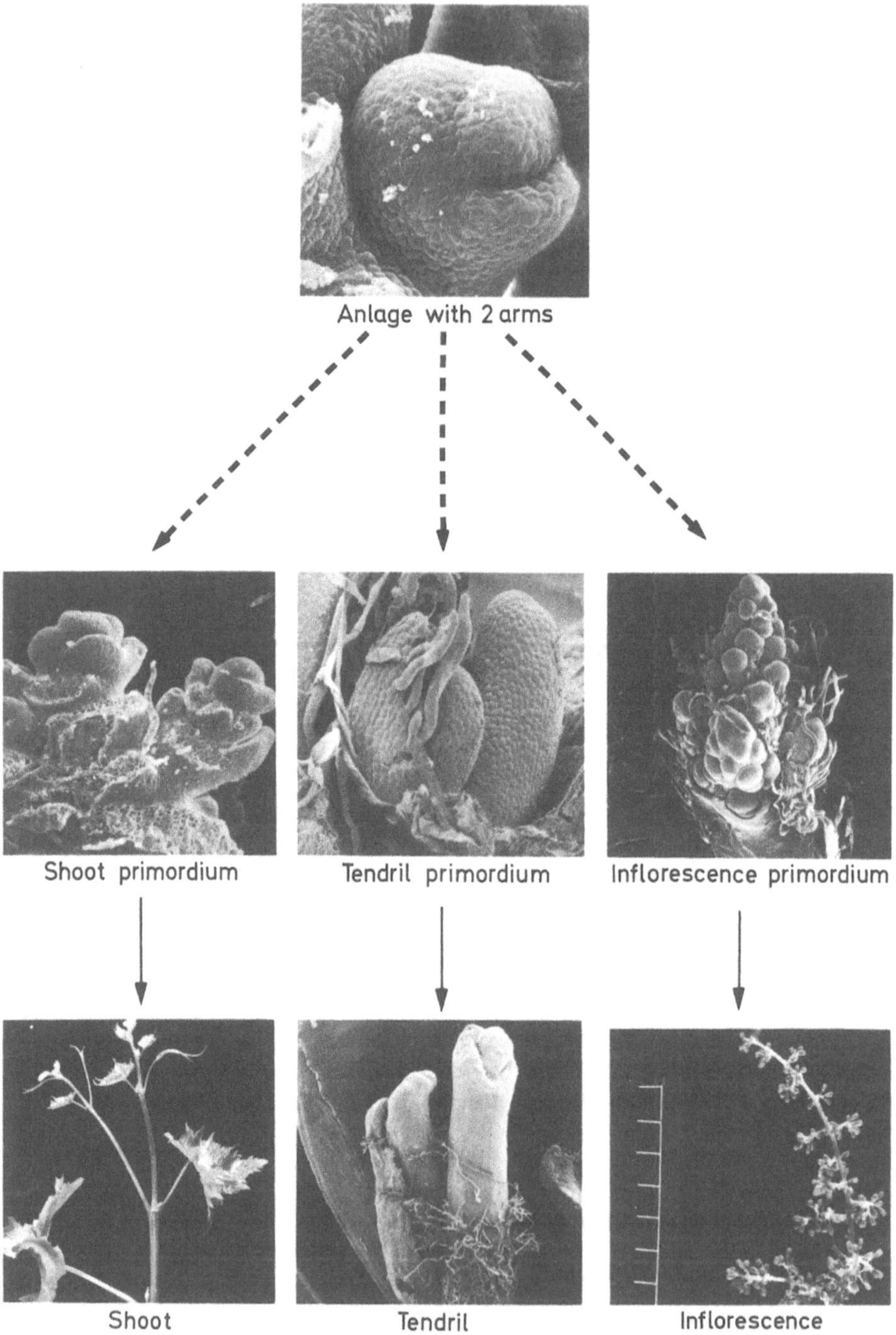

Fig. 2. Pathways of Anlagen development. Anlage with 2 arms, × 125; shoot primordium, × 35; tendril primordium, × 75; inflorescence primordium, × 20; shoot, × 0.15; tendril, × 10; inflorescence, × 0.25

Anlagen which undergo repeated branching give rise to inflorescence primordia (2) but those which produce only two or three branches give rise to tendrils (28). It follows that the control of inflorescence formation in grapes hinges upon the control of branching of Anlagen (or of tendrils). This process has been studied by growing isolated apices and tendrils in aseptic culture with various growth substances (29) and by application of growth substances to the apices and tendrils of container-grown plants (30, 31).

Isolated tendrils were induced to branch profusely, and they grew into inflorescences or inflorescence-like structures when cultured in vitro with the cytokinins benzyladenine (BA), 6-(benzylamino)-9-(2-tetrahydropyranyl)-9H purine (PBA), or zeatin riboside. There was normal development in vitro of calyx, corolla (calyptra) stamens and pistils, but micro- and megasporogenesis were absent. With normally grown plants inflorescences and fruits with viable seeds were produced in place of tendrils when shoot apices of a *Vitis vinifera* cultivar (Muscat of Alexandria) and of a hybrid grape *(V. vinifera* x *V. rupestris)*, were given repeated applications of PBA (50–200 μM).

The ability of cytokinin to transform tendrils into inflorescences is not restricted to cultivars of the grapevine but is applicable also to grapevine seedlings. By repeated application of cytokinin (BA or PBA) to the first-formed tendrils of grape seedlings, flower initials have been induced to form within four weeks of germination. These results, which are still preliminary, raise the prospect of greatly reducing the generation time in grapevine breeding.

To summarize, inflorescence formation in the grapevine appears to involve a weakening of apical dominance in Anlagen or in young tendrils. The effects of cytokinins on branch formation which are observed in vitro and in vivo are similar in character to responses to cytokinins in other correlative phenomena (19).

Environmental Factors and Inflorescence Formation

The effects of environmental factors on inflorescence formation in the grapevine have been studied in the field and in plants grown in controlled conditions (4). There are several reports that the mode of development of Anlagen is regulated by temperature conditions during the differentiation of latent buds. In the cultivar Muscat of Alexandria (syn. Muscat Gordo Blanco), for example, high temperatures (30°C) favor the formation of inflorescence primordia by Anlagen but low temperatures favor the formation of tendril primordia (5). The mechanism by which temperature affects inflorescence formation in *Vitis* is unknown, but knowledge of the effects of exogenous growth substances indicates that temperature may exert its effect on flowering by modifying the levels of endogenous cytokinin(s). This view is supported by evidence of increasing cytokinin production with increasing temperature (1) and by qualitative differences in cytokinins in the xylem sap of Sultana vines which were grown at either 20°C or 30°C (25).

Differentiation of Flowers

When dormant latent buds are activated in spring the inflorescence primordia which were formed in the previous summer undergo a period of rapid development. Each branch of the inflorescence primordium divides many times to give groups of flower primordia. Depending on the cultivar the flower primordia are formed in groups of 3 or 5 (24, 28).

Flower formation from the inflorescence primordium is a cytokinin-controlled process. There is high cytokinin activity in the xylem (bleeding) sap of the grapevine during bud burst and flowering (18, 26) and there is evidence that the cytokinin produced by roots is involved in the regulation of flower development. Inflorescence primordia in hardwood cuttings of the grapevine atrophy if bud burst precedes the emergence of adventitious roots, but normal inflorescences are formed when cuttings are propagated by a technique which ensures root formation in advance of bud burst (14). Rootless cuttings require exogenous cytokinin (BA or PBA) for normal differentiation of inflorescences (15, 16).

Other effects of exogenous cytokinins in *Vitis* include the induction or promotion of pistil development in male vines (17) and the promotion of fruit set in grape cultivars (35). An effect of cytokinin and of roots in promoting femaleness in dioecious species has been reported recently by Chailakhyan and Khryanin [(10); see also the following paper in this volume].

The Control of Flowering

The concept of a single trigger for flowering is inappropriate in the grapevine because inflorescence formation is regulated at two levels – that of formation and that of differentiation of Anlagen. The theory that the floral stimulus involves two complementary stimuli, as suggested by Carr (7), Chailakhyan (8, 9), and Evans (11) is a more acceptable working hypothesis.

According to Thimann (32), flowering is simply a developmental process under the control of the interplay of hormones. Zeevaart (36) made a distinction between annual and perennial plants by proposing that the requirement for a specific balance of hormones for flower formation is readily applicable to woody perennials but less attractive as a hypothesis for control of flower formation in herbaceous plants. Evidence is accumulating that in the grapevine gibberellin and cytokinin are the principal regulators of flowering.

Chailakhyan (9) has suggested that gibberellins are involved in the formation and growth of floral stems or inflorescence axes. The responses of grapes to exogenous GA and chlormequat are consistent with this view. Gibberellin is necessary for the formation of inflorescence axes (initiation of Anlagen) and for the growth of inflorescence axes (two-branched stage of the Anlagen).

The observation that gibberellins are inhibitors of flowering in many fruit species (12) requires qualification in the case of *Vitis.* The role of GA in flowering in grapevines varies with the stage of development of the latent bud. At an early stage GA is a promoter of flowering because Anlagen formation is a GA-requiring process. Later, GA acts as an inhibitor of flowering because it directs the Anlagen to form tendrils.

Cytokinins are implicated in the control of many aspects of reproduction in the grapevine (15–17, 28–31), including the branching of Anlagen to form inflorescence primordia and the differentiation of flowers. The mechanism by which cytokinins exert these effects is unknown but may be associated with the partitioning of assimilates. It has been demonstrated in many plants, including *Vitis vinifera,* that cytokinins are strong mobilizers of assimilates to the site of application. In another perennial, *Bougainvillea,* Tse et al. (33) showed that PBA-induced accumulation of ^{14}C-assimilates was followed by inflorescence formation, and several other authors have suggested that redistribution of metabolites is involved in the regulation of flowering (22, 23).

The effects of exogenous gibberellin and cytokinin on flowering in the grapevine are unequivocal and indicate that flowering is regulated by a gibberellin: cytokinin interaction. The origin and action of endogenous gibberellin and cytokinin in flowering in the intact plant is less clear. The xylem sap of the grapevine contains gibberellin (25) and has a high cytokinin activity during bud burst (18, 26). There is a specific effect of roots on inflorescence development and it is likely that the root system is an important source of the growth substances which are involved in the control of flowering. This has yet to be proved, however, and it is noteworthy that recent work by Wang and Wareing (34) has called into question the view that correlative phenomena such as apical dominance are regulated by cytokinin which originates in roots.

References

1. Atkin, R.K., Barton, G.E., Robinson, D.K.: J. Exp. Bot. *79,* 259–266 (1973)
2. Bugnon, F., Bessis, R.: In: Biologie de la vigne: Acquisitions récentes et problèmes actuels, pp. 160. Paris: Masson 1968
3. Buttrose, M.S.: Vitis *9,* 121–125 (1970)
4. Buttrose, M.S.: Hortic. Abstr. *44,* 319–325 (1974)
5. Carolus, M.: Thèse, 3me Cycle, Université de Bordeaux, France
6. Carolus, M.: Conn. Vigne Vin. *2,* 163–173 (1971)
7. Carr, D.J.: Ann. N.Y. Acad. Sci. *144,* 305–312 (1967)
8. Chailakhyan, M.Kh.: Annu. Rev. Plant Physiol. *19,* 1–36 (1968)
9. Chailakhyan, M.Kh.: In: Plant Growth Regulation. Pilet, P.E. (ed.), pp. 258–272. Berlin, Heidelberg, New York: Springer 1977
10. Chailakhyan, M.Kh., Khryanin, V.N.: Planta *142,* 207–210 (1978)
11. Evans, L.T.: Annu. Rev. Plant Physiol. *22,* 365–394 (1971)
12. Jackson, D.I., Sweet, G.B.: Hortic. Abstr. *42,* 9–24 (1972)
13. Lang, A.: Annu. Rev. Plant Physiol. *21,* 537–570 (1970)
14. Mullins, M.G.: Nature (Lond.) 209, 419–420 (1966)
15. Mullins, M.G.: J. Exp. Bot. *18,* 206–214 (1967)
16. Mullins, M.G.: J. Exp. Bot. *19,* 532–543 (1968)
17. Negi, S.S., Olmo, H.P.: Science *152,* 1624–1625 (1966)
18. Nitsch, J.P., Nitsch, C.: Bull. Soc. Bot. Fr. *112,* 11–19 (1975)
19. Phillips I.D.J.: Annu. Rev. Plant Physiol. *26,* 341–367 (1975)
20. Pratt C.: Am. J. Enol. Viticult. *22,* 92–109 (1971)
21. Pratt, C.: Am. J. Enol. Viticult. *25,* 131–150 (1974)
22. Sachs, R.M.: HortScience *12,* 220–222 (1977)
23. Sachs, R.M., Hackett, W.P.: Acta Hortic. *68,* 29–49 (1976)

24. Scholefield, P.B., Ward, R.C.: Vitis *14*, 14–19 (1975)
25. Skene, K.G.M.: Planta *74*, 250–262 (1967)
26. Skene, K.G.M.: In: The Development and Function of Roots. Torrey, J.G., Clarkson, D.T. (eds.), pp. 36–96. London, New York: Academic Press 1975
27. Skene, K.G.M., Kerridge, G.H.: Plant Physiol. *42*, 1131–1139 (1967)
28. Srinivasan, C., Mullins, M.G.: Ann. Bot. *38*, 1079–1084 (1976)
29. Srinivasan, C., Mullins, M.G.: Plant Physiol. *61*, 127–130 (1978)
30. Srinivasan, C., Mullins, M.G.: Planta *145*, 187–192 (1979)
31. Srinivasan, C., Mullins, M.G.: Ann. Bot. *45*, 439–446 (1980)
32. Thimann, K.V.: Plant Physiol. *54*, 450–453 (1974)
33. Tse, A.Y.T., Ramina, A., Hackett, W.P., Sachs, R.M.: Plant Physiol. *54*, 404–407 (1974)
34. Wang, T.L., Wareing, P.F.: New Phytol. *82*, 19–28 (1979)
35. Weaver, R.J., Van Overbeek, J., Pool, R.M.: Nature (Lond.) *206*, 952–953 (1965)
36. Zeevaart, J.A.D.: Annu. Rev. Plant Physiol. *27*, 321–348 (1976)

Hormonal Regulation of Sex Expression in Plants

M.Kh. CHAILAKHYAN [1] and V.N. KHRYANIN [2]

Introduction

The sex of plants is encoded in their genetic material and can be explained on the basis of the chromosome theory of sex inheritance. In most dioecious plants the female individuals have a pair of identical X chromosomes while the males have a pair of heterochromosomes, X and Y. In two typical dioecious plants, hemp *(Cannabis sativa)* and spinach *(Spinacia oleracea),* the main genes that determine sex are located in these chromosomes (1). Thus, the genetic control of sex determination in these plants is clear and beyond doubt.

However, phenotypical sex expression in plants both in the natural environment and under experimental conditions is subject to drastic fluctuations caused by various natural and introduced factors. The most important such environmental factors are daylength, light intensity, temperature, mineral nutrition, and ambient atmosphere (2, 3, 4). Among the applied factors, exogenous phytohormones have proven to exert the most striking effects on sex expression. It was shown in 1949 (4a) that treatment of cucumber plants with auxins greatly increases the ratio of female to male flowers. Later it was discovered that auxin in this case acts indirectly, by increasing the production of ethylene by the plants, and that ethylene is the acutal sex-modifying agent. This work with auxin was the first clear report on effects of a phytohormone on sex expression in a monoecious plant. Somewhat later, it was shown that auxin treatment of hemp, a dioecious plant, increased the number of female individuals (5, 6). Auxin effects on sex expression were also found in maize (7, 8). Later it was shown in cucumber (9) and hemp (10) that spraying with ethrel (2-chloroethylphosphonic acid), a compound releasing ethylene, changed sex expression in the female direction, whereas in contrast, treatment of hemp with GA caused a change in the male direction (10–13). Furthermore, this latter effect was not obtained when the plants were treated with abscisic acid at the same time (14). Results with other physiologically active substances have been less clear or even contradictory. One reason for such different results probably is that different investigators used plants at different stages of development.

---

1 K.A. Timiriazev Institute of Plant Physiology, Academy of Sciences of USSR, Botanicheskaya ul. 35, Moscow, USSR

2 V.G. Belinsky State Pedagogical Institute, Penza, USSR

Materials and Methods

During the past five years the Laboratory of Growth and Development of the K.A. Timiriazev Institute of Plant Physiology, Academy of Sciences of USSR, has conducted studies of the regulation of sex expression in plants. In accord with our long-standing rule in investigating any phenomenon, the program was designed to provide answers to three questions: (1) *when* (at what age of the plant) does the phenomenon take place, (2) *where* (in what organs of the plant) does it take place, and (3) *what* is taking place (which internal changes bring about the phenomenon).

We conducted experiments with the dioecious short-day plant, hemp *(Cannabis sativa)* strain US-6, and the dioecious long-day plant, spinach *(Spinacia oleracea)* cv. "Victoria". In another series of experiments two monoecious plants, cucumber *(Cucumis sativus)* cv. "Nyerosimiye" and maize *(Zea mays)* cv. "Voronezhskaya" and cv. "Odesskaya-10", were used. The plants were grown in a greenhouse and in controlled-environment cabinets at the Timiriazev Institute.

Experiments carried out to resolve the first question showed that differentiation of the shoot apices characteristic of male and female plants begins quite early: in hemp, when the third leaf pair becomes visible (15); in spinach, at the time of formation of the third leaf. These experiments showed that in order to affect sex expression, phytohormones and inhibitors should be applied to the plant at a very early stage of growth. We therefore treated seeds before germination; or applied plant hormones and other substances through the roots of seedlings; or germinated excised embryos on White's nutrient medium (16, 17) with and without added phytohormones.

In experiments aimed at resolving the second question (in which organs do the processes influencing sex expression take place) we used de-rooted young plants grown in Knop's nutrient solution. The plants were divided into three groups: (1) with leaves and with newly regenerating adventitious roots, (2) with leaves but without adventitious roots (the roots were removed as soon as they appeared), (3) without leaves (only the 2–3 uppermost young ones were allowed to remain) but with adventitious roots.

To answer the third question (what changes in the plant cause changes in sex expression), plants of the three groups were treated for a short time with phytohormones or inhibitors. For this purpose, the roots were cut off and the plants were placed with their cut bases in the test solutions, usually for 24–28 h. We call this entire system (which permits us to study the role of different organs, and of the hormones they produce) the integral model of sex expression in plants.

To answer the third question we also determined the content of natural cytokinins and gibberellins. When flower buds appeared, samples of leaves and roots were fixed in liquid nitrogen and lyophilized. Cytokinins were extracted and purified according to techniques previously described (18–20); cytokinin activity was estimated by an *Amaranthus* seedling bioassay, measuring the optical density of betacyanin produced under the influence of cytokinins (21, 22). Extraction and purification of gibberellin-like substances were carried out utilizing a method by Lozhnikova et al. (23) which measures GA activity in terms of growth of dwarf pea seedlings, *Pisum sativum* cv. "Pioneer" (24).

Results and Discussion

Effects of Phytohormones and Inhibitors on Sex Expression in Dioecious Plants when Applied Through the Root System

Treatment of hemp seeds before germination resulted in the same shifts in sex expression as observed by other investigators in earlier experiments in which the plants had been sprayed with solutions of phytohormones and inhibitors (25). When the substances were introduced into young hemp or spinach seedlings through the roots, the effect on the ratio of male and female plants was very marked. In these experiments the seeds were germinated in the dark for three days. Then seedlings, selected for roots of the same length, were placed for 24–28 h with their roots in solutions of phytohormones, while control seedlings were placed in water. The treatments were as follows: (1) water control, (2) gibberellic acid (GA_3), (3) indole-3-acetic acid (IAA), (4) 6-benzylaminopurine (6-BAP), (5) abscisic acid (ABA). After treatment, the hemp seedlings were grown in Knop nutrient solution in a controlled environment cabinet with a short (8 h) day regime; the spinach seedlings were grown in soil in a greenhouse in long (18 h) days (natural light extended with light from xenon-arc lamps). The treatment of the seedlings with phytohormones introduced through the roots caused marked changes in growth and flowering and a very pronounced shift in sex expression in both hemp and spinach. Application of GA_3 resulted in prevalence of male plants, whereas applications of either 6-BAP or IAA induced predominantly female sex expression in the treated plants [(26–29), Table 1].

Table 1. Effect of phytohormones absorbed by the roots on sex expression in hemp and spinach. Plants were treated when they had developed three leaf pairs (hemp) or three leaves (spinach)

	Percent of plants, ± standard error of the mean					
Treatment	Hemp			Spinach		
	Male	Female	Intersexes	Male	Female	Intersexes
Control	29 ± 1.2	37 ± 2.3	34 ± 1.1	48 ± 1.7	52 ± 2.3	–
GA_3	84 ± 2.6	7 ± 0.4	9 ± 0.7	79 ± 3.2	16 ± 0.6	5 ± 0.3
6-BAP	–	47 ± 1.8	53 ± 3.2	11 ± 0.8	87 ± 2.9	2 ± 0.0
IAA	–	40 ± 2.0	60 ± 2.8	21 ± 0.1	76 ± 2.2	3 ± 0.6
ABA	20 ± 3.0	39 ± 1.4	41 ± 1.8	29 ± 1.4	71 ± 1.8	–

These results confirm the assumption that regulation of sex expression by phytohormones is effective only at an early stage in the development of a plant. All subsequent experiments are based on this concept.

The Effect of Different Organs and Phytohormones in Sex Expression in Dioecious Plants

To elucidate the role of different plant organs in sex expression, i.e., to answer our second question, hemp and spinach plants were grown in soil in a greenhouse on long days. The hemp plants were grown until the third pair of leaves became visible and the spinach plants until the appearance of the third leaf, when the shoot apices were still vegetative. The plants were then cut off at the root-shoot junction, placed with their cut bases in Knop solution, and divided into two groups. In the first group (control) adventitious roots were allowed to develop; in the second group they were systematically removed, while the leaves were left on all plants. The number of male plants was much greater (up to 80%–90%) both in hemp and spinach when the roots were removed; when roots were allowed to develop the number of female plants increased to as much as 80%–90%.

Thus, leaves and roots play important roles in the sex of plants: roots enhance female sex expression whereas leaves seem to promote male sex expression. It should be noted here that Molotkovsky (7, 8), from earlier investigations of the polarity of maize plants and taking into consideration the relative position of the female and male inflorescences (the cobs being inserted relatively close to the roots, and the tassels at the shoot apex), concluded that the roots were determinants of the gynoecium and the shoots determinants of the androecium.

To answer the third question – which physiologically active substances in roots and leaves are responsible for their role in sex expression – we kept in mind both ideas expressed earlier by other investigators and our own data. Since treatment of the plants with 6-BAP introduced through the roots had considerably increased female sex expression in hemp and spinach seedlings, and bearing in mind the concepts of Sabinin (30) and Mothes (31, 32) that roots synthesize hormones, particularly cytokinins, a solution of 6-BAP (15 mg/l) was introduced for 28 h through the cut ends of the stems of hemp and spinach seedlings with their leaves intact but with any adventitious roots removed. This treatment markedly enhanced the number of female plants (Table 2). It may be concluded that in hemp and spinach the role of the roots in female sex expression is associated with their synthesis of cytokinins (27, 29, 33, 34).

Since male expression was considerably enhanced when gibberellin was applied to the roots of seedlings, and proceeding from our concept (35, 36) that gibberellins are

Table 2. Influence of roots and of 6-BAP on sex expression in hemp and spinach. Plants treated at age shown in Table 1

Treatment	Hemp				Spinach			
	Male		Female		Male		Female	
	No.	%	No.	%	No.	%	No.	%
With leaves and roots	8	9	83	91	7	15	96	85
With leaves without roots	81	90	9	10	85	85	15	15
With leaves without roots + 6-BAP	9	19	39	81	16	16	82	84

synthesized in the leaves, seedlings of hemp and spinach were supplied with gibberellin, by placing them for 28 h with the ends of the cut stems into a solution of GA_3 (25 mg/l); these seedlings were defoliated but were allowed to develop adventitious roots. This treatment considerably enhanced the number of male plants (Table 3). Evidently, the role of the leaves in male sex expression in hemp and spinach is associated with their synthesis of gibberellins (37, 38).

Table 3. Influence of leaves and of GA_3 on sex expression in hemp and spinach. Plants treated at age shown in Table 1

Treatment	Hemp				Spinach			
	Male		Female		Male		Female	
	No.	%	No.	%	No.	%	No.	%
With roots and leaves	15	18	70	82	16	15	89	85
With roots, without leaves	18	20	72	80	14	14	87	86
With roots, without leaves, + GA_3	68	81	16	19	76	77	23	23

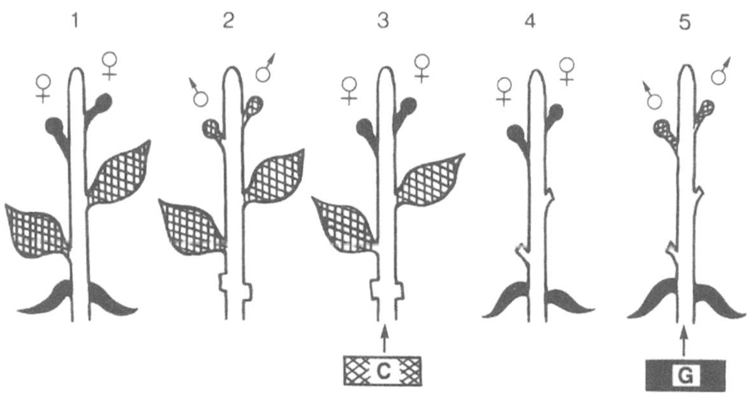

Fig. 1. Role of organs and of phytohormones they synthesize in sex expression in dioecious plants. *From left to right: 1* plants with leaves and roots; *2* with leaves, whithout roots; *3* with leaves, without roots + 6-BAP treatment; *4* without leaves, with roots; *5* without leaves, with roots + GA_3 treatment. *Hatched* GA_3 ; *solid black* 6-BAP

A general scheme in Fig. 1 illustrates the roles of roots and leaves in sex expression in dioecious plants and the importance of the formation of certain phytohormones in these organs for sex expression in plants.

The Roles of Phytohormones and Organs in Sex Expression in Monoecious Plants

The consistent effects of different phytohormones on sex expression found in the two dioecious plants led us to study two monoecious plants, cucumber and corn. The experiments were conducted in the same way as with the dioecious plants. Those with cucumber showed that if GA_3 was introduced through the roots of seedlings, a great majority of the flowers on the developing plant became male. It had been shown (39) that when plants of a female line of cucumber were treated with gibberellin they produced both male and female flowers. When we applied 6-BAP to cucumber seedlings the ratio of male to female flowers became about 1:1, whereas in the controls it was 4:1. Treatment with IAA or abscisic acid (ABA) did not markedly alter the ratio of male to female flowers from that in the controls (40).

In experiments on the roles of roots and leaves in sex expression in cucumber we found that removing the former and leaving the latter resulted in predominant formation of male flowers, whereas removing the leaves and allowing adventitious roots to develop resulted in predominant formation of female flowers. 6-BAP treatment of plants with their roots removed enhanced the formation of female flowers (up to 45.7%), whereas GA_3 treatment of plants with their leaves removed promoted formation of male flowers (to as high as 98.6%). These results are strong indirect evidence that in cucumber, as in the dioecious plants studied, the promotive effect of leaves on male sex expression is associated with their synthesis of gibberellins, and the promotive effect of roots on female sex expression is associated with their synthesis of cytokinins (41).

In experiments with maize (42), 3-day-old seedlings, cv. Voronezhskaya, selected for roots of the same length, were treated with phytohormone solutions for 28 h while controls were placed in water; then the plants were transferred to soil in a greenhouse

Fig. 2. Effect of cytokinin (6-BAP), applied both through the roots and to the shoots, on sex expression in maize plants. Note formation of cobs within the tassels

kept on an 18-h day. The GA_3 treatment hastened the appearance of tassels by four and also enhanced their growth while at the same time it retarded the formation of female inflorescences by five days. Both 6-BAP and IAA treatments retarded tassel formation by five days while formation of a second cob was accelerated. In an experiment with cultivar Odesskaya-10, 6-BAP (15 mg/l) was also introduced through the roots, but in addition, the tips of the plant were sprayed with 6-BAP solution after the formation of the third leaves. In this case 28 out of a total of 36 plants formed female flowers in the tassels (Fig. 2). A smaller such shift toward female sex expression had been observed in maize plants sprayed with IAA solution at the two leaf stage of development (43).

Thus, our experiments with monoecious as well as dioecious plants demonstrate that gibberellin enhanced male and cytokinin enhanced female sex expression.

Effects of Phytohormones and Growth Inhibitors, Applied Individually or in Combination

It is well known that substances other than cytokinins and gibberellins may markedly affect sex expression in plants (1, 44). Therefore, the effects of various phytohormones and growth inhibitors and the antagonisms between these in sex expression were studied in hemp. In one series of plants with adventitious roots allowed to develop while all but the two youngest leaves were removed, we tested the effects of GA_3 and CCC, separately and in combination. In another series in which the leaves were allowed to develop while secondary (adventitious) roots were systematically removed, we tested the effects of 6-BAP, IAA, ABA, CCC, and ethrel individually, and the effect of 6-BAP and ABA in combination.

The results (Table 4) show that a pronounced shift towards male sex expression was induced by GA_3 and towards female sex expression by 6-BAP. In the ethrel treatment, about half the plants were female, the other half were intersexes, i.e., possessed male, female and bisexual flowers, but male plants were entirely absent. The retardant CCC shifted sex expression toward femaleness; in plants with the leaves removed and

Table 4. Sex expression in hemp plants treated with phytohormones and growth inhibitors

Treatment	Percent of plants, ± standard error of the mean		
	Male	Female	Intersexes
1. Control	19 ± 1.2	81 ± 3.3	–
2. GA_3	85 ± 2.7	15 ± 0.2	–
3. CCC	25 ± 0.9	75 ± 1.9	–
4. GA_3 + CCC	71 ± 2.3	29 ± 1.5	–
5. 6-BAP	16 ± 0.8	84 ± 3.0	–
6. IAA	28 ± 1.7	67 ± 2.8	5 ± 0.2
7. ABA	39 ± 1.1	58 ± 1.6	3 ± 0.5
8. CCC	32 ± 1.4	68 ± 2.6	–
9. Ethrel	–	54 ± 1.5	46 ± 1.3
10. 6-BAP + ABA	34 ± 1.8	66 ± 2.2	–

The plants were treated when having formed three leaf pairs. Those in treatments 2–4 had roots but no leaves; those in treatments 5–10 had leaves but no roots

the adventitious roots present this trend was more pronounced than in plants with leaves present and adventitious roots removed. The effect of GA_3 was considerably reduced when it was applied together with CCC. ABA reduced the effect of 6-BAP on female sex expression.

Thus, the results once more confirm the assumption that the major phytohormones which promote male and female sex expression in hemp are gibberellins and cytokinins respectively; their effects are realized even when other phytohormones or inhibitors are present (45).

Biological Activities of Cytokinins and Gibberellins in Dioecious Plants

To obtain a more precise idea of the roles of different organs, and of the phytohormones synthesized therein, in sex expression in dioecious plants, we determined the biological activities of cytokinins and gibberellins in leaves and roots of hemp and spinach plants at the stage when sex differentiation in the apices had already occurred. We found that in intact plants grown in soil, cytokinin activity both in leaves and primary roots was greater in female than in male plants (Fig. 3). It was known (46) that

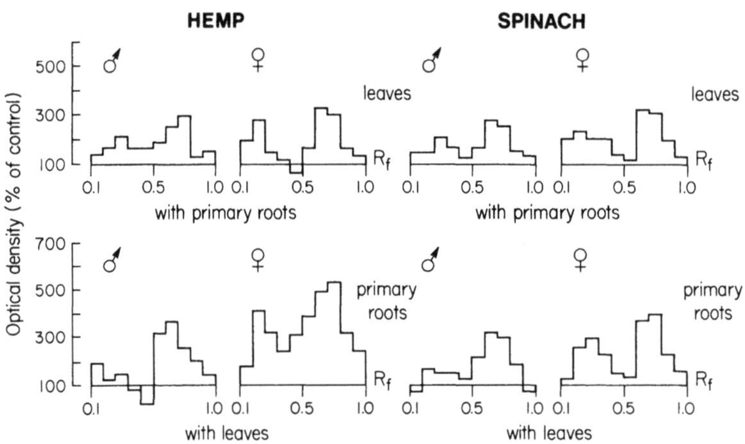

Fig. 3. Biological activity of cytokinins in leaves and roots of intact hemp and spinach plants, grown in soil. *Abscissa* Rf zones; *ordinate* betacyanin activity as percent of controls

there are more cytokinins in the inflorescences of female than of male plants. High cytokinin activity was found in the leaves of female hemp and spinach plants which had formed adventitious roots, but especially striking was the content of cytokinins in the roots themselves, both in plants with and without leaves (Fig. 4). It can be concluded that the regeneration of adventitious roots rich in cytokinins and their subsequent activity determined the sharp shift toward female sex expression.

Histograms of the content of gibberellin-like substances in leaves (Fig. 5) show that there is more GA activity in male than in female hemp and spinach plants. Intact plants grown in soil showed this difference. In the leaves of male, derooted plants

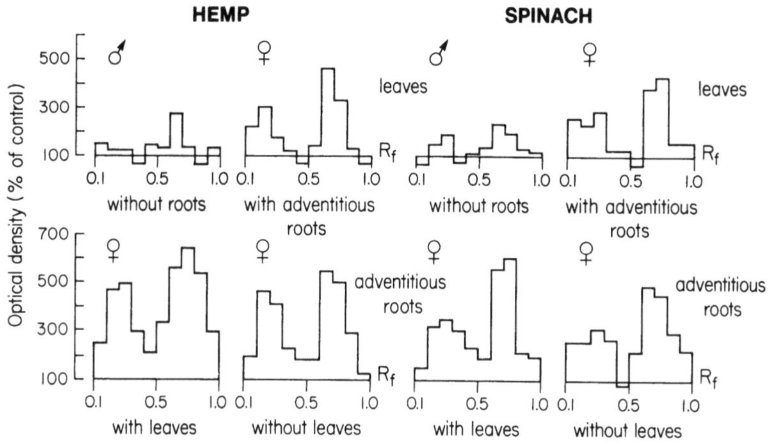

Fig. 4. Biological activity of cytokinins in leaves and adventitious roots of hemp and spinach plants grown in nutrient solution and with leaves or roots removed. *Abscissa* and *ordinate* as in Fig. 3

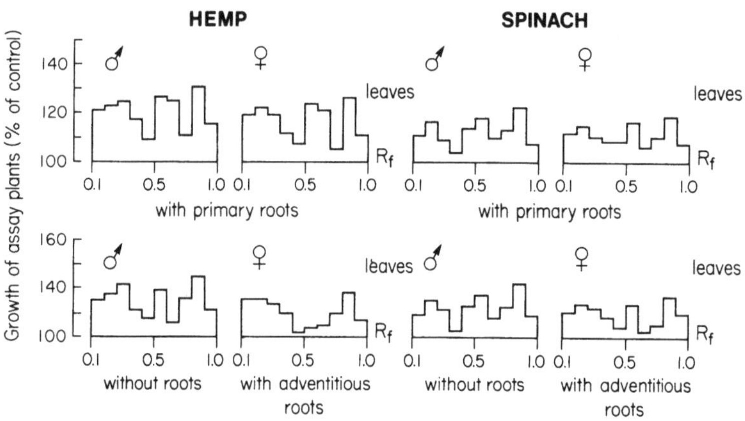

Fig. 5. Biological activity of gibberellins in leaves and roots of intact hemp and spinach plants, and plants with leaves or roots removed. *Abscissa* Rf zones; *ordinate* growth of assay plants (dwarf pea seedlings) as percent of controls

there were more gibberellins than in the leaves of female plants with adventitious roots (Fig. 5.). The cytokinin content was low in the leaves of derooted male plants of hemp and spinach (Fig. 4). Most likely, the high content of natural gibberellins and low content of cytokinins in the leaves of derooted plants are the major cause of the shift of sex expression in the male direction.

Thus, the effects of exogenous phytohormones on sex expression and the variations in the endogenous phytohormone content in the plants are analogous, supporting our conclusion about the roles of these hormones in sex expression in plants (47).

General Idea of Sex Expression in Dioecious and Monoecious Plants

Our general idea of sex expression in dioecious plants is presented schematically in Fig. 6. We propose that the leaves synthesize gibberellins which are transported to the terminal buds where they affect metabolism so as to favor male sex expression. The roots, on the other hand, produce cytokinins which are also transported to the buds and promote female sex expression.

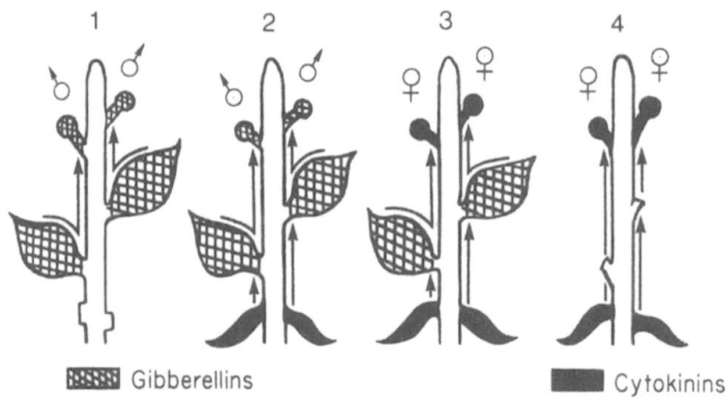

Fig. 6. Hormonal and genetic regulation of sex expression in plants. From left to right: *1* plants with leaves, without roots; *2, 3* with leaves and roots; *4* without leaves, with roots. *Hatched* gibberellins; *solid black* cytokinins

Under favorable natural conditions, when a certain hormonal balance is present, sex expression is controlled essentially by the genetic material, and the number of male and female plants is approximately the same (48).

The mechanism of phytohormone action in sex expression in monoecious plants is somewhat more complex than in dioecious ones since male and female flowers are formed on one and the same individual, but often in a definite sequence. For example, in cucumber at first only male flowers are formed, then both male and female flowers, and finally almost only female flowers are produced. Most probably this sequence is due to the fact that in the first stage the foliage is relatively better developed than the root system and more gibberellins are accumulated in the plant. In the second stage the foliage and root system are both well developed and both gibberellins and cytokinins are present in sufficient amounts. Finally, in the third stage, decreased activity in the aging leaves results in a predominance of cytokinins in the plant. In maize, in accordance with Molotkovsky's concept (7, 8), one may assume that the formation of female inflorescences (cobs) in the lower zone of the shoot, is related to the better supply of this part of the plant with cytokinins from the roots, whereas the formation of male inflorescences (tassels) at the shoot tip is related to a better supply of gibberellins from the leaves to this region.

Regulation of Sex Expression in Isolated Embryos of Hemp

To study the immediate effects of phytohormones in sex expression we carried out experiments with isolated hemp embryos grown on White's nutrient agar in test tubes under controlled conditions (20°C–22°C, 80% relative humidity, and light from fluorescent lamps). For 7 days the cultures were exposed to a 16-h day and then until the end of the experiment to an 8-h short day. In three separate experiments, it was found that inclusions of GA_3 (12.5 mg/l) in the medium enhanced the growth of the embryos, accelerated flowering of the resulting miniature plants, and 95.5%–100% of the latter became male. Inclusion of 6-BAP (5 mg/l) reduced growth and delayed flowering, and 93%–98% of the plantlets became female [(49); Fig. 7]. Thus, gibberellin and cytokinin, when supplied to excised embryos, have an even greater effect on sex expression in the resulting plants than when they are supplied to seedlings via the roots.

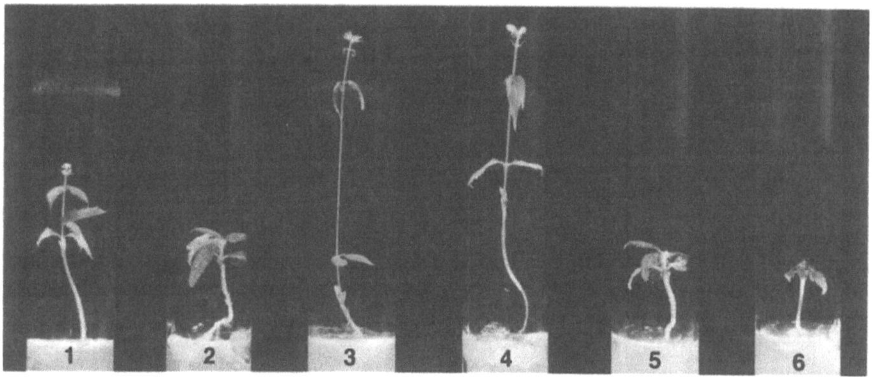

Fig. 7. Regulation of sex expression in hemp plants grown from isolated embryos. Pairs of plants from left to right: controls; treated with GA_3; treated with 6-BAP

Effects of Phytohormones and of Inhibitors of Nucleic Acid and Protein Metabolism on Sex Expression in Hemp

In current studies of the roles of phytohormones, much attention is paid to the molecular mechanism of their action in promoting synthesis of nucleic acids and proteins (50, 51). We have attempted to understand the molecular mechanism of regulation of sex expression in hemp by gibberellin and cytokinin, by determining the responses of plants to GA_3 and 6-BAP in the presence of the antimetabolites mitomycin, actinomycin D, and puromycin, which are known to inhibit DNA replication, transcription, and translation, respectively. The antimetabolites were applied at a concentration of 0.5 mg/l; 1 mg/l had been found to affect sex expression without affecting growth or causing visible damage to the plants. The plants were treated after having formed the third leaf pair, and were grown with the cut bases of the shoots in Knop solution. They were treated first with phytohormone for 28 h, and then with antimetabolite for the next 28 h.

In the experiments with GA_3, the leaves of the plants were removed after regeneration of adventitious roots had started, and the roots were allowed to develop. It was found that the shift toward male sex expression typical of exogenous GA_3 was prevented by actinomycin D, whereas the other two antimetabolites were much less effective in this regard (Table 5). We may infer from this result that gibberellin affects sex expression at the transcriptional level.

Table 5. Sex expression in hemp seedling plants treated with phytohormones and with inhibitors of nucleic acid and protein metabolism

	Percent of plants, ± standard error of the mean	
Treatment	Male	Female
Control	26 ± 0.8	74 ± 2.9
GA_3	81 ± 3.4	19 ± 0.9
GA_3 + mitomycin	67 ± 3.3	33 ± 1.2
GA_3 + actinomycin D	29 ± 1.3	71 ± 2.8
GA_3 + puromycin	65 ± 2.6	35 ± 1.4
6-BAP	23 ± 0.5	77 ± 3.1
6-BAP + mitomycin	73 ± 2.3	27 ± 1.5
6-BAP + actinomycin D	22 ± 0.4	78 ± 2.6
6-BAP + puromycin	76 ± 3.2	24 ± 0.8

The treatments (28 h with hormone followed where appropriate by 28 h with antimetabolite) were applied when the plants had formed three leaf pairs. Those treated with GA_3 had no leaves but had roots; those treated with 6-BAP had leaves but no roots. Counts made 30 days after start of treatment

In the experiments with 6-BAP the leaves were left on the plants, but all regenerating roots were removed as soon as they appeared. The shift toward female sex expression normally elicited by 6-BAP was prevented by mitomycin and puromycin but not by actinomycin D (Table 5). Hence, cytokinin seems to affect sex expression at the replicational and translational levels. These results provide only indirect evidence which needs to be confirmed by direct analysis. Nevertheless, they suggest that the mechanism of phytohormone action in sex expression involves changes in the activity of the plants' genetic material (52).

Conclusions

The experimental results and their interpretation summarized in this report allow us to propose the general idea that in plants with unisexual flowers, sex is determined by the genetic information of the individual but that actual, phenotopic sex expression depends, within a broad range, on the balance of hormones and inhibitors which is established in the plant under natural conditions or as a result of experimental treatments. Thus, elucidation of the exact mechanisms of interaction between genetic and hormonal regulation in sex expression appears to us, as we have said before, to be "one of the most fascinating areas of plant biology" (53, 54).

References

1. Frankel, R., Galun, E.: Pollination Mechanisms, Reproduction and Plant Breeding. Berlin, Heidelberg, New York: Springer 1977
2. Minina, E.G.: Modification of Sex in Plants under the Influence of Factors of the External Environment. (In Russian.) Moscow: Acad. Sci. USSR 1952
3. Dzhaparidze, L.I.: Sex in Plants, Pt. 1, 2. (In Russian.) Tbilisi, USSR: Akad. Nauk Gruz SSR/ Metsniereba. Engl. transl. 1968, Jerusalem: Israeli Program Scient. Transl. 1963/65
4. Maurinya, Kh.A., Berzinya-Berzite, R.V.: Nauchn. Tr. Vses. Sel. Genet. Inst. *11*, 116–121 (1974)

4a. Laibach, F., Kribben, F.J.: Ber. Dtsch. Bot. Ges. *62*, 53–55 (1949)

5. Heslop-Harrison, J.: Physiol. Plant. *4*, 588–597 (1956)
6. Heslop-Harrison, J.: Biol. Rev. *32*, 38–90 (1957)
7. Molotkovsky, G.Kh.: Byull. Mosk. Ova. Ispyt. Prir. *65* (6), 65–77 (1960)
8. Molotkovsky, G.Kh.: Polarity of Development and Physiological Genetics in Plants. (In Russ.) Chernovitsy: Univ. Press 1968
9. Heslop-Harrison, J.: Brookhaven Symp. Biol. *16*, 109–122 (1963)
10. Mohan Ram, H.Y., Jaiswal, V.S.: Experientia *26*, 214–216 (1970)
11. Atal, C.K.: Curr. Sci. *28*, 408–409 (1959)
12. Köhler, D.: Ber. Dtsch. Bot. Ges. *78*, 275–281 (1964)
13. Zhukov, M.S., Chailakhyan, M.Kh., Kochankov, V.G., Sazhko, M.M.: In: Gibberellins and their effect on plants. Chailakhyan, M.Kh. (ed.), pp. 261–269. (In Russ.) Moscow: Acad. Sci. USSR 1963
14. Mohan Ram, H.Y., Jaiswal, V.S.: Planta *105*, 263–266 (1972)
15. Khryanin, V.N., Milyaeva, E.L.: Dokl. Akad. Nauk SSSR *234*, 982–984 (1977). Engl. transl. in: Dokl. Bot. Sci. *232–234*, 62–64
16. White, P.R.: Growth *10*, 231–289 (1946)
17. Butenko, R.G.: The Culture of Isolated Tissue and Physiology of Plant Morphogenesis. (In Russ.) Moscow: Nauka 1964. . Engl. transl. Jerusalem: Israeli Program Sci. Transl. 1968
18. Letham, D.S., Williams, M.W.: Physiol. Plant. *22*, 925–936 (1969)
19. Van Staden, J.: J.S. Afrik. Bot. *39*, 261–263 (1973)
20. Konopskaya, L.N.: Dokl. Akad. Nauk SSSR *236*, 1270–1272 (1977). Engl. transl. in: Dokl. Bot. Sci. *235–237*, 113–115
21. Bigot, C.: C.R. Acad. Sci. *266* D, 349–352 (1968)
22. Mazin, V.V., Shashkova, L.S., Andreev, L.N., Komizerko, E.I., Zhloba, N.M., Kefeli, V.I.: Dokl. Akad. Nauk SSSR *231*, 506–509 (1976). Engl. transl. in: Dokl. Bot. Sci. *229–231*, 132–134
23. Lozhnikova, V.N., Khlopenkova, L.P., Chailakhyan, M.Kh.: In: Methods of Determination of Phytohormones, Growth Inhibitors, Defoliants and Herbicides, pp. 50–58. (In Russ.) Moscow: Nauka 1973
24. Muromtsev, G.S., Rusanova, N.V.: In: Methods of Determination of Growth Regulators and Herbicides, pp. 89–96. (In Russ.) Moscow: Nauka 1966
25. Khryanin, V.N., Kochankov, V.G., Chailakhyan, M.Kh.: Fiziol. Rast. *25*, 698–703 (1978). Engl. transl. in: Sov. Plant Physiol. *25*, 545–550
26. Chailakhyan, M.Kh., Khryanin, V.N.: Dokl. Akad. Nauk SSSR *236*, 268–271 (1977). Engl. transl in Dokl. Bot. Sci. *235–237*, 113–115
27. Chailakhyan, M.Kh., Khryanin, V.N.: Dokl. Akad. Nauk SSSR *239*, 1262–1264 (1978). Engl. trans. in: Dokl. Bot. Sci. *238–240*, 16–19
28. Chailakhyan, M.Kh., Khryanin, V.N.: Planta *138*, 181–184 (1978)
29. Chailakhyan, M.Kh., Khryanin, V.N.: Planta *142*, 207–210 (1978)
30. Sabinin, D.A.: On the Role of the Root System in the Vital Activity of Plants. (In Russ.) Timiryazev Lectures No. IX. Moscow: Akad. Nauk SSSR 1949
31. Mothes, K., Engelbrecht, L., Kulajewa, O.: Flora (Jena) *147*, 445–447 (1959)
32. Mothes, K.: In Regulateurs Naturels de la Croissance Vegetale (Colloq. Int. C.N.R.S. No. 123), pp. 597–609. Paris: C.N.R.S. 1964

33. Chailakhyan, M.Kh., Khryanin, V.N.: Dokl. Akad. Nauk SSSR *236*, 509–512 (1977). Engl. transl. in: Dokl. Bot. Sci. *235–237*, 106–109
34. Chailakhyan, M.Kh., Khryanin, V.N.: Planta *138*, 185–187 (1978)
35. Chailakhyan, M.Kh.: Fiziol. Rast. *18*, 348–357 (1971). Engl. transl. in: Sov. Plant Physiol. *18*, 289–295
36. Chailakhyan, M.Kh.: In: Plant Growth Substances 1970. Proc. 7th Int. Conf. Plant Growth Substances. Carr, D.J. (ed.), pp. 745–757. Berlin, Heidelberg, New York: Springer 1972
37. Chailakhyan, M.Kh., Khryanin, V.N.: Dokl. Akad. Nauk SSSR *239*, 1497–1500 (1978). Engl. transl. in: Dokl. Bot. Sci. *238–240*, 20–23
38. Chailakhyan, M.Kh., Khryanin, V.N.: Planta *144*, 205–207 (1979)
39. Galun, E.: Phyton (Buenos Aires) *13*, 1–8 (1959)
40. Khryanin, V.N., Chailakhyan, M.Kh.: Dokl. Vses. Akad. Skh. Nauk *1*, 10–13 (1979)
41. Khryanin, V.N., Chailakhyan, M.Kh.: Dokl. Vses. Akad. Skh. Nauk *4*, 9–12 (1979)
42. Khryanin, V.N., Chailakhyan, M.Kh.: Fiziol. Rast. (in press, 1980)
43. Molotkovsky, G.Kh.: Fiziol. Biokhim. Kult. Rast. *8*, 384–390 (1976)
44. Sidorsky, A.G.: Usp. Sovrem. Biol. *85* (1), 111–124 (1978)
45. Khryanin, V.N., Chailakhyan, M.Kh.: Fiziol. Rast. *26*, 455–458 (1979)
46. Engelbrecht, L.: Pr. Inst. Sadow. Skierniewice (Ser. E *3*, 389–392 (1973)
47. Khryanin, V.N., Chailakhyan, M.Kh.: Fiziol. Rast. *26*, 1008–1015 (1979)
48. Chailakhyan, M.Kh., Khryanin, V.N.: Dokl. Akad. Nauk SSSR *242*, 493–496 (1978)
49. Chailakhyan, M.Kh., Khryanin, V.N.: Dokl. Akad. Nauk SSSR *244*, 1037–1039 (1979)
50. Kulaeva, O.N.: Cytokinins, Their Structure and Function. (In Russ.) Moscow: Nauka 1973
51. Muromtsev, G.S., Agnistikova, V.N.: Plant Hormones. Gibberellins. Moscow: Nauka 1973
52. Chailakhyan, M.Kh., Khryanin, V.N.: Dokl. Akad. Nauk SSSR *243*, 1341–1344 (1978). Engl. transl. in: Dokl. Bot. Sci. *241–243*, 87–90
53. Chailakhyan, M.Kh.: Fiziol. Rast. *25*, 952–974 (1978). Engl. transl. in: Sov. Plant Physiol. *25*, 757–775
54. Chailakhyan, M.Kh.: Am. J. Bot. *66*, 717–736 (1979)

Growth Substances: Roles in Fertilization and Sex Expression

T.-H. TSAO [1]

This report summarizes studies on reproductive development done in our laboratory. It includes early work, some of which was published in Chinese and therefore not readily available to our Western colleagues. Our primary interest has been in the involvement of plant growth substances in fertilization, pollen tube extension, and sex expression.

Fertilization

It was demonstrated some years ago by Hsiang that "pollination effects" in orchids could be mimicked by the application of IAA or NAA to the stigma, resulting in (1) increases in catalase activity and respiratory rate, (2) an increase in transpiration followed by wilting of the perianth, and (3) mobilization of nitrogenous and phosphorus-containing compounds from the perianth to the column and the swelling of the latter. The effects of auxin treatment diminished with time, whereas in normal pollination these processes showed a progressive increase (1, 2). Thus, according to Linskins, pollination may involve the transfer of auxin from pollen to the gynoecium (3).

Our laboratory has studied respiratory rates in the tomato during pollination and fertilization, particularly as affected by hybridization. We have found three respiratory peaks occurring 2–4, 35–48, and 96 h after pollination. These correspond to germination of the pollen and growth of the pollen tubes, gametic union, and the division of the endosperm nucleus. The respiratory rate was highest in intravarietal crosses and lowest in intergeneric crosses, the crosses decreasing in order as follows: intravarietal > intervarietal > intersubspecific > intergeneric. Thus, the post-fertilization respiratory rise seemed to be a measure of the affinity between the crossing partners. This finding may have practical value in plant breeding (4).

The content of inhibitors in tomato and tobacco ovaries was followed, but no significant changes were found (5). We are currently following changes in isozyme patterns during pollination and fertilization.

Gibberellins and Pollen

We have found gibberellin-like substances in tomato and tobacco stamens and in the pollen grains of four varieties of corn, determined on the basis of rice seedling bioassays

1 Laboratory of Plant Development, Biology Department, Peking University, China

and paper chromatography. The activity was located at Rf 0.1–0.2, whereas our sample of GA_3 gave an Rf value of 0.7–0.9 (solvent system: isopropyl alcohol/ammonia water 24%–25%, 5/1, v/v). In terms of GA_3, the gibberellin content of stamens and pollen was 3–9 μg/g fresh weight, which is a relatively high content compared to that of other organs (6). Our values compare well with those of others. The Rf values for gibberellin-like substances in bamboo shoots as reported by Koshimizu, using the same solvent system, were identical to ours (8). Barendse obtained values of the same order of magnitude for gibberellin in *Lilium* (8).

The experimental data of Mitchell and Whitehead (9) on the crude extracts of corn pollen have been re-examined by us (6). We are confident on the basis of our results that the active substances obtained by these workers in 1941 were indeed gibberellins, and believe that they should be so credited.

Subsequent to our report of factors which induce a chemotropic response in pollen (1), there have been other reports (11). Later, studying snapdragon, we found Rf value of 0.4–0.5 for the active factor in a *n*-butanol/acetic acid/water solvent system (4/1/2). The eluate from this region contained a substance which inhibited the growth of *Avena* coleoptile sections.

Sex Expression

Sex expression constitutes a specific aspect of the physiology of flowering and provides a good subject for the study of differentiation. Therefore, our laboratory has devoted much attention to this topic since 1975.

CO treatment was found to remarkably affect sex expression in cucumber. Treatments of 24-day-old seedlings with CO gas at 1%, 0.5%, and 0.3% for 161 h not only altered the sequence in which male and female flowers appeared, but also changed the absolute and relative numbers of flowers of each sex as follows: control plants, 216 male and 5 female flowers; plants treated with 0.3% CO, 52 male and 22 female flowers (13). According to present knowledge, this effect of CO may be due to the suppression of IAA oxidase activity, resulting in a rise in the auxin content (14). To investigate this problem further, we cultured excised stem tips under aseptic conditions (15). Galun et al. (16, 17) reported work on undifferentiated floral buds of cucumber, and Prakash and Kumar (18) have used the stem tip culture method for the study of sex expression in *Cucumis melophant.*

Recently Wand and Tsao (19) have shown that treatment with CCC causes the exclusive formation of female flowers on normally monoecious cucumber plants, while GA_3 has the opposite effect, causing the formation of exclusively male flowers. In an extreme case, as many as 26 female flowers were produced on a single plantlet with as many as 11 flowers in a single leaf axil.

To study the interaction of CCC and GA_3 on sex expression, these chemicals were applied alone and in combination. As can be seen in Table 1, GA_3 alone at a concentration of 2×10^{-4} M had little effect of the number of male flowers produced and CCC reduced this number only when applied simultaneously at a fourfold higher concentration, i.e., at 8×10^{-4} M. CCC, when supplied alone at this concentration, promoted female flower formation and suppressed male flower formation, averaging 20 female to 8 male flowers as compared to a ratio of 5 female to 20 male for the control plants. The addition of GA_3 at 2×10^{-4} M completely suppressed female flower formation, even in the presence of the fourfold higher concentration of CCC.

Table 1. Effect of CCC and GA_3 applied singly or in combination, on the sex expression of excised shoot apex

CCC	GA_3	Av. no. of node at which the 1st ♀ fl. appeared	No. explants	Av. no. ♀ fls/plant	Av. no. ♂ fls/plant	♀ : ♂
0	0	2.2	10	5.1	20.4	1:4.0
8×10^{-4} M	0	1.3	10	20.3	7.9	1:0.4
0	2×10^{-4} M	–	9	0	19.9	0:19.9
8×10^{-5} M	2×10^{-4} M	–	10	0	18.7	0:18.7
4×10^{-4} M	2×10^{-4} M	–	10	0	18.0	0:18.0
8×10^{-4} M	2×10^{-4} M	–	10	0	10.0	0:10.0

Treatments with IAA, NAA, and *trans*-cinnamic acid all increased the number of female flowers, decreased the number of male flowers, and lowered the node at which the first female flowers appeared. As would be expected, auxin antagonists such as TIBA and MH lowered the female/male ratio, as did salicylic acid.

An interesting correlation exists between sex expression and root growth. In all cases when IAA, NAA, and *trans*-cinnamic acid promoted femaleness, they promoted root formation and root growth; when TIBA, MH, and salicylic inhibited the differentiation of female flowers, they inhibited root formation and growth. Better root growth was associated with femaleness and vice versa. Similar results have been found by Chailakhyan (20–23) in studies of hemp. We agree with Chailakhyan's suggestion that a better root system may synthesize more cytokinins, which may be transported to and influence development of aerial parts.

Since the resumption of our studies in 1977 we have been interested in comparing the hormonal control of sex expression in plants belonging to different photoperiodic types. A long-day plant, spinach, and a short-day plant, hemp, were chosen. The work on spinach is near completion, while the work on hemp has just been initiated. In spinach, foliar application of 50 ppm of GA_3 at the seedling stage raised the female/male ratio from 0.97 (control) to 1.67. Ethylene at 200 ppm decreased the ratio to 0.65. When the two hormones were used in combination, the ratio become 1.10.

In hemp, our results to date indicate that ethylene increases the number of female flowers, while GA_3 suppresses all formation of female flowers.

Peroxidase isozyme patterns of the young leaves of plants of different sexes have been compared by our group, using both polyacrylamide disk electrophoresis and starch gel electrophoresis methods modified from Smithes (24). Results from both methods showed consistent differences in the peroxidase isozyme patterns of GA_3-treated plants and controls; in male plants a specific band was missing. Further work along these lines is in progress.

We believe that sex expression involves differential gene activity, resulting in selective protein synthesis at the translational level. We plan to test this possibility in dioecious plants such as hemp.

References

1. Hsiang, T.T.: Plant Physiol. *26,* 441 (1951)
2. Hsiang, T.T.: Plant Physiol. *26,* 708 (1951)
3. Linskens, H.F. (ed.): Fertilization in Higher Plants. Amsterdam: North Nolland Publ. Co. 1974
4. Tsao, T.H., Liu K.J., et al.: Proc. Symp. 1st Meet. Chin. Soc. Plant Physiol. (1963)
5. Chang, H.C.: Thesis, unpublished
6. Sun, F.C., Tsao, T.H.: Acta Phytophysiol. Sin. *3* (1), 1–7 (1966)
7. Koshimizu, K., Zawamura, H., Mitsui, T., Ogawa, Y.: Nature (Lond.) *198* (4887), 1306–1307 (1963)
8. Barendse, G.W.M. et al.: Acta Bot. Neerl. *19* (1), 175–186 (1970)
9. Mitchell, J.W., Whitehead, D.C.: Bot. Gaz. *102,* 770 (1941)
10. Tsao, T.H.: Plant Physiol. *24,* 494–504 (1949)
11. Mascarenhas, J.P.: Bot. Rev. *41* (3), 259–314 (1975)
12. Tsao, T.H., Tan, C.Y., Wang, T.F.: Proc. Symp. 1st Meet. Chin. Soc. Plant Physiol. (1964)
13. Tsao, T.H., Li, C.K., Chin, I.F., Wu, C.M.: Acta Sci. Nat. Univ. Pekinensis *3* (2), 233–246 (1957)
14. Heslop-Harrison, J., et al.: Proc. R. Soc. Edinburgh Ser. B *66,* 424–434 (1957)
15. Wang, B.L., Tsao, T.H.: Plant Physiol. Newslett. *3,* 1–6 (1963)
16. Galun, E., Jung, Y., Lang, A.: Nature (Lond.) *194,* 596–598 (1962)
17. Galun, E., Jung, Y., Lang, A.: Dev. Biol. *6,* 370–387 (1963)
18. Prakash, G., Kumar, V.: In: 4th Int. Congr. Plant Tissue Culture, p. 45. 1978
19. Wang, B.L., Tsao, T.H.: Proc. Symp. Plant Tissue Culture Peking, pp. 511–516 (1978)
20. Chailakhyan, M.Kh., Khryanin, V.N.: Planta *138,* 185–187 (1978)
21. Chailakhyan, M.Kh., Khryanin, V.N.: Planta *138,* 181–184 (1978)
22. Chailakhyan, M.Kh., Khryanin, V.N.: Planta *142,* 207–210 (1978)
23. Chailakhyan, M.Kh., Khryanin, V.N.: Planta *144,* 205–207 (1979)
24. Smithes, C.: Biochem. J. *61,* 629 (1955)

Hormonal Regulation of Morphogenesis

Chairman: D. E. FOSKET

The Hormonal Regulation of Morphogenesis in Mosses

M. BOPP [1]

Introduction

This report is a short summary of present knowledge and recent work on the morphogenesis of the moss protonema. The clear definition of chloronema and caulonema as the two main stages in the development of the moss protonema by Sironval (55) in 1947 established the mosses as a tool for morphogenetic studies. Subsequent important discoveries for this purpose were the finding by Gorton and Eakin (26) that kinetin can induce bud formation in the protonema of *Tortella caespitosa* and the demonstration by Johri and Desai (38) that exogenous auxin induces the transition from chloronema to caulonema in a suspension culture of *Funaria hygrometrica.* Previously only an inhibitory effect of auxin on protonemal growth was well documented (5).

These results are the background for all further research: they show not only that the moss protonema is a "morphogenetic system" (8) in which hormonal control of morphogenesis is much clearer than in many other systems, in that distinct morphogenetic steps are discernable. Morphogenesis in moss protonema can be reduced to one single event in each case.

Germination of Spores

The development of a moss protonema starts with the germination of a spore or occasionally with the outgrowth of gemmae (20). The first step after swelling of the spore is the emergence of a single filament, either a primary rhizoid or a chloronema (34), and this is also the first step in morphogenesis. In an electric field the spores tend to form the primary rhizoids towards the positive electrode, and it has been stated that the pronounced field response of *Funaria* spores is mediated by active calcium ion uptake into that part of the cell. This hypothesis is supported by experiments with the ionophore A 23187 which promotes entry of calcium ions across the membrane into the germinating spores, showing that the formation of rhizoids is predominately towards the side of higher ionophore concentration (15). However, subsequent growth of both rhizoids and chloronemata is directed towards the negative electrode. We found, in collaboration with Dr. Weisenseel, that in growing caulonema filaments it is possible to measure an electric current entering the tip of each growing cell of the main filaments of a protonema and also entering the tips of side branches. In the context of our interest these are the only growing regions in a moss protonema.

1 Botanisches Institut, Universität Heidelberg, 6900 Heidelberg, FRG

Chloronema Formation

After germination on a medium like agar, development continues with the formation of chloronemata, whereby the filaments grow in all directions along the substrate surface. This growth habit is regulated by substances produced by the protonemal filaments themselves. Therefore, the morphogenetic system of a protonema (8) is independent of its origin from one single spore, from a few, or from many spores (Fig. 1). This alternative differentiation of protonema is also seen in suspension cultures where a low cell density after a certain time results in the formation of several caulonema cells (36), whereas in a high cell density culture the chloronema cells simply continue to proliferate.

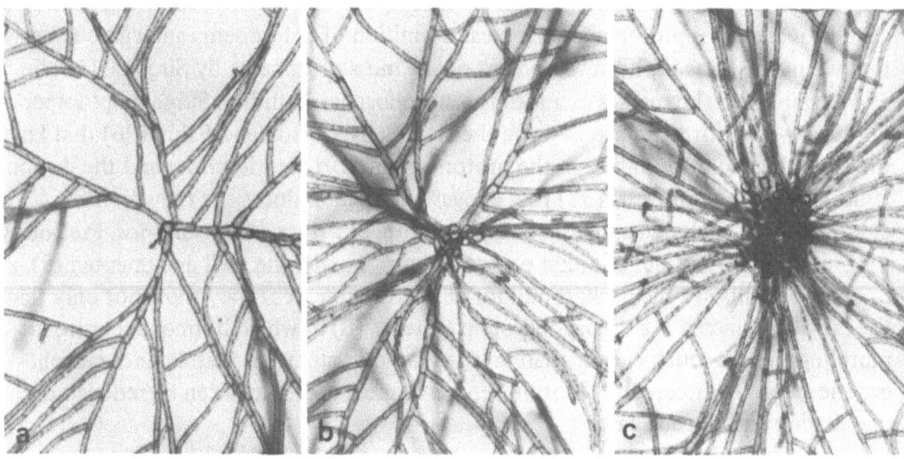

Fig. 1a–c. Protonema of *Funaria hygrometrica* 6 days after germination between the agar surface and a cellophane layer. **a** From one single spore, **b** from several spores germinating together, **c** from many spores. In each case the germinating filaments form one single "morphogenetic system"

The substances responsible for this cellular regulation are not yet known. They are produced by the growing protonema, secreted into the substrate, and act from there to regulate the development of all filaments. It is possible that one of the substances is auxin, that others are enzymes like phenolase (40) and very probably IAA oxidase, and therefore one finds in a cell suspension culture a high correlation between cell density and the amount of auxin-like activity; the denser the suspension, the lower the auxin activity. The total enzyme activity in the solution may also be dependent on pH. In a chloronema suspension culture the pH increases from about 5 to 7 in two weeks (21).

Caulonema Formation

Exogenous auxin can induce the next step in development, the transition from chloronema to caulonema (37). The caulonema is best characterized by a comparison of the walls between the cells in the filament. In every filament tip of primary chloronemata, as well as in "secondary" chloronemata which form as a result of deficiency conditions (40, 42), the cell walls are strongly and regularly perpendicular. In caulonemata the walls between the cells are always oblique (Fig. 2). The oblique cell walls are initiated

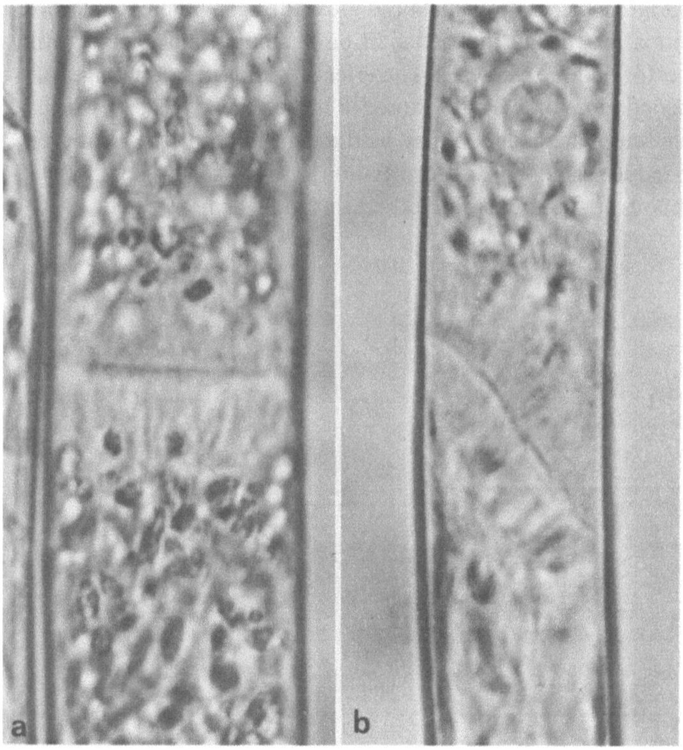

Fig. 2. **a** Perpendicular cell wall in a regenerated chloronema cell of *Funaria hygrometrica.* Both daughter cells are identical. **b** Oblique cell wall in a growing caulonema cell. Same stage as **a**. The tip cell above. In both cases the microtubules forming the spindle are clearly seen

as perpendicular; that is, the original orientation of the spindle is parallel to the filament axis, but the orientation changes after reconstitution of the daughter nuclei. The turning of the spindle and the phragmoplast is due to the activity of microtubules, which can be seen during the movement around the phragmoplast. Colchicine, which prevents the assembly of tubulin into microtubules, also inhibits the turning of the spindle, so that under its influence the new cell walls remain perpendicular (52–54). This demonstrates the active participation of the microtubules in orienting the cell wall. The oblique cell wall, however, can be considered the most distinctive characteristic of the caulonema stage of the protonema. In cell suspension cultures a change in cell wall orientation was seen 20 h after IAA treatment (37), even though the uptake of IAA is very slow into these cells (50) as compared with higher plant suspension cultures (48). The effect of auxin is antagonized by cAMP (32, 37), but because phosphate deficiency can enhance caulonema formation very strongly, it is not yet clear whether or not the cAMP effect is specific, especially as interaction between cytokinin and cAMP has been reported for the moss species *Pylaisiella* (56)

In suspension cultures auxin induction of caulonema formation is striking and reproducible (21, 38). In cultures on a solid medium like agar or cellophane, however, the differentiation to caulonema takes place spontaneously, and auxin only inhibits growth. Under low light intensities protonemata can grow for a long time as chloronemata without differentiation (6). In this case addition of auxin stimulates the formation of caulonemata, i.e., of oblique cell walls. The effective concentration ranges of different auxins correspond more or less with those for higher plants (Fig. 3), which suggests that the mechanism of auxin action may be the same in both cases. But even though it is possible to induce caulonema differentiation with exogenous auxin, the participation of endogenous auxin in this process remains to be demonstrated.

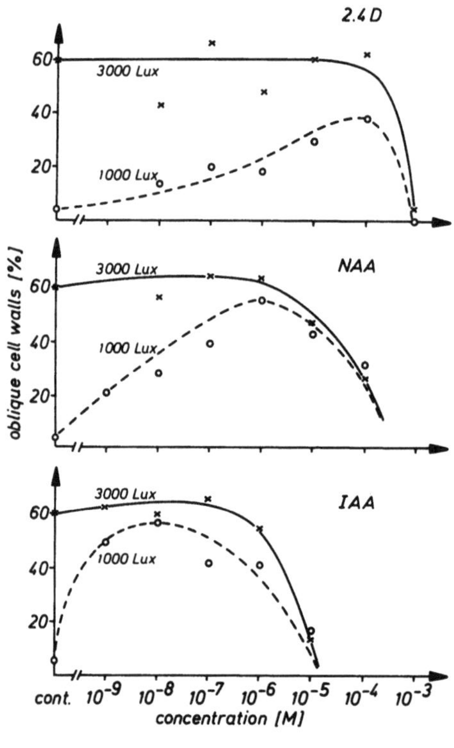

Fig. 3. Protonema of *Funaria hygrometrica* cultivated in low (1000 lux) and high (3000 lux) light intensity on agar. At the higher light intensity auxins inhibit the growth and also the formation of oblique cell walls. In lower light intensity the formation of oblique cell walls is enhanced by auxin. The number of the cells with oblique cell walls was counted when in the control protonema 60% of the walls were oblique. *NAA* α naphthalene acetic acid; *IAA* indole-3-acetic acid; *2,4 D* 2,4-dichlorophenoxy acetic acid

Effect of Auxin and Antiauxin

Several tests are possible to confirm the assumption that endogenous auxin may regulate caulonema morphogenesis: 1. Direct measurement of auxin content in the protonema; such experiments are in progress. 2. Selection of mutants which are altered in their response to auxin, as first performed by Hatanaka-Ernst (33) and recently by Cove and co-workers (1–3, 27, 28). They found mutants which differ in their response to auxin, and some of them are thought to be unable to produce auxin (2). 3. Use of an antiauxin which can regulate either the endogenous auxin level or the effect of endogenous auxin. One such antiauxin is *p*-chlorophenoxyisobutyric acid (PCIB) (29, 58, 59).

Either PCIB or IAA inhibits moss growth when applied in high concentrations (58), but the inhibition by IAA can be reduced with PCIB. PCIB application leads quickly to several effects. There is a temporary cessation of growth (Fig. 4), and the higher the concentration the longer this inhibition lasts. Cell length is also reduced as

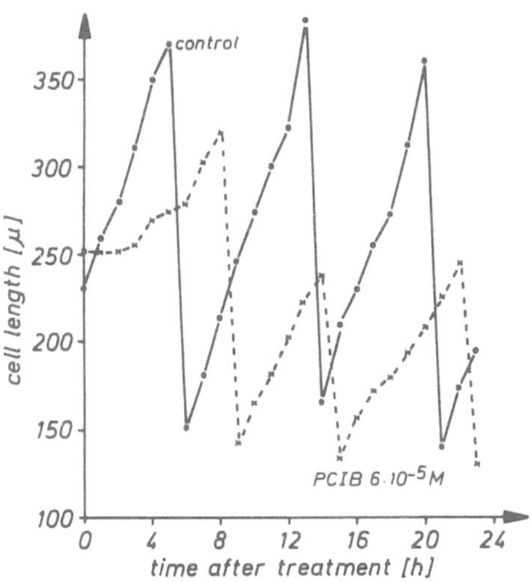

Fig. 4. Growth of a control caulonema of *Funaria hygrometrica* and of a PCIB (6 × 10^{-5} M) treated caulonema. The growth with PCIB stops temporarily. Growth rate is diminished, but the cell division rate is not changed. Therefore the cell length is reduced. *PCIB* Para-chlorophenoxyisobutyric acid

a function of PCIB concentration, and cell diameter is decreased, but the frequency of cell division (cell cycle) is not changed. Chlorophyll formation is strongly enhanced. Chloroplasts in the inhibited tip cells move to the apical end, later the original "tip body" in the cell is restored, and the first new-formed cell wall remains perpendicular (Fig. 5). It can be concluded that one of the clearest effects of auxin is the regulation of the position of the cell wall.

All events described so far transform the caulonema filaments into chloronemata. Since cytokinin-induced bud formation takes place in caulonemata, this process is no longer possible in PCIB-treated protonemata after 16 h (Fig. 6). We may conclude that the formation of caulonemata as well as the maintenance of this stage of differentiation must be dependent on the endogenous auxin.

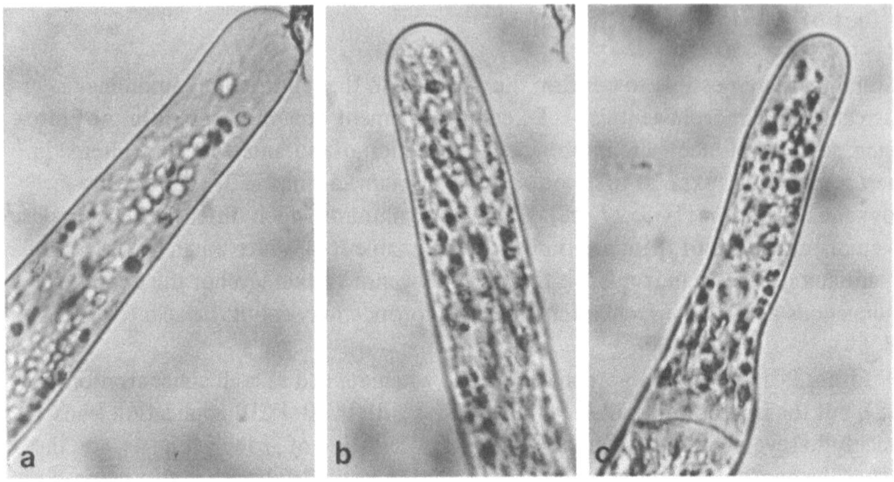

Fig. 5a–c. Caulonema tip of *Funaria hygrometrica* treated with 10^{-4} M PCIB. **a** Untreated control; **b** after 6 h of treatment; **c** after 12 h of treatment. After reconstitution of the tip, the filament grow as a chloronema with a perpendicular cell wall

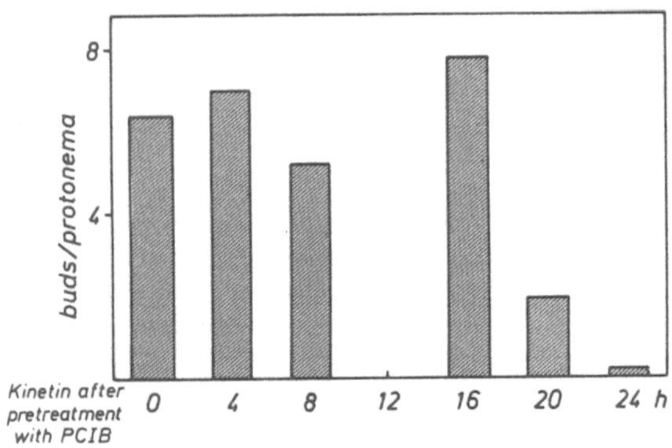

Fig. 6. Bud formation of PCIB (10^{-4} M) treated protonema. The protonema was pretreated with PCIB and after different times transferred to a medium containing PCIB (10^{-4} M) and kinetin (10^{-5} M). The number of buds is counted 48 h after treatment

Experiments designed to show how PCIB works in mosses have demonstrated that the uptake of ^{14}C-labeled IAA is greatly reduced. This may be important if we assume that a part of the auxin effect is mediated through the substrate, as can be concluded from experiments with high and low cell densities in *Funaria hygrometrica* suspension cultures (37). If the uptake of auxin is inhibited, the tip cell of a caulonema does not get enough auxin to remain in this stage of differentiation. We failed to find in vivo stimulation of IAA oxidation by PCIB as had been found in vitro (25); on the contrary, the degradation of ^{14}C-labeled IAA was reduced. Therefore, we believe that PCIB can block IAA at different sites in the cell, such as those responsible for uptake, metabolism, and the hormonal effect of auxin. All the data seem to demonstrate that caulonema formation is dependent on a high level of endogenous auxin activity.

It is possible that malate plays a role in the causal chain between auxin and the morphogenetic effect, because the malate content is high in light-grown caulonema-forming cultures and very low in cultures under low light intensities, in which caulonemata do not appear.

The caulonema is characterized by other unusual features, one of which is continuous DNA synthesis after cell division (41, 44), so that the cells become diploid, tetraploid and even octoploid (43, 51); this never happens in chloronemata. Another feature is the strong polarity of the cells, which can be changed by several external factors (54)

Bud Formation

The next step in differentiation, bud formation, occurs on caulonemata formed from protonemata under the influence of exogenous or endogenous auxin (2, 47, 58). Cytokinin is one of the factors which induce buds. Another one is factor H (39) which has been known for many years, but which contrary to frequent reports (2, 37), is neither a cytokinin nor a cytokinin-like molecule. Cytokinin is present in mosses, however, as was demonstrated in the one and only existing moss callus (which was produced by cross-breeding *Funaria hygrometrica* and *Physcomitrium pyriforme*), and this "bryokinin" has been identified as N^6-(Δ^2-isopentenyl)adenine (4).

Evidence for the existence of endogenous cytokinin is provided by the work of Ashton et al. (3), who isolated at least two different mutants of *Physcomitrella patens* that overproduce gametophores in the absence of exogenous cytokinins. Alternative explanations for this behavior are discussed by the authors (3): (a) overproduction of the normal endogenous cytokinin, (b) production of an altered cytokinin more active than the normal hormone, (c) increased sensitivity to the normal endogenous cytokinin, (d) absence of a cytokinin requirement for gametophyte development and (e) underproduction of a cytokinin antagonist. Because cross-feeding between mutants and wild-type strains through the culture medium is possible, the most likely explanation is that the mutants are cytokinin overproducers (3) as was demonstrated by quantitative cytokinin analysis of mutant and wild-type strains (64). Furthermore, even though the moss callus may be a cytokinin-overproducer, it may lack target cells capable of bud formation. However as far as we know, both the uptake and metabolism of kinetin or BAP in the caulonema and chloronema are more or less the same (23). Exogenous cytokinin must be metabolized very quickly in the protonema, because it has no "after effect"; its effect ceases as soon as it is removed from the medium (7, 14, cf. 64).

Chloronema cells show only one response to cytokinin treatment, a higher rate of cell division (61), whereas caulonema cells show at least two responses, stimulated cell division and bud formation, the latter cells therefore are the target cells with respect to the second response. It was shown by cytokinin treatment that the isolated single caulonema cell itself, freed from side branches, is a target cell. The pattern of distribution of buds among the cells isolated in sequence from a filament is identical, regardless of whether side branches are intact or removed during cytokinin treatment. When the cells are separated from one another, the sixth cell of a filament, always produces the greatest number of buds (Fig. 7).

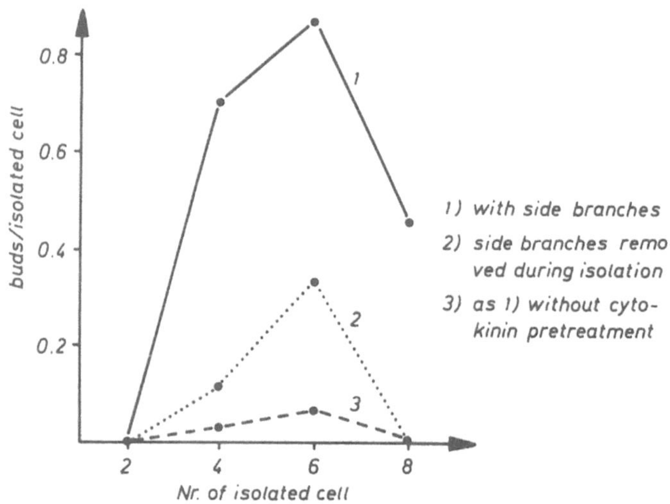

Fig. 7. Response of isolated single cells to a kinetin treatment. Always the 6th cell shows the strongest response to kinetin, independent of whether the buds forming side branches are present or absent during treatment. *1, 2* the filaments are isolated after a 15 h pretreatment with kinetin, *3* no pretreatment

We assume that the target character of the caulonema cells is associated with certain high molecular weight proteins which occur only in the peripheral caulonema cells of older protonemata that are otherwise ripe to form buds (10) and are absent or nonfunctional in the chloronema cells. We call these caulonema-specific-proteins (CSP) (11, 22). However, it seems that they are also present in young protonemata which have not yet attained the status of target cells for bud formation. Therefore, the attainment of the critical size – the smallest protonema able to form buds under certain conditions – is not a consequence of the formation of the CSP. At the time of bud formation one can clearly distinguish between the CSP-containing zone in which the buds are formed, and a zone without these proteins where no buds can be induced (22).

During regeneration of caulonema cells, when the ability to form buds is lost (with a half-life of 5 h), the CSP are no longer synthesized and disappear. Hence, they may be important but are not solely responsible for the target character of the cells. The CSP are certainly not the primary cytokinin receptor sites, and no clearly defined such protein has yet been found in the caulonema cells (25a, Knoop, unpubl.).

To summarize: the effect of cytokinin is very specific. All cytokinins have the same effect (9, 62). Nearly all mosses respond in the same way (74), and the effect of cytokinins disappears immediately after they are removed, in which case buds already induced grow out as chloronema filaments (14a).

Bud Growth

The next question is what constitutes bud formation in terms of morphogenesis? The response to cytokinin treatment takes place in the very young side branches of the caulonema. In *Physcomitrella*, cells are already committed to produce either buds or filaments when the caulonema cells start to enlarge (13), while in *Funaria* the side branches can still shift from filament to bud formation during the early stages of development.

In general, buds may be viewed as transformed side branches. In cytokinin-treated protonemata bud formation starts within 10 h [see (36)]. In the light microscope one can see a change in cell shape (11, 48); the new cell wall material is no longer incorporated in the direction of the longitudinal axis of the filament but perpendicularly to it in the tip region. This results in a triangular incorporation of wall material precursors as indicated in autoradiographs. The important and only morphogenetic step in bud formation, therefore, is the change in direction of deposition of wall material (13).

Additional steps in growth regulation result in the formation of a trisected apical cell. These steps are not very well understood (19). Under the influence of high concentrations of cytokinin the buds commonly fail to develop into leafy stems, and tend to retain a reduced spherical or moruloid shape (60), which is also found in the "cytokinin-overproducing" mutants (3). Sometimes callus is also produced in response to cytokinin (24, 47, 56).

High concentrations of auxin also inhibit normal development of buds, either all cells grow out as protonema, forming a "spiky" appearance (2, 5, 26) or in somewhat older buds, short, thick, leafless stems appear (2, 5, 45). Such structures are reminiscent of gemmae in some mosses. For example, in *Bryum cornatum,* gemmae are formed with a few rudimentary leaves under the same conditions where *Leptobryum* produces normal gametophores (17, 18). With low concentrations of kinetin, gemmae are formed in *Bryum klinggraeffii,* whereas with high concentrations inhibited buds appear at the same loci. Therefore, in these cases, gemma formation can be regarded as being analogous to the budding response of the caulonema. Both kinds of experiment suggest that in the induction of buds and their subsequent development into gametophores two distinct processes are involved.

Conclusions

The elements of the "morphogenetic system" of the moss protonema are very labile, as stages in differentiation endure only under the conditions which induced them. Induction and maintenance of the different stages are strikingly influenced by exogenous hormones (auxin, cytokinin) which suggests that endogenous hormone regulation also exists. Evidence for this also comes from studies of effects of hormone antagonists, and of environmental conditions which modify the response to exogenously applied hormones, and, finally, from the direct demonstration of the presence of hormones in mosses and of excessive hormone production by certain mutants.

Acknowledgments. Our research was supported by the Deutsche Forschungsgemeinschaft. Thanks to my co-workers B. Lehnert, S. Rose, and S. Zimmermann for valuable cooperation.

References

1. Ashton, N.W., Cove, D.J.: Mol. Gen. Genet. *154*, 87–95 (1977)
2. Ashton, N.W., Grimsley, N.H., Cove, D.J.: Planta *144*, 427–435 (1979)
3. Ashton, N.W., Cove, D.J., Featherstone, D.R.: Planta *144*, 437–442 (1979)
4. Beutelmann, P., Bauer, L.: Planta *133*, 215–217 (1977)
5. Bopp, M.: Z. Bot. *41*, 1–16 (1953)
6. Bopp, M.: Planta *53*, 178–197 (1959)
7. Bopp, M.: J. Linn. Soc. London Bot. *58*, 305–309 (1963)
8. Bopp, M.: Ber. Dtsch. Bot. Ges. *78*, 44–54 (1965)
9. Bopp, M.: Proc. 8th Int. Conf. Plant Growth Substances, pp. 934–944. Tokyo: Hirokawa Publ. 1974
10. Bopp, M., Brandes, H.: Planta *62*, 116–136 (1964)
11. Bopp, M., Diekmann, W.: Planta *74*, 86–96 (1967)
12. Bopp, M., Erichsen, U., Nessel, M., Knoop, B.: Physiol. Plant. *42*, 73–78 (1978)
13. Bopp, M., Fell, J.: Z. Pflanzenphysiol. *79*, 81–87 (1976)
14. Brandes, H.: Rev. Bryol. Lichenol. *35*, 158–164 (1967)
14a. Brandes, H., Kende, H.: Plant Physiol. *43*, 827–837 (1968)
15. Chen, T.H., Jaffe, L.F.: Planta *144*, 401–406 (1979)
16. Chopra, R.N., Kumar, P.K., Rekhi, A.: Curr. Sci. *47*, 735–737 (1978)
17. Chopra, R.N., Rawat, M.S.: Bryologist *80*, 655–661 (1977)
18. Chopra, R.N., Rawat, M.S.: Beitr. Biol. Pflanz. *53*, 353–357 (1977)
19. Chopra, R.N., Rashid, A.: Z. Pflanzenphysiol. *61*, 192–198 (1969)
20. Correns, C.: Untersuchungen über die Vermehrung der Laubmoose durch Brutorgane und Stecklinge. Jena: Fischer 1899
21. Dirlam, J.: Diplomarb. Fak. Biol. Heidelberg 1978
22. Erichsen, J., Knoop, B., Bopp, M.: Planta *135*, 161–168 (1977)
23. Erichsen, U., Knoop, B., Bopp, M.: Plant Cell Physiol. *19*, 839–850 (1978)
24. Fell, J.: Diplomarb. Fak. Biol. Heidelberg 1975
25. Frenkel, C.H., Haard, N.F.: Plant Physiol. *53*, 380–384 (1973)
25a. Gardner, G., Sussmann, M.R., Kende, H.: Planta *143*, 67–73 (1978)
26. Gorton, B.S., Eakin, R.E.: Bot. Gaz. *119*, 31–38 (1957)
27. Grimsley, N.H., Ashton, N.W., Cove, D.J.: Mol. Gen. Genet. *154*, 97–100 (1977)
28. Grimsley, N.H., Ashton, N.W., Cove, D.J.: Mol. Gen. Genet. *155*, 103–107 (1977)
29. Grotha, R.: Planta *129*, 235–238 (1976)
30. Hahn, H., Bopp, M.: Planta *83*, 115–118 (1968)
31. Hahn, H., Bopp, M.: Z. Pflanzenphysiol. *68*, 19–29 (1972)
32. Handa, A.K., Johri, M.M.: Nature (Lond.) *259*, 480–482 (1976)
33. Hatanaka-Ernst, M.: Z. Pflanzenphysiol. *55*, 259–277 (1966)
34. Heitz, E.: Ber. Dtsch. Bot. Ges. *60*, 17–27 (1942)
35. Hildebrand, R., Kotte, J., Winkler, S.: Beitr. Biol. Pflanz. *54*, 1–12 (1978)
36. Idzikowska, K., Szweykowska, A.: Protoplasma *94*, 41–52 (1978)
37. Johri, M.M.: Frontier of Plant Tissue Cultures 1978, pp. 27–36. Proc. 4th Int. Congr. Plant Tissue and Cell Culture 1978
38. Johri, M.M., Desai, S.: Nature (Lond.) *245*, 223–224 (1973)
39. Klein, B.: Planta *73*, 12–27 (1967)
40. Knoop, B.: Z. Pflanzenphysiol. *70*, 22–33 (1973)
41. Knoop, B.: Planta *121*, 75–80 (1974)
42. Knoop, B.: Z. Pflanzenphysiol. *77*, 350–358 (1976)

43. Knoop, B.: Protoplasma *94*, 307–314 (1978)
44. Knoop, B., Hahn, H., Bopp, M.: Planta *88*, 288–292 (1969)
45. Kofler, L.: Rev. Bryol. Lichenol. *28*, 1–202 (1959)
46. Kottke, J., Krisch, T., Winkler, S.: Beitr. Biol. Pflanz. *51*, 407–415 (1976)
47. Menon, M.K.C., Lal, M.: Planta *115*, 319–328 (1974)
48. Nebel, B.J., Naylor, A.W.: Am. J. Bot. *55*, 33–37 (1968)
49. Pfrüner, H., Bentrup, F.W.: Planta *143*, 213–223 (1978)
50. Rose, S.: Diplomarb. Fak. Biol. Heidelberg 1979
51. Ripetzky, R.T., Demkiv, O.T., Lavrinkh, T.V.: Ukr. Bot. Zh. *29*, 315–320 (1972)
52. Schmiedel, G.: Diss. Fak. Biol. Heidelberg 1978
53. Schmiedel, G., Schnepf, E.: Protoplasma *100*, 367–383 (1979); *101*, 47–59 (1979)
54. Schnepf, E., Schmiedel, G.: Proc. R. Microsc. Soc. *14*, 234 (1979)
55. Sironval, C.: Bull. Soc. R. Bot. Belg. *79*, 48–78 (1947)
56. Spiess, L.D.: Bryologist *82*, 47–53 (1979)
57. Spiess, L.D., Lippincott, B.B., Lippincott, J.A.: Bot. Gaz. *138*, 35–40 (1977)
58. Sood, S., Hackenberg, D.: Z. Pflanzenphysiol. *91*, 385–397 (1979)
59. Stange, L.: Planta *135*, 289–295 (1977)
60. Szweykowska, A.: Acta Soc. Bot. Pol. *31*, 553–557 (1962)
61. Szweykowska, A.: Tissue Culture and Plant Sci. 461–475 (1974)
62. Szweykowska, A., Korcz, J., Jsakiewicz-Mroczkowska, B., Metelska, E.: Acta Soc. Bot. Pol. *41*, 401–409 (1972)
63. Vaarama, A., Taren, N.: Bot. Not. *112*, 481–488 (1959)
64. Wang, T.C., Beutelmann, P., Featherstone, D.R., Cove, D.J.: Abstr. 10th Int. Conf. Plant Growth Substances 1979, p. 14

Hormonal Control of Morphogenesis in Cultured Tissues

D.E. FOSKET [1]

Introduction

Several aspects of the hormonal control of morphogenesis have been reviewed recently (1–5). As a result, a comprehensive review of this topic would appear to be redundant at the present time. A better use for these pages would appear to be the presentation of a critical evaluation of the idea that two hormones, auxin and cytokinin, represent the primary morphogens in higher plants. To narrow the topic somewhat, I intend to confine my attention to studies which have been conducted with cultured tissues, where much of the important work has been done. Although it is by no means an established fact, the bulk of this work supports the hypothesis that auxin and cytokinin represent the primary signals for organogenesis and cytodifferentiation in higher plants.

One of the more interesting ideas that has been put forth in recent years to explain why development takes place where it does is Wolpert's (6) concept of cellular positional information. He has proposed that concentration gradients of morphogens exist within organisms and that the developmental fate of any given cell is determined by its position within these gradients. Botanists should be particularly receptive to Wolpert's hypothesis since the significance of a cell's position within the plant in predicting its subsequent development has been acknowledged for over 100 years (7). Furthermore, auxin and cytokinin, which are known to elicit striking morphogenetic effects in higher plants, could be expected to establish gradients of concentrations within the plant body. In the simplest case, one can envision concentration gradients of these morphogens extending through a tissue. Since the sources and transport mechanisms of these two hormones differ, the shape of the gradient would be different in each case. Cells within the tissue would respond in a characteristic manner according to their position within these gradients.

Regulation of Organogenesis

Organogenesis in Cultured Tobacco Tissues

Over 30 years have passed since Skoog and his co-workers first demonstrated that organogenesis could be regulated by chemical treatments in cultured tobacco tissues (8, 9). After the subsequent discovery of the cytokinins in the same laboratory (10),

1 Department of Developmental and Cell Biology, University of California, Irvine, CA 92717, USA

the idea that morphogenesis could be manipulated by chemical treatments was refined to suggest that it was the auxin/cytokinin ratio that controlled development in cultured tissues (11). This concept has been repeatedly and elegantly verified with tobacco where high auxin/cytokinin ratios promote root formation, high cytokinin/auxin ratios lead to shoot initiation, while unorganized callus growth occurs with intermediate levels of the two hormones (12–15). Both hormones are required for the growth of tobacco pith tissue in culture (14), while the hormonal balance will determine whether the growth is organized or not, as well as the kind of organization exhibited by the newly formed tissue. It should also be noted that each hormone appears to exhibit both a positive and a negative morphogenetic effect, as suggested by the fact that, if a given level of auxin and cytokinin will support bud formation, raising the auxin concentration while keeping the cytokinin level constant can suppress bud initiation. Similarly, high cytokinin can suppress auxin-induced rooting.

Control of Organogenesis in Other Species

In recent years tissue culture has become a widely used method for the clonal propagation of many commercially important plants (16, 17). Frequently this approach involves the sequential induction first of callus and then of organs. Growth and morphogenesis are brought about by the transfer of tissues to specific inductive media where hormones play a significant, if not dominant, role in determining the developmental events. In many of these studies with dicot species, the roles of auxin and cytokinin in controlling growth and organogenesis have been shown to be similar to the story that has unfolded for tobacco (18–26). Nevertheless, at the present time it would be premature to conclude that organogenesis universally is regulated by an interaction between auxin and cytokinin in higher plants.

One dicot species that exhibits a significantly different type of developmental behavior in response to auxin and cytokinin is alfalfa *(Medicago sativa).* Bingham found that sequential treatment with two different media was required to bring about morphogenesis. Unorganized callus tissue was produced when alfalfa explants were cultured on a medium containing 2,4-D and a cytokinin. Organogenesis occurred when the callus tissue subsequently was transferred to a medium lacking both hormones. No buds were produced on the second medium unless the first medium contained 2,4-D, while cytokinin, although stimulatory, was not required for bud formation (27, 28). Recently Walker and co-workers (29) demonstrated that roots were formed from alfalfa callus upon transfer to hormone-free medium if the callus was produced on a medium containing a high cytokinin/auxin ratio. Conversely, when the alfalfa callus was initiated on a medium with a high auxin/cytokinin ratio, it produced shoots upon transfer to the hormone-free medium. Thus, although these results suggest that organogenesis is regulated by an interaction between auxin and cytokinin in this species, the specific role of the hormones seems to be the opposite of that found in tobacco and other dicots. High auxin levels stimulated shoot formation while high cytokinin levels stimulated root initiation.

Organogenesis in Monocots

Another major difficulty for the hypothesis is that the factors controlling organogenesis in cultured tissues of monocot species are poorly understood. In most cases an auxin has been found to be necessary for the initiation of callus growth, but frequently the addition of cytokinin to the medium has little effect on either growth or morphogenesis. In several cereal grasses and in onion, where a potent auxin such as 2,4-D has been used to evoke cell proliferation, organogenesis occurs when the tissues are transferred either to a medium lacking an auxin, or to a medium containing a less potent auxin such as IAA (30–34). It has been suggested that the cells of cereal tissues are already programed to differentiate, and that the function of the strong auxin is not only to initiate cell proliferation, but to suppress the tendency toward organized growth (35). Since as noted above both hormones may be thought to have both inhibitory and stimulatory morphogenetic effects, these results do not necessarily argue against the hypothesis that organogenesis is controlled by an auxin-cytokinin interaction in monocots. We simply don't have enough data to evaluate this possibility at the present time.

The Role of Endogenous Hormones

One of the difficulties in evaluating those cases where organogenesis does not appear to be regulated by an auxin-cytokinin interaction is that very little is known about the endogenous hormone levels of cultured tissues or how these levels change in response to cultural conditions. However, there are known examples of changes in the apparent morphogenetic controls in a given species which are associated with developmental alterations that might be expected to modify endogenous hormone levels. For example, Bonnett and Torrey (36) observed that auxin alone controlled organogenesis in cultured segments of *Convolvulus* roots. Continuous exposure to medium containing IAA evoked the formation of lateral root primordia, whereas treatment with the same auxin concentration for a few days, followed by the transfer of the segments to medium lacking auxin, evoked bud initiation. These results are in striking contrast to those of Earle and Torrey (37), who found that bud formation in *Convulvulus* callus of root origin could be induced predictably by providing the proper auxin-cytokinin balance in the culture medium. Both hormones interacted to control bud initiation in the callus tissue of this species, but root segments, which might be expected to contain comparatively high levels of cytokinins, only responded to exogenous auxin.

The available evidence, although circumstantial, suggests that endogenous hormone levels, particularly cytokinins, may also play a significant role in regulating adventive embryogenesis in cultured carrot tissues. It is now well known that somatic cells of wild or domestic species of carrots are capable of undergoing normal embryological development in culture (38–40). Auxin has been shown to be a particularly important factor in determining whether or not embryos are formed, but its specific regulatory role is complex. Reinert found that 2,4-D inhibited embryo formation when carrot callus was initiated and maintained on White's medium. The same 2,4-D concentration that inhibited morphogenesis was optimal for callus growth. However, embryos were produced in the presence of 2,4-D when callus was initiated and maintained on Murashige-Skoog medium (41). Cytokinin was unable to induce embryogenesis when it was added to the noninductive White's medium, but it did suppress embryo formation when it was added to inductive Murashige-Skoog medium at a concentration that did not affect the growth rate of the tissues (42).

As carrot tissue is subcultured, it appears to lose its capacity to form embryos in response to an inductive medium (42–44). A similar loss of morphogenetic capacity upon prolonged subculture has been noted in other systems where it is correlated with the accumulation of cells with chromosomal abnormalities (45, 46). However, one must be cautious in attributing the loss of morphogenetic capacity directly to the chromosomal abnormalities, particularly in the light of the work of Sacristan and Melchers (47, 48), who regenerated plants from very old, habituated tobacco callus cultures exhibiting a high degree of aneuploidy. Although abnormal, sterile, aneuploid plants were formed, this nevertheless demonstrates that it is possible to induce organized development in such tissues. If this is the case, it seems appropriate to ask if the loss of morphogenetic capacity in carrot cultures might not be due to some physiological or epigenetic change which is potentially reversible.

Although there are few data to support this suggestion, the observation of Wochok and Wetherell (49) that the morphogenic capacity of old carrot cultures can be restored, at least temporarily, by adding cytokinin to the inductive medium may be an important step toward our understanding of the control mechanisms in this system. Why does this tissue suddenly become responsive to cytokinin? One possible explanation is that adventive embryogenesis is a response of cultured carrot tissues to a specific auxin/cytokinin ratio. Linstedt and Reinert (50) have demonstrated that carrot callus, cultured on inductive Murashige-Skoog medium, contains substantial levels of a cytokinin-like substance. Perhaps aging carrot cultures simply lose the ability to produce cytokinins and thus their ability to undergo morphogenesis in response to exogenous auxin.

Certainly there are several known examples of callus tissues, capable of growing on medium lacking a cytokinin, which have been shown to produce their own cytokinins (51–53). Furthermore, there is evidence that cytokinin biosynthesis may be activated by high levels of auxin some tissues (54, 55). There is even evidence that, under some circumstances, auxin biosynthesis may occur in response to cytokinin (14, 56). If nothing else, these observations mean that the endogenous hormone levels must be known before conclusions are made about the factors controlling morphogenesis in a given tissue.

Other Factors Affecting Organogenesis

Unfortunately, even if a tissue's auxin and cytokinin levels were known, it might not be possible to predict its developmental behavior in culture. It has been well established that a plethora of additional factors can enhance, modify, or negate the expected response from tissues in which organogenesis is elicited by specific hormonal regimens. These factors include inorganic ions such as phosphate, iron, and ammonium (8, 57–59); amino acids such as tyrosine (60); the nucleic acid base adenine (9, 61); other hormones such as abscisic acid (62, 63) and gibberellic acid (64–69); as well as physical and cultural parameters such as light (70) and liquid versus solid medium. Furthermore, the physiological and developmental state of the explanted tissue can have a profound influence on the response obtained. For example, in wheat, callus formation occurred readily from embryos and seedling nodal segments, but not at all from leaf tissue (34). Also, in tobacco, explants from flowering or fruiting stems not only are capable of a qualitatively different morphogenic response from that of vegetative explants (production of floral instead of vegetative buds in culture), but also exhibit an altered sensitivity to exogenous auxin and cytokinin (15, 71).

One factor affecting organogenesis which has been relatively little investigated is the genetic background of the plant in question. It is widely accepted that members of certain families, such as the Solanaceae, tend to exhibit morphogenesis in culture much more readily than do members of other families, such as the Leguminosae. Presumably such differences have a genetic component. Linsmaier and Skoog (72) tested the response of five different tobacco cultivars to the same media. At a given auxin/cytokinin ratio, some of the cultivars produced many shoots, some produced few shoots, while others did not initiate shoots at all. A genetic basis for differences in morphogenetic behavior was demonstrated by Bingham et al. (73), who observed that alfalfa cultivars differed in their capacity to undergo organogenesis. They were able to breed a cultivar with a greatly increased frequency of organogenesis in culture by crossing fertile plants which had been regenerated from callus tissue.

Cytodifferentiation

An interaction between auxin and cytokinin has been shown to play a role in the regulation of cytodifferentiation, principally of tracheary elements, in various cultured tissues. Although auxin has been shown to be the principal factor limiting tracheary element formation in *Coleus* stem tissue (74, 75) and in some callus cultures (76–79), cytokinin has been shown to interact with auxin to control tracheary element formation in soybean callus tissue (80), in pea root cortical explants (81, 82) and in pea root callus (83). Furthermore, cytokinin has been shown to stimulate xylem element formation in a number of systems where auxin appears to exert the primary control over cytodifferentiation (84–86).

Some of the problems encountered in demonstrating that cytokinin plays a role in the regulation of tracheary element formation are illustrated by the work of Mizuno (87), who found that tracheary elements were formed in carrot taproot explants when they where cultured in the light on a medium containing an auxin, but lacking a cytokinin. However, tracheary elements did not form when the explants were cultured on the same medium in the dark. Recently Mizuno and Komamine (88) isolated and identified a cytokinin-like substance from light-grown carrot explants which would induce tracheary elements in dark-grown explants when it was added to the culture medium together with an auxin. Furthermore, they identified the active substances as zeatin and zeatin riboside and demonstrated that these cytokinins were absent from the dark-grown explants.

As noted above, Linstedt and Reinert (50) have shown that carrot callus produces cytokinins when cultured on medium containing 2,4-D. The work of Mizuno and Komamine (88) indicates that, at least in the carrot cultivar they used for their studies, this 2,4-D-induced cytokinin biosynthesis is activated by light, which may help explain why light has been found to be highly stimulatory to tracheary element differentiation in some systems (89).

As was found to be the case with organogenesis, a number of additional factors are known to modify the response of a given tissue to an otherwise inductive stimulus for cytodifferentiation. Perhaps the most dramatic of these is ethylene which, at relatively low levels that did not prevent cell division, completely blocked tracheary element formation in cultured lettuce explants (90). Both gibberellic acid and abscisic acid have been shown to influence tracheary element formation in various experimental systems. In cultured Jerusalem artichoke tuber disks, gibberellic acid can be stimulatory (84) or inhibitory (85) to differentiation, apparently depending upon how the tubers were stored before experimentation, while abscisic acid strongly inhibits cytodifferentiation in the same system (85). Of particular significance is the role of sucrose, which not only seems to be necessary for tracheary element formation as an energy source (91), but also has been suggested to play a direct role in the induction of differentiation as well (78, 79).

A greater problem to the hypothesis that tracheary element formation is controlled by a cytokinin-auxin interaction is the work of Clutter (92) and Earle (93). Both of these workers studied tracheary element formation in excised, cultured pith tissues. In both cases auxin was the only variable examined, but it clearly induced tracheary element differentiation in the absence of exogenous cytokinin. Furthermore, in the case of Clutter's (92) studies with tobacco, this freshly excised pith tissue would not be expected to contain endogenous cytokinins (11). However, Clutter noted that tracheary elements were produced only after a prolonged period in culture, and that the auxin treatment slowly induced a limited amount of cell proliferation, with tracheary elements forming from some of the derivatives produced in culture. Since cell division is controlled by an auxin-cytokinin balance in tobacco (4, 11), this may be another example of auxin-induced cytokinin biosynthesis with both cell division and cytodifferentiation occurring in response to exogenous auxin and endogenous cytokinin. Certainly this possibility needs to be examined.

References

1. Roberts, L.W.: Cytodifferentiation in Plants. Xylogenesis as a Model System. New York: Cambridge University Press 1976
2. Kohlenbach, H.W.: In: Plant Tissue Culture and its Biotechnical Application. Barz, W., Reinhard E., Zenk, M.H. (eds.), pp. 355–366. Berlin, Heidelberg, New York: Springer 1977
3. Narayanaswamy, S.: In: Plant Cell, Tissue, and Organ Culture. Reinert, J., Bajaj, Y.P.S. (eds.), pp. 179–206. Berlin, Heidelberg, New York: Springer 1977
4. Fosket, D.E.: In: Mechanism and Control of Cell Division. Rost, T.L., Gifford, E.M., Jr. (eds.), pp. 62–91. Stroudsburg, Pa.: Dowden, Hutchison and Ross 1977
5. Street, H.E.: In: Plant Cell and Tissue Culture. Principles and Applications. Sharp, W.R., Larson, P.O., Paddock, E.F., Raghavan, V. (eds.), pp. 123–153. Columbus: Ohio State University Press 1979
6. Wolpert, L.: J. Theoret. Biol. *25*, 1–47 (1969)
7. Vöchting, H.: Über Organbildung im Pflanzenreich, p. 241. Bonn: Max Cohen 1878
8. Skoog, F., Tsui, C.: Am. J. Bot. *35*, 782–787 (1948)
9. Skoog, F., Tsui, C.: In: Plant Growth Substances. Skoog, F. (ed.), pp. 263–286. Madison: University of Wisconsin Press 1951

10. Miller, C.O., Skoog, F., Okumura, F.S., von Saltza, M.H., Strong, F.M.: J. Am Chem. Soc. *78*, 1345–1350 (1956)
11. Skoog, F., Miller, C.O.: Symp. Soc. Exp. Biol. *11*, 118–131 (1957)
12. Vasil, V., Hildebrandt, A.C.: Science *150*, 889–892 (1965)
13. Vasil, V., Hildebrandt, A.C.: Planta *75*, 139–151 (1967)
14. Skoog, F.: In: Les Cultures de Tissus des Plantes. Colloq. Int. C.N.R.S. *193*, 115–136 (1971)
15. Van, M.T.T., Dien, N.T., Chlyah, A.: Planta *119*, 149–159 (1974)
16. Murashige, T.: Annu. Rev. Plant Physiol. *25*, 135–166 (1974)
17. Murashige, T.: In: Plant Tissue Culture and Its Biotechnological Applications. Barz, W., Reinhard, E., Zenk, M.H. (eds.), pp. 392–403. Berlin, Heidelberg, New York: Springer 1979
18. Kartha, K.K., Gamborg, O.L., Constabel, F.: Physiol. Plant. *31*, 217–220 (1974)
19. Gunay, A.L., Rao, P.S.: Plant Sci. Lett. *11*, 365–372 (1978)
20. Hervey, A., Robbins, W.J.: In Vitro *14*, 294–299 (1978)
21. Hiraoka, N., Tabata, M.: Phytochemistry *13*, 1671–1675 (1974)
22. Pillai, S.K., Hildebrandt, A.C.: Am. J. Bot. *56*, 52–58 (1969)
23. Kartha, K.K., Gamborg, O.L., Shyluk, J.P., Constabel, F.: Z. Pflanzenphysiol. *77*, 292–301 (1976)
24. Rao, P.S., Handro, W., Harada, H.: Physiol. Plant. *28*, 458–463 (1973)
25. Bapat, V.A., Rao, P.S.: Z. Pflanzenphysiol. *85*, 403–416 (1977)
26. Kamat, M.B., Rao, P.S.: Plant Sci. Lett. *13*, 57–65 (1978)
27. Saunders, J.W., Bingham, E.T.: Crop Sci. *12*, 804–808 (1972)
28. Saunders, J.W., Bingham, E.T.: Am. J. Bot. *62*, 850–855 (1975)
29. Walker, K.A., Yu, P.C., Sato, S.J., Jaworski, E.G.: Am. J. Bot. *65*, 654–659 (1978)
30. Fridborg, G.: Physiol. Plant. *25*, 436–440 (1971)
31. Carter, O., Yamada, Y., Takahashi, E.: Nature (Lond.) *214*, 1029–1030 (1967)
32. Nishi, J., Yamada, Y., Takahashi, E.: Nature (Lond.) *219*, 508–509 (1968)
33. Rangan, T.S., Z. Pflanzenphysiol. *72*, 456–459 (1974)
34. O'Hara, J.F., Street, H.E.: Ann. Bot. *42*, 1029–1038 (1978)
35. King, P.J., Potrykus, I., Thomas, E.: Physiol. Veg. *16*, 381–399 (1978)
36. Bonnett, H.T., Torrey, J.G.: Plant Physiol. *40*, 1228–1236 (1965)
37. Earle, E.D., Torrey, J.G.: Am. J. Bot. *52*, 891–899 (1965)
38. Reinert, J.: Planta *53*, 318–333 (1959)
39. Halperin, W., Wetherell, D.F.: Am. J. Bot. *51*, 274–283 (1964)
40. Steard, F.C., Mapes, M., Kent, A., Holsten, R.: Science *143*, 20–27 (1964)
41. Reinert, J., Tazawa, M.: Planta *87*, 239–248 (1969)
42. Reinert, J.: In: Plant Growth Substances, Carr, D.J. (ed.), pp. 686–694. Berlin, Heidelberg, New York: Springer 1970
43. Reinert, J., Backs, D.: Nature (Lond.) *220*, 1340–1341 (1968)
44. Smith, S.M., Street, H.E.: Ann. Bot. *38*, 223–241 (1974)
45. Murashige, T., Nakano, R.: Am. J. Bot. *54*, 963–970 (1967)
46. Torrey, J.G.: Physiol. Plant. *20*, 265–275 (1967)
47. Sacristan, M.D., Melchers, G.: Mol. Gen. Genet. *105*, 317–333 (1969)
48. Sacristan, M.D., Melchers, G.: Mol. Gen. Genet. *152*, 111–117 (1977)
49. Wochok, Z.S., Wetherell, D.S.: Experientia *28*, 104–105 (1972)
50. Linstedt, D., Reinert, J.: Naturwissenschaften *62*, 238–239 (1975)
51. Miura, G.A., Miller, C.O.: Plant Physiol. *44*, 1035–1039 (1969)
52. Short, K.C., Torrey, J.G.: J. Exp. Bot. *23*, 1099–1105 (1972)
53. Einset, J.W., Skoog, F.: Proc. Natl. Acad. Sci. USA *70*, 658–660 (1973)
54. Witham, F.H.: Plant Physiol. *43*, 1455–1457 (1968)
55. Einset, J.W.: Plant Physiol. *59*, 45–57 (1977)
56. Jordan, W.R., Skoog, F.: Plant Physiol. *48*, 97–99 (1971)
57. Halperin, W., Wetherell, D.F.: Nature (Lond.) *205*, 519–520 (1965)
58. Tazawa, M., Reinert, J.: Protoplasma *68*, 157–173 (1969)
59. Singh, J.P.: Biol. Plant. *20*, 436–439 (1978)
60. La Motte, C.E.: Ph.D. Dissertation, University of Wisconsin, Madison 1960

61. Nitsch, J.P., Nitsch, C., Rossini, L.M.E., Bui, B.H.: Phytomorphology *17*, 446–453 (1967)
62. Thorpe, T.A., Meier, D.D.: Physiol. Plant. *29*, 121–124 (1973)
63. Haddon, L., Northcote, D.H.: J. Cell Sci. *20*, 47–55 (1976)
64. Murashige, T.: Science *134*, 280 (1961)
65. Murashige, T.: Physiol. Plant. *17*, 636–643 (1964)
66. Thorpe, T.A., Murashige, T.: Science *160*, 421–422 (1968)
67. Besemer, J., Harden, U., Reinert, J.: Z. Pflanzenphysiol. *60*, 123–134 (1969)
68. Thorpe, T.A., Meier, D.D.: Physiol. Plant. *27*, 365–369 (1972)
69. Engelke, A.L., Hamzi, H.Q., Skoog, F.: Am. J. Bot. *60*, 491–495 (1973)
70. Seibert, M.: In Vitro *8*, 435 (1973)
71. Wardell, W.L., Skoog, F.: Plant Physiol. *44*, 1402–1406 (1969)
72. Linsmeier, E.M., Skoog, F.: Physiol. Plant. *18*, 100–127 (1965)
73. Bingham, E.T., Hurley, L.V., Kaatz, D.M., Saunders, J.W.: Crop Sci. *15*, 719–721 (1975)
74. Jacobs, W.P.: Am. J. Bot. *39*, 301–309 (1952)
75. Jacobs, W.P.: Am. Nat. *88*, 327–333 (1954)
76. Wetmore, R.H., Sorokin, S.: J. Arnold Arbor. Harv. Univ. *36*, 305–317 (1955)
77. Wetmore, R.H., Rier, J.P.: Am. J. Bot. *50*, 418–430 (1963)
78. Jeffs, R.A., Northcote, D.H.: Biochem. J. *101*, 146–152 (1966)
79. Jeffs, R.A., Northcote, D.H.: J. Cell Sci. *2*, 77–88 (1967)
80. Fosket, D.E., Torrey, J.G.: Plant Physiol. *44*, 871–880 (1969)
81. Shininger, T.L., Torrey, J.G.: In: Mechanism of Regulation of Plant Growth. Bieleoki, R.L., Ferguson, A.R., Cresswell, M.M. (eds.), pp. 721–728. R. Soc. N.Z. Wellington 1974
82. Shininger, T.L.: In Vitro *14*, 31–50 (1978)
83. Torrey, J.G.: In: Biochemistry and Physiology of Plant Growth Substances. Wightman, F., Setterfield, G. (eds.), pp. 843–855. Ottawa: Runge Press 1968
84. Dalessandro, G.: Plant Cell Physiol. *14*, 1167–1176 (1973)
85. Minocha, S.C., Halperin, W.: Planta *116*, 319–331 (1974)
86. Phillips, R., Dodds, J.H.: Planta *135*, 207–212 (1977)
87. Mizuno, K., Komamine, A., Shimokoriyama, M.: Plant Cell Physiol. *12*, 823–830 (1971)
88. Mizuno, K., Komamine, A.: Planta *138*, 59–62 (1978)
89. Kleiber, H., Mohr, H.: Planta *76*, 85–92 (1967)
90. Zobel, R.W., Roberts, L.W.: Can. J. Bot. *56*, 987–990 (1978)
91. Fosket, D.E., Roberts, L.W.: Am. J. Bot. *51*, 19–25 (1964)
92. Clutter, M.E.: Science *132*, 548–549 (1960)
93. Earle, E.D.: Am. J. Bot. *55*, 302–305 (1968)

Agricultural Uses of Plant Growth Regulators

Chairman: P. W. MORGAN

Agricultural Uses of Plant Growth Substances: Historical Perspective

P.W. MORGAN [1]

After 30 years the International Conference on Plant Growth Substances has returned to Madison, and after the same number of years attention has again been given to the practical uses of the science that interests us all. Perhaps this 30 years should be called on to supply the historical perspective to help evaluate the past development and the future of plant growth substances in agriculture.

Many of the details of the historical development of agricultural applications of plant growth substances have been published (1–3); therefore, this introduction to the following papers on plant growth regulators will deal with only a few of the factors that have strongly influenced the present state of the industry.

In 1949 P.W. Zimmermann (4) reviewed "plant hormones in practice" and listed the following agricultural uses as in existence or prospect: rooting, parthenocarpic fruit set, thinning fruit, inhibiting sprouting, inhibiting bud growth, inducing flowering, defoliation, preventing preharvest fruit drop, and killing weeds. All of these uses involved auxins exclusively, except that induction of flowering in pineapple could be done with acetylene generated from calcium carbide and nursery stock was defoliated by exposure to ethylene. It is obvious that the recognition of new classes of plant hormones has been one of the major influences on the chemical industry interested in plant growth substances. The effect has not just been to add new compounds to test but, possibly more important, it has expanded our understanding of the degree to which plants can be manipulated. Consequently, if new classes of natural growth substances are discovered – for example, flowering regulators, auxin transport inhibitors, or growth retardants – the resultant knowledge may have profound effects on practical applications.

One important factor that has changed since 1949 is the agricultural chemical industry itself. Excluding fertilizers, the modern chemical approach to agriculture owes its origin to two classic compounds, DDT and 2,4-D, which came into production and extensive use following World War II. Compared to present-day circumstances the agricultural chemical industry was in its infancy in 1949. All of the experience, successful products, marketing organizations and grower acceptance that have developed in recent years have provided a base for future developments that was essentially nonexistent in the 1930's and 1940's. This state of industrial experience, plus the more extensive basic understanding of the physiology of several crop plants, may be largely responsible for the multiple growth-regulator applications that are practiced in several crops such as sugar cane, apples, and cherries.

1 Department of Plant Sciences, Texas A&M University, College Station, Texas 77843, USA

It should also be noted that one of the historical events that shaped the modern agricultural chemicals industry – the discovery of the selective herbicidal activity of 2,4-D – grew out of research on plant hormones and related growth regulators. While the modern herbicide industry has come far since 2,4-D, its roots in auxin research should be remembered and appreciated.

One significant historical episode, often overlooked in discussions of growth regulators, was the development of chemicals to defoliate cotton in order to facilitate mechanical harvesting (5). This effort represented an early example of success of the empirical approach to agricultural chemistry. A specific problem was recognized (the need to remove leaves), several means were developed to test for active compounds, an organization was formed of people interested in the problem, which facilitated cooperation and exchange of information (Cotton Defoliation Conference organized by the National Cotton Council), and financial support was provided by industrial firms to university scientists to underwrite studies of defoliants. The result was the discovery, development, marketing, and extensive use of an array of chemical defoliants and desiccants that were relatively effective, economical, and safe, by the standards of the time. Elements of this empirical pattern have been repeated many times since the early 1950's, and the pattern is seen in current efforts, for example the citrus fruit harvest-aid program in Florida. F.T. Addicott, H.C. Carns, and W.C. Hall were involved in the early cotton defoliation group. Addicott and Carns contributed to the discovery of ABA (6, 7), and Hall to the recognition of the hormonal roles of endogenous ethylene (8, 9).

The empirical approach to development of commercial plant growth regulators may also be linked to one of the current failings. Specifically, a compound capable of significantly increasing basic productivity or yield of a major field crop – for example, wheat, corn, rice, or soybeans – is not yet at hand. In fact, an empirical search for a yield-enhancer is probably a task too complex to be accomplished. To empirically identify the right crop and the proper compound (or compounds), to then apply the proper amount in the proper manner at the right stage of development, and then weeks or months later to measure an increased yield simply defies success. This is especially true since the entire procedure must be done under slow and costly field conditions due to the unavailability of bioassays to reliably predict yield-increasing abilities of chemicals. Successes in promoting yields have almost always occurred when a specific goal was recognized that would subsequently lead to increased yield. Current practices to increase sugar content of cane, promote flow of latex in *Hevea,* and prevent lodging of small grains are good examples of such specific goals. The recognition of specific developmental and metabolic changes that will improve productivity seems to be an essential component, absent in some efforts, of successful programs to develop yield-enhancers.

The intense efforts to commercialize gibberellic acid during the late 1950's and the rather abrupt termination of interest by some of the firms involved probably had an effect on subsequent interest in development of growth regulators. The gibberellin experience certainly re-emphasized a real gap that exists between discovery of strong biological activity and a successful commercial compound. Later ABA and the cytokinins failed to generate any major commercial products, and growth inhibitors such as maleic hydrazide and chlormequat chloride came to make up much of the volume and value of plant growth substances which were used commercially. A causal inspection of the current array of products and uses shows that most of them are growth inhibitors or involve restriction of growth. Historically, we have done much less to increase growth than we have to obtain benefits by restricting it. What this means for the future remains to be seen.

In an area closer to my own experience, it seems as if recognition of the hormonal roles of ethylene, as facilitated by the application of H_2 flame ionization gas chromatography to the study of fruit ripening (10) and by the rediscovery of auxin-mediation of ethylene synthesis, were important (11, 12). The resultant strong interest in ethylene (13–16) probably contributed to the rapid development of 2-chloroethyl phosphonic acid as a commercial growth regulator and opened up all the options to manipulate ethylene-sensitive developmental events (17, 18). These options include inhibiting ethylene synthesis and action as well as the current practices of supplying ethylene (via ethephon) and removing ethylene (via reduced pressure).

Another important factor that has influenced the commercialization of plant growth substances is governmental regulations and the accompanying cost of registration. Obviously, no one desires to see unsafe products marketed, and the costs involved probably do not prevent development of good products for major markets. However, the registration procedure probably does keep a substantial number of products off the market that could be used in rather small markets. The result is that experience and information which might be useful to the total science of growth regulation is unavailable while such limited-market compounds rest on company shelves. As means to circumvent this problem become available, many new growth regulators may be disclosed.

This series of observations is not intended to be complete but simply to set the stage for the papers in this section. In 1949 Zimmermann (4) made some remarkable predictions regarding plant growth substances. He predicted: herbicides for specific crops, pre-emergence herbicides, bud-inducing chemicals (cytokinins), shoot-inducing substances (gibberellins), and substances to synchronize and concentrate fruit ripening (daminozide, ethephon). The papers that follow will discuss the fulfilment of some of these predictions, and they will also talk about the future. Perhaps the past 30 years have brought us to the point that we can go beyond manipulation of developmental events and actually increase productivity of major field crops. Historically, this seems to be an idea whose time has come.

References

1. Tukey, H.B. (ed.): Plant Regulators in Agriculture. New York: Wiley 1954
2. Weaver, J.R.: Plant Growth Substances in Agriculture. Freeman 1972
3. Wittwer, S.H.: Outlook Agric. *6*, 205–217 (1971)
4. Zimmerman, P.W.: In: Plant Growth Substances. Skoog, F. (ed.), pp. 81–89. University of Wisconsin Press 1951
5. Addicott, F.T.: In: Plant Regulators in Agriculture. Tukey, H.B. (ed.), pp. 99–116. New York: Wiley 1954
6. Addicott, F.T., Carns, H.R., Lyon, J.L., Smith, O.E., McMeans, J.L.: In: Régulateurs Naturels de la Croissance Végétale, pp. 687–703. Paris: C.N.R.S. 1964

7. Ohkuma, K., Addicottt, F.T., Smith, O.E., Thiessen, W.E.: Tetrahedron Lett. *29,* 2529–2535 (1965)
8. Hall, W.C.: Bot. Gaz. *113,* 310–322 (1952)
9. Hall, W.C., Truchelut, B.B., Leinwelkr, C.L., Herrero, F.A.: Physiol. Plant. *10,* 306–317 (1957)
10. Burg, S.P., Burg, E.A.: Plant Physiol. *37,* 179–189 (1962)
11. Morgan, P.W., Hall, W.C.: Physiol. Plant. *15,* 420–427 (1962)
12. Morgan, P.W., Hall, W.C.: Nature (Lond.) *201,* 99 (1964)
13. Burg, S.P.: Annu. Rev. Plant Physiol. *13,* 265–302 (1962)
14. Pratt, H.K., Goeschl, J.D.: Annu. Rev. Plant Physiol. *20,* 541–584 (1969)
15. Abeles, F.B.: Annu. Rev. Plant Physiol. *23,* 259–292 (1972)
16. Lieberman, M.: Annu. Rev. Plant Physiol. *30,* 533–591 (1979)
17. Morgan, P.W.: Acta Hortic. *34,* 41–54 (1973)
18. Morgan, P.W.: Adv. Chem. *159,* 42–53 (1977)

Applications of Gibberellins in Agriculture

L. RAPPAPORT [1]

Introduction

Since their introduction as plant growth regulators heralded by full page newspaper advertisements and editorials about 10-foot cabbages *(Brassica oleracea),* GAs have found a niche in specialized areas of agriculture. The image of towering plants yielding a crop cornucopia excited public investigators as well as commercial companies in the late 1950's and early 1960's. While the predictions may have been exaggerated, several uses have emerged, and it is likely that further investigation will result in new applications.

This report is intended to update information on the current applications of the GAs as well as those that appear close to adoption, to examine reasons for the relatively slow development of commercial applications, and to consider the potential uses of GA s in agriculture.

Uses of GAs on apples and sugar cane, and of growth inhibitors are reviewed elsewhere in this volume. They will receive only limited attention here. Information about the applications of gibberellins (GA s) in agriculture have been considered primarily in the context of commercial sources (1), general treatments of growth regulators (2–8), in specific reviews (9–12), and in specialized articles about particular crops (13–16).

Current Applications

Control of Vegetative Growth

The bolting response of rosette plants to GAs , first reported by Lang (17), was attributed to promotion of cell division and enlargement in the subapical meristem (18). This spectacular effect has been observed in numerous species and in caulescent plants pretreated with inhibitors of GA biosynthesis; here GA reverses the inhibition. The most prominent, although minor, use of this treatment is promotion of seedstalk elongation and increased seed yield in firm-headed lettuce *(Lactuca sativa)* (19). During formation of the "head" the infolding leaves entrap the emerging seedstalk. Applications of GA_3 at the appearance of the 8th and 12th leaves result in release of 85% to 90% of the seedstalks, compared with only 20% in the untreated controls (J.W. Chaney, pers. comm.).

---

1 Department of Vegetable Crops/Plant Growth Laboratory, University of California, Davis, California 95616, USA

Growth of sugar cane *(Saccharum officinarum)* is enhanced by GA_3, a result associated with an increase in total sugar (14, 4). This application is marginal in view of the cost of the hormone for the treatment of sugar cane and the increased availability of compounds that act as "ripening" agents which promote sucrose accumulation (14, 4).

Freshly harvested potato *(Solanum tuberosum)* tubers exhibit a rest period which must be broken before the tubers are replanted as seed. This may be accomplished by treatment with GA_3 (20, 21) which is also used by seed certification agencies to promote sprouting in order to detect virus infection in seedlots before they are planted the next season; infected lots are discarded.

De Angelis (22) and Snyder et al. (23) found that foliar applications of GA_3 to artichokes *(Cynara scolymus)* accelerated peduncle elongation and bud development, especially at low temperatures which limit bud growth. Gibberellin application is popular in Cyprus where the peduncle is eaten as well as the artichoke bud (Fig. 1). Treatment may result in smaller buds and, on occasion, reduced total yields.

Fig. 1. The effect of foliage sprays of GA_3 (32 mg/L) on peduncle elongation in globe artichokes grown in Cyprus. GA_3 is used to accelerate artichoke bud development, especially at low temperatures. (Photo courtesy of Dr. George Elson, ICI Ltd., England)

Wittwer and Bukovac (9) and Takatori et al. (24) observed that GA_3 promoted elongation and increased fresh weight of petioles of certain celery *(Apium graveolens)* cultivars. There is a limited use of GA_3 to attain marketable petiole length, however, undesirable seedstalk elongation and reduced petiole quality may result.

Application of GA_3 to tart cherry *(Prunus cerasus)* trees allows an interesting avoidance of the effects of a pollen-transmitted virus disease called sour cherry yellows (25, 26). This virus actually promotes blossom bud development in terminal and lateral shoots, but causes reduction in yield because vegetative growth ceases. Since GA s inhibit flowering in stone fruit trees and, at optimal concentrations, promote vegetative growth (25), application of the hormone enhances new growth and production of spurs which do not exhibit the infection.

The use of compounds that inhibit specific stages of GA biosynthesis, thereby limiting elongation, deserves attention here. Application of chlormequat (CCC) has proved highly effective in inhibiting lodging of tall cultivars of wheat (27, 28) (Fig. 2). Yield is increased directly due to maximum recovery of grain from upright plants and indirectly because treatment permits application of high levels of N. In northern Europe, CCC is applied to about 80% of the wheat crop for these purposes. Control of plant height by growth retardants is a standard practice in the nursery industry. Potted plants of many ornamental species are treated with such compounds as ancymidol, chlormequat, phosphon and daminozide in order to reduce stem height.

Fig. 2. The response of 'Opal' wheat *(Triticum aestivum)* plants to concentrations of 0 to 3000 mg/L of (2-Chloroethyltrimethyl) ammonium chloride applied as a foliage spray. (Photo courtesy of J. Jung, BASF, West Germany)

Control of Flowering

Although GA_3 promotes flowering in some species, there are today few commercial applications of this knowledge. The major use of GA s in regulation of flowering is to enhance male flower formation in monoecious and gynoecious cucumbers, an excellent example of hormone application in plant breeding. It is based on the observations that GA s promote staminate flower formation in gynoecious cucumbers (29–31), that GA s delay or inhibit female flower formation (32, 33), and that higher endogenous GA levels correspond to the expression of the monoecious rather than the gynoecious condition (34). Today, GA_3, and a more effective mixture of GA_4 and GA_7 ($GA_{4/7}$, GA_{4+7}) (35) are used commercially for producing pollen parents in gynoecious cucumber fields. Tolla and Peterson (36) determined that silver nitrate, an antagonist of ethylene action, is much more effective than $GA_{4/7}$, in promoting male flower formation. Ethylene is thought to be the primary determinant of female sex expression (37). Field tests are required before it will be possible to ascertain the agricultural value of silver nitrate treatment.

Regulation of reproductive capacity is a goal of scientists interested in hormonal regulation. Therefore, the discovery by Phatak et al. (38) that GA_3 applied before anthesis induces fertility in otherwise sterile tomato anthers has met with interest, especially from seed companies in Europe. In addition, GA_3 has been shown to induce male sterility in onion which has also proved of value for hybrid seed production (J. Rudich, pers. comm.).

One of the most spectacular effects of GA s is induction of flowering in certain coniferous species (15). Repeated applications of GA_3 and $GA_{4/7}$ promoted flowering within a year in Arizona cyprus, *Cupressus arizonia,* western red cedar, *Thuja plicata,* coastal and giant redwoods, *Sequoia sempervirens* and *Sequoia gigantea* (39, 40). These results indicate the possibility of enhancing breeding programs designed to improve yield and quality of trees for high quality lumber. In the family Pinaceae, in which flowering may not occur for 10 years and longer, it was found that several auxins, applied jointly with GA_3, A_4, A_5, A_7, A_9 or A_{14}, promoted formation of strobili of both sexes within a year (15, 41). The effect is enhanced by girdling (42).

In Douglas fir, *Pseudotsuga menziesii,* the most effective formulation for promoting male strobilus formation was a mixture of naphthaleneacetic acid (25 μg/branch/application), $GA_{4/7}$ (200 + 300 μg/shoot application) and GA_9 (40 μg/application). The necessity for determining optimal hormone combinations, concentrations and times of applications was underlined in these studies. Pharis (41) reported that the auxin/GA treatment also increased seed yield 30- to 50-fold. This treatment could permit increased production of seed from genetically desirable trees.

Gibberellins inhibit or delay flowering in many species, especially fruit trees (2), and this effect has some practical advantage, especially where freezing is a danger. Dennis et al. (43) showed up to a 90% reduction in flower bud formation in "Bartlett" pears.

Control of Fruit Set, Development and Quality

Unquestionably, the major applications of the GAs have been to fruit crops. These uses stem from the ability of the GAs to induce parthenocarpic fruit set (2, 44–47), promote fruit enlargement (48–50), alter fruit color, shape and size (43, 51, 52) and to overcome a variety of physiological diseases (53, 54) and surface and internal defects (51, 55, 56).

Gibberellin-induced parthenocarpic fruit set in pear *(Pyrus communis),* reported by Gil et al. (57), is of value in regions where frost injury causes damage to the ovules or fruits set parthenocarpically for lack of pollination (2, 58, 59; S. Sansivini, pers. comm.). In Japan, grape-flower clusters of 'Delaware', a seeded cultivar, are dipped in a solution of GA_3 prior to bloom and two weeks later to induce parthenocarpic fruit set and accelerate apparent fruit maturity (47); berry size is sometimes increased by treatment. As many as 12,000 hectares of 'Delaware' grapes are treated in this way in Japan (Luckwill, pers. comm.). In seeded cultivars, such as 'Delaware', berry size is directly correlated with seed number. Berries with but one seed tend not to enlarge to full size. Application of a gibberellin results in attainment of full size, irrespective of seed number (60).

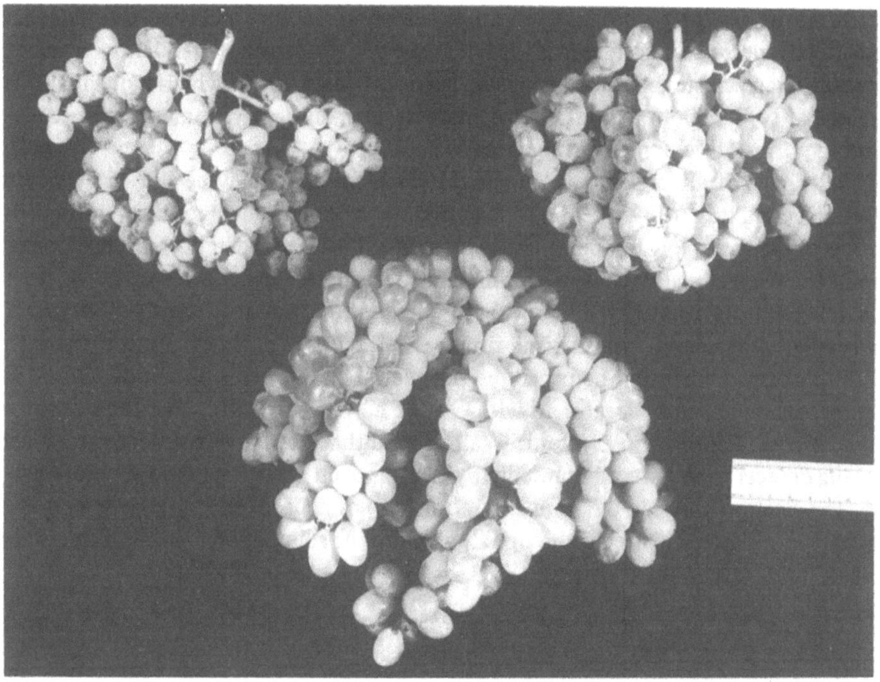

Fig. 3. Effect of GA_3 in promoting berry enlargement of 'Thompson' seedless grapes. *Upper left* unsprayed, ungirdled, control; *upper right* ungirdled, sprayed with 50 mg/L. *Below* girdled, sprayed with 50 mg/L. (Photo courtesy of Dr. R.J. Weaver, University of California, Davis)

The remarkable effect of GA_3 in stimulating berry enlargment in seedless grapes (Fig. 3) such as 'Thompson Seedless' and 'Perlette' (48) has led to near universal use of the hormone on such cultivars for table use. Gibberellin A_3 applied to the flower clusters during bloom increases rachis length, thereby loosening the cluster (61). However, a second and third application of GA_3 post-bloom generally increases berry size. Because of resulting competition for space it is necessary to thin about a third of the cluster, and total yield of the crop is often increased, especially when the canes are girdled to limit phloem transport of photosynthate (49). The treated berries tend to be elongate and greener in color than the untreated ones.

Since the berries enlarge rapidly, there is concern that the clusters will be harvested before the sugar content is adequate to impart good flavor. However, harvest is based on sugar content, not size, and rate of maturity is related to cropping load. Singh et al. (50) found that early harvest resulted in berries with significantly lower sugar content; however, a one-week delay in harvest allowed sugar content of the treated berries to equal that of the earlier harvested controls. Acidity was not altered in these experiments.

Unlike the response with seedless grapes, prebloom GA_3 treatment of seeded cultivars with high concentrations generally has undesirable effects, and may even be toxic to the vines (62). However, in seeded cultivars which produce very compact clusters, GA_3 applied pre-bloom successfully thins the berries and elongates the rachis. The berries on the resulting looser clusters are less likely to rupture and become infected. The effectiveness of fungicides is also increased by the loosening. Frequently, after berry set, grape clusters produce shriveled berries, a condition in which individual berries lose turgor about a month before harvest. Post-anthesis treatments with GA_3 result in normal development of such berries.

Fruit shape of 'Red Delicious' apples *(Malus domestica)* may be altered by sprays containing a mixture of GA_{4+7} and the cytokinin N^6-benzyladenine (63). During periods of relatively low temperature, the treatment is effective in promoting elongation of the characteristic lobes to produce the ideal shape for the cultivar.

Gibberellin A_3 applied 4–6 weeks before harvest to 'Rainier' cherries *(Prunus avium)* slightly increased fruit size and firmness and reduced anthocyanin content with a resulting improvement in the processed product. Ascorbic acid content of 'Bing' cherries was increased by treatment with GA_3 (51). Working with 'Van' and 'Lambert' cherries, Looney and Lidster (56) also found that postharvest surface pitting and bruising defects were reduced markedly by GA_3 applied about one month before harvest. The relation between GA_3 concentration, size of cherries and surface defects is illustrated in Fig. 4. Treatment did not alter acidity or soluble solids in the expressed juice.

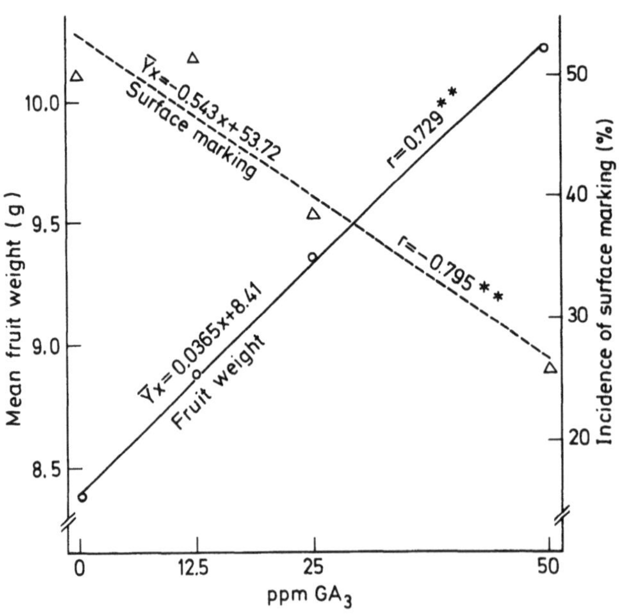

Fig. 4. Effect of GA_3 applied as a foliage application on fruit weight and development of surface defects on 'Lambert' cherries (Looney and Lidster, 1979)

Gibberellin treatments have a number of beneficial effects on citrus species. Fruit set of 'Fino' Clementine mandarin *(Citrus reticulata)* was increased dramatically with 100 mg/l GA_3 applied in the period between anthesis and two weeks post-anthesis (64). They found that in 'Monreal' Clementine, a self-compatible cultivar, GA_3 increased total fruit number and the number of seedless fruits, in agreement with results of Soost and Burnett (65). In 'Fino' Clementine, however, fruit size was decreased in relation to concentration, but the yield of marketable fruit was increased significantly. The results appear to vary with the location, but there is a good chance the treatment will be adopted where it improves yield.

In lemon, Coggins et al. (66) found that application of the hormone prior to degreening results in an extension of the storage life of the fruit up to three months. The fruits continue to increase in size during the post-application period. Figure 5 shows

Fig. 5. GA_3 treatment of lemon fruits results in delayed degreening and prolonged storage life. *Right* untreated; *left* fruit treated with 10 mg/L in October and photographed in March. (Photos courtesy of C.W. Coggins, University of California, Riverside)

delay in rind senescence in the treated lemons. Similarly, ripening in limes (67), "Navel" oranges (68), and grapefruit (69) is delayed by a preharvest spray application. The hormone is often mixed with 2,4-D to delay fruit abscission.

Table 1. Influence of GA_3 on rind staining of 'Navel' oranges at different levels of N. The hormone was applied December 15, 1964 and the observations were made on February 8, 1965. (Data provided by C.W. Coggins, University of California, Riverside)

Nitrogen applied	0	1 lb/tree in Feb.	1 lb/tree split:Feb., Aug.	3 lb/tree in Feb.	3 lb/tree split:Feb., Aug.
			% Fruit defective		
No GA_3	9	14	17	13	35
100 mg/L GA_3	3	4	3	5	8

Surface defects are markedly reduced in citrus by treatment with GA_3. In Israel, the hormone sprayed in June or July is used to reduce "creasing" of 'Valencia' oranges, the result of cracking of the albedo (13, 53). Later application, however, results in undesirable delayed development of carotenoids and chlorophyll degradation. In California, GA_3 is used by 'Navel' orange growers to reduce surface defects associated with rind senescence. Such physiological disorders as "rind staining" (54), "puffy rind", "sticky rind" and "water spot" are reduced measurably by application of GA_3 (Fig. 6; Tables 1 and 2).

Fig. 6. Physiological rind defects of 'Navel' orange which are alleviated by foliage sprays of GA_3 (10 mg/L) applied in October. *Top* "Water spot". Treated fruit on the *left*. The condition results from penetration of surface water into the rind, especially after application of insecticidal oil sprays. Subsequently, the area dessicates giving the symptoms seen in this figure. GA_3 is applied together with insecticidal oil sprays to overcome the disorder. *Bottom* "Sticky rind", preharvest disease which increases in severity after harvest. The sticky exudate is not water soluble. GA_3 application *(left)* essentially eliminates this condition. Flecking is due to soil particles. (Photos courtesy of C.W. Coggins, University of California, Riverside)

Table 2. "Sticky rind" surface defect in "Navel" orange is reduced by GA_3. The hormone was applied January 19 and fruit was harvested on June 18, 1965. (Data supplied by C.W. Coggins, University of California, Riverside)

GA_3 (mg/L)	0	5	10	20
% Fruit with sticky rind	87	26	21	6

Malting

Scientists in the brewing industry early recognized the possible advantages of using GA for promoting enzyme activity in malting barley (70); see review by Paleg (71). The hormone promotes synthesis, release, or both, of α-amylase, proteinases, and cellulases which break down starch, proteins and walls of the cells of the endosperm into a mixture of sugars, polypeptides and amino acids. This is the malt used in fermentation. Addition of GA_3 to the steep water or during early germination shortens the germination period, thereby accelerating the malting process. Treatment increases the yield of malt extract, reducing losses and producing a uniform product more suitable for fermentation (1).

Limitations to Commercial Application

According to recent predictions (72), the future of growth regulator use is bright. A 60% worldwide increase in sales of plant growth regulators is predicted by 1984, exclusive of herbicides, defoliants, and dessicants. Yet, relative to those excluded compounds, the amount of plant growth regulators sold is very small. One important factor limiting use of GA's is the relatively high cost. Thus GA_3 is effective in preventing internal browning in Italian prunes, but the cost of application is prohibitive (Proebsting, pers. comm.).

Morgan (3), in an interesting perspective, presented a number of cogent reasons for the slower adoption of plant growth regulators than most other agricultural chemicals. He pointed out that whereas insecticides or herbicides are expected to produce a rapid and qualitative effect, plant growth substances generally are expected to act over a long period of time. The effects of genotype and environment on action of the regulator are, therefore, much more pronounced than with an herbicide. Another reason is that the physiology of the processes being regulated is generally not understood. Therefore, the ideal time of application and the concentration required are usually not known and are difficult to ascertain because of the vagaries of climate and genotypic response. The side effects, such as inhibition of flowering in fruit trees for a number of years after treatment, may therefore be very undesirable. Still another problem is the fact that endogenous hormones are generally metabolized at a rapid rate (73); there is a great need for stable analogs of GA s and cytokinins. Finally, results obtained under controlled conditions are not easily reproduced in the field.

Applications are most effective when (1) the endogenous hormone seems to be limiting, (2) expression of desirable growth or form is prevented because of genetic characteristics, or (3) because the crop is grown under environmental constraints which GAs may overcome.

The effect of exogenous GAs on enzyme induction in the isolated cereal aleurone layer is a clear-cut example of the absence of a hormone at its natural target site. Normally, the embryo provides the GAs necessary for activation of enzyme syntheses in the aleurone layer of cereal seeds.

Limited endogenous GA was reported to be related to the rosette condition in radish, *Raphanus sativus* (74). Exposure to favorable vernalization and photoperiodic treatment resulted in an increase both in endogenous GA s and bolting. Seedstalk formation, but not flowering, was suppressed by a growth retardant, and GA_3 reversed the inhibition imposed by the retardant; applied GA s promoted both bolting and flowering under unfavorable conditions. Thus environmental constraints, low temperature and short day, were overcome by exogenous GA s.

Even with a relatively reproducible process, such as GA-induced elongation, environmental interactions powerfully influence expression. The effects of GAs on three genotypes of tomato grown at different temperatures illustrate this problem (Fig. 7a, b). As Morgan (3) states, it is pointless to expect that a growth regulator would elicit the same response under vastly different environmental conditions.

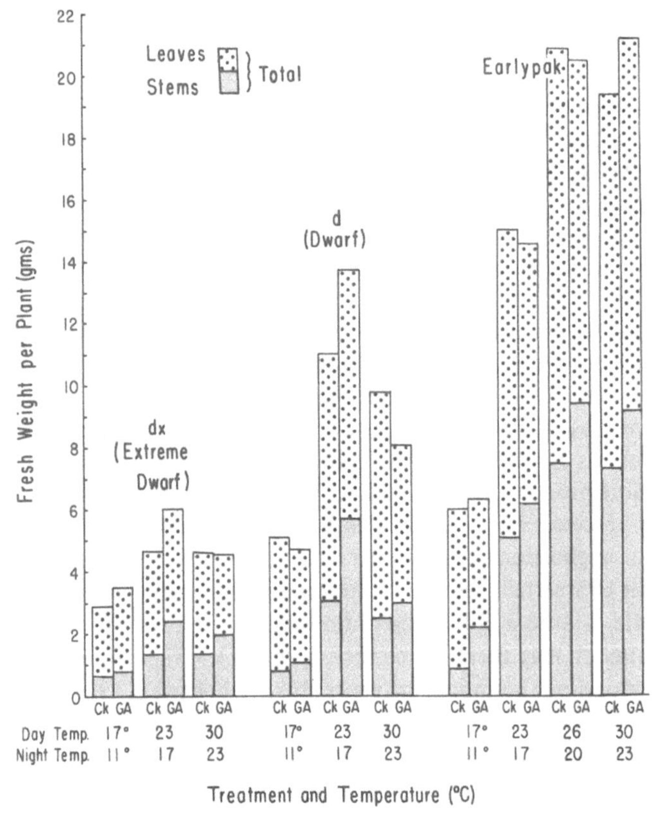

Fig. 7a. The effect of GA_3 (100 μg/plant) applied to the apex on weight of dx, d and 'Earlypak' tomato plants

Bolting induced in head lettuce by GA_3 illustrates circumvention of a formative effect resulting from breeding. Gibberellin-induced bolting permits escape of the seedstalk before it is trapped by the leaves. Alteration of fruit shape in 'Delicious' apples is another example (63), however, the public makes the final decision on the acceptability of a treatment. Thus, in pears which are treated because of poor pollination, GA s produce an undesirable outgrowth on the "blossom end" of the fruits. Lavee (pers. comm.) stated this altered shape was found acceptable in Israel and there is growing interest in the treatment in Europe.

While the nature of GA-inhibition of surface defects on cherries (51, 56) and inhibition of internal browning in Italian prunes (55) has not been explained, it is likely that application of GA_3 slows some aspects of fruit maturation (softening, color development) by providing a limiting factor. Browning reactions which are related to surface defects may be reduced as a result of GA inhibition of polyphenoloxidase (75).

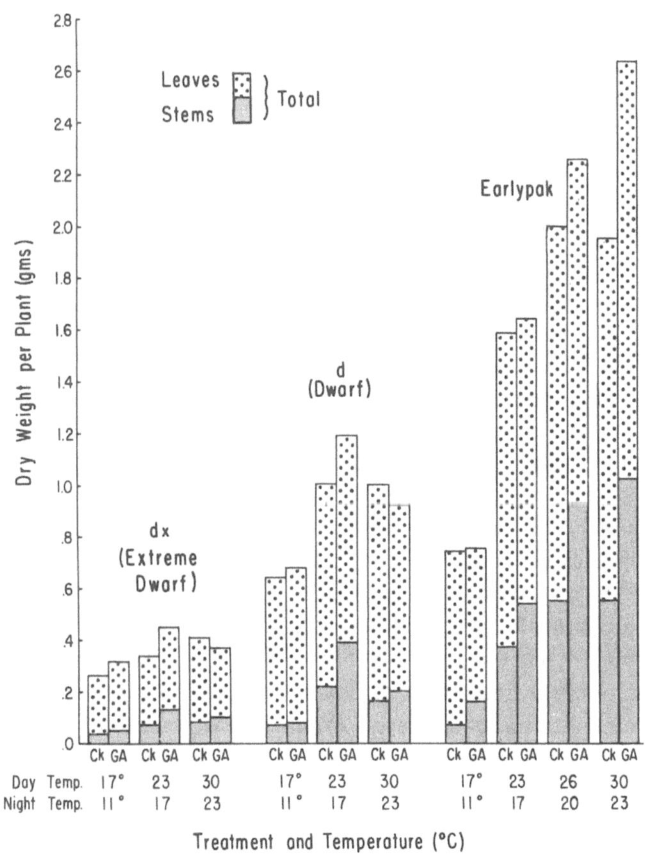

Fig. 7b. Plants were grown in temperature controlled greenhouses. (L. Rappaport, unpublished data)

Potential Applications

Future applications of the GA s will depend on the development of new information on the processes which appear to be most limiting. Special attention must be paid to concentration and timing of application. The subtle effects of very low concentrations need to be examined closely since they may elicit desirable results without producing detrimental side effects. The ability of GA s to enhance certain processes should be investigated further. For example, the hormone might be used effectively to enhance bolting and flowering in certain biennial species in regions in which temperatures required for vernalization are marginally low. Alternatively, the hormone should be tested to induce stem elongation in slow-bolting biennials as an aid in selecting plants with greater resistance to bolting. Seed from such plants might be used in regions in which low temperature-induced precocious bolting is a serious problem.

The potential of GA s applied the previous summer for thinning fruits by inhibiting flower differentiation requires further investigation. Growers are fearful of using GA s for this purpose because the spray applied the previous summer may totally inhibit flowering. It would appear that adjustment of cropping load, the choice of the correct GA, timing of application and appropriate concentration for a given cultivar could be accomplished with the major advantage of reducing the energy loss of producing so many unwanted fruits (Lavee, pers. comm.).

Another consideration affecting new discoveries is the plethora of known GA s, (currently fifty-six). Regardless of the significance of these many forms (76, 77), it is a fact that only GA_3 and a mixture of GA_4/A_7 are in commercial use. Many workers have speculated about the possibilities of finding specific GA s for specialized applications, however, there are only a few instances of species specificity. An example is the greater growth and reproductive response of cucurbits to the 13-deoxyGA s (e.g., GA_4, A_7, A_9) as compared with the 13-hydroxyGA s (e.g., GA_1, GA_3) (32). Coombe (78) showed that GA_{32} isolated from apricot tissue was more effective than GA_3 in promoting apricot fruit enlargment. Similarly, Bearder et al. (79) and Martin et al. (80) found GA_{45} isolated from pear seeds most actively increased fruit size, fresh weight and length/diameter ratio. Perhaps other examples of specificity will be discovered, but it will require the availability of GA s on a large scale for adequate testing.

A new trend in research and formulation may extend the limits of utilization of GA s in agriculture. The approach is to mix GA s with other hormones or inorganic constituents, especially Ca^{2+}. A mixture of GA_3 and CCC – which by itself is effective in setting fruit (81) – promoted berry enlargement in certain grape cultivars in Australia (Coombe, pers. comm.), and GA_3 and 2,4-D are used both to delay degreening and to inhibit abscission of citrus fruits (66, 68). Young and Edgerton (82) found that phytotoxic effects of ethephon on several peach *(Prunus persica)* cultivars (leaf chlorosis, accelerated abscission and "gummosis") were reduced or eliminated by simultaneous application with GA_3. Pharis (41) and Pharis and Ross (16) showed that sex expression in conifers is markedly affected by combinations of different GA s and auxins. Interestingly, male flower formation is enhanced by high auxin levels, contrary to results with angiosperms. Already available commercially is a mixture of GA_3 and N^6-benzyladenine to promote characteristic fruit shape in pears and apples under unfavorable environmental conditions. Gibberellins may be mixed with nonhormonal compounds to extend their effectiveness. For example, GA_3 is mixed in the insecticidal oil applied to 'Valencia' orange, both to alleviate the water spot injury caused by the oil and to reduce the cost of application. It may be anticipated that mixtures of GA s with other substances will be tested extensively in the coming years.

Finally, it is a fact that GA s have found use primarily in horticultural crops. These applications have generally developed as a result of repeated empirical tests at numerous locations. It is likely that progress will continue to be made, and the range of applications extended, especially as our understanding of plant processes becomes more sophisticated. The growing interest in source/sink relationships and their regulation is intensifying interest not only in endogenous hormonal controls, but also in control by exogenous growth substances. In addition to these eminently researchable areas, it may be anticipated that increasing attention will be paid to the role of GAs, and GAs in combination with other substances, on such diverse processes as water and nutrient uptake, stress effects, thinning, abscission, senescence of plant canopies, photosynthesis, storage organ formation, flowering, fruit set and development, and senescence of fruits and vegetables.

Thus, it is predictable that there are many uses of gibberellins yet to be discovered.

Acknowledgment. Drs. J. García-Martínez, S. Lavee, and J. Rudich are acknowledged for their constructive reviews of the manuscript.

References

1. Anonymous: Gibberellic acid. A Handbook of User Information and Technical Data. Fernhurst, England: ICI Plant Protection Ltd. 1974
2. Dennis, F.G.: Acta Hortic. *34,* 251–259 (1972)
3. Morgan, P.: AAAS Symp. Plant Growth Regul. Houston. January (1979)
4. Nickell, L.G.: Chem. Eng. News *9,* 8–19 (1978)
5. Rappaport, L.: In: Research for the World Food Crisis. AAAS, pp. 147–180 (1970)
6. Wittwer, S.H.: Outlook Agric. *6,* 205–217 (1971)
7. Weaver, R.J.: Plant Growth Substances in Agriculture. San Francisco: Freeman & Co. 1972
8. Wareing, R.F.: Outlook Agric. *9,* 42–45 (1976)
9. Wittwer, S.H., Bukovac, M.J.: Econ. Bot. *12,* 213–255 (1958)
10. Rappaport, L., Singh, I.J.: Ind. J. Hortic. *18,* 230–244 (1961)
11. Stuart, N.W., Cathey, H.M.: Annu. Rev. Plant Physiol. *12,* 369–394 (1961)
12. Morgan, P.W.: Adv. Chem. Ser. *159,* 42–53 (1977)
13. Monselise, S.P., Goren, R.: Hort. Science *13,* 134–138 (1978)
14. Nickell, L.G.: Outlook Agric. *9,* 57–61 (1976)
15. Pharis, R.P., Kuo, C.G.: Can. J. For. Res. *7,* 299–325 (1977)
16. Pharis, R.P., Ross, S.D.: Outlook Agric. *9,* 82–87 (1976)
17. Lang, A.: Annu. Rev. Plant Physiol. *21,* 537–570 (1970)
18. Sachs, R.M., Lang, A.: In: 4th Int. Conf. Plant Growth Regul., pp. 567–578. Iowa State Press 1961
19. Harrington, J.F.: Proc. Am. Soc. Hortic. Sci. *75,* 476–479 (1960)
20. Rappaport, L.: Calif. Agric. *10* (12), 4, 11 (1956)
21. Rappaport, L., Lippert, L.F., Timm, H.: Am. Potato J. *34,* 254–260 (1957)
22. de Angelis, J.G.: Isr. J. Agric. Res. *20,* 149–157 (1970)

23. Snyder, M., Welch, N.C., Rubatzky, V.E.: HortScience *6*, 484–485 (1971)
24. Takatori, F.H., Lorenz, O.A., Zink, F.W.: Calif. Agric. *13* (7), 3–4 (1959)
25. Fliegel, P., Parker, K.G., Edgerton, L.J.: Plant Dis. Rep. *50*, 240–242 (1966)
26. Parker, K.G., Edgerton, L.J.: Am. Fruit Grow. *84*, 25–28 (1964)
27. Chrominski, A.: J. Agric. Food. Chem. *15*, 109–112 (1967)
28. Jung, J.: In: Plant Regulation and World Agriculture. Scott, T.K. (ed.), pp. 279–307. New York: Plenum Publ. Co. 1979
29. Galun, E.: Phyton (Buenos Aires) *13*, 1–8 (1959)
30. Peterson, C.E., Anhder, L.D.: Science *131*, 1673–1674 (1960)
31. Pike, L.M., Peterson, C.E.: Euphytica *18*, 106–109 (1969)
32. Bukovac, M.J., Wittwer, S.H.: Adv. Chem. Ser. *28*, 80–88 (1961)
33. Friedlander, M., Atsmon, D., Galun, E.: Plant Cell Physiol. *18*, 261–269 (1977)
34. Rudich, J., Halevy, A.H., Kedar, N.: Plant Physiol. *50*, 585–590 (1972)
35. Clark, R.K., Kenney, D.S.: J. Am. Soc. Hortic. Sci. *94*, 131–132 (1969)
36. Tolla, G.E., Peterson, C.E.: HortScience *14*, 542–544 (1979)
37. Beyer, E.M., Jr.: HortScience *11*, 195–196 (1976)
38. Phatak, S.C., Wittwer, S.H., Honma, S., Bukovac, M.J.: Nature (Lond.) *209*, 636–637 (1966)
39. Pharis, R.P., Morf, W.: Bioscience *19*, 719–720 (1969)
40. Pharis, R.P., Ruddat, M.D.E., Glenn, J.L., Morf, W.: Can. J. Bot. *48*, 653–658 (1970)
41. Pharis, R.P.: US Patent No. 4,110,102. Aug. 29 (1978)
42. Ross, S.D., Pharis, R.P.: Physiol. Plant. *36*, 182–186 (1976)
43. Dennis, F.G., Edgerton, L.J., Parker, K.G.: HortScience *5*, 158–160 (1970)
44. Rappaport, L.: Plant Physiol. *32*, 440–444 (1957)
45. Wittwer, S.H., Bukovac, M.J., Sell, H.M., Weller, L.E.: Plant Physiol. *32*, 39–41 (1957)
46. Crane, J.C., Primer, P.E., Campbell, R.C.V.: Proc. Am. Soc. Hortic. Sci. *75*, 129–137 (1960)
47. Itakura, T., Kosaki, I., Machida, Y.: Bull. Hortic. Res. Stn. Minist. Agric. For. Ser. A (Hiratsuka) *4*, 68–95 (1965)
48. Weaver, R.S., McCune, S.B.: Hilgardia *28*, 297–350 (1959)
49. Weaver, R., McCune, S.B.: Hilgardia *28*, 421–456 (1959)
50. Singh, K., Weaver, R.J., Johnson, J.O.: Am. J. Enol. Vitic. *29*, 258–262 (1978)
51. Drake, S.R., Proebsting, E.L., Jr., Carter, G.H., Nelson, J.W.: J. Am. Soc. Hortic. Sci. *103*, 162–164 (1978)
52. Martin, G.C., Nishijima, C.: HortScience *13*, 255–256 (1978)
53. Monselise, S.P., Weiser, M., Shafir, N., Goren, R., Goldschmidt, E.E.: J. Hortic. Sci. *51*, 341–351 (1976)
54. Coggins, C.W., Eaks, I.L., Hield, H.Z., Jones, W.W.: Proc. Am. Soc. Hortic. Sci. *82*, 154–157 (1963)
55. Proebsting, E.L., Mills, H.H.: Proc. Am. Soc. Hortic. Sci. *89*, 135–139 (1966)
56. Looney, N.E., Lidster, P.D.: J. Am. Soc. Hortic. Sci. *105*, 130–134 (1980)
57. Gil, G.F., Griggs, W.H., Martin, G.C.: HortScience 7 (6), 559–561 (1972)
58. Luckwill, L.C.: Annu. Rep. Long Ashton Agric. Hortic. Res. Stn. *1961*, 61–66 (1962)
59. Varga, A.: Fruittelt *55*, 1506–1507 (1965)
60. Lavee, S.: Nature (Lond.) *185*, 395–396 (1960)
61. Miele, A., Weaver, R.J., Johnson, J.: Am. J. Enol. Vitic. *29*, 79–82 (1978)
62. Weaver, R., McCune, S.B.: Hilgardia *30*, 425–444 (1961)
63. Williams, M.W., Stahly, E.A.: J. Am. Soc. Hortic. Sci. *14*, 17–19 (1969)
64. García-Martínez, J.L., Garcia-Papi, M.A.: Sci. Hortic. *10*, 285–293 (1979)
65. Soost, R.K., Burnett, R.H.: Proc. Am. Soc. Hortic. Sci. *77*, 194–201 (1961)
66. Coggins, C.W., Hield, H.Z., Boswell, S.B.: Proc. Am. Soc. Hortic. Sci. *76*, 199–207 (1960)
67. Burns, R., Rosedale, D.O., Pehrson, J.E., Jr., Coggins, C.W., Jr.: Calif. Agric. *18* (7), 14–15 (1964)
68. Coggins, C.W., Hield, H.Z.: Proc. Am. Soc. Hortic. Sci. *81*, 227–230 (1963)
69. Gilfillan, I.M., Kockmoer, W., Stevens, J.: Citrus Sub-Trop. Fruit J. *458*, 5–7 (1972)
70. Prentice, N., Dickinson, A.D., Burkhart, B.A., Standridge, N.N.: Cereal Chem. *40*, 208–220 (1963)

71. Paleg, L.G.: Annu. Rev. Plant Physiol. *16*, 291–322 (1965)
72. Anonymous: A look at world pesticide markets. Farm Chem. *14*, 38–43 (1977)
73. Lang, A.: Annu. Rev. Plant Physiol. *21*, 537–570 (1970)
74. Suge, H., Rappaport, L.: Plant Physiol. *43*, 1208–1214 (1968)
75. Knapp, F.W., Hall, C.B., Buchanan, D.W., Biggs, R.H.: Phytochemistry *9*, 1453–1456 (1970)
76. Hedden, P., Phinney, B.O., MacMillan, J.: Annu. Rev. Plant Physiol. *29*, 149–192 (1978)
77. Rappaport, L., Adams, D.: Philos. Trans. R. Soc. London Ser. B *284*, 521–539 (1978)
78. Coombe, B.: Science *172*, 856–857 (1971)
79. Bearder, J.R., Dennis, F.G., MacMillan, J., Martin, G.C., Phinney, B.O.: Tetrahedron Lett. *9*, 669–670 (1975)
80. Martin, G.C., Dennis, F.G., Jr., MacMillan, J., Gaskin, P.: J. Am. Soc. Hortic. Sci. *102*, 16–19 (1976)
81. Coombe, B.G.: J. Hortic. Sci. *45*, 415–425 (1970)
82. Young, E., Edgerton, L.J.: HortScience *14* (6), 713–714 (1979)

Ethylene and Ethylene Physiology

D.R. DILLEY [1]

Ethylene, indeed, is a busy gas. So busy, in fact, that seemingly, each month when the professional journals are published, some new involvement is discovered or a new application for ethylene in agriculture is developed. Prior to the 1950's, about the only commercial use of ethylene in agriculture was gassing mature green bananas and tomatoes to make them ripen. Limited application was made of the use of ethylene-saturated water to induce flowering in pineapple. Today there is an ever-growing long list of uses and potential uses for ethylene (10, 14), making it one of the most widely used plant growth regulators in agriculture. This has come about through a better understanding of ethylene's role in plant physiology (6, 12) and through the development and use of agricultural chemicals such as ethephon, which when applied to crop plants evolve ethylene gas. Ethrel and CEPA are commercial formulations of ethephon in widespread use today.

The widespread interest in ethylene as a plant growth regulator stems from the fact that so many physiological processes are either stimulated or inhibited in the presence of the gas (1, 14). Moreover, the direction of the ethylene effect can sometimes change depending upon concentration and/or stage of plant development at the time the application is made.

While there is a wide variety of plant responses favorably affected by ethylene, there are numerous examples where the effects may be adverse or premature in terms of crop utilization. This has led to interest in developing means to attenuate ethylene biosynthesis, inhibit ethylene action, or remove ethylene from the atmosphere (1). Ethylene is a natural component of the atmosphere; clean ambient air contains 5–20 ppb, crop canopy levels of 80 ppb are recorded, and levels in excess of 300 ppb are commonly found in urban and industrial areas as a pollutant from fossil fuel combustion (1). These ambient ethylene levels become important when one considers that some plant responses are seen at ethylene levels as low as 20 ppb and most plant responses are half-maximally affected at a concentration of 100 ppb. The concentration of ethylene as a dissolved gas in water at room temperature in equilibrium with 100 ppb ethylene in the atmosphere is approximately 1×10^{-10} M. This is several orders of magnitude lower than the concentration at which most other plant growth regulators are active.

In many, but not all, of the processes ethylene has been found to stimulate or attenuate, it does so as an endogenous plant hormone. These range from stimulating seed germination to causing fruit ripening and senescence. The manifold effects include:

1 Horticulture Department, Michigan State University, East Lansing, Michigan 48824, USA

Breaking of seed and bud dormancy	Flower initiation
Root initiation	Modification of flower sex
Hypocotyl hook formation and strengthening	Flower and fruitlet thinning
Stem strengthening	Fruit growth stimulation
Internode shortening	Fruit ripening, abscission, and senescence
Lateral branching	Fruit and leaf degreening
Leaf epinasty	Leaf abscission and senescence

Ethylene effects thus span all phases of plant development. Examples of important applications of ethylene in agriculture involving effects in different developmental phases will be discussed briefly.

Seed Germination and Breaking of Dormancy

Dormancy of certain seeds requiring light for germination can be broken by ethylene treatment. Ethylene also helps overcome high-temperature inhibition of germination of some seeds. Apart from breaking of dormancy, an absolute requirement for ethylene in germination has not been demonstrated. Bud dormancy of certain fruit trees requiring chilling is broken by ethylene. This is sometimes seen prematurely following heat or drought stress which induces ethylene synthesis, and causes some trees and shrubs to bloom in the fall rather than remain dormant until spring. Dormancy of buds on rhizomes of weeds such as quackgrass, Johnsongrass, and nutgrass is broken by ethylene, and this has some practical value in stimulating early growth for more effective herbicide application (15). Rooting of cuttings of some plants can be improved if the plants or cuttings are first treated with ethylene. Ethylene does not seem to play a direct role in these responses but rather may exert its effects by altering the distribution of auxin.

Seedling Development

Ethylene may inhibit or stimulate stem growth (17). It causes hypocotyl hook formation and maintenance. Mechanical impedance of the growth of the seedling by a soil crust causes an increase in ethylene production. This slows elongation and promotes radial growth which strengthens the stem and allows the seedling to break through the soil crust. Stem elongation of certain aquatic plants is stimulated by ethylene. Ethylene accumulates in the hollow stem of these submerged plants and the stem continues to elongate until the frond reaches the surface, at which time the gas is released and stem elongation ceases. Ethylene acts as a modulator of growth in these instances.

Vegetative Growth

There are numerous examples of ethylene effects in the vegetative growth phase (17). These include: growth inhibition, leaf epinasty, induction of rooting, release of apical dominance, promotion of tillering and rhizome growth, stem stiffening, and latex flow. Some effects are important agronomically or horticulturally. Inhibition of terminal growth improves the appearance and value of plants such as azalea and poinsettia by increasing lateral branching and making a more compact flowering plant. Lodging in small grains can be attenuated by ethylene. This is an important commercial use of ethephon. Release from apical dominance by ethylene allows lateral and basal buds of Kentucky bluegrass turf plants to grow, helping to improve the stand.

Latex flow in rubber trees is stimulated by applying ethephon to the bark below the tapping cut (4). This markedly increases rubber yields by increasing latex flow, saving labor by reducing the tapping frequency, and by prolonging the life of the rubber tree by allowing shorter tapping cuts. This application has become the most important commercial use of Ethrel.

Flowering and Fruiting

Flowering of certain plants is promoted by ethylene. This is true of pineapple (18) and some ornamental bromeliads. In the case of pineapple, treatment of fields with ethephon induces uniform flower formation which is essential for scheduling harvest (10). Flower bud formation is indirectly stimulated in young fruit trees by suppressing vegetative growth with an ethephon application made early in the growing season. This helps to bring orchards into bearing as much as a year or two earlier then normal. By promoting abscission of flowers and fruits, ethylene can be used as a thinning agent for fruits and to remove nuisance fruits in lawns and gardens (10).

Female sex expression in cucurbits is promoted by ethylene. In fact, ethylene appears to play a definitive role in this respect (9). Promotion of female flowers and elimination of male flowers by ethephon is important in hybridizing to prevent self-pollination of some crops. In commercial cucurbit production the increased femaleness can also result in earlier and increased yields (10, 14).

Postharvest aspects of ethylene physiology are particularly important and have been widely investigated. While most of what has gone before dealt with effects of ethylene that contribute in a positive way to plant growth and development and accrue positively with respect to productivity or quality, most of the postharvest aspects of ethylene affect the commodity in a negative sense from the handler's standpoint. This is a natural consequence of ethylene's role in promoting senescence of plants and plant organs: a role that may have survival value to the species but may interfere in postharvest commodity preservation. A number of approaches are potentially useful and some are commercially employed to attenuate the effect of ethylene after harvest.

The role of ethylene in accelerating senescence of fruits, vegetables, and flowers has been recognized for many years (1, 12). Many of the storage and handling procedures that have been developed for these commodities directly or indirectly subdue the synthesis or action of ethylene in its natural role as a ripening or senescence promoting agent (11). These include the use of controlled atmosphere storage at refrigerated temperatures, which diminishes the rate of ethylene production by lowering the O_2 level to below 3%, and inhibits the action of ethylene by elevating the CO_2 in the storage atmosphere to 3%–5%. This technique is used to prolong the storage life and delay ripening of over 30 million bushels of apples annually in the USA and Canada alone. Inhibition of ethylene action by CO_2 is widely recognized in cut flowers such as carnations.

Recognition of the central role of ethylene has led to past attempts to reduce the concentration of ethylene surrounding stored commodities by ventilation, absorption, or destruction. However, the concentration of ethylene within the tissue is governed by its production rate and the diffusion rate from the tissue. Although the external ethylene concentration can be reduced to negligible levels, sufficient gas remains in most tissues to cause a response, since only as little as 50 ppb of ethylene is required to cause a half-maximal effect in some tissues (6).

Hypobaric or Low Pressure Storage

Studies of the gas exchange process in fruits by Burg and Burg (7) led to the observation that ethylene or CO_2 produced within the organ escapes through an air-filled pore or opening, such as a stoma or lenticel, more readily when the commodity is subjected to a low-molecular-weight gaseous atmosphere. Although the rate of gas production was not affected, gas diffusion was facilitated because the density of the media through which diffusion occurred was lowered. Gas diffusion coefficients are inversely proportional to the molecular weight of the diffusion media and to atmospheric pressure. Thus, by lowering the atmospheric pressure to 0.1 atmosphere, ethylene, CO_2, and other volatiles escape from the tissue 10 times more rapidly than into air at atmospheric pressure. This, together with the lowered O_2 supply, reduces the equilibrium concentration of ethylene and some other gasses produced within tissues by at least a factor of 10. These studies led to the development of hypobaric storage (8). This subject was recently reviewed (13).

In actual practice, and on a limited commercial scale, the commodity is placed in a vacuum-tight, refrigerated container and evacuated by a vacuum pump to the desired low pressure which, depending upon the commodity, may vary from 10 mm Hg to 76 mm Hg. The oxygen level varies in direct response to the absolute pressure. When the desired low pressure is obtained, fresh air is admitted to the chamber through a pressure regulator and then humidified. In the continuously ventilated partial vacuum, carbon dioxide, ethylene, and waste volatile by-products of metabolism rapidly diffuse out of the commodity and are flushed from the storage chamber. Hypobaric storage technology is being developed commercially by Grumman Allied Industries. Inc.

There are several consequences of ventilating a commodity at hypobaric pressure: (i) Oxygen supply is reduced and this reduces the respiration rate and ethylene synthesis. (ii) Ethylene, which the tissues produce and which causes them to ripen or senesce, is removed from the tissue and flushed out of the storage. Hence, deterioration is delayed. (iii) Other volatile substances produced by the fruit such as carbon dioxide, acetaldehyde, acetic acid, esters, and α-farnesene (a terpene) are swept from the fruit. Acetic acid and acetaldehyde have been implicated in the development of internal breakdown and α-farnesene has been related to superficial scald of some fruits.

Inhibitors of Ethylene Synthesis and Action

Another approach to attenuate ethylene effects is to reduce or inhibit the synthesis of ethylene in the tissue specifically or to block the action of the gas. Aminoethoxyvinylglycine (AVG), an analog of the antibiotic rhizobitoxine, is a rather specific inhibitor of ethylene biosynthesis from methionine (12). It blocks the conversion of s-adenosylmethionine to 1-aminocyclopropane-1-carboxylic acid (ACC), which is the immediate precursor of ethylene in plants (2). Processes such as fruit ripening and flower senescence are retarded by treatment with AVG, but commercial use of AVG should be approached with caution because of the broad range of other reactions that may be affected.

Ag^+ has been found to be a very potent inhibitor of ethylene action (3). It is effective at low concentrations when applied as a dip or foliar spray to the tissue, and it prevents ethylene action in a broad range of responses including fruit ripening and flower fading. Cost as well as potential toxicity to man will limit practical uses of Ag^+. A substituted benzothiadiazole also inhibits ethylene action in a number of responses (19). It apparently inhibits ethylene action (9), although ethylene synthesis can be stimulated by the compound.

In summary, ethylene is the gaseous plant hormone that is active in physiological and morphological processes at virtually all stages of plant development from seed germination through senescence. Environmental, mechanical, and chemical factors may attenuate or stimulate ethylene synthesis or action. Numerous important agronomic and horticultural practices have evolved and continue to be developed with increasing recognition of the role of ethylene in plant growth and development.

References

1. Abeles, F.B.: Ethylene in Plant Biology. London, New York: Academic Press 1973
2. Adams, D.O., Yang, S.F.: Proc. Natl. Acad. Sci. USA *76*, 170–174 (1979)
3. Beyer, E.M.: Plant Physiol. *58*, 268–271 (1976)
4. Blencoe, J.W.: World Crops *23*, 126–132 (1971)
5. Bukovac, M.J.: HortScience *6*, 385–388 (1971)
6. Burg, S.P.: Proc. Natl. Acad. Sci. USA *70*, 591–597 (1973)
7. Burg, S.P., Burg, E.A.: Physiol. Plant. *18*, 870–884 (1965)
8. Burg, S.P., Burg, E.A.: Science *153*, 314–315 (1966)
9. Byers, R.E., Baker, L.R., Sell, H.M., Herner, R.C., Dilley, D.R.: Proc. Natl. Acad. Sci. USA *69*, 717–720 (1972)
10. de Wilde, R.C.: HortScience *6*, 364–370 (1971)
11. Dilley, D.R.: J. Food Biochem. *2*, 235–242 (1979)
12. Lieberman, M.: Annu. Rev. Plant Physiol. *20*, 533–591 (1979)
13. Lougheed, E.C., Muir, D.P., Berard, Luce: HortScience *13*, 21–27 (1978)
14. Morgan, P.W.: Acta Hortic. *34*, 41–54 (1973)
15. Morgan, P.W.: In: Herbicides: Physiology, Biochemistry, Ecology. Audus, L.J. (ed), vol. I, pp. 255–280. London, New York: Academic Press 1976, 608 pp.
16. Poovaiah, B.W., Leopold, A.C.: Crop Sci. *13*, 755–758 (1973)
17. Pratt, H.K., Goeschl, J.D.: Annu. Rev. Plant Physiol. *20*, 541–584 (1969)
18. Traub, H.P., Cooper, W.C., Reece, P.C.: J. Am. Soc. Hortic. Sci. *37*, 521–525 (1939)
19. Van Daalen, J.J., Daams, J.: Naturwissenschaften *8*, 395 (1970)

Applied Uses of Growth Substances – Growth Inhibitors

G.L. STEFFENS [1]

Introduction

Plant growth regulating chemicals are used in agriculture to increase the efficiency of production. Production efficiency "standards" change as economic conditions, consumer needs, and desires change. When considering the present and future uses of plant growth regulators (PGR's), this fact becomes important. Yield of a crop per unit area is certainly important but yields must be considered also in relation to cost of production, which changes with demand and improved technology. The total world market for PGR's (including defoliants and desiccants) is projected to be about $ 150 million by 1984 – up from about $ 59 million in 1974 (1). Even though this 150% increase seems impressive, it represents only a little over 1% of the world pesticide sales which are projected to be almost $ 10 billion by 1984. In 1974 the total cost of treatment for plant defoliation or regulation in the U.S. was nearly $ 35 million (2). Hardy (3) associates several factors with the slow development of PGR's, but he also indicates that a number of factors suggest an improved future of them.

An evaluation of the PGR's most widely used in agriculture (3–5) shows that indeed a great many are used on horticultural crops (high-value crops) and most affect plant morphology rather than cause increased yields per se. Many are plant growth inhibitors – they inhibit growth of the whole plant or plant parts. Some plant growth inhibitors increase yields, but usually in an indirect way. Nevertheless, they all are designed to increase crop production efficiency.

Major Agricultural Uses of Growth Inhibitors

The seven chemicals or chemical types which constitute the major plant growth inhibitors in terms of economics, usage, or acreage treated are shown in Fig. 1. These make up a major portion of all PGR's used worldwide in agriculture. In addition, there are a number of chemicals with relatively specialized uses, and new ones are being evaluated.

---

1 Plant Hormone and Regulators Laboratory, U.S. Department of Agriculture, SEA-AR, Beltsville Agricultural Research Center, Beltsville, Maryland 20705, USA

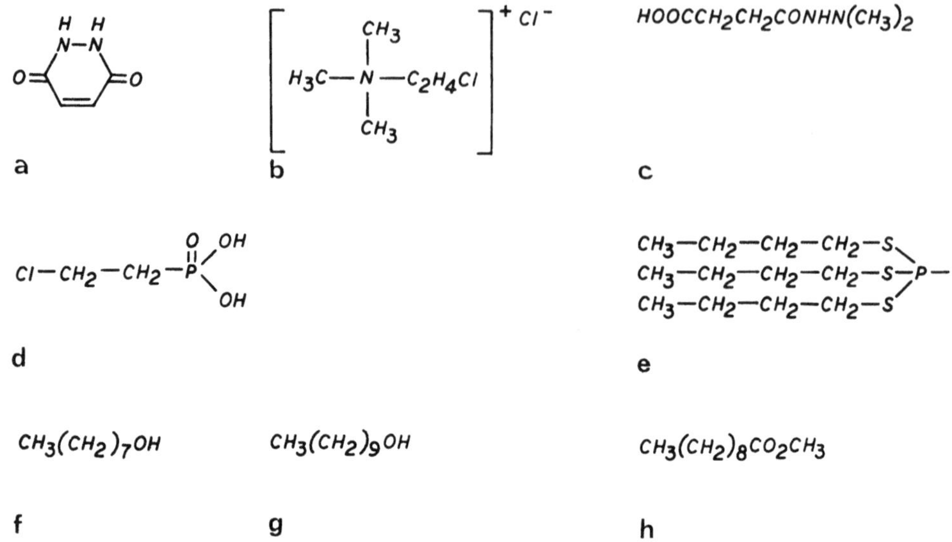

Fig. 1. Plant growth inhibitors. **a** Maleic hydrazide (MH) – 1,2-dihydro-3,6-pyridazinedione; **b** chlormequat – (2-chloroethyl)trimethylammonium chloride; **c** daminozide – butaneodioic acid mono(2,2-dimethylhydrazide); **d** ethephon – (2-chloroethyl)phosphonic acid; Organic Phosphate Defoliant, **e** tributyl phosphorotrithioite; Contact Sucker Control Agents, **f** 1-octanol, **g** 1-decanol; and Chemical Pinching Agent, **h** decanoic acid, methyl ester

Maleic Hydrazide (MH)

This is a systemic plant growth inhibitor (Fig. 1), first reported to be biologically active in 1949 (6) and by the late 1950's had begun to be extensively used. In 1976, nearly 4 million pounds of MH were produced in the U.S. (2). Over 3 million pounds were used to inhibit the growth of axillary buds or suckers on tobacco. The remainder, about 0.8 million pounds, was used to inhibit the sprouting of potatoes and onions during storage and to inhibit growth of trees, shrubs, and grasses; the largest portion (0.7 million pounds) was for potato sprout control.

The terminal flower or bud is usually removed (topping) in commercial tobacco production to increase size of upper leaves. This operation triggers the development of suckers. To maintain tobacco quality, producers removed suckers by hand, an activity which was costly and disagreeable. Tobacco producers now use MH as a systemic sucker control chemical to inhibit growth. About 80% to 90% of U.S.-grown tobacco is treated with MH and it is also used in many other tobacco-producing countries of the world. For the inhibition of potato sprouting, MH is sprayed on the vines 2 to 3 weeks after full bloom and is translocated to the tubers so that sprouting can be controlled for up to 11 months in storage. As many as 250,000 acres of potatoes are treated with MH annually, especially those grown for chipping and for processing. To inhibit the sprouting of onions in storage, MH is applied to the foliage when bulbs are mature but when leaves are still green, to allow translocation to the bulbs. A large portion of the onions grown for storage are treated with MH so that they can be kept in good condition for 6 months. Limited amounts of MH are used along with traditional trimming methods to restrict the size of mature trees and shrubs under power lines, along city streets and highways, in parks and in other landscaped areas. Also, MH is used to inhibit the growth of turf grasses along highways, especially in areas difficult to mow.

When applied as a spray to plants, MH is taken up by the leaves and is readily translocated in both the xylem and phloem tissue resulting in distribution throughout the plant (7–9). Recent [^{14}C]-MH labeling studies (10) show that MH moves without conversion. Because residues of MH in treated crops may be high in some instances, the risks and benefits of MH are being evaluated by the U.S. Environmental Protection Agency under the RPAR (rebuttable presumption against registration) process (11).

Even though the mode of action of MH has been studied for nearly 30 years, the biochemical events through which it affects plant development are still not well understood (7–9, 12). Soon after MH was found to be an effective plant growth inhibitor, it was shown to inhibit cell division. MH may inhibit DNA and RNA synthesis (13), inhibit mitosis by reaction with sulfhydryl groups (14), and inhibit uracil uptake into cells as well as become incorporated into RNA (15). Other workers (8, 9, 16) have shown that such physiological processes as transpiration, photosynthesis, and respiration are inhibited by MH. It is strongly bound to protein, and affects a number of enzyme systems.

Chlormequat

Chlormequat is a quaternary ammonium plant "growth retardant" with the structure shown in Fig. 1. The growth retardants inhibit internode elongation in plants without inhibiting the function of the apical meristem and without causing loss of apical dominance (17, 18). Growth retardants usually do not cause formative effects, whereas chemical inhibitors which disrupt cell division and affect apical dominance usually do. The first plant growth retardants were found in about 1950 (19, 20), and Tolbert reported on the activity of chlormequat in 1960 (21). Data on the total amount of chlormequat used annually in the U.S. or in world agriculture are difficult to obtain. In the U.S., federal law prohibits publishing such statistical data when only one or a few companies are involved in manufacture (2). However, Nickell (4) indicates that chlormequat is among the plant growth regulators most widely used in the world.

A major use of chlormequat is on cereals to prevent lodging, especially under heavy nitrogen fertilization and rainfall (22–25). Namokar (24), in a recent review on the potential agricultural uses of chlormequat, estimates that 20% to 25% of the wheat in Germany and Austria is treated for this purpose, whereas Nickell (4) estimates that 50% is treated. Shortening and strengthening of the stems, due to increased cell wall thickness and increased stem diameter, prevents lodging and decreases yield losses. Chlormequat is more active on wheat and oats than on rye and barley. It is not used on cereals in the U.S. because climatic conditions and fertilization practices in the cereal-producing areas are not conducive to lodging. In the U.S., chlormequat has been widely used for height reduction of poinsettias, azaleas, and other ornamentals. Of 88 ornamental species evaluated, Cathey (26) found that 21 responded to chlormequat sprays or soil treatment.

Most of the research on uptake and metabolism of chlormequat dates to the 1960's. It is only slowly taken up by wheat leaves (24) and is broken down in moist soil by microorganisms. Decomposition rate is affected by temperature; little decomposition occurs below 4°C or above 40°C (27). The mechanism of action of the plant growth retardants, including chlormequat, has been associated with the inhibition of synthesis or action of gibberellin. Early work (28, 29) showed that chlormequat blocked gibberellin biosynthesis in the fungus *Fusarium moniliforme* and in higher plants. In most cases application of gibberellins prevents inhibition of shoot elongation by chlormequat and even reverses its effects. More recent work indicates that the enzyme kaurene synthetase may be involved (30, 31) and that chlormequat changes the proportion of free to bound forms of gibberellin (32). Chlormequat may affect IAA metabolism (33, 34), stimulate ethylene production (35), and affect sterol synthesis (36).

Daminozide

Daminozide (Fig. 1), another growth retardant, was first reported to be active by Riddell et al. in 1962 (37). The growth retardant effects obtained with daminozide are similar in many respects to those found with chlormequat. It is mainly used on tree fruits, ornamentals, and to a certain extent on peanuts. No data are available on amounts manufactured, but about 30% to 35% of the apples produced in the U.S. and a large percentage in England and European countries are from treated trees. Large numbers of greenhouse-grown azaleas, poinsettias, and chrysanthemums are treated, as are other nursery and bedding plants.

Daminozide effectively reduces vegetative growth and promotes flower bud initiation in apples (22, 38) and can delay fruit maturity, aid in control of fruit drop, increase firmness of fruit, reduce fruit size, and prolong storage life. In the U.S. it is used on cherries, peaches, nectarines, and on several varieties of grapes to increase yield and reduce growth rate. See also the report by N.E. Looney in this volume on uses of PGR's in apple production.

With respect to ornamentals, daminozide is used in the U.S. on pot chrysanthemums to produce compactly branched plants and to shorten flower stem for cut flowers; on azaleas to promote development of buds in greenhouse plants, and to produce compact nursery plants; and on poinsettias to reduce elongation. Cathey (26) found that of 88 plant species evaluated, 44 responded to daminozide, compared to 22 for chlormequat. It is used in the U.S. on tomatoes prior to transplanting and on peanuts to shorten vines and make them more erect and green, thus increasing yields. On nursery trees such as apple, pear, cherry, and plum, daminozide reduces growth and induces terminal bud formation without suppressing root growth.

Daminozide is water-soluble; its uptake is rapid over the first few hours and then declines sharply after about 8 h (39). As discussed by Dicks (25), daminozide is transported in both the phloem and xylem, and is not degraded within the plant (40). Studies suggest a relationship between daminozide concentration in stem tissue and stem growth inhibition, although Sachs and Mock (41) indicate an insignificant relationship between daminozide tissue level in elongating branches and inhibition of stem elongation. Decreases in concentration of daminozide in growing tissue are probably due to dilution by new tissue growth since it is highly persistent. In soil, daminozide is broken down rather rapidly by microbial action.

Initially daminozide was thought to act via its hydrolysis to 1,1-dimethylhydrazine (42), which could affect auxin concentration. Daminozide, like other growth retardants, was thought to block gibberellin biosynthesis in a manner similar to that of chlormequat (44). It was not active in the α-amylase barley endosperm assay (44), but its effect in retarding shoot growth could be at least partially overcome by gibberellins (18, 45). Based on work by Sachs and coworkers (41, 46), Dicks (25) considers that daminozide may be involved in the later stages of gibberellin biosynthesis, although it may also be involved in auxin metabolism (45).

Ethephon

The structure of ethephon is shown in Fig. 1; it releases ethylene at pH > 4 (47). It was first evaluated as a plant growth regulator in the mid-1960's and was found to cause responses similar to those produced by ethylene itself. This discovery made it practical to treat field-grown plants with ethylene. There are many physiological responses to ethylene (48–50), and a number of responses to ethephon have been reviewed (51). (See also the review by Dilley in this volume.)

Because of the many plant responses to ethephon, the concept of chemical plant growth regulation became a real possibility in the minds of many scientists, both basic and applied. At present, there are a large number of practical or potential uses for ethephon. Some are inhibitory processes, but others are not. Only a few will be considered here. In the U.S. ethephon is used for delaying spring bloom of sweet cherries and, when used on young, nonbearing apple trees, it reduces vegetative growth and increases bud development during the season, so increased flowering occurs the following spring. Ethephon, in combination with chlormequat, has shown promise for inhibiting plant height and lodging of cereal grains (51, 52) and is being used in some European countries for that purpose. Obviously, ethephon is widely used in many ways, especially to increase efficiency of mechanized handling and harvesting.

Organic Phosphate and Inorganic Defoliants

These are represented here by tributyl phosphorotrithioite (Fig. 1) and sodium chlorate ($NaClO_3$) respectively. Of the 13 million acres of cotton grown in the U.S., an estimated 5 to 6 million acres are treated with the phosphorotrithio-type defoliants (trithioite and trithioate), and about 2 million acres are treated with $NaClO_3$ (4, 5). The need for effective removal of leaves from cotton plants became important with the widespread use of mechanical cotton pickers. The defoliant harvest aids are usually applied to cotton 1 or 2 weeks before harvest. They induce leaf fall to facilitate the use of the rotating spindle type pickers which take only cotton from open bolls. Sodium chlorate, a strong oxidizing agent, is highly inflammable, so formulations usually include borates to reduce fire hazards. In the U.S. the organic phosphate cotton defoliants were developed in the late 1950's and early 1960's. They are now used in other countries where mechanical harvesting is practical and also to advantage where hand picking is still used.

Sodium chlorate can be absorbed through the leaves as well as the roots and is carried downward through the xylem, since it kills phloem tissue (53). The organic phosphate defoliants are more effective at higher temperatures and some difficulty has been encountered with effectiveness on drough-stressed plants, rapidly growing plants, and plants under cool conditions (54). Only those leaves are affected on which spray droplets fall. After about 2 to 6 h, the organic phosphates penetrate epidermal leaf cells, causing swelling and blistering (55), and within a few hours the chemical diffuses to the palisade cells. The mode of action of the cotton defoliants involves mechanisms that are associated with leaf abscission such as leaf injury, increased ethylene production, reduced auxin levels, and water loss (22, 56–58). There is little structural similarity among chemicals which cause defoliation, but many are oxidizing agents, and all cause leaf injury which probably initiates the abscission process. However, the exact mechanism of leaf abscission is not known.

Arsenic Acid Desiccant

H_3AsO_4 is used in the U.S. as a desiccant annually on over 2 million acres of cotton. It came into wide usage in areas of Texas and Oklahoma in the mid-1950's on cotton harvested by mechanical strippers. This type of machine removes or strips all the leaves. bolls, bracts, and side branches from the main stem of the plant. The desiccants cause the leaves to dry in 1 to 3 days after application so that the mechanical stripper (via metallic fingers, steel rollers or brushes) can remove the cotton along with "trash" in a once-over harvesting operation. To prevent deterioration before clean-up and ginning, the moisture content of the stripped burr cotton must be low. Also, to remove plant parts efficiently from the seed cotton, the trash must be dry as it goes through a series of clean-up steps prior to ginning.

Because of possible exposure to arsenic by applicators and gin workers, in addition to its possible environmental effects, the use of arsenic acid as a cotton desiccant is now undergoing evaluation by the U.S. Environmental Protection Agency under the RPAR process (59).

Arsenic acid applied to cotton leaves causes such rapid drying that no abscission layer is formed and the leaves remain on the plants. Miller (60, 61) has determined that an internal leaf-blade concentration of 100 ppm of arsenic is required for desiccation to take place. Only about 4% of the arsenic acid directed toward the cotton plant actually reaches the leaf cells to cause desiccation.

Contact Sucker Control and Pinching Agents

The effectiveness of hydrocarbons as sucker control agents in tobacco was found in 1949 by D. Anderson and F. Skoog at North Carolina State University. They used mineral oil as a carrier for NAA and other auxins in field tests of auxin as chemical desuckering agents and observed that the oil itself killed axillary buds within a few hours without injuring the stem or leaf tissue. Still more active chemicals are represented by the C_{10} fatty acid methyl ester and the C_8 and C_{10} fatty alcohols (Fig. 1), first shown to be effective contact inhibitors of sucker growth on tobacco in the mid-

1960's (62, 63) and then as pinching agents for ornamental and fruit crops (64). Of the fatty alcohols evaluated as tobacco sucker control agents, the C_8 and C_{10} alcohols were among the most effective (65). About 8 to 9 million pounds of the C_8 and C_{10} fatty alcohols are used in worldwide production of tobacco and of that total, about 5 million pounds are used in the U.S. The use of methyl esters as pinching agents has been limited to azaleas and certain other woody species (25).

Fatty alcohol emulsions are applied to tobacco plants as sprays and must drain down the stalk to contact and kill the immature suckers. The emulsions are phytotoxic to young buds but cause little or no injury to more mature tissue (66). The type and amount of surfactant are important in controlling selectivity (67). Because secondary suckers soon develop after initial application, a second application of the fatty alcohol or a systemic chemical such as MH is made (68). In the U.S. the sequential method of sucker control (contact followed by systemic) is widely used on flue-cured tobacco. In countries where MH is not used, the fatty alcohol contact-type agents are extensively used. To pinch ornamentals and woody species effectively, the fatty acid esters must kill terminal buds which stimulate axillary meristem development. The fatty acid derivatives are not translocated, and their action is probably by localized destruction of cell membranes and tissue kill. Fatty acid residues on tobacco were not detected 26 days after treatment in one study (69) and were found only at about 1 ppm in another (70).

Older Chemicals Which Have Limited Usage

A number of inhibitors have been developed over the years which have had rather specific uses or which have mostly been replaced by newer chemicals that are more effective, have a broader range of activity, or are easier to use. These older chemicals include chlorpropham (isopropyl *m*-chlorocarbanilate) (71); chlorphonium chloride [tributyl(2,4-dichlorobenzyl)phosphonium chloride] (17); and various forms of NAA (1-naphthalene-acetic acid) (22).

Newer Plant Growth Inhibitors

As a result of increased interest in the development of plant growth regulators by the chemical industry in the 1960's, quite an array of new chemicals has been evaluated or marketed since about 1970 (3). Most of them are inhibitors and some are chemically related or produce responses similar to those already on the market.

Chlorflurenol (morphactin)

Methyl 2-chloro-9-hydroxyfluorene-9-carboxylate is one of a group of related structures which was first developed as a plant growth inhibitor in the mid-to late-1960's (72). Such inhibitors are mainly used in California to retard growth of grasses and weeds in hard-to-mow areas and to retard hedges, shrubs, ground covers, and trees under utility lines, near highways, and in parks (73). These compounds inhibit apical growth and produce a wide range of morphological effects (25, 74).

Ancymidol

α-Cyclopropyl-α-(4-methoxyphenyl)-5-pyrimidinemethanol is a growth retardant developed in the early 1970's, and is used on chrysanthemums, lilies, poinsettias, and other ornamentals. Ancymidol has the widest spectrum of activity of any retardant and was active on 68 of 88 species evaluated (26). Its inhibitory activity can be reversed by gibberellin, and it was found to inhibit gibberellin-regulated growth. Hence it is thought to block gibberellin action (75), but it can also block gibberellin biosynthesis (76). Its effects are now being evaluated on other plants, including agronomic crops.

UBI-P293

2,3-Dihydro-5,6-diphenyl-1,4-oxathiin is a growth inhibitor released for evaluation in about 1973. Cathey (77) indicates that it acts as a localized blocker of cell division and cell expansion. It inhibits stem growth and may inhibit both auxin- and gibberellin-induced growth (78). It is being tested for pinching and shortening ornamentals, delaying flowering in peaches, and for controlling growth of other plants such as sugar beets, sugarcane, potatoes, apple trees, and beans.

Melfluidide

N-[2,4-Dimethyl-5-[[(trifluoromethyl)sulfonyl]amino]phenyl] acetamide has been under development since about 1974, and received U.S. label registration in 1978. Melfluidide is used for regulating the growth of turf grasses and broadleaf vegetation. It reduces mowing requirements by regulating growth and suppressing seedhead formation (79). It is also being evaluated in the field as a sugarcane ripener as well as for the inhibition of growth of trees and woody ornamentals after trimming.

Dikegulac

2,3:4,6-bis-*O*-(1-methylethylidene)-α-*L*-xylo-2-hexulofuranosonic acid was first reported to be a growth retardant in 1974 (80). It acts on a wide range of plants and has received label registration in the U.S. for systemic chemical pinching and growth regulation of ornamentals. It has been widely evaluated in European countries and is used there on a number of plant species (81). Dikegulac has been shown to interact with gibberellins, auxins, kinetin, and ethylene (82) and to inhibit DNA synthesis, but its initial effects may be on cell membranes (83).

Piproctanylium Bromide

1-Allyl-1-(3,7-dimethyloctyl)-piperidinium bromide, a quaternary ammonium growth retardant, was first evaluated in 1976 on several greenhouse-grown ornamentals (84). It is used in the U.S. on chrysanthemums, bedding plants, and other ornamental species and is being used in Europe for these purposes. It affects certain enzyme systems, increases chlorophyll and protein levels, and likely is involved in gibberellin metabolism (85, 86).

Mepiquat Chloride

1,1-Dimethylpiperidinium chloride is also a piperidinium compound introduced in the mid-1970's (87–89) as an inhibitor of vegetative growth of cotton. This chemical reduces vertical and lateral growth by reducing internode length on main stems and on side branches. Height of cotton under some conditions was reduced 30%, and treated plants matured as much as 10 days earlier than controls. In many cases these effects result in increased yields and reduced boll rot. It is being further evaluated in the U.S. on cotton in the Experimental Use Permit Program. Mepiquat chloride treatment has caused increased calcium and potassium levels in leaves, and increased CO_2 fixation. Increased chlorophyll and leaf thickness and reduced area per leaf have also been reported. It is quite active on cotton; application dosages are only 10 to 40 g per acre.

Fosamine

Ethyl hydrogen (aminocarbonyl)phosphonate is used to control and suppress growth of a number of woody brush species on rights-of-way of railroads, utilities, pipelines, and other noncrop land areas (90). When applied during the 2-month period before fall coloration of leaves, it is absorbed by foliage, stems, and buds with no apparent effect until the following spring. Bud development is either inhibited or completely prevented so that only very small leaves are produced. Very susceptible woody species do not produce new leaves and eventually die (91). The chemical is rapidly decomposed in soil by microorganisms.

CGA-41065

N-Ethyl-*N*-(2-chloro-6-fluorobenzyl)-2′,6′-dinitro-4′-trifluoromethylaniline is being evaluated for the control of axillary bud growth on tobacco (92, 93) and is effective for this purpose over a wide range of environmental conditions and tobacco types. CGA-41065 acts as a contact-systemic agent, since buds must be contacted for inhibition of growth, but they are not usually killed.

UBI-N252

2,3-Dihydro-5,6-dimethyl-1,4-dithiin 1,1,4,4-tetraoxide is being evaluated as a defoliant on cotton, grapes, beans, and nursery stock and as a vine desiccant on potatoes. It was first evaluated in about 1975 and has been effective on cotton (94) and potatoes (95).

Thidiazuron

N-Phenyl-N′-1,2,3-thiadiazol-5-ylurea has been evaluated since about 1977 as a defoliant and a regrowth inhibitor on cotton. It is especially interesting because it does both. From mode of action studies in progress it appears to be rather specific for cotton. Further studies are being conducted in the U.S. under the Experimental Use Permit Program (G.W. Cathey, Stoneville, MS, pers. comm.).

Etacelasil

2-Chloroethyl-tris (2′-methoxy-ethoxy)-silane is an ethylene-releasing agent evaluated for promotion of abscission of olives in Europe (27, 96). Its effects are similar in many respects to those obtained with ethephon. Its release of ethylene gas is pH dependent.

Nitrapyrin

2-Chloro-6-(trichloromethyl)pyridine selectively inhibits *Nitrosomonas* (3, 53), thus inhibiting nitrification. It is applied with various ammonium fertilizers.

There are a number of new synthetic and natural growth regulators that have not yet been evaluated thoroughly enough to judge their effectiveness. The chemistry of most of these has not been revealed. Because of development costs, especially of long-term toxicology studies, chemical companies are making an effort to expand the uses of chemicals already on the market. Mixtures of active chemicals are being investigated to some extent. New plant growth regulators that will significantly and consistently increase yields of major agronomic crops are still the goal of this search.

References

1. Farm Chem. *140* (9), 38–43 (1975)
2. Fowler, D.L., Mahan, J.N. (Preparers): The Pesticide Review 1977. U.S. Dep. Agric. ASCS 1978
3. Hardy, R.W.F.: In: Plant Regulation in World Agriculture. Scott, T. (ed.), pp. 165–206. New York: Plenum Press 1979
4. Nickell, L.G.: Chem. Eng. News *56*, 18–34 (1978)
5. Morgan, P.W.: Bot. Gaz. (in press, 1980)
6. Schoene, D.L., Hoffman, O.L.: Science *109*, 588–590 (1949)
7. Nooden, L.D.: Physiol. Plant. *32*, 157–160 (1969)

8. Zukel J.W. (Compiler): MHIS No. 8, Naugatuck Chem. Div., U.S. Rubber Co., Naguatuck, CN 1957
9. Zukel, J.W. (Compiler): A Literature Summary on Maleic Hydrazide 1957–1963. Naugatuck Chem. Div., U.S. Rubber Co., Naugatuck, CN 1963
10. Frear, D.S., Swanson, H.R.. J. Agric. Food Chem. *26*, 660–666 (1978)
11. U.S. Environmental Protection Agency: Federal Register 42 (No. 208, Part IV): 56920–56935 (1977)
12. Bush, L.P., Sims, J.L.: Physiol. Plant. *32*, 157–160 (1974)
13. Nooden, L.D.: Plant Cell Physiol. *13*, 609–621 (1972)
14. Hughes, C., Spragg, S.P.: Biochem. J. *70*, 205–212 (1958)
15. Coupland, D., Peel, A.J.: Planta *103*, 249–253 (1972)
16. Baker, J.E.: Physiol. Plant. *14*, 76–88 (1961)
17. Cathey, H.M.: Annu. Rev. Plant Physiol. *15*, 271–302 (1964)
18. Sach, R.M., Hackett, W.P.: HortScience *7*, 440–447 (1972)
19. Mitchell, J.W., Wirwille, J.W., Weil, L.: Science *110*, 252–254 (1949)
20. Wirwille, J.W., Mitchell, J.W.: Bot. Gaz. *111*, 491–494 (1950)
21. Tolbert, N.E.: J. Biol. Chem. *235*, 475–479 (1960)
22. Weaver, R.J.: Plant Growth Substances in Agriculture. San Francisco: Freeman 1972
23. Wareing, P.F.: Outlook Agric. *9*, 42–45 (1976)
24. Namokar, J.: Pesticides *11*, 53–57 (1977)
25. Dicks, J.W.: Outlook Agric. *9*, 69–75 (1976)
26. Cathey, H.M.: HortScience *10*, 204–216 (1975)
27. Abdel-Rahman, M. (ed.): Plant Growth Regulator Handbook. Plant Growth Regulator Working Group 1977
28. Kende, H., Ninnemann, H., Lang, A.: Naturwissenschaften *50*, 599–600 (1963)
29. Harada, H., Lang, A.: Plant Physiol. *40*, 176–183 (1965)
30. Fall, R.R., West, C.A.: J. Biol. Chem. *246*, 6913–6928 (1971)
31. West, C.A., Fall, R.R.: In: Plant Growth Substances 1970. Carr, D.J. (ed.), pp. 133–142. Berlin, Heidelberg, New York: Springer 1972
32. Snir, I., Kessler, B.: Planta *125*, 73–75 (1975)
33. Kuraishi, S., Muir, R.M.: Plant Physiol. *38*, 19–24 (1963)
34. Kabarity, A., Khodary, S.: Cytologia *41*, 429–444 (1976)
35. Abdel Gawad, H.A., Ketellapper, H.J.: Plant Physiol. Suppl. *47*, 16 (1971)
36. Wain, R.L.: Proc. R. Soc. London Ser. B *191*, 335–352 (1975)
37. Riddell, J.A., Hageman, H.A., J'Anthony, C.M., Hubbard, W.L.: Science *136*, 39 (1962)
38. Luckwill, L.C.: Outlook Agric. *9*, 46–51 (1976)
39. Dicks, J.W.: Ann. Appl. Biol. *72*, 313–326 (1972)
40. Peisker, K., Sternkopf, G.: Arch. Gartenbau *24*, 327–333 (1976)
41. Sachs, R.M., Mock, T.: Bull. Plant Growth Regul. Work.Group *2*, 36–37 (1974)
42. Reed, D.J., Moore, T.C., Anderson, J.D.: Science *148*, 1469–1471 (1965)
43. Ninnemann, H., Zeevaart, J.A.D., Kende, H., Lang, A.: Planta *61*, 229–235 (1964)
44. Paleg, L., Kende, H., Ninnemann, H., Lang, A.: Plant Physiol. *40*, 165–169 (1965)
45. Audus, L.J.: Plant Growth Substances, Vol. 1: Chemistry and Physiology. New York: Barnes and Noble 1972
46. Ryugo, K., Sachs, R.M.: J. Am. Soc. Hortic. Sci. *94*, 529–533 (1969)
47. Yang, S.F.: Plant Physiol. *44*, 1203–1204 (1969)
48. Pratt, H.K., Goeschl, J.D.: Annu. Rev. Plant Physiol. *20*, 541–584 (1969)
49. Abeles, F.B.: Ethylene in Plant Biology. London, New York: Academic Press 1973
50. Proc. Symp. Ethylene-Fruit Abscission 1971. HortScience *6*, 354–392 (1971)
51. deWilde, R.C.: HortScience *6*, 364–370 (1971)
52. Abdel-Rahman, M.: Proc. 4th Annu. Meet. Plant Growth Regul. Work.Group, 1977, p. 29
53. Thompson, W.T.: Rev. Agricultural Chemicals, Book II. Fesno, CA: Thompson 1977
54. Thompson, W.T.: Rev. Agricultural Chemicals. Book III. Fresno, CA: Thompson 1976/77
55. Miller, C.S., Aboul-Ela, M.M.: Proc. Cotton Defoliation and Physiol. Conf. 1965, p. 93
56. Cooper, W.C., Rasmussen, G.K., Rogers, B.J., Reece, P.C., Henry, W.H.: Plant Physiol. *43*, 1560–1576 (1968)

57. Morgan, P.W., Beyer, E.M., Jr., Lipe, J.A., McAfee, J.A.: Beltwide Cotton Prod. Res. Conf. Proc. 1971, pp. 42–43
58. Beyer, E.M., Wittenbach, V.A., Sweetser, P.B.: Beltwide Cotton Prod. Res. Conf. Proc.1975, p. 52
59. U.S. Environmental Protection Agency 1978. Federal Register *43*, 48157 (1978)
60. Miller, C.S.: Beltwide Cotton Prod. Res. Conf. Proc. 1968, p. 62
61. Miller, C.S.: Beltwide Cotton Prod. Res. Conf. Proc. 1969, p. 126
62. Tso, T.C.: Nature (Lond.) *202*, 511–512 (1964)
63. Tso, T.C., Steffens, G.L., Engelhaupt, M.E.: J. Agric. Food Chem. *13*, 78–81 (1965)
64. Cathey, H.M., Steffens, G.L., Stuart, N.W., Zimmerman, R.H.: Science *153*, 1382–1383 (1966)
65. Steffens, G.L., Tso, T.C., Spaulding, D.W.: J. Agric. Food Chem. *15*, 972–975 (1967)
66. Nelson, P.V., Poteet, L.Z.: Florists Rev. *145* (3752), 19, 50–53 (1969)
67. Steffens, G.L., Cathey, H.M.: J. Agric. Food Chem. *17*, 312–317 (1969)
68. Collins, W.K., Hawks, S.N., Kittrell, B.U.: Tob. Sci. *14*, 65–68 (1970)
69. Tancogne, J.: Le Tabac Ann., Sect. 2 *11*, 231–238 (1974)
70. Tso, T.C., Chu, H., DeJong, D.W.: Beitr. Tabakforsch. *8*, 241–245 (1975)
71. Marth, P.C., Schultz, E.S.: Am. Potato J. *29*, 268–272 (1952)
72. Schneider, G., Erdmann, D., Lust, S., Mohr, G., Niethammer, K.: Nature (Lond.) *208*, 1013 (1965)
73. Backhaus, R.A., Sachs, R.M., Hield, H.Z.: HortScience *11*, 578–580 (1976)
74. Schneider, G.: Annu. Rev. Plant Physiol. *21*, 499–533 (1970)
75. Leopold, A.C.: Plant Physiol. *48*, 537–540 (1971)
76. Coolbaugh, R.C., Hamilton, R.: Plant Physiol. *57*, 245–248 (1976)
77. Cathey, H.M.: J. Am. Soc. Hortic. Sci. *101*, 599–604 (1976)
78. Purohit, A., Meudt, W.J., Shanks, J.B.: Proc. 5th Annu. Meet. Plant Growth Regulat. Work. Group 1978, p. 260
79. Nielsen, A.P., Wakefield, R.C.: Proc. NE Weed Sci. Soc. *29*, 403–408 (1975)
80. Bocion, P.F., deSilva, W.H., Huppi, G.A., Szkrybalo, W.: Nature (Lond.) *258*, 142–144 (1975)
81. Bocion, P.F., deSilva, W.H., Walther, H.R., Grof, H.R.: Proc. Joint Br. Crop Prot. Conf. Br. Plant Growth Regul. Group Symp. 1978, pp. 195–201
82. Bocion, P.F., deSilva, W.H.: In: Plant Growth Regulation. Pilet, P.E. (ed.)., pp. 189–198. Berlin, Heidelberg, New York: Springer 1977
83. Zilkah, S., Gressel, J.: Planta *145*, 273–278 (1979)
84. Huppi, G.A., Bocion, P.F., deSilva, W.H.: Experientia *32*, 37–38 (1976)
85. Knypl, J.S.: Experientia *33*, 725–726 (1977)
86. Menhenett, R.: Ann. Appl. Biol. *87*, 451–463 (1977)
87. Willard, J.I., Kupelian, R.H., Schott, P.E.: Beltwide Cotton Prod. Res. Conf. Proc. 1977, p. 69
88. Cothren, J.T., Nester, P.R., Stutte, C.A.: Proc. 4th Annu. Meet. Plant Growth Regul. Work. Group 1977, p. 204
89. Gausman, H.W., Rittig, F.R., Namken, L.N., Rodriguez, R.R., Escobar, D.E., Garza, M.V.: Proc. 5th Annu. Meet. Plant Growth Regul. Work. Group 1978, pp. 137–145
90. Hernandez, T.J., Hudson, W.H., Rushing, T.T.: Proc. NE Weed Sci. Soc. *29*, 325 (1975)
91. Weigel, R.C., Beyer, E.M., Jr., Riggleman, J.D.: Proc. 5th Annu. Meet. Plant Growth Regul. Work.Group 1978, pp. 250–251
92. Wilcox, M., Chen, I.Y., Kennedy, P.C., Li, Y.Y., Kincaid, L.R., Helseth, N.T.: Proc. 4th Annu. Meet. Plant Growth Regul. Work.Group 1977, pp. 194–200
93. Kennedy, P.C., Seltmann, H., Atkinson, W.O., Whitty, E.B., Wilcox, M.: Proc. 5th Ann. Meet. Plant Growth Regul. Work. Group 1978, pp. 172–177
94. McIntire, S., Lambert, B., Brewer, A.D.: Proc. 4th Annu. Meet. Plant Growth Regul. Work. Group 1977, p. 202
95. Bohne, P.: Proc. 4th Annu. Meet. Plant Growth Regul. Work.Group 1977, p. 172
96. Fory, W., Fischer, H.P., Geiger, M., Furer, R.: 4th Int. Congr. Pest. Chem. Abstr. II-455 (1978)

Growth Regulator Use in Commercial Apple Production

N.E. LOONEY [1]

Introduction

Modern apple production is an intensive and expensive agricultural pursuit. It occupies valuable land which could be used for many other agricultural or nonagricultural purposes; establishing the orchard requires a considerable capital outlay for trees, usually an expensive irrigation system, and a wide range of specialized orchard equipment; and inevitably, several years of orchard maintenance precede the first economic return. Because most orchard systems assume a long productive orchard life, very important decisions must be made before planting and early in the life of the orchard. Decisions about tree spacing, rootstocks, cultivars, and the distribution of pollinizers will reward, plague, and limit for many years.

Perhaps the relatively rapid development and the widespread use of growth regulator techniques by scientists and orchardists relate to this long production cycle. Orchard managers are understandably reluctant to remove producing trees even if they begin to crop biennially, exhibit poor fruit quality, or develop a preharvest fruit drop problem. Growth regulators permit quicker solutions.

An impressive array of growth regulator techniques is used in commercial apple production world-wide. Tree propagators and nurserymen use auxins to promote rooting of hardwood and softwood cuttings, abscission-promoting compounds to accelerate autumn defoliation of nursery stock, and growth retardants and branching agents to modify tree shape. Growth regulators are used in young commercial orchards to encourage tree branching, to discourage unwanted sucker growth, and to promote flowering and fruit set. Mature trees are treated to thin the crop, to prevent harvest drop, and to improve fruit color, shape and internal quality; harvest date can be advanced or delayed; and root suckers and "water sprouts" are controlled with auxinic paints or sprays.

This report will briefly describe the growth regulator techniques presently important in commercial apple production and others that are in advanced stages of development. Also, some unsolved production and fruit quality problems are discussed within the framework of new growth regulator development opportunities.

---

1 Agriculture Canada Research Station, Summerland, British Columbia, Canada, VOH 1Z0

Tree Propagation and Nursery Management

Rooting Hardwood Cuttings

Commercial apple cultivars are propagated asexually by grafting or budding a scion piece onto an apple seedling or onto a size-controlling clonal rootstock. Producing the clonal rootstock layers has traditionally involved a stool bed layering technique. However, considerable progress has been made in recent years toward reliable rooting of hardwood cuttings, taken in early spring from a vigorous hedge of the rootstock cultivar (1, 2), or from young nonbranched trees of commercial scion varieties treated the previous summer with the growth retardant, daminozide (3). Reduced cost and greater flexibility in meeting the continuously changing demand for various clones are important advantages of this system.

The procedure involves dipping the proximal end of pencil-thick shoots about 60 cm long into an alcoholic solution of about 2500 ppm of 4-(indolyl-3) butyric acid (IBA) (1, 2). These shoots are then planted in beds with carefully controlled bottom heat (3), and rooting is achieved within two to four weeks. The rooted cuttings are grown under protection for a period of time and eventually lined out in a nursery.

Propagation by Meristem-tip Culture

Excellent progress is now being made towards achieving rapid in vitro propagation of apple rootstocks (4) and rooted trees of several commercial cultivars (5, 6). The technique appears destined to become commonplace within a few years. It promises to revolutionize nursery tree production and may dramatically increase orchard productivity levels by making the high-density planting concept practical for more orchardists. For example, McIntosh Wijcik, a super compact mutant of McIntosh, does not need a dwarfing rootstock to keep tree size small. If self-rooted trees could be obtained cheaply enough, growers would undoubtedly experiment with ultra-high-density planting systems such as the meadow orchard (7, 8) or other short cycle systems.

The procedure described by Lane (6) is to culture surface-sterilized shoot meristem tips on a Murashige and Skoog (9) medium containing about 5×10^{-6} M benzyladenine. The cytokinin promotes shoot proliferation of this mother culture, and these shoots are "harvested" periodically and planted on a similar medium containing 1×10^{-5} M naphthaleneacetic acid (NAA) to stimulate root initiation. A third transfer onto a growth regulator-free medium optimizes root growth and the complete plants are potted and grown into saleable trees using normal nursery procedures. Desirable clones can be rapidly multiplied at any time of year from mother cultures maintained in a small growth cabinet.

Defoliating Nursery Trees

Finished nursery trees intended for fall or spring orchard planting are dug in the autumn after the trees are dormant and defoliated, and before the soil freezes. Since nursery trees are encouraged to grow vigorously, natural leaf fall is delayed. This annual problem has led to the search for an effective and reliable chemical defoliant. Potassium iodide (10) and bromodine (11) have been used by some nurserymen, but these often are only marginally effective.

The most promising recent development is the use of a surfactant or wetting agent (Dupont WK: DWK) with ethephon (12). Three sprays are applied within a two week period in late October. A spray of 1% DWK plus 100 ppm ethephon is followed by two sprays of 1.5% DWK plus 200 ppm ethephon. Defoliation is advanced several critical weeks by this procedure but uneven results and occasional tree damage have been reported. The search for a safe and effective nursery tree defoliant continues.

Managing the Young Orchard

Chemical Pruning and Training

A subject of recent research interest in Europe and North America aims at encouraging young orchard trees or nursery trees to develop more lateral branches with wider crotch angles (13–15). Well-branched trees are reported to be more productive early in the life of the orchard (16, 17).

Encouraging early results were obtained with cytokinin or gibberellin plus cytokinin treatment (15, 18, 19), but materials such as Off-Shoot-0 (Proctor and Gamble Inc.) and NC 9634 (Fisons Ltd.) are proving to be more effective (20, 21). Off-Shoot-0 is a proprietary mixture containing methyl esters of C_6 to C_{12} straight-chain fatty acids whereas NC 9634 is [(3-phenyl-1,2,4-thiadiazol-5-yl)thiol]acetic acid. These and other materials (21), applied early in the growing season, selectively kill or temporarily check the growth of actively growing shoot tips. They interfere with apical dominance and lateral branches develop.

Another problem is to encourage the lateral branches to develop where they can effectively serve as permanent scaffold limbs. A procedure developed and used in England involves spraying the bottom 60 cm of young trees with a non-translocated growth retardant (Amex A-820, Amchem Inc.). This treatment, applied in early June, results in improved branching above 60 cm (22).

Growth regulators which promote branching and spur development may also help to correct certain growth habit problems in mature apple trees such as bare, unproductive branch sections. A chemical to encourage the development of small lateral branches, which will become fruiting spurs, could be very useful.

Promoting Flower Initiation

Flowering and fruiting of young apple trees is either encouraged or discouraged, depending upon tree age and tree spacing. Excessive crops on small trees adversely affect tree shape and growth rate. At present this situation is avoided by maintaining a high level of tree vigor. There is currently no growth regulator technique to effectively suppress flowering of apple.

However, the more common problem is that highly vigorous young trees fail to flower and fruit reliably. Particularly in high-density plantings, early cropping is desired to help slow tree growth and to begin repaying the high orchard establishment cost. Several growth regulators promote flowering of apple trees (Table 1), and daminozide and ethephon are registered for this purpose in the United States.

Table 1. Plant growth regulators which promote flowering of apple trees

Chemical	References
Daminozide; succinic acid, 2,2-dimethylhydrazide	23–26
Ethephon; (2-chloroethyl)phosphonic acid	24, 25
TIBA; 2,3,5-triiodobenzoic acid	27
Zeatin; 6-(4-hydroxy-3-methyl-2-butenylamino)purine	26
Benzyladenine; 6-benzylaminopurine	26

Since flower initiation on spurs occurs early in the growing season of the year preceding anthesis, spray timing is important. Daminozide is applied about two weeks after full bloom. Ethephon is applied four to five weeks after full bloom (as it can otherwise cause excessive fruit thinning). Daminozide is often applied with one of the post-bloom, fruit-thinning sprays. Daminozide and ethephon are occasionally combined on non-cropping trees.

Promoting Fruit Set

The postbloom period of fruit abscission commonly referred to as the "June drop" is often severe enough on young vigorous trees to reduce final crop load excessively. In England a mixture of GA_3 and NAA is registered for use on Cox's Orange Pippin apples to improve set and yield. Research at Wye College has shown that a three-chemical mixture of GA_3, N'N'-diphenylurea (DPU) and 2-naphthoxyacetic acid (NOXA) is even better (28), but registration has not yet been achieved.

Early results in Washington suggest that aminoethoxyvinylglycine (AVG), a rhizobitoxin analog, can essentially eliminate the June drop of Delicious apples (M.W. Williams, pers. comm.). This result is consistent with the findings of Bangerth (29) with mature apples.

Crop Management in the Mature Orchard

Chemical Thinning

Williams (3) has recently reviewed the literature on chemical thinning of apples. This subject could easily constitute a graduate course in pomology and only a brief introduction can be provided here.

Selective removal of blossoms and small fruits with chemical sprays is used by apple growers in all producing regions but is perhaps most highly developed in western North America. Growers in these areas produce primarily dessert-quality fruit which is transported to distant markets. Early thinning increases the size of the remaining fruits and improves other aspects of fruit quality such as sugar and acid levels and general fruit appearance by balancing crop load with leaf area. It also adjusts the crop to the physical tree framework to maximize light exposure and to facilitate spraying and hand harvesting.

Another, and often the most important reason for flower and fruit thinning, relates to the biennial bearing tendency of many apple cultivars. For example, Golden Delicious trees blooming profusely and cropped excessively in 1980 will initiate very few flowers for the 1981 crop. The rather profound influence of chemical thinning agents on alleviating biennial bearing of apples can be seen by examining national or regional production records before and after thinning practices came into widespread use. Before 1949 biennial bearing was a problem throughout the United States. After 1949 it largely disappeared.

To maintain annual cropping, a proportion of the fruiting spurs on an apple tree must "rest" each year. Thus, a "snowball" bloom, where essentially every spur flowers, is a crisis situation in crop management. The growth regulator program used by a western apple grower when this situation occurs may include all of the following sprays:

a) Liquid or dry powder formulations of sodium 4,6-dinitro-ortho-cresylate (DNOC) are applied at or just before full bloom to reduce the number of flowers which are fertilized. The action is caustic or herbicidal to the male and female flower parts.

b) NAA or naphthaleneacetamide (NAAm) thinning sprays are applied about 10 days after full bloom. These materials accentuate the abscission of weak fruits during the normal "June drop" period. NAAm is applied at several times the NAA rate because of its attenuated auxinic activity.

c) Carbaryl, 1-naphthyl(N-)methylcarbamate, is also an effective apple thinner when applied within about three weeks of bloom. It is used in preference to NAAm on Delicious apples since the use of NAAm on that cultivar results in "pygmy" apples being retained until harvest. Occasionally carbaryl and NAAm are applied together.

d) Daminozide, a growth retardant, does not thin but has flower promoting activity. It is often added to one postbloom thinning spray as additional insurance against biennial bearing. Daminozide has other uses in the mature orchard which will be discussed later.

Preventing Preharvest Fruit Drop

Most apple cultivars ripening in the early fall require protection against preharvest fruit abscission. NAA and 2,4,5-trichlorophenoxypropionic acid (fenoprop) have been used for this purpose for more than 30 years (31). Of the two, fenoprop is the more effective, but its use can lead to tree injury in some climates. Thus, NAA has been the preferred material in regions where humidity is high.

These auxinic chemicals are applied just before fruit drop is expected and can give protection for up to three weeks. They appear to act by delaying senescence of the fruit abscission zone. However, they also promote fruit softening and thus shorten storage life because of the ethylene generated by the treated fruit (32, 33).

Daminozide applied early to mid-season also effectively controls preharvest fruit drop and does so without advancing fruit ripening (23, 32). In fact, harvest can be safely delayed. Because of this and other advantages, daminozide is the treatment preferred by many growers. The fact that it can be applied months ahead of harvest is particularly important in close-planted orchards that can become quite inaccessible nearer harvest time.

Daminozide appears to regulate apple fruit abscission by suppressing ethylene production by the mature fruit (34). Early results with AVG suggest a similar action (29), but whereas daminozide treatment also reduces fruit size, AVG does not appear to have this side effect.

Control of harvest drop in the future will likely be accomplished as one of the several effects of more sophisticated crop and tree management chemicals.

Improving Fruit Quality and Regulating Ripening

Pomologists have long known that growth regulators used for various other reasons can also affect fruit quality. However, the idea of developing growth regulators specifically to improve fruit quality is quite recent.

An important development in this regard was the introduction of daminozide to the apple growing industry in the mid-1960's. Applied in early to mid-summer it improves fall coloration of red apples (35, 36), reduces the development of water core (37) and superficial scald (36) on Delicious apples, and increases fruit firmness at harvest and to a lesser extent after extended cold storage (35). Ripe fruit flavor components such as aroma and juiciness are delayed in their development but not reduced. Sugar and acid levels are not altered.

The appearance of ethephon in the late 1960's presented new possibilities for regulating fruit ripening and improving certain aspects of fruit quality. Applied with NAA or fenoprop to control fruit drop, ethephon dramatically improves the red coloration of some apple cultivars and rapidly transforms "green" fruit to ripe fruit for early marketing (32, 38). The aroma, flavor, and color of ethephon-treated apples are indistinguishable from fruits harvested several weeks later. Thus, consumers purchasing first-of-the-season apples are presented with a more palatable product; growers are able to extend their marketing season in the "early" direction; and, where fruit color is an important grade consideration, growers benefit by the fruit coloring effect of this treatment.

The complete early-ripening procedure used for McIntosh apples in British Columbia also involves daminozide. Growers are encouraged to apply daminozide in early summer to help counter the fruit softening and fruit abscission effects of ethephon (38). The ethephon-auxin spray is applied in mid-to-late August and, depending somewhat upon weather conditions, the treated fruit is harvested 10 days later. This combination of growth regulators results in ripe apples firmly attached to the tree at the time of picking. They are highly colored, firm, and of excellent eating quality. After being graded, waxed, and packaged, they are sold out of cold storage within about two months of harvest. The main-crop McIntosh apples are harvested in mid-September and sometimes held until April in controlled atmosphere storage before they are sold.

The most recently introduced growth regulator treatment developed to improve an aspect of fruit "quality" aims at increasing the length to diameter ratio of Delicious apples (39–41). Marketed as Promalin by Abbott Laboratories, this proprietary mixture of benzyladenine (BA) and gibberellins A_4 plus A_7 (GA_{4+7}) is applied at full bloom at a rate of about 20 ppm each of BA and GA_{4+7}. The result is a Delicious apple which better meets market expectations with respect to shape.

The long-term marketing benefits of such an obviously cosmetic treatment are difficult to predict. However, this, or similar mixtures of growth regulators may prove to have other beneficial effects. For example, fruit size, color and yields may be increased (41) and the growth habit of certain problem cultivars improved (15).

Vegetation Management in Apple Orchards

Eliminating Unwanted Shoot Growth

Highly vigorous shoots (sprouts or suckers) arising from adventitious buds on roots, trunks or major scaffold branches are generally undesirable and must be removed by pruning or prevented from growing with growth regulator sprays or paints. The treatment described by Raese (42) for mature trees and marketed as Trehold by Amchem Products is a 1% solution of the ethyl ester of NAA which is either sprayed or painted (with 50% white latex) on the tree parts where unwanted shoots may occur. A single treatment controls sprout growth for several years. Others have used similar NAA treatments to control unwanted growth on younger trees or as an aid to tree shaping (43,44).

Retarding Shoot Growth

Both daminozide and ethephon are used to shorten internode length and thus reduce shoot extension growth of apple trees (24,45). However, to control tree size effectively with either chemical requires early application at high rates which, depending on the chemical used, can excessively reduce fruit size, crop load, and fruit quality.

Thus, apple growers are still awaiting a chemical treatment which effectively slows shoot growth without reducing yield or fruit quality. Potential benefits of such a treatment include reduced pruning costs, less shading of the tree interior because of less peripheral shoot growth, and the ability to tailor tree size to a particular orchard spacing or management system.

Opportunities for Development

It almost goes without saying that many other opportunities exist for developing new growth regulator techniques beneficial to apple producers. Morgan (46) recently outlined several research areas where the potential for new development is high and several of his points are particularly relevant to apple production. These include the development of growth regulators to modify plant response to the environment, to regulate nitrogen and mineral uptake and metabolism, and to modify plant metabolism to to improve quality.

Protection Against Climatic Hazards

By selecting appropriate cultivars, rootstocks, and management systems, apples can be successfully grown over quite a wide geographic range. However, insufficient winter chilling becomes the limiting factor in warm regions and excessive winter cold limits production to truly temperate zones. Spring frosts are a hazard in almost all apple producing areas.

Growth regulators have been used for decades to help apple growers overcome the problems associated with insufficient winter chilling but the results are characteristically erratic (47). A chemical treatment which reliably induced spring bud break of apple (and many other deciduous fruits) would be welcomed enthusiastically.

Similarly, a growth regulator treatment to protect against deep winter cold would be of great benefit in colder apple growing regions. Several cryoprotectants have shown promise in this regard (48), and we may soon see a commercial product developed. Growth regulators will probably find use in this area by affecting plant dormancy. Thus, a reasonable expectation might be that growth regulator treatments will be found to extend the period of natural cold resistance and that cryoprotectants will be used to guard against periods of exceptional cold. This subject has recently been reviewed by Howell and Dennis (49). A great deal of research activity is presently being directed towards regulating cold hardiness in perennial plants.

Finally, a logical approach to protecting flowers from spring frosts would be to delay flowering with growth regulators. While little success has been reported for apples (49, 50), some other tree species do appear amenable to this type of regulation.

The challenge is clear. New growth regulators are needed if we are to extend the climatic range suitable for apple production and effectively protect existing trees from some serious climate-related hazards.

Improving the Quality of Stored Apples

New growth regulator techniques are needed which delay senescence of apples held in cold storage. We appear to be approaching the limits to which fresh product life can be extended by manipulating storage temperature and atmosphere. Fresh approaches are needed.

Much is now known about the importance of tissue calcium levels to long storage life in apple (51, 52). Thus, one avenue worth investigating may be to try to increase Ca levels with plant growth regulators (53).

Alternatively, an approach aimed at more directly delaying tissue or organ senescence with hormones may be more successful. Auxins, gibberellins, and cytokinins are known to have senescence-delaying activity with some plant tissues. Perhaps these effects will eventually be exploited with whole organs.

An important key to delaying senescence of apple fruits appears to be the control of ethylene synthesis and action (54). The use of a powerful inhibitor of ethylene synthesis such as AVG, together with one or more of the above hormones, may lead to profound improvements in the quality of stored apples and perhaps even permit storage at higher temperatures. We are verging on important breakthroughs in this area.

Conclusions

Plant growth regulators help solve a wide range of apple production problems. In fact, it is difficult to imagine profitable orcharding without some use of these versatile chemicals. Some chemicals are used for several purposes. NAA is used to encourage abscission of young fruits, to prevent abscission of mature fruits, and to prevent unwanted shoot growth. Daminozide promotes flower initiation, retards shoot growth, prevents harvest drop, delays fruit ripening and improves fruit color and firmness. Aminoethoxyvinylglycine does not yet have a registered use, but it promises to promote lateral bud breaks, to prevent both the "June drop" and harvest drop, and to prolong the storage life of mature fruit by inhibiting ethylene production.

Apple producers are accomplished growth regulator experimenters. They make good use of new techniques as these are developed, and they often discover innovative new uses on their own. Growth regulator use in commercial apple production is something of a success story in which the plot is still developing and more characters will be introduced.

References

1. Howard, B.H.: J. Hortic. Sci. *43*, 23–31 (1968)
2. Howard, B.H., Nahlawi, N.: Rep. E. Malling Res. Stn. 1969, 91–94 (1970)
3. Child, R.D., Hughes, R.F.: Acta Hortic. *79*, 43–48 (1978)
4. Jones, O.P., Hopgood, Margaret E., O'Farrell, D.: J. Hortic. Sci. *52*, 235–238 (1977)
5. Jones, O.P.: Sci. Hortic. *30*, 44–48 (1979)
6. Lane, W.D.: Plant Sci. Lett. *13*, 281–285 (1978)
7. Luckwill, L.C., Child, R.D.: Acta Hortic. *34*, 213–220 (1973)
8. Luckwill, L.C.: Acta Hortic. *65*, 237–243 (1978)
9. Murashige, T., Skoog, F.: Physiol. Plant. *15*, 473–497 (1962)
10. Larsen, F.E.: Proc. Am. Soc. Hortic. Sci. *88*, 690–697 (1966)
11. Larsen, F.E.: J. Am. Soc. Hortic. Sci. *95*, 213–233 (1970)
12. Larsen, F.E.: J. Am. Soc. Hortic. Sci. *98*, 34–36 (1973)
13. Quinlan, J.D., Preston, A.P.: Acta Hortic. *34*, 123–128 (1973)
14. Baldini, E., Sansavini, S., Zocca, A.: J. Hortic. Sci. *48*, 327–337 (1973)
15. Williams, M.W., Billingsley, H.D.: J. Am. Soc. Hortic. Sci. *95*, 649–651 (1970)
16. Preston, A.P.: J. Hortic. Sci. *43*, 17–22 (1968)
17. Van Oosten, H.J.: Acta Hortic. *65*, 123–127 (1978)
18. Kender, W.J., Carpenter, S.: J. Am. Soc. Hortic. Sci. *97*, 377–380 (1972)
19. Luckwill, L.C.: J. Hortic. Sci. *43*, 91–101 (1968)
20. Quinlan, J.D., Preston, A.P.: J. Hortic. Sci. *53*, 39–43 (1978)
21. Quinlan, J.D.: Acta Hortic. *80*, 39–48 (1978)
22. Quinlan, J.D.: ARC Res. Rev. *5*, 3–5 (1979)

23. Batjer, L.P., Williams, M.W., Martin, G.C.: Proc. Am. Soc. Hortic. Sci. *85*, 11–16 (1964)
24. Williams, M.W.: J. Am. Soc. Hortic. Sci. *97*, 210–212 (1972)
25. Byers, R.E., Barden, J.A.: HortScience *11*, 506–507 (1976)
26. Ramirez, H., Hoad, G.V.: Br. Plant Growth Regul. Group. Monogr. *2*, 37–47 (1978)
27. Bukovac, M.J.: Am. Fruit Grow. *88*, 24 (1963)
28. Goldwin, G.K.: Acta Hortic. *80*, 115–121 (1978)
29. Bangerth, F.: J. Am. Soc. Hortic. Sci. *103*, 401–404 (1978)
30. Williams, M.W.: Hortic. Rev. *1*, 270–300 (1979)
31. Edgerton, L.J.: Acta Hortic. *34*, 333–343 (1973)
32. Looney, N.E.: Acta Hortic. *34*, 397–406 (1973)
33. Southwick, F.W., Demoranville, I.E., Anderson, J.F.: Proc. Am. Soc. Hortic. Sci. *61*, 155–162 (1953)
34. Looney, N.E.: Plant Physiol. *43*, 1133–1137 (1968)
35. Faust, M.: Acta Hortic. *34*, 407–420 (1973)
36. Williams, M.W., Batjer, L.P., Martin, G.C.: Proc. Am. Soc. Hortic. Sci. *85*, 17–19 (1964)
37. Batjer, L.P., Williams, M.W.: Proc. Am. Soc. Hortic. Sci. *88*, 76–79 (1966)
38. Looney, N.E.: J. Am. Soc. Hortic. Sci. *96*, 350–353 (1971)
39. Williams, M.W., Stahly, E.A.: J. Am. Soc. Hortic. Sci. *94*, 17–19 (1969)
40. Stembridge, G.E., Morrell, G.: J. Am. Soc. Hortic. Sci. *97*, 464–467 (1972)
41. Unrath, C.R.: J. Am. Soc. Hortic. Sci. *99*, 381–384 (1974)
42. Raese, J.T.: HortScience *10*, 396–398 (1975)
43. Dozier, W.A., Jr., Hollingsworth, M.H.: HortScience *11*, 392–393 (1976)
44. Elfving, D.C., Forshey, C.G.: J. Am. Soc. Hortic. Sci. *102*, 418–423 (1977)
45. Byers, R.E., Barden, J.A.: HortScience *11*, 506–507 (1976)
46. Morgan, P.W.: Proc. 6th Annu. Meet. Plant Growth Regul. Work. Group (in press, 1980)
47. Erez, A., Lavee, S., Samish, R.M.: J. Am. Soc. Hortic. Sci. *96*, 519–522 (1971)
48. Ketchie, D.O., Murren, C.: J. Am. Soc. Hortic. Sci. *101*, 57–59 (1976)
49. Howell, G.S., Jr., Dennis, F.G., Jr.: In: Analysis and Improvement of Cold Hardiness in Plants. Olein, C.R. (ed.). Cleveland, Ohio, CRC Press 1980, in press
50. Dennis, F.G., Jr : J. Am. Soc. Hortic. Sci. *101*, 241–245 (1976)
51. Shear, C.B.: Commun. Soil Sci. Plant Anal. *6*, 233–244 (1975)
52. Sharples, R.O., Johnson, D.S.: Ann. Appl. Biol. *85*, 450–453 (1977)
53. Marcelle, R., Clijsters, H.: Acta Hortic. *80*, 353–360 (1978)
54. Jerie, P.H., Hall, M.A., Zeroni, M.: Acta Hortic. *80*, 325–332 (1978)

Uses of Plant Growth Substances in the Production of Sugarcane: A Practical Case History

L. G. NICKELL[1]

Introduction

In 1974, Wittwer (32) made the statement that "growth regulants are emerging as significant in the production of major food crops." It is fitting to point out that some of the earliest applied work on growth substances was conducted in Hawaii by our chairman, Professor Skoog, and in Puerto Rico by Dr. J. van Overbeek (30), two famous names in plant physiology and especially in plant growth substances.

Plant growth regulators have been used in Hawaii to increase the production of sucrose in sugarcane for over two decades (17, 18). The first commercial success was in the prevention of flowering, followed by the application of gibberellic acid for the increase of stalk elongation which ultimately resulted in increased sugar production. Currently, the greatest interest centers around the use of chemical compounds for the control of maturation – the so-called chemical "ripeners". All of these uses are now on a commercial basis and will be discussed in some detail. In addition, research has been successful in affecting both germination of the vegetative "seed pieces" used for propagation and on the tillering of young plants of sugarcane. Because of the importance of and the success with ripening control, these latter two uses have not been pursued rigorously during the last few years. Consequently, they are not at the commercial stage at this time.

This brief introduction shows that the sugar industry in many places throughout the world and especially in Hawaii is using chemicals of the plant growth regulator type at almost every stage of the development of the crop – "from the cradle to the grave" as it were. The results with this crop amply confirm Dr. Wittwer's prediction.

Rather than discuss the various uses for plant growth regulators in historical order, I would prefer to present them sequentially, as they would be used in the development of the crop, i.e., from planting through harvest.

Germination

Because of the problems involved in emergence of sugarcane after planting (including such factors as depth of planting, angle of the bud, adverse weather conditions – particularly excess moisture and cold – and fungal or bacterial infection) (19), it has al-

1 Research & Development, Velsicol Chemical Corporation, 341 E. Ohio Street, Chicago, Ill. 60611, USA

ways been the aim of sugarcane planters to try to obtain a "good stand". In order to get a good stand, one has to have germination and rapid early growth, i.e., emergence of the first green leaf. In an empirical search through many compounds looking for those which might stimulate the rate of germination or early growth of the young sugarcane plant, it was found that the amino acid arginine caused a considerable stimulation of germination (20). This amino acid was also shown to cause an increase in the growth rate of the sugarcane plant at later stages, resulting in increased production of both cane and sugar (19). Additional studies showed that this effect occurs at the cellular level – that is, cells of sugarcane grown in suspension culture respond rather dramatically to this amino acid (11, 21). Subsequently, considerable work has been done with arginine and its effects on sugarcane growth and metabolism (12, 19, 22). The dramatic results obtained with arginine suggest that, if ways to exploit these findings commercially could be devised, they could easily lead to one of the greatest increases in sugar production yet achieved by the application of chemicals. In order to achieve this, much more whole-plant physiology and field work must be conducted.

Tillering

Of the many variables involved in the production of sugar from the cane plant, probably the most significantly related factor is the number of stalks per acre of land at harvest (16, 19). In order to obtain the maximum number of stalks supportable by an acre of cane, it is necessary to induce branching at an early stage. A search for compounds that would induce tillering in sugarcane was started several years ago. Several compounds, such as ethephon, have been found to be active (26). To date, very little has been done to carry these studies beyond the screening stage.

Stalk Elongation

During the 1950's when experimental amounts of gibberellic acid and related gibberellins became available, they were applied to sugarcane, and stalk elongation was demonstrated (5, 6). Because of the high cost of crystalline gibberellic acid and its unavailability in quantity, more than a decade passed before such materials were evaluated to determine whether or not they would increase crop production on a commercial scale. During the early studies of this class of plant hormones, more and more potential uses became apparent. Fermentation companies consequently started making larger amounts of gibberellins, and eventually unrefined fermentation liquids became available for evaluation, at greatly reduced costs. These products were shown to increase the cane tonnage per unit area, particularly when applied during the cold or winter months when growth is slowed down considerably (28). Later, field work with several types of products including broths, semipurified and crystalline gibberellins showed on average, under Hawaiian plantation conditions, a gain of about 5 tons of cane per acre at harvest, with a gain in sugar from 0.2 to 0.5 tons per acre from the application of 2–3 oz. per acre of gibberellin (29). Most of the early work was done by application during the second season of the two-year crop cycle in Hawaii. Subsequent to the registration of gibberellin products with regulatory agencies, additional studies have been carried out with first season applications and these also have proved to be effective, indicating that maximum response is obtained with applications made to young sugarcane plants (3). More recent studies have shown that the maximum response is a function of: the cultivar treated, the amount of gibberellic acid applied, the number of applications, the time interval between applications, and the allowance of sufficient time before harvests for maturation of the gibberellin-produced growth (13, 14, 15).

Flowering

The flowering process in sugarcane is extremely sensitive to the environment. This is true of flower initiation, emergence, and pollen fertility. The optimum photoperiod appears to be about 12 1/2 h, and most commercial varieties respond to this throughout the world. The 12 1/2 h day occurs about September 2 in Hawaii, and night interruption experiments indicate this is the approximate date on which initiation begins in that state. For most commercial varieties studied thus far, initiation is usually complete by September 20. The highly reproducible short period of floral initiation makes possible the suppression of flowering in young cane on a commercial scale (4). Experiments in Hawaii have established that flower inhibition by night lights, chemical sprays, or the withdrawal of water can be expected to increase yields by 10%–20% under conditions of heavy flowering (4).

After determining that night interruption from September 1–20 would inhibit flowering (commonly referred to as tasseling or arrowing in the sugar industry) of the varieties worked with in Hawaii in the 1940's, field experiments were conducted to determine the quantitative effects tasseling might have on the yield of sucrose. The results in 1949 of a replicated field test of 10 paired plots using variety H32-8560 supported the belief, long held by sugar growers, that tasseling reduced sugar yields. In this field test, the average gain from suppression of tasseling was 15%, or the equivalent of 1.3 tons of sucrose per acre.

In the years immediately following these results, the factors affecting tasseling, as well as methods for preventing its occurrence, were extensively studied. The effective ways found to prevent tasseling included: night interruption with light, lowered temperature, leaf and spindle trimming, withdrawal of water, and the application of chemicals. Because temperature cannot be controlled in the field and because leaf trimming and light interruption on a commercial scale were not operationally feasible, emphasis was placed on water withdrawal and the application of chemicals. Withdrawal of water, possible only on irrigated plantations, had other operational problems, so that the use of chemicals eventually became the standard practice in Hawaii.

The first potentially useful commercial chemical was maleic hydrazide which, at best, gave about 60% control. Rapid development led to the establishment of monuron as the chemical of choice. These studies later included diuron. When either chemical was properly applied at 4 lbs of active ingredient per acre, virtually complete control of heavy-tasseling varieties used in the 1950's and 1960's in Hawaii could be obtained. A continuation of testing for more active chemicals to prevent tasseling led to the discovery that diquat, applied from the air at rates as low as 0.125 lbs/acre of the cation, was as active as monuron at 4 lbs/acre, making diquat one of the most active compounds yet evaluated for this purpose. The resulting cut in cost per unit area for control was substantial (27). Positive effects with chemical control of flowering, with diquat the compound of choice, have been obtained in Guyana (7), Mexico (10), the Philippines (2), and Taiwan (33).

Ripening

Ripening is considered one of the most important aspects of sugarcane production – from both a research and an operational point of view. To say that the phenomenon of cane ripening is extremely complex would be, at best, a gross understatement. Studies on the use of plant growth regulators have appeared in the literature spasmodically since 1949. The first material reported to be effective was 2,4-D (1). This was followed by studies with maleic hydrazide, triiodobenzoic acid, dalapon, CMU, DCMU, EDTA, Trysben, Pesco 1850 (a mixture of MCPA and Trysben), as well as a number of enzyme inhibitors and metabolic inhibitors. No large-scale program was launched, however, until basic studies on the effects of defoliation on translocation in sugarcane had furnished a solid basis for such a program (8, 9). The screening test used is a very simple one, consisting of adding the test material by pipette or by needle and syringe into the whorl of leaves at the top of the sugarcane stalk, which is field grown and almost at the stage for normal maturation. At a specified time or times (4,5, and/or 8 weeks) after application of the test material, 5–10 stalks are harvested, analyzed, and compared with an untreated group of stalks. The effectiveness of a test compound as a ripener is based on its ability to increase the quality of the treated stalks in two major parameters for sugar production (juice purity and sugar as a percent of field cane weight) (24).

After more than a decade of screening and evaluation, a surprising number of compounds of diverse chemical structure have been found which can be used to increase the sugar content of the cane crop at harvest. A program similar to that originally started in Hawaii (24) was initiated in Australia but was subsequently curtailed. Less extensive programs, generally limited to field evaluation of compounds reported to have activity, have been carried out in Florida, Louisiana, Mauritius, the Philippines, Puerto Rico, Rhodesia, South Africa, Taiwan, Trinidad, and other sugar-producing countries.

At the present time, only one compound is registered in the United States as a commercial product for the purpose of increasing the sucrose content of sugarcane at harvest. This material is glyphosine [N,N-bis-(phosphonylmethyl)glycine], common name Polaris, which has been evaluated over a period of several years on close to 100,000 acres in Hawaii and other sugar producing areas (23) and has given substantial gains – about 10%–15% increase in yields – which is an increase of 1 ton per acre or more when applied to certain varieties grown on the rainy coasts of the Island of Hawaii. More recent work has shown that varieties previously thought to be nonresponsive to this ripener have been found to respond positively when surfactants are added to the formulation (17). Similarly, it has been found to be effective on irrigated lands when surfactants are added. Glyphosine treatment results in a reduced rate of terminal cane growth, but how this relates to its mode of action has not yet been established (23).

Four other chemicals: Ripenthol, chlormequat, Embark, and MON-8000, have been registered under experimental labels in the United States for field evaluation as commercial ripeners.

Ripenthol, the monoamine salt of Endothall, was one of the first materials found to have significant activity on sugarcane in Hawaii (24). Numerous relatives of this compound were tested in the early screening stages, and it was found that, while the acid itself had very low activity, amine salts were more active than the disubstituted amines. Ripenthol (also known as Hydrothol) has considerable phytotoxic activity and, because of this, care must be taken in its application, especially to avoid drift when applied by air.

Chlormequat (2-chloroethyl-trimethylammonium chloride), also known as Cycocel, is among the most widely used plant growth regulators in the world (31). It has been evaluated on more than 1000 acres in Hawaii, but preliminary results suggest that its activity might be too low to be commercially successful.

Embark (common name, mefluidide) is being tested at the present time under an experimental label in Hawaii, the Philippines, and certain other countries.

MON-8000 is the sodium salt of N-phosphonylmethyl glycine. The isopropylamine salt of this compound is the well-known herbicide, Round-Up. MON-8000 is being evaluated in Hawaii under an experimental label. It is much more active than Polaris. The recommended rate for Polaris is 4 lbs active ingredient per acre, whereas MON-8000 is being evaluated in the 1/2 to 1 lb per acre range.

Ethephon (Ethrel) has been known to be active as a sugarcane ripener for many years, but for commercial reasons, was not developed as such in the United States. Rostron (25) finds ethephon to be much more effective than Polaris in southern Africa, whereas the reverse was found to be true in Hawaii and other places where the two have been compared.

Concluding Remarks

Long-term research by the sugar industry in Hawaii and other tropical and semi-tropical countries has supported a multi-pronged investigation of the activity of chemicals on most of the steps of cane development, from germination through ripening and harvest. While the use of plant growth regulators is still in its infancy, success to date in sugarcane, in terms of yield increases greater than 10%, substantiates the belief that the regulation of crop growth and metabolism may result in one of the most important quantitative gains yet achieved in agriculture. The monumental task of producing raw materials to supply the world's food and to supplement its energy requirements may depend to a large degree on achievements of this magnitude in a wide range of crops.

Hawaii, with its high costs of operation and high yields of sugarcane, nonseasonal environmental conditions, and the necessity to harvest the year-round, can afford high-priced chemicals. In other cane-producing countries this may not be the case. For example, Australian investigators were among the first to study the use of chemical rieners. In fact, during the early 1960's, there was a cooperative program between Hawaiian and Australian workers on this subject. This was not pursued when it was realized that Australian conditions, with cool and dry weather at harvest, were for the most part conducive to excellent natural ripening. In other cane-growing areas in the world, however, as more is learned about the relationship of a given chemical to the process which it affects, as sugarcane agronomy improves and its economy becomes more favorable, the use of chemical ripeners will undoubtedly become more widespread; essentially the same can be said for the use of gibberellins and chemicals for flower control.

Historically, agricultural research has been primarily concerned with improvement of total crop yield by the removal of obstacles to optimal production. Now that many of these obstacles can be overcome with herbicides, pesticides, fertilizers, irrigation, and improved management practices, the stage is set for further yield increases by the use of sophisticated techniques of physiological manipulation of the plant and its metabolism. Evidence already in hand suggests that this is more than just wishful thinking.

References

1. Beauchamp, C.E.: Proc. 23rd Annu. Meet. Assoc. Tech. Azucareros Cuba, 55–87 (1949)
2. Benedicto, F.: Victorias Milling Co. Exp. Stn. Bull., March–April, p. 3 (1967)
3. Buren, L.L.: 1971 Rep. Hawaii Sugar Technol. 104–108 (1972)
4. Burr, G.O., Hartt, C.E., Brodie, H.W., Tanimoto, T., Kortschak, H.P., Takahashi, D., Ashton, F.M., Coleman, R.E.: Annu. Rev. Plant Physiol. *8,* 275–308 (1957)
5. Coleman, R.E.: Sugar J. *20,* 23–26 (1958)
6. Coleman, R.E., Todd, E.H., Stokes, I.E., Coleman, O.H.: Sugar J. *23,* 11–21 (1960)
7. Evans, H., Bates, J.F.: Proc. Br. West Indies Sugar Technol. (1966)
8. Hartt, C.E.: 1963 Rep. Hawaii Sugar Technol. 151–167 (1963)
9. Hartt, C.E., Kortschak, H.P., Burr, G.O.: Plant Physiol. *39,* 15–22 (1964)
10. Humbert, R.P., Lima, M., Goveas, J.: Proc. 13th Int. Soc. Sugar Cane Technol. Taiwan, 462–467 (1968)
11. Maretzki, A., Nickell, L.G.: 7th Int. Congr. Biochem. Abstr. Vol. J-207 (1967)
12. Maretzki, A., Nickell, L.G., Thom, M.: Physiol. Plant. *22,* 827–839 (1969)
13. Moore, P.H.: Proc. 5th Annu. Meet. Plant Growth Regul. Work. Group 158–162 (1978)
14. Moore, P.H., Buren, L.L.: Crop Sci. *17,* 443–446 (1978)
15. Moore, P.H., Ginoza, H.: Crop Sci. *20,* 78–82 (1980)
16. Nickell, L.G.: 1967 Rep. Hawaii Sugar Technol. 147–155 (1968)
17. Nickell, L.G.: Bull. Plant Growth Regul. *2,* 51–54 (1974)
18. Nickell, L.G.: Outlook Agric. *9,* 57–61 (1976)
19. Nickell, L.G.: In: Ecophysiology of Tropical Crops. Alvim, P. deT., Kozlowski, T.T. (eds.), pp. 89–111. London, New York: Academic Press 1977
20. Nickell, L.G., Kortschak, H.P.: Hawaii Planters Rec. *57,* 230–236 (1964)
21. Nickell, L.G., Maretzki, A.: Physiol. Plant *22,* 117–125 (1969)
22. Nickell, L.G., Maretzki, A.: Proc. IV Int. Ferment. Symp. Ferment. Technol. Today 681–688 (1972)
23. Nickell, L.G., Takahashi, D.T.: 1971 Rep. Hawaii Sugar Technol. 73–82 (1972)
24. Nickell, L.G., Tanimoto, T.T.: 1965 Rep. Hawaii Sugar Technol. 152–166 (1966)
25. Rostron, H.: Proc. 16th Congr. Int. Soc. Sugar Cane Technol. Brazil 1977 *2,* 1605–1618 (1978)
26. Takahashi, D.: 1969 Annu. Rep. Exp. Stn. Hawaii Sugar Planters Assoc., p. 50 (1969)

27. Tanimoto, T., Nickell, L.G.: Proc. 12th Int. Soc. Sugar Cane Technol. Puerto Rico, 1965, 113–116 (1967)
28. Tanimoto, T., Nickell, L.G.: 1966 Rep. Hawaii Sugar Technol. 184–190 (1967)
29. Tanimoto, T., Nickell, L.G.: 1967 Rep. Hawaii Sugar Technol. 137–146 (1968)
30. van Overbeek, J., Davila Olivo, G., de Vazquez, E.M.S.: Bot. Gaz. *106*, 440–451 (1945)
31. Wittwer, S.H.: Outlook Agric. *6*, 205–217 (1971)
32. Wittwer, S.H.: Bull. Plant Growth Regul. *2*, 5–8 (1974)
33. Yang, P.C., Pao, T.P., Ho, F.W.: Taiwan Sugar *19*, 21–27 (1972)

Plant Growth Substances in Commercial Uses of Tissue Culture

TOSHIO MURASHIGE[1]

Introduction

The aseptic culture of plant cells and organs, known generically as plant tissue culture, has emerged as a versatile research procedure and a significant commercial practice. And, as Haberlandt discovered in his pioneering experiments (33), the successes have depended on the availability of suitable growth regulating chemicals that could be included in the nutrient medium. Thus, non-tumorous callus was not culturable until IAA (indole-3-acetic acid), the first auxin according to Kögl et al. (48), was used as supplement. Even then, culturability remained restricted to tissues that required only auxin as an exogenous growth substance. More widespread successes were made possible by the discovery of kinetin, the first cytokinin (56). Kinetin, added together with IAA or other auxins, replaced the need for coconut endosperm, the callus-promoting factor discovered earlier (91). Manipulations involving auxin and cytokinin provisions continue to serve as the basis of nearly all of today's plant tissue culture.

The use of plant tissue culture to pursue practical objectives is not a recent idea. Embryo culture has been a tool of plant hybridizers for 50 years (34). The many rare orchids in culture are due to the aseptic seed germination technique introduced in 1922 by Knudson (47). Plant pathologists and horticulturists have utilized shoot apex cultures to obtain specific-pathogen-free (SPF) plants since Morel and Martin (59) demonstrated the feasibility of freeing dahlia plants from dahlia mosaic virus in 1952. Cloning of plants through tissue culture began in the 1960's, following Morel's (58) suggestion that 4 million orchid plants would be obtainable annually from a single shoot apex explant.

Currently, the commercial sector uses plant tissue culture for profit as well as in research. The former includes tissue culturing as a process that can generate immediately marketable commodities and as an effort directed at increasing production efficiency and improving product quality. Cloned plants and secondary metabolities are examples of immediately marketable commodities, whereas the recovery of SPF plants and development of new varieties are tissue culture activities aimed at enhancing production efficiency and product quality. This report summarizes the status of activities that currently appear commercially significant.

---

1 Department of Botany and Plant Sciences, University of California, Riverside, California 92521, USA

Marketable Commodities

Rapidly Cloned Plants

The most extensive commercial use of plant tissue culture is in rapid clonal progapation. Although orchid multiplication rates of the magnitude predicted by Morel (58) remain unattained, the method has utility for species and hybrids of at least 23 genera of the Orchidaceae (61, 88). It is noteworthy that nearly all successes with orchids have been accomplished with nutrient media lacking hormonal supplements. Thus, improvement of multiplication and growth rates and extension to additional genera are highly probable through suitable provisions of growth substances.

Propagation of other crops has now surpassed the orchids and about 100 private laboratories employ tissue culture to reproduce a large variety of flowers, foliage plants, and other ornamentals. A few also utilize it for vegetables, fruits, field crops, forest trees and shrubs, and other economic plants. The extent to which diverse crops are currently propagatable by tissue culture is shown in Table 1, which does not denote actual commercial use, but includes species for which only a feasibility has been reported.

The plant multiplication procedure of most laboratories relies on mass production of shoots, either adventitiously from a variety of explants or by enhanced axillary branching of excised shoot tips. Complete plants are obtained by ultimately rooting individual.shoots. The method of adventitious shoots is considerably faster than that of axillary branching. However, its usefulness is limited to a few species and it is frequently associated with a high incidence of aberrant plants, i.e., plants that fail to reproduce the cultivar characteristics. Both methods of shoot multiplication require media with suitably belanced levels of auxin and cytokinin. They are based on physiological concepts delineated 20 years ago, for adventitious organogenesis by Skoog and Miller (86) and for axillary branching by Wickson and Thimann (93). Adventitious shoots, whether

Table 1. Extent to which diverse crops are currently propagatable clonally via tissue culture[a]

Crop types	No. species reported
Flowers (excluding orchids)	64
Ferns	20
Foliage plants	25
Succulents	12
Landscape ornamentals	26
Vegetables and condiment plants	40
Field crops	33
Fruits and nuts	22
Forest trees and shrubs	34
Medicinal and related plants	18
Total	294

[a] From (3, 9, 12, 13, 14, 15, 16, 18, 19, 20, 26, 35, 39, 40, 42, 43, 45, 46, 52, 60, 62, 66, 67, 72, 73, 75, 80, 83, 84, 87, 90, 92, 97)

arising directly from explant or through intermediary callus, occur when auxin and cytokinin are provided in moderate and about equal levels, e.g., 2–3 mg/l each of IAA and kinetin. A medium very high in cytokinin, e.g., 30 mg/l of N^6-isopentenyladenine, and low in auxin, e.g., 0–0.3 mg/l IAA, has been more suitable for enhanced axillary shoot development in shoot tip explants.

The shoot multiplication methods have been satisfactory for highly heterozygous cultivars, the propagation of which must be accomplished asexually, even if only very slowly. They are inadequate and uneconomical for crops that are establishable in fairly uniform stands, and inexpensively, from seeds, e.g., many annual vegetables and flowers. The cloning of seed-propagated crops requires methods that can generate large numbers of plants very quickly and at extremely low unit costs. Those presently under development are focused on exploiting somatic cell embryogenesis. Apparently somatic embryogenesis is widespread among plants; a recent survey (89) disclosed its occurrence, naturally or in vitro, in 150 species belonging to over 70 families. Under optimum conditions a few grams of cultured cells can produce thousands of embryos. Pelletization of the embryos, furnishing them with a protective and nourishing coat, is under investigation. The pellets of "clonal seeds" might then be stored, transported, and mechanically sowed in soil. Also under study is the transplanting of pre-germinated embryos, directly from test-tube to soil, via the fluid drill process.

Secondary Products

Although American interest in tissue culture as an alternate source of medicinals and other plant products has waned considerably since the 1950's, European and Asian efforts seem revitalized. Frequent symposia, focused specifically on the subject, have been held recently (1, 10). A large variety of secondary constituents is reportedly obtainable from cell cultures of diverse plants, and numerous patents have been awarded (1, 10, 22). Nevertheless, pharmaceutical firms have not yet begun to displace traditional sources of chemicals with tissue cultures, perhaps because of relatively low yields. But there are many examples of productivity by cultured cells that equal or surpass the progenitor plant or organ (Table 2). Yields of secondary products from some organ cultures have also been found to be higher than from plants or cell cultures; thus, they might also be tapped for certain constituents.

Increasing Production Efficiency and Improving Product Quality

Recovering Specific-Pathogen-Free Plants

It would be reasonable to assume that crops that have been propagated by traditional asexual methods are systematically infected with viruses and similar pathogens, unless the pathogens have been detected and eliminated. Infection by a pathogen does not invariably result in death of the plant. Nevertheless, it can reduce yield and/or quality of the marketed commodity. Yield increases averaging 30%, and up to 300%, have been observed following replacement of virus-infected plants with SPF plants.

The available in vitro methods are adequate to enable establishment of SPF stocks of practically any crop from known plant pathogens, including viruses, viroids, and

Table 2. Examples of secondary metabolites that are produced by cell cultures in levels as high or higher than plant or organ (1)

Secondary substance	Plant cell culture
Anthocyanins	*Haplopappus and Daucus*
Betalains	*Beta*
Anthraquinones	*Morinda citrifolia, Lithospermum erythrorhizon, Cassia tora* and *Galium* sp.
Ubiquinone	*Nicotiana tabacum*
L-dopa	*Mucuna pruriens*
Serpentine	*Catharanthus roseus*
Ajmalicine	*Catharanthus roseus*
Ginseng saponins	*Panax ginseng*
Glycyrrhizin	*Glycyrrhiza glabra*
Rosmarinic acid	*Coleus blumei*
Diosgenin	*Dioscorea deltoidea*

mycoplasmas (24, 25, 62). The methods have also been proven safe when used in transporting genetic stocks internationally and are amenable to cryogenic storage of germplasm.

Shoot apex cultures have been most often used to obtain SPF plants. When properly executed, the explant is comprised of the apical meristem and one to a few primordial leaves, and the whole measures 0.1 to 0.3 mm tall. Shoot apices of the dimensions used in rapid clonal propagation are too large and risk including cells or tissues containing viruses, viroids, mycoplasmas, and even some bacteria and fungi.

Meristem culture, or culture of the true stem apical meristem, has rarely been used. By definition, the explant of a meristem culture is confined to the 0.05 to 0.1 mm tall dome. (Sometimes the tissue is considerably less than 0.05 mm tall and can even appear depressed). Its successful isolation requires special dexterity and experience. The survival rate is low and plant development among the surviving meristem explants proceeds very slowly. More important, the probability of excluding pathogens by meristem culture may not be substantially better than by a properly executed culture of the larger shoot apex. The hormonal needs of the excised apical meristem are evidently distinct from those of the shoot apex with subjacent leaf primordia attached. Meristems have been unable to develop in media lacking auxin and cytokinin. In contrast, the larger, leaf-primordium-containing explants of many cultivars have produced rooted plants in unsupplemented media. It might thus be concluded that the source of hormones in the shoot apex is the developing leaves, and not the apical meristem as once believed (81).

The immediate aim of any method of obtaining SPF plants is to establish a rooted plant quickly, so that prescribed pathogen tests can be performed expeditiously. When the excised shoot apex is unrootable, it can be grafted aseptically to produce the complete plant. This method has been especially advantageous with woody perennials and other difficult-to-root plants.

Sometimes pathogen-free plants can be generated in cell cultures established from infected plants or explants. But the method has serious risks: too often it yields variant plants. Its applicability is also limited to cultivars in which plant regeneration by adventitious processes can be achieved relatively easily in cultures originating in explants from mature plants.

Developing New Varieties

Aids to Traditional Breeding Methods. Embryo culture, the method commonly used when embryogenesis of rare hybrids falls short of completion in vivo, has been accomplished with over 40 plant families (62). To be more useful, however, consistently successful culturing must include the smaller globular isolates, or the embryos that are in the stage when most developmental failures set in. The very young embryos probably have complex nutritional requirements, fulfillment of which is sometimes achievable by using endosperm fluids, high sugar or high osmotic pressure. A suitable balance involving auxin, cytokinin and other growth substances has also been implicated (74). It is also possible that the intractability is simply due to excessive injury inflicted during the isolation steps.

When development fails before the embryo reaches isolable dimensions, the alternative has been to culture ovules or ovularies excised shortly after pollination. Cultures of ovules and ovularies from unpollinated flowers have also been used to circumvent some types of incompatibility. To date, the application of pollen to excised ovules or ovularies in vitro has resulted in viable seeds of these genera: *Agrostemma, Antirrhinum, Argenome, Brassica, Dianthus, Eschscholzia, Melandrium, Narcissus, Nicotiana, Papaver, Petunia,* and *Zea* (7, 62).

Asparagus seeds of distinctly superior quality, particularly uniform reproduction of desired characteristics by their seedlings, are being produced commercially from parent plants chosen for progeny behavior and cloned in large numbers through tissue culture (21). Seed producers are now exploring the procedure's extendability to other vegetable crops, such as broccoli and lettuce.

Many variant plants usable in variety development have been obtained by selections performed with cell cultures (Table 3), some of which have already contributed to new cultivars. Frequently the variants are not mutants, but simply epigenetic manifestations; thus, it is probable that their original characteristics will reappear if progagated extensively by asexual methods.

Haploid plants and cell cultures simplify considerably the process of obtaining mutants. The isogenic plants can also be utilized to reduce substantially the time spent in breeding to create new varieties. Anther and miscrospore cultures have thus far enabled the attainment of haploid and isogenic plants of other ploidy levels of over 100 species of plants, belonging to more than 40 genera in 15 families (2, 36, 38, 57, 62, 71, 78). Using plants originating in anther cultures, the Chinese have succeeded in developing 3 new varieties of tobacco, 3 of rice and 2 of wheat (38).

The consideration of in vitro techniques as aids in plant breeding might also include germplasm storage, especially that of asexually propagated plants. Cryopreserved tissue cultures could be the alternative to variety collections and other extensive field plantings. Cell cultures, somatic cell embryos and shoot apices of at least 16 species (5, 6, 8, 23, 27, 32, 44, 29, 50, 64, 65, 70, 76, 79, 82, 94, 95, 96) have been stored successfully at liquid nitrogen temperature. The cryopreserved shoot apices (6, 32, 44, 79)

Table 3. Examples of variant plants obtained through cell cultures

Crop	Variant	References
Sugar cane	Resistance to *Helminthosporium sacchari*	(67)
Sugar cane	Resistance to Fiji virus	(67)
Sugar cane	Resistance to *Sclerospora sacchari*	(67)
Sugar cane	Drought resistance	(67)
Sugar cane	Temperature tolerance	(67)
Sugar cane	Yield increase	(51)
Potato	Resistance to *Alternaria solani*	(53)
Geranium	"Velvet Rose" cultivar	(85)
Tobacco	Resistance to IPC herbicide	(4)
Tobacco	Polyploids	(63)
Kale	Polyploids	(37)

and embryos (23, 94) are particularly noteworthy, inasmuch as plant regeneration is virtually assured on their return to culture.

Novel Breeding Methods. Protoplast fusion that culminates in a somatic hybrid plant is a quick method of increasing a crop's genetic diversity. Fusion between protoplasts of plants belonging to different families, and even between plant protoplasts and animal cells, has been accomplished. However, successful development of fused cells into hybrid plants has been limited to fairly closely related plants, e.g., intra- and interspecific hybrids within *Nicotiana, Datura, Daucus,* or *Petunia* (11, 17, 30, 31, 41, 55, 98). There is reason for optimism, nevertheless, because plants have been obtained recently in two cases where the comparable hybrids have not been attainable by sexual means: *Arabidopsis thaliana* x *Brassica campestris* (28, 29) and *Lycopersicon esculentum* x *Solanum tuberosum* (54).

Gene transfers in crop plants might also be manipulable through protoplasts. Inasmuch as foreign DNA in the unprotected state is degraded quickly and not integrated into the receipient cell's genome, studies in several laboratories are focused on plasmids, organelles, and lipid vesicles as possible vehicles for the gene transfer process.

Summary and Outlook

Plant tissue culture, or aseptic culture of plant cells and organs, has gained considerable prominence in research and commercial applications. The successes have clearly depended on the use of growth-regulating substances, especially auxins and cytokinins. Current commercial applications include tissue culture as a method of rapid clonal propagation, an alternative source of secondary plant products, a means of recovering specific-pathogen-free plants, and as an aid in developing new varieties.

Virtually all marketable orchids are clonable through tissue culture. About 100 private laboratories can now propagate nearly 300 other crop plants by the method, utilizing the processes of adventitious and axillary shoot formation. Cloning of normally seed-propagated plants will probably involve exploitation of somatic cell embryogenesis, combined with pelletization or fluid drill processes.

Many plant constituents are now obtainable from cell cultures in levels as high or higher than the progenitor plant or organ. Hence, tissue culture may soon become a major commercial source of secondary plant products.

Extensive use is expected of in vitro methods to obtain specific-pathogen-free plants, especially in view of the utility of SPF plants to recover crop productivity and enhance germplasm storage and transport.

Tissue culture methods that are currently applicable to plant breeding are embryo, ovule, ovulary, anther or microspore, and cell cultures. Somatic hybrid plants have been obtained recently following fusion of tomato and potato protoplasts, and *Brassica* and *Arabidopsis* protoplasts. These rare hybrids provide a unique opportunity to confirm their value in enhancing genetic diversity of crop plants. Soon we should also learn whether plasmids, organelles, lipid vesicles, or other carriers will help to accomplish gene transfer into crop plants.

References

1. Alfermann, A.W., Reinhard, E. (eds): In: Production of Natural Compounds by Cell Culture Methods. Munich: Federal Ministry for Research and Technology. 361 pp. 1978
2. Amos, J.A., Scholl, R.L.: Z. Pflanzenphysiol. *90*, 33–43 (1978)
3. Arnold S.V., Eriksson, T.: Physiol. Plant. *45*, 29–34 (1979)
4. Aviv, D., Galun, E.: Z. Pflanzenphysiol. *83*, 267–273 (1977)
5. Bajaj, Y.P.S.: Physiol. Plant. *37*, 263–268 (1976)
6. Bajaj, Y.P.S.: Crop Improv. *4*, 48–53 (1977)
7. Balatková, V., Tupý, J., Hrabětová, E.: Plant Sci. Lett. *8*, 17–21 (1977)
8. Bannier, L.J., Steponkus, P.L.: HortScience *7*, 194 (1922)
9. Barnes, L.R.: In: Propagation of Higher Plants Through Tissue Culture. Hughes, K.W., Henke, R., Constantin, M. (eds), p. 250. Oak Ridge: U.S. Dep. of Energy 1978
10. Barz, W., Reinhard, E., Zenk, M.H. (eds) Plant Tissue Culture and Its Bio-Technological Application. 419 pp. Berlin-Heidelberg-New York: Springer 1977
11. Belliard, G., Pelletier, G., Vedel, F., Quetier, F.: Mol. Gen. Genet. *165*, 231–237 (1978)
12. Bhansali, R.R., Kumar, A., Arya, H.C.: Curr. Sci. *47*, 551–552 (1978)
13. Bychenkova, F.A., David, A.: Fiziol. Rast. *25*, 274–282 (1978)
14. Cailloux, M.: In: Propagation of Higher Plants Through Tissue Culture. Hughes, K.W., Henke, R., Constantin, M. (eds). p. 248. Oak Ridge: U.S. Dep. of Energy 1978
15. Cheah, K.T., Sagawa, Y.: HortScience *13*, 661–662 (1978)
16. Chen, C.H., Lo, P.F., Ross, J.G.: Crop Sci. *19*, 117–118 (1979)
17. Cocking, E.C.: In: Frontiers of Plant Tissue Culture. Thorpe, T.A. (ed.). pp. 151–158. Calgary: Int. Assoc. Plant Tissue Culture 1978
18. David, H., Isemukali, K., David, A.: C.R. Acad. Sci. *278*, 245–248 (1978)
19. Davies, M.E., Dale, M.M.: Z. Pflanzenphysiol. *92*, 51–56 (1978)
20. Debergh, P., Hargaoui, Y.: In: Propagation of Higher Plants Through Tissue Culture. Hughes, K.W., Henke, R., Constantin, M. (eds.). p. 250. Oak Ridge: U.S. Dep. of Energy 1978
21. Doré, C.: Acta Hortic. *78*, 89–93 (1977)
22. Dougall, D.K.: In: Plant Cell and Tissue Culture. Sharp, W.R., Larson, P.O., Paddock, E.F., Raghavan, V. (eds.). pp. 727–743. Columbus: Ohio State University Press 1979
23. Dougall, D.K., Wetherell, D.F.: Cryobiology *11*, 410–415 (1974)
24. Duron, M.: C.R. Acad. Sci. *284*, 183–185 (1977)
25. Duron, M., Morand, J.C.: Ann. Phytopathol. *10*, 371–374 (1978)
26. Ellyard, R.K.: HortScience *13*, 662–663 (1978)
27. Finkle, B.J., Ulrich: Plant Physiol. *63*, 598–604 (1979)
28. Gleba, Y.Y., Hoffmann, F.: Mol. Gen. Genet. *165*, 257–264 (1978)

29. Gleba, Y., Kohlenbach, H.W., Hoffmann, F.: Naturwissenschaften *65*, 655–656 (1978)
30. Gleba, Y.Y., Piven, N.M., Komarnitskii, I.K., Sytnik, K.M.: Dokl. Akad. Nauk SSSR *240*, 1223–1226 (1978)
31. Glimelius, K., Eriksson, T., Grafe, R., Müller, A.J.: Physiol. Plant. *44*, 273–277 (1978)
32. Grout, B.W.W., Henshaw, G.G.: Ann. Bot. *42*, 1227–1229 (1978)
33. Haberlandt, G.: Sitzungsber. Mat. Nat. Kl. Kais Akad. Wiss. Wien *111*, 69–92 (1902)
34. Hannig, E.: Bot. Zg. *62*, 45–80 (1904)
35. Handley, L.W., Chambliss, O.L.: HortScience *14*, 22–23 (1979)
36. Heilungkiang Institute of Forestry, Tree Improvement Laboratory: Sci. Sin. *18*, 22–23 (1979)
37. Horak, J., Lustinec, J., Mesicek, J., Kaminek, M., Polackova, D.: Ann. Bot. *39*, 571–577 (1975)
38. Hu, H., Hsi, T.Y., Tseng, C.C., Ouyang, T.W., Ching, C.K.: In: Frontiers of Plant Tissue Culture. Thorpe, T.A. (ed.). pp. 123–130. Calgary: Int. Assoc. Plant Tissue Culture 1978
39. Hughes, K.W., Barni, B.: In: Propagation of Higher Plants Through Tissue Culture. Hughes, K. W., Henke, R., Constantin, M. (eds.). pp. 248–249. Oak Ridge: U.S. Dep. Energy 1978
40. Hui, L.H., Zee, S.-Y.: Z. Pflanzenphysiol. *89*, 77–80 (1978)
41. Izhar, S., Power, J.B.: Plant Sci. Lett. *14*, 49–55 (1979)
42. Johnson, B.B.: HortScience *13*, 596 (1978)
43. Jones, O.P., Hopgood, M.E.: J. Hortic. Sci. *54*, 63–66 (1979)
44. Kartha, K.K., Leung, N.L., Gamborg, O.L.: Plant Sci. Lett. *15*, 7–15 (1979)
45. Khanna, R., Chopra, R.N.: Phytomorphology *27*, 267–274 (1977)
46. Khoder, M., Villemur, P., Jonard, R.: C.R. Acad. Sci. *288*, 323–326 (1979)
47. Knudson, L.: Bot. Gaz. *73*, 1–25 (1922)
48. Kögl, F., Haagen-Smit, A.J., Erxleben, H.: Z. Physiol. Chem. *228*, 90–103 (1934)
49. Latta, R.: Can J. Bot. *49*, 1253–1254 (1971)
50. Leddet, C.: C.R. Acad. Sci. *282*, 2083–2085 (1976)
51. Liu, M.C., Chen, W.H.: Euphytica *27*, 273–282 (1978)
52. Lowe, K.W., Conger, B.V.: In: Propagation of Higher Plants Through Tissue Culture. Hughes, K.W., Henke, R., Constantin, M. (eds.). p. 256. Oak Ridge: U.S. Dep. Energy 1978
53. Matern, U., Strobel, G., Shepard, J.: Proc. Natl. Acad. Sci. USA *75*, 4935–4939 (1978)
54. Melchers, G., Sacristan, M.D., Holder, A.A.: Carlsberg Res. Commun. *43*, 203–218 (1978)
55. Menczel, L., Lazar, G., Maliga, P.: Planta *143*, 29–32 (1978)
56. Miller, C.O., Skoog, F., Okumura, F.S., von Saltza, M.H., Strong, F.M.: J. Am. Chem. Soc. *78*, 1375–1380 (1956)
57. Mokhtarzadeh, A., Constantin, J.J.: Crop Sci. *18*, 567–572 (1978)
58. Morel, G.M.: Am. Orchid Soc. Bull. *33*, 473–478 (1964)
59. Morel, G., Martin, C.: C.R. Acad. Sci. *235*, 1324–1325 (1952)
60. Mukhopadhyay, A., Bhojwani, S.S.: Z. Pflanzenphysiol. *88*, 263–268 (1978)
61. Murashige, T.: Annu. Rev. Plant Physiol. *25*, 135–166 (1974)
62. Murashige, T.: In: Frontiers of Plant Tissue Culture. Thrope, T.A. (ed.). pp. 15–26. Calgary: Int. Assoc. Plant Tissue Culture 1978
63. Murashige, T., Nakano, R.: J. Hered. *57*, 114–118 (1966)
64. Nag, K.K., Street, H.E.: Physiol. Plant. *34*, 254–260 (1975)
65. Nag, K.K., Street, H.E.: Physiol. Plant. *34*, 261–265 (1975)
66. Nessler, C.L., Mahlberg, P.G.: Can. J. Bot. *57*, 675–685 (1977)
67. Nickell, L.G.: Crop Sci. *17*, 717–719 (1977)
68. Nickerson, N.L.: HortScience *13*, 698 (1978)
69. Pence, V.C., Hasegawa, P.M., Janick, J.: J. Am. Soc. Hortic. Sci. *104*, 145–148 (1979)
70. Quatrano, R.S.: Plant Physiol. *43*, 2057–2061 (1968)
71. Radojevic, L.: Protoplasma *96*, 369–374 (1978)
72. Radojevic, L.: Z. Pflanzenphysiol. *91*, 57–62 (1978)
73. Ramawat, K.G., Bhansali, R.R., Arya, H.C.: Phytomorphology *29*, 303–307 (1977)
74. Raghavan, V., Torrey, J.G.: Am. J. Bot. *50*, 540–551 (1963)
75. Russo, S.L., Varnell, R.J.: Soil Crop Sci. Soc. Fla. Proc. *37*, 34–36 (1977)
76. Sakai, A., Sugawara, Y.: Plant Cell Physiol. *14*, 1201–1204 (1973)
77. Sala, F., Cella, R., Rollo, F.: Physiol. Plant. *45*, 170–176 (1979)

78. Sehgal, C.B.: Z. Pflanzenphysiol. *88*, 349–352 (1978)
79. Seibert, M.: Science *191*, 1178–1179 (1976)
80. Sen, B, Gupta, S.: Physiol. Plant. *45*, 425–428 (1979)
81. Shabde, M., Murashige, T.: Am. J. Bot. *64*, 443–448 (1977)
82. Siminovitch, D.: Plant Physiol. *63*, 722–725 (1979)
83. Singh, J.P.: Biol. Plant. *20*, 436–439 (1978)
84. Sinha, M., Chaturvedi, H.C.: Curr. Sci. *48*, 176–178 (1978)
85. Skirvin, R.M., Janick, J.: HortScience *11*, 61–62 (1976)
86. Skoog, F., Miller, C.O.: Symp. Soc. Exp. Biol. *11*, 118–130 (1957)
87. Smith, R.H., Nightingale, A.E.: HortScience *14*, 20 (1979)
88. Teo, C.K.H.: Am. Orchid Soc. Bull. *47*, 1028–1030 (1978)
89. Tisserat, B., Esan, E.B., Murashige, T.: Hortic. Rev. *1*, 1–78 (1979)
90. Tomes, D.T.: Can. J. Bot. *57*, 137–140 (1978)
91. Van Oberbeek, J, Conklin, M.E., Blakeslee, A.F.: Science *94*, 350–351 (1941)
92. Walkey, D.G.A., Matthews, K.A.: Plant Sci. Lett. *14*, 287–290 (1979)
93. Wickson, M.E., Thimann, K.V.: Physiol. Plant. *11*, 62–74 (1958)
94. Wickson, M.E., Thimann, K.V.: Physiol. Plant. *11*, 62–74 (1958)
94. Withers, L.A.: Plant Physiol. *63*, 460–467 (1979)
95. Withers, L.A., Street, H.E.: Physiol. Plant. *39*, 171–178 (1977)
96. Withers, L.A., Street, H.E.: In: Plant Tissue Culture and Its Bio-technological Applications. Barz, W., Reinhard, E., Zenk, M.H. (eds.). pp. 226–244. Berlin-Heidelberg-New York: Springer 1977
97. Wu, L., Antonovics, J.: Am. J. Bot. *65*, 268–271 (1978)
98. Zelcer, A., Aviv, D., Galun, E.: Z. Pflanzenphysiol. *90*, 397–407 (1978)

Symposium on Plant Movements

Chairman: A. W. GALSTON

Circumnutation, Rhythms and Light-Regulated Movements in Plants[1]

A.W. GALSTON[2]

Circumnutation as the Basis for Tropistic Movements

Of all Charles Darwin's contributions to the study of plant movements, none is more fundamental than his suggestion that circumnutation is the basic process underlying the tropistic responses of stems and roots to unilateral stimuli such as light and gravity. Like Pfeffer before him (1), Darwin appreciated that cells around the perimeter of cylindrical organs do not all grow at the same rate, and that the rhythmic changes in these patterns of unequal growth cause the organs to oscillate around the vertical as they grow up and down, respectively. Observing the displacement of stem and root tips relative to a fixed point, Darwin was able to plot on a horizontal glass plate the approximately elliptical pattern described by the tips of such asymmetrically growing organs (2). The true three-dimensional pattern described by the growing tip was of course that of an elliptical spiral; Darwin's use of a flat recording surface fixed above the tip precluded recording in the vertical plane.

When a unilateral or asymmetric stimulus such as light was applied to such a circumnutating system, the elliptical pattern became asymmetric, and the tip was progressively displaced from the center. In three-dimensional terms, this meant that the elongating apical region bent toward or away from the direction of the greatest force of the stimulus. Thus, tropistic curvatures result from the conversion of a basically symmetrical elliptical oscillation into an asymmetric pattern, although the original symmetrical circumnutation pattern is itself the integrated result of a series of momentary displacements based on asymmetric growth.

Studies on circumnutation have not been numerous in the last century, and we still do not understand much of the physiology underlying the spiral growth observed by Darwin and others. Two theories currently contend: one says that circumnutation results from an endogenously rhythmic oscillation in growth rates that itself rotates around the stem (3). If this is true, then the rate and amplitude of circumnutation should change in response to various applied perturbations that are known to alter rhythms. The other main theory says that circumnutation results from successive overshoots of the geotropic correction of an originally randomly displaced tip (4). If this latter explanation is true, then certain mathematical relations between angular displace-

1 Paper presented as part of the Centennial Symposium honoring the publication of Charles Darwin's "The Power of Movement in Plants".

2 Department of Biology, Yale University, P.O. Box 6666, New Haven, Connecticut 06511, USA

ment and circumnutation should exist (5), and these relations are currently being tested. At the moment, the results seem somewhat ambiguous, in that they do not permit a clear choice to be made between the two theories.

To these two theories, it occurs to me that it should be possible to add a third more economical theory based entirely on known structural attributes of plant cells. It is well known that a spring caused to stretch by the fall of an attached weight is subjected not only to tension but also to torque. This can be easily visualized by attaching an arrow to the falling weight and watching its rotation, especially near the bottom of the extension path of the spring, where the change in length is minimal, but the angular rotation still considerable. The twisting effect arises solely from the spiral orientation of the body being deformed, and from the fact that tension and torque are inevitably linked in such a structure. Now let us transfer this model to a plant cell, in which the cellulose microfibrils strengthening the stretchable primary wall are spirally oriented. The force causing distention of this system is of course the internally generated turgor pressure. As in the case of the spring, the effect of stretching the spirally strengthened wall will be to generate a torque. Since the cells below the extension zone are more rigid and fixed in space, the deformation due to the torque will be manifested closer to the tip, where the walls are undergoing this plastic deformation. If the cells are mechanically tightly coupled to one another, and if the direction of the spiral is the same in all cells, then the effect will be that of twisting the end of a deformable rod. In such a case, the angle through which the rod is twisted,

$$\phi = \frac{2\,lL}{n\pi r^4}$$, where l = length of the rod, r = radius of the rod, L = torque, and n = modulus of rigidity.

This mechanism should lead to a screw-like motion which, when coupled to a more rigid basal (posterior) structure, could lead to a spiral displacement of the tip, i.e., circumnutation. Like the other two hypotheses, this one requires considerable testing before it becomes applicable to the living system, but it has the virtue of logical economy in that it does not invoke other poorly understood processes (i.e., rhythms, geotropism) to explain the phenomenon under consideration.

Whatever the mechanism by which they are produced, circumnutational movements seem linked to the appearance of tropistic curvatures, as Darwin appreciated. Indeed, on the basis of published information, it seems appropriate to propose a generalization "Ohne Cirkumnutation kein Tropismus", similar to Went's dictum "Ohne Wuchsstoff kein Wachstum" (6) relating growth to the necessary presence of auxin. The test of this generalization should help unravel the physiology underlying both processes. (See *Addendum,* page 443).

Alteration of Circumnutational Patterns by Light

Circumnutational patterns are altered by light absorbed by either phytochrome or a blue-light-absorbing photoreceptor. In the latter case, depending on the light intensity, the symmetrical oscillation is altered to an asymmetric oscillation tending toward or away from the region of higher photon flux, or even to a straight line displacement of the tip from the center, without signs of residual circumnutation. So far as we know, unilateral blue light does not change the frequency of the basic oscillatory pattern, only its relative amplitude on the two sides of the organ.

By contrast, red light perceived by phytochrome alters both the frequency and the amplitude of the circumnutational oscillation (7). In etiolated peas we found the frequency and the amplitude of the oscillation to increase 15–22 hours after the onset of photomorphogenically active weak red light. When bright white light was given, the effect was first an immediate damping of all oscillatory activity, correlated with an inhibition of growth, and then, some hours later, an enhancement of circumnutational amplitude and frequency above the control (unirradiated) value. These effects are clearly seen in a time-lapse photographic study made with photomorphogenically inactive green "safelights" (7). Whether the basic rhythm underlying the circumnutation is of the circadian type and has been rephased by the red light has not been determined.

Phototropism

When an asymmetric blue light field is imposed on a circumnutating cylindrical plant organ, the circumnutational pattern becomes asymmetric and the organ bends. At least three questions can be posed with respect to this phenomenon. (1) What is the nature of the photoreceptor pigment? (2) How is this light energy transduced into chemical or physical effects within the plant? (3) In view of the small energies required to elicit curvature, how is the effect of each absorbed photon amplified?

The action spectrum for phototropism should be useful in inferring the nature of the receptor pigment. Early determinations (8) showed peaks in the blue, near 480 nm, 440 nm, and 420 nm. The general resemblance of this action spectrum to the absorption spectrum for β-carotene and other carotenoids led, for about 20 years, to an uncritical acceptance of the idea that carotenoids were, in fact, the photoreceptors for this process. The fact that carotenoids occurred in oat coleoptiles and are near the tip, where photoreception was known to be maximal, and that they decreased in concentration down the coleoptile, where photoreceptivity was lower, was taken to bolster the hypothesis of carotenoid photoreception (9, 10). The discovery of the riboflavin-sensitized photooxidation of the growth hormone indoleacetic acid (11) and related compounds, and the similarity of the action spectrum for this process to that for phototropism (12) led to the proposal that flavins be considered as serious candidates for the photoreceptor role (13). In support of flavins over carotenoids, it was pointed out that several mutant strains of maize lacking or almost completely lacking carotenoids gave approximately normal phototropic response.

Critical proof on the nature of the photoreceptor has not yet come, almost 30 years later, although the weight of pro-flavin evidence has been growing steadily. Action spectra for phototropism extended into the ultraviolet showed peaks near 370 nm, where flavins and flavoproteins have an absorption peak but carotenoids generally do not (14). Flavins in nonpolar solvents were shown to develop the definite peaks near 420 and 440 nm that were only suggested as shoulders in aqueous solvents (15). Contrariwise, carotenoids in acetone-water mixtures were shown to develop the 370 nm peak normally lacking in nonpolar solvent solutions (16) and cis-carotenoids of a type not shown to occur naturally in plant tissues also showed this peak (17). Clearly, action spectra alone could not be used to distinguish between flavins and carotenoids as photoreceptors.

Recent interest in flavin photoreception has been heightened by the discovery of an apparently ubiquitous photoreaction in which light activation of a yet uncharacterized flavoprotein causes reduction of a b-type cytochrome (18). Originally studied in fungi (19), this reaction has also been found in higher plants, especially maize coleoptiles (20, 21) and even in human cells in tissue culture (22). In etiolated peas, the flavin-sensitized reduction of cytochrome is shown well in preparations from buds, but the stem tissue shows instead a bleaching attributable to a flavin (23), the electrons for the reduction of riboflavin to leucoflavin presumably coming from the EDTA used in the homogenization medium. An anaerobic IAA-mediated flavin bleaching had previously been described (11). It is important to note that no firm evidence links either the flavin-mediated cytochrome reduction, the photobleaching of flavin, or the flavin-sensitized oxidation of IAA to any physiological reaction. Any or none of these reactions may be linked to physiological reactions such as phototropism and only future work will help to fill in the details.

The original flavin-sensitized oxidation of IAA (11) had been the basis of a proposal (12) that photodestruction of IAA might be the cause of the observed inequities of auxin known (6) to occur on the light versus the dark sides of a unilaterally illuminated coleoptile. The subsequent proof (24) that light-mediated lateral transport of auxin, rather than auxin destruction, is involved in coleoptile phototropism led to an unwarranted wholesale rejection of the flavin hypothesis of photoreception (25). In effect, the baby had been thrown out with the bathwater. It is necessary to reassert that even if lateral auxin transport is the sole basis for phototropism in green plants (a point by no means established), then either flavins or carotenoids could still be the photoreceptors for this process. While the photoreceptor for phototropism is not yet known with certainty, even investigators unaware of the earlier work on flavin photoreception (26) are increasingly regarding the flavins as the more likely candidates for this role (27, 28).

The transduction of asymmetrically supplied light energy into curvature seems to involve effects on growth itself. In both grass coleoptiles (29) and *Phycomyces* sporangiophores (30), where blue light causes curvature, symmetrically supplied blue light causes a *light-growth reaction.* In *Avena,* a transient decrease in growth rate is followed by a supranormal overshoot; in *Phycomyces,* a transient increase in growth rate is followed by an abnormally low undershoot. In both cases, the final result is a temporally disturbed pattern, but quantitatively unaltered net elongation over the long run. Blue light also causes altered rates of cytoplasmic streaming (31) and changes in cytoplasmic viscosity (32) and facilitates subsequent thigmonastic response of pea tendrils (33); the action spectra suggest that the same pigment may be involved in all these processes, as well as in phototropism.

What is the chemical basis for these varied effects? It is important to note that while lateral auxin transport has been accepted as the basis for phototropism in grass coleoptiles, it is almost certainly not the answer for *Phycomyces,* in which auxin produces no effects on growth. Also, phototropism in radish and other dicot seedlings seems to involve auxin synthesis or longitudinal transport rates, rather than lateral transport (34); phototropism in sunflower stems may involve the photoproduction of a carotenoid-type growth inhibitor, xanthoxin (35), and in roots, phototropic curvature may involve the release and transport of another growth inhibitor, abscisic acid, from the root cap to the elongating zone (36). At the moment, a unitary mode of action of light in causing phototropic curvature is not at hand; it may be that following the initial excitation of the receptor pigment, diverse metabolic pathways are involved in different organisms.

In this brief analysis of the metabolic events underlying phototropic curvature, it remains only to point out that some amplification mechanism is necessary to account for the large growth effects resulting from the unilateral application of relatively few quanta. With regard to lateral auxin transport, I have previously calculated that each photon would have to cause the lateral migration of more than 3000 IAA molecules (13). Since such high quantum yields are virtually unknown in aqueous solutions, it seems clear that amplification must be effected either by (a) photoactivation of some enzyme with a high turnover number, (b) change in properties of membranes, or (c) gene activation. Which of these is involved in tropistic curvature is yet unknown, but gene activation would seem to be least likely on the basis of the rapidity of both the light-growth reaction and tropistic response itself, especially in *Phycomyces*.

Nyctinasty of Leaves

Darwin (2) also studied the sleep movements or *nyctinasty* of leaves. Noting the prevalence of such movements in plants of tropical origin, he attributed to the nocturnal closure, or folding together, the function of minimizing radiative heat loss to the open sky. A later proposal (37) emphasizes that the closing and subsequent tilting of the folded leaves or leaflets serves to minimize the possible interference by bright moonlight of the photoperiodic signals given by the sun during the day. Neither theory has as yet been fully substantiated.

Just as tropistic behavior seems linked to circumnutation, so nyctinasty seems linked with endogenously rhythmic events in leaf bases, especially in pulvini. In several species of legumes, especially *Phaseolus* (38), *Albizzia* and *Samanea* (39), and *Mimosa* (40), it seems clear that light-dark cycles initiate opening and closing movements that are then sustained for at least several cycles under constant conditions. These latter movements have all the requirements of true circadian rhythms (41). Light absorbed both by phytochrome (42) and blue-absorbing pigment (43) affects leaf movement. Phytochrome in the P_{fr} state at the end of the light period predisposes the leaf to subsequent closure, while a brief exposure to far-red light at this point either prevents or inhibits closure. P_{fr} phytochrome is also effective in phase-shifting the endogenous rhythm (44), the effect of red light being nicely annulled by immediately subsequent far-red. Blue light acts generally to open the leaflets; an action spectrum for this effect (45) resembles action spectra for phototropism and the other blue-light-mediated events already described.

In several nyctinastic plants, it has been shown conclusively that leaf movement results from massive shuttling of ions, especially K^+ and Cl^-, from one group of motor cells to another. Sufficiently large quantities are moved to effect the massive osmotic movement of water; this in turn affects turgor pressure, the size of motor cells and the ultimate orientation of the leaf. Both diffusive and active ion movements seem to be involved in the process (46), the former in part through leakage channels. Considerable evidence indicates that both phytochrome and blue absorber act in nyctinasty through control of membrane-localized events. In view of the evidence that phytochrome may be localized in membranes (47), some workers have proposed that blue- and red-light-absorbing pigments reside in membranes, and that their photoactivation changes ion flux by altering the physical state of membrane pores and/or permeases. The manner in which light signals affect rhythmic processes is unknown, but it is significant that considerable evidence also links rhythmic events to membrane-based phenomena (48). This, together with the apparent absence of circadian events in prokaryotic cells, emphasizes the possibility that membrane barriers within the cell are involved in generating rhythmically changing events, and that the modification of these events by absorbed photons leads to light responses.

The motor cells of *Samanea* and *Phaseolus* pulvini show endogenous circadian rhythms in transmembrane potential that may be linked to the circadian changes in salt content of their cells (49, 50). The state of phytochrome can affect the magnitude of this potential at appropriate times in the circadian cycle. It is not known whether blue light can have similar effects, but in grass coleoptiles exposed to unilateral blue light, a transverse surface electrical polarization can be detected (51) which, depending on the site of application, can act together with or in opposition to auxin. Since unambiguous evidence (52) shows that asymmetric auxin distribution precedes rather than follows the development of transverse electrical potentials, current thinking downplays the significance of the electrical data in phototropism. For most workers, the transverse potentials are merely a sign that asymmetric auxin distribution has occurred.. Yet the development of the transverse potential could have significance for the subsequent events, such as ion and water movement controlling curvature.

Many details of the mechanisms of phototropism and nyctinasty remain to be worked out, but their occurrence in organs with basic oscillatory movements underscores the possible organic linkage between the two kinds of processes.

Acknowledgment. The author's work has been supported by grants from the National Science Foundation and recently from the National Aeronautics and Space Administration as well. Drs. R.L. Satter and S.J. Britz made numerous suggestions for improvement of earlier versions of this paper.

References

1. Pfeffer, W.: Die periodische Bewegungen der Blattorgane. Leipzig 1875
2. Darwin, C., Darwin, F.: The power of movement in plants. New York: Appleton-Century-Crofts 1881
3. Heathcote, D.G., Aston, T.J.: J. Exp. Bot. *21*, 997–1002 (1970)
4. Andersen, H., Johnsson, A.: Physiol. Plant. *26*, 44–61 (1972)
5. Brown, A.H., Chapman, D.K.: Plant Physiol. *59*, 636–640 (1977)
6. Went, F.W.: Rec. Trav. Bot. Néerl. *25*, 1–116 (1929)
7. Galston, A.W., Tuttle, A.A., Penny, P.J.: Am. J. Bot. *51*, 853–858 (1964)
8. Johnston, E.S.: Smithson. Inst. Miscell. Coll. *92*, 1–17 (1934)
9. Wald, G., duBuy, H.G.: Science *84*, 247 (1936)

10. Bünning, E.: Planta *27*, 148–158 (1937)
11. Galston, A.W.: Proc. Natl. Acad. Sci. USA *35*, 10–17 (1949)
12. Galston, A.W., Baker, R.S.: Am. J. Bot. *36*, 773–780 (1949)
13. Galston, A.W.: Encycl. Plant Physiol. *XVII* (1), 492–529 (1959)
14. Curry, G, Gruen, H.E.: Proc. Natl. Acad. Sci. USA *45*, 797–804 (1959)
15. Harbury, H.A., LaNoue, K.F., Loach, P.A., Amick, R.M.: Proc. Natl. Acad. Sci. USA *45*, 1708–1717 (1959)
16. Hager, A.: Planta *91*, 38–53 (1970)
17. Inhoffen, H.H., Bohlmann, F., Rummert, G.: Liebigs Ann. Chem. *571*, 75–83 (1950)
18. Muñoz, V., Butler, W.A.: Plant Physiol. *55*, 421–426 (1975)
19. Poff, K., Butler, W.A.: Plant Physiol. *55*, 427–429 (1975)
20. Brain, R.D., Freeberg, J.A., Weiss, C.V., Briggs, W.R.: Plant Physiol. *59*, 948–952 (1977)
21. Goldsmith, M.H.M., Caubergs, R., Briggs, W.R.: Plant Physiol. Suppl. *63*, 155 (1979)
22. Pereira, O.M., Smith, J.R., Packer, L.: Photochem. Photobiol. *24*, 237–242 (1976)
23. Galston, A.W., Britz, S.J., Briggs, W.R.: Carnegie Inst. Yearb. *76*, 293–295 (1977)
24. Briggs, W.R., Tocher, R.D., Wilson, J.F.: Science *125*, 210–212 (1957)
25. Thimann, K.V., Curry, G.M.: Comp. Biochem. *1*, 243–309 (1960)
26. Galston, A.W.: Photochem. Photobiol. *25*, 503–504 (1977)
27. Song, P.S., Moore, T.A.: Photochem. Photobiol. *19*, 435–441 (1974)
28. Presti, D., Delbrück, M.: Plant, Cell Environ. *1*, 81–100 (1978)
29. Blaauw, A.H.: Meded. Landbouwhogesch. Wageningen *15*, 89–204 (1919)
30. Blaauw, A.H.: Rec. Trav. Bot. Néerl. *5*, 209–372 (1909)
31. Bottelier, H.P.: Proc. K. Akad. Wet. *36*, 790–795 (1933)
32. Virgin, H.: Physiol. Plant. *1*, 147–155 (1948)
33. Shotwell, H., Jaffe, M.J.: Photochem. Photobiol. *29*, 1153–1156 (1979)
34. Van Overbeek, J.: Rec. Trav. Bot. Néerl. *30*, 537–626 (1933)
35. Thompson, A.G., Bruinsma, J.: J. Exp. Bot. *28*, 804–810 (1977)
36. Wilkins, M.B.: In: Plant Growth Regulation. Pilet, P.E. (ed.). pp. 199–207. Berlin-Heidelberg-New York: Springer 1977
37. Bünning, E., Moser, I.: Proc. Natl. Acad. Sci. USA *62*, 1018–1022 (1969)
38. Bünning, E., Moser, I.: Proc. Natl. Acad. Sci. USA *70*, 3387–3389 (1973)
39. Satter, R.L., Galston, A.W.: BioScience *23*, 407–416 (1973)
40. Fondéville, J.C.: C.R. Soc. Biol. *157*, 1291–1294 (1963)
41. Bünning E.: Cold Spring Harbor Symp. Quant. Biol. *25*, 249–256 (1960)
42. Fondéville, J.C., Borthwick, H.A., Hendricks, S.B.: Planta *69*, 357–364 (1966)
43. Jaffe, M.J., Galston, A.W.: Planta *77*, 135–141 (1967)
44. Simon, E., Satter, R.L., Galston, A.W.: Plant Physiol. *58*, 421–425 (1976)
45. Evans, L.T., Allaway, W.G.: Austr. J. Biol. Sci. *25*, 885–893 (1972)
46. Satter, R.L., Geballe, G.T., Applewhite, P.B., Galston, A.W.: J. Gen. Physiol. *64*, 413–430 (1974)
47. Haupt, W.: Annu. Rev. Plant Physiol. *16*, 267–290 (1965)
48. Njus, D., Sulzman, F., Hastings, J.W.: Nature (London) *248*, 116–120 (1974)
49. Racusen, R.H., Satter, R.L.: Nature (London) *225*, 408–410 (1975)
50. Scott, B.I.H., Gulline, H.F., Robinson, G.R.: Austr. J. Plant Physiol. *4*, 193–206 (1977)
51. Schrank, A.R.: Plant Physiol. *21*, 362–365 (1946)
52. Grahm, L., Hertz, C.H.: Physiol. Plant. *15*, 96–114 (1962)

Addendum. Recent work by S.J. Britz at Yale shows that circumnutation and geotropism *can* be separated. Etiolated pea seedlings do not circumnutate, yet geotrope as well as preilluminated seedlings, which circumnutate vigorously. Also, the latter geotrope equally parallel or perpendicular to the plane of the hook, yet circumnutation is much greater in the parallel plane.

Phototropism as a Phenomenon of Inhibition

J. BRUINSMA, J.M. FRANSSEN, and E. KNEGT[1]

Introduction

The concept of hormonal regulation of plant growth, gradually developed during the first quarter of this century, eventually led to the Cholodny-Went theory of tropistic movements. According to this theory the curving response of stems and roots to unilateral photo- and geotropic stimuli results from an uneven lateral distribution of auxin. It was earlier reported (2) that the evidence in favor of the Cholodny-Went theory of phototropism is only indirect, and that physicochemical analysis of the distribution of endogenous indoleacetic acid (IAA) in the phototropically curving hypocotyl of green sunflower seedlings could not establish any lateral auxin gradient. As an alternative, an inhibitor gradient was suggested, with xanthoxin as a possible candidate, analogous to the probable function of abscisic acid in the geotropism of roots. The present paper provides further evidence for the view that phototropism is a phenomenon of growth inhibition, of course requiring a growing tissue to take place, but independent of the rate of elongation of that tissue.

Materials and Methods

Plant material was the 6-day-old sunflower seedling, *Helianthus annuus* L., grown in a 14-h photoperiod at 25°C; for details see (1). This seedling responds to unilateral illumination with white light with a curvature that reaches a maximum value after about one hour (Fig. 1). Owing to geotropic stimulation of the inclining upper part of the hypocotyl, the curvature decreases again and an undulating movement develops, also because of the interference of a slow but appreciable circumnutation.

For analysis of endogenous growth substances only the curving parts of the hypocotyls were sampled. The determination of IAA was performed spectrophotofluorometrically (4). Xanthoxin had to be determined with a bioassay, because insufficient pure reference material was available for a gas-chromatographic determination until recently (3). The cress seed germination test as described by Thompson and Bruinsma (5) is much more sensitive to xanthoxin than to abscisic acid. The two inhibitors are easily separated because of the neutral character of xanthoxin.

---

1 Department of Plant Physiology, Agricultural University, Wageningen, The Netherlands

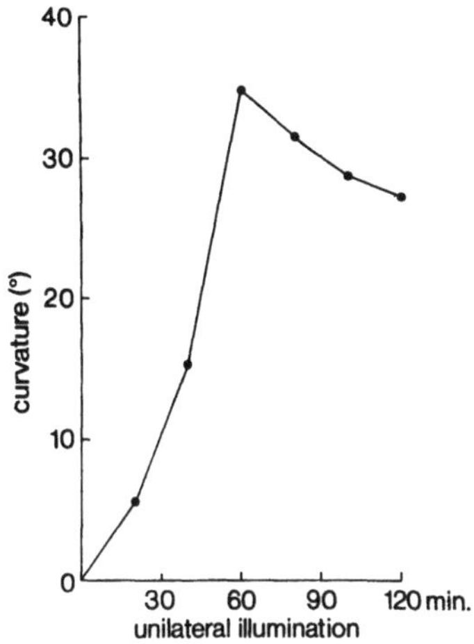

Fig. 1. Phototropic response of a typical sunflower seedling during unilateral illumination with white light. Conditions as described in (1)

Results and Discussion

Lateral Distribution of IAA

At the onset of unilateral illumination, both sides of the straight hypocotyl contained the same amount of IAA (Table 1). Because of variation between experiments, the data are expressed as percentages; on an average, the tissue contained 35 ng IAA per g fresh weight.

Table 1. The lateral distribution of IAA during phototropic bending. Data from 5 experiments. n = number of seedlings per experiment

0 min.		half 1	half 2		n
	exp 1	49.5	50.5		55
	exp 2	48.5	51.5		60
30–45 min.	exp 3	54	46	curvature	n
		46	54	← hν	
		56	44	17°	415
240 min.		shaded half	light half	curvature	n
	exp 4	48	52	21°	44
	exp 5	49	51	23°	45

Table 2. The lateral distribution of xanthoxin during phototropic stimulation. Data from 6 experiments, n = number of seedlings per experiment

0 min.		half 1	half 2		n
	exp 6	48	52		208
	exp 7	49	51		235
30–45 min.		shaded half	light half	curvature	n
	exp 8	28	72	14°	395
	exp 9	39	61	17°	475
60–80 min.		shaded half	light half	curvature	n
	exp 10	38	62	35°	422
	exp 11	44	56	34°	433

In the course of the first bending period the distribution of IAA did not change. In experiment 3, after an average curvature of 17° had been reached, auxin analysis of the illuminated and shaded sides of three portions of the bending zone confirmed that auxin did not accumulate on the shaded side. This is one typical example of several experiments that always gave the same result [see (2)]. During continued bending, no change in the auxin distribution occurred.

Lateral Distribution of Xanthoxin

The lateral distribution of xanthoxin during phototropic bending turned out to be much more interesting than that of IAA: the equal distribution in the straight hypocotyl rapidly changed upon unilateral illumination (Table 2). When the curving process was fully underway, the illuminated side contained far more xanthoxin than the shaded side so that, in spite of an equal distribution of auxin, cell elongation on the illuminated side would be considerably inhibited, causing the hypocotyl to curve toward the light source. Since the amounts of xanthoxin were similar to those of IAA, 22 ng per g fresh weight on average, such an inhibition is quite feasible. After further curvature the unequal distribution of xanthoxin persisted, although it was perhaps less distinct.

Xanthoxin Content and Phototropic Sensitivity

At that time, a possible role of xanthoxin in phototropism was indicated by the finding that etiolated sunflower seedlings, which are unable to give a phototropic response, are practically devoid of xanthoxin (5). A further indication that xanthoxin is involved

pretreatment in	ng Xa/g fr.w. (t=o) exp 1	exp2
white light	39.5	33.3
blue light	38.3	28.0
red light	40.7	19.0
darkness	15.6	13.9

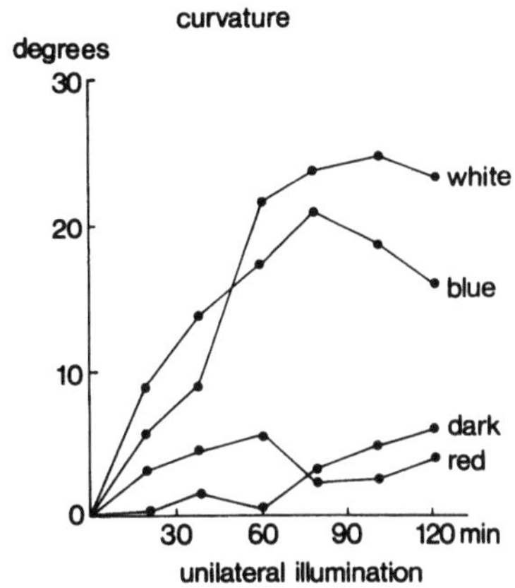

Fig. 2. Xanthoxin content and phototropic responsiveness after 18 h pretreatment under different illumination conditions. Data from samples of 100 hypocotyls

was derived from experiments in which light-grown seedlings were pretreated under different illumination conditions prior to exposure to unilateral white light (Fig. 2).

A pretreatment of 18 h darkness strongly reduced both the phototropic sensitivity and the xanthoxin content of the hypocotyls. Since the dose-response curves of such inhibiting substances are linear rather than logarithmic, a decrease to about one third must have considerable physiological consequences. In contrast, pretreatment with blue light was phototropically almost as effective as white light, and did not influence the amount of xanthoxin. Pretreatment with red light was almost as adverse as darkness to the phototropic sensitivity of the sunflower hypocotyl. In one experiment this was accompanied by a considerable drop in the xanthoxin content, but in another it was not. Results of experiments on the possible influence of phytochrome in the phototropic response pointed towards changes in the intracellular distribution of xanthoxin; red light might affect this distribution.

Phototropism and Elongation Growth

Various experiments indicated that the phototropic sensitivity of the hypocotyl tissue is relatively independent of the rate of its elongation growth. Firstly, seedlings in blue light grow faster than those in white light (1.3 $mm.h^{-1}$ versus 1.0 $mm.h^{-1}$), yet their phototropic response is often slightly less than that of white-grown seedlings (Fig. 2). Secondly, dark-grown seedlings de-etiolate far better in red light than in blue light: the cotyledons become darker green and open wider, the elongation growth of the hypocotyl is more inhibited and its radial extension is enhanced. Nevertheless, the phototropic sensitivity of these seedlings is very poor (Fig. 2). Apparently, elongation growth is a phytochrome-directed phenomenon whereas the development of phototropic sensitivity is not.

That phototropism and elongation growth are relatively independent phenomena was also demonstrated in experiments in which the elongation growth was retarded by the addition of chlormequat (CCC) to the soil. Ten and 50 mM CCC caused growth in-

Table 3. Reduction of elongation growth by the addition of chlormequat to the soil of sunflower seedlings. n = number of seedlings per sample

Treatment	Length on day 5 (cm)	n
Control	4.4 ± 1.4	137
10 mM CCC	3.6 ± 1.1	92
50 mM CCC	2.5 ± 0.8	95

hibitions of 28% and 40%, respectively (Table 3). Yet, when seedlings grown to the same length at these different growth rates were subjected to identical phototropic stimuli, they gave identical responses (Fig. 3).

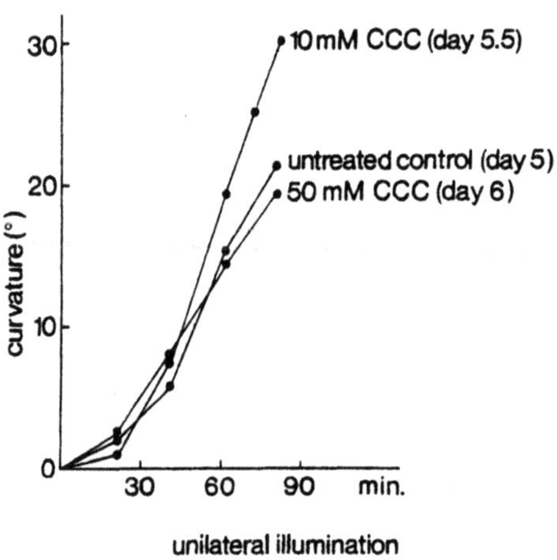

Fig. 3. Phototropic response of 4–5 cm long seedlings with different rates of elongation

Conclusions

Phototropism in the sunflower seedling is a process of growth inhibition, relatively independent of the rate of elongation growth.

Xanthoxin is involved as a growth-inhibiting substance because the degree of phototropic sensitivity is generally related to the xanthoxin content and, moreover, the lateral distribution of xanthoxin becomes unequal, particularly during phototropic curvature, in such a way that it can account for the curvature.

The synthesis, translocation, and metabolism of xanthoxin require further investigations.

References

1. Bruinsma, J., Karssen, C.M., Benschop, M., van Dort, J.B.: J. Exp. Bot. *92,* 411–418 (1975)
2. Bruinsma, J.: In: Plant Growth Regulation. Pilet, P.E. (ed.). pp. 218–225. Berlin-Heidelberg-New York: Springer 1977
3. Franssen, J.M., Knegt, E., Bruinsma, J.: Z. Pflanzenphysiol. *94,* 155–158 (1979)
4. Knegt, E., Bruinsma, J.: Phytochemistry *12,* 753–756 (1973)
5. Thompson, A.G., Bruinsma, J.: J. Exp. Bot. *28,* 804–810 (1977)

Hormonal Control of Root Georeaction: Some Light Effects

P.E. PILET[1]

Introduction

In darkness, when kept in horizontal position, primary roots of a few plants do not georeact (4) or only bend slowly (18), whereas exposed to light they respond promptly. This is because light induces the formation (17, 31, 32, 33), in the cap, of growth inhibitors which then accumulate in the lower part of a horizontally oriented root, decreasing its elongation relative to the upper part (11, 16) and thus causing downward curvature. While these inhibitors are produced only in light, the actual georeaction can occur in the dark (14, 15, 29). According to unpublished data of G.S.B. Gibbons and M.B. Wilkins (32), the light-induced positive geotropic sensitivity of maize roots is only temporary.

On the other hand, an inhibitory effect of white light on the elongation of primary roots, respectively of wheat, rice, lentil, and pea, has been previously reported (24, 21, 1, 6, 8). In maize roots this growth inhibition was found, by experiments involving removal of the root cap, to be dependent on the perception of light by the root cap (14, 15, 25, 27).

Pilet and Ney (20) observed a strong and rapid inhibition of root elongation when a microbeam of light was directed at the cap, while when the growing zone was illuminated, roots continued to grow at the same rate as before the illumination.

The objectives of the present report are to discuss (1) some biological properties of the growth inhibitors formed, in light, in the root cap, (2) the light effects on georeaction and the importance of the cap in light perception, (3) the localized light action, (4) the effects of light pretreatment on the magnitude and the duration of the light-induced geosensitivity in roots, and (5) the chemical nature of these inhibiting substances, more particularly the implication of abscisic acid (ABA) as one of the growth inhibitors in the georeaction controlled by light.

Materials and Methods

The conditions for preparing maize caryopses (several varieties – which will be indicated for each assay – have been used) have been previously described (14, 17). When the primary radicles, which elongated vertically, reached 15 ± 3 mm in length, they

---

1 Institute of Plant Biology and Physiology of the University, 6, Place de la Riponne, 1005 Lausanne, Switzerland

were selected in green light (530 ± 20 nm; 1.2 $\mu W\ cm^{-2}$ at the root level) and only entirely straight apical segments, 10 ± 0.2 mm long, were cut and mounted in plastic frames with their basal ends in contact with buffered (pH 6.1) moist filter paper. The frames were then placed (17) in temperature-controlled boxes (22 ± 1°C) in which a humid atmosphere (90% ± 5%) was maintained. For light exposure, a white fluorescent lamp (13 W/25-ky X; 0.90 ± 0.06 $J\ m^{-2}\ s^{-1}$: Philips, Eindhoven, The Netherlands) was used. Curvature was recorded by means of shadow photographs. Standard errors of the mean have been calculated and significant differences assessed by the *t*-test.

Results and Discussion

Some Biological Properties of the Growth Inhibitors

A few data will briefly be discussed concerning the *production* and the *transport* of the growth-inhibiting substances in apical maize (cv. Kelvedon 33) root segments exposed to light.

At first (Fig. 1) segments were horizontally placed (A), and after 6 h developed a strong curvature. Their caps were removed. Simultaneously, vertical segments were decapped (B). Geostimulated caps were then placed (C) on the apical cut ends of these vertical segments. Significant bending occurred after 4 h, indicating that the elongation of the stump was less on the side in contact with the originally lower half of the cap. This clearly indicates (16) that the lower half of the cap from a geostimulated root produced growth inhibitor(s) in larger amount(s) than the upper half.

In a second series of assays (Fig. 2), half-decapitated segments (A) developed a curvature toward the side on which the remaining half-tip was located. The insertion of a mica barrier into the side of the root without the half-tip (B) did not change the reaction. In contrast, mica inserted into the side with the half-tip present (C) caused an inversion of the curvature. It is concluded (11–13, 17) that inhibitors move basipetally from the cap to the extending and georeacting zone of roots and that they can also be transported laterally inside the root apex.

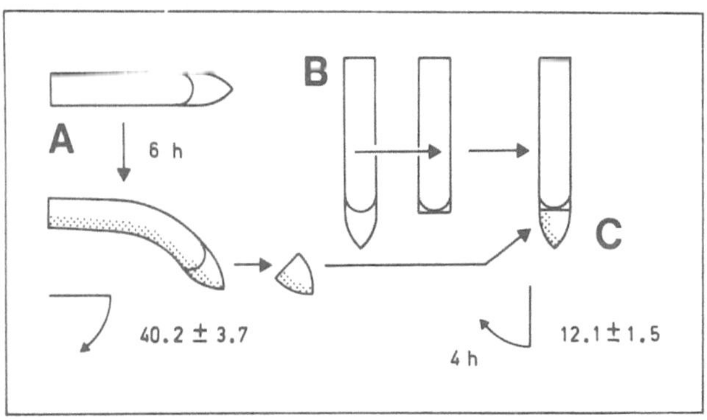

Fig. 1. Curvature in degrees (± standard error) of horizontally placed root segments after 6 h (A) and of vertical segments after 4 h, the caps of which were removed (B) and replaced by caps from 6-h geostimulated roots, caps covering exactly the apical cut sections (C). Segments not to scale

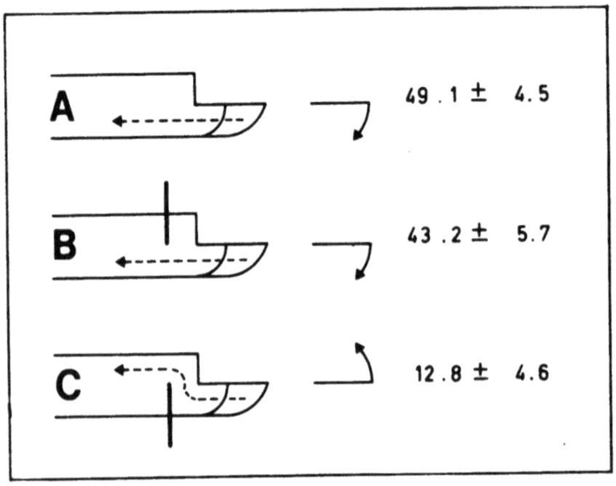

Fig. 2. Curvature in degrees (± standard error) after 14 ±0.5 h of half-decapitated root segments, horizontally placed, with or without a mica barrier. For further details see text

In another set of experiments (Fig. 3), the mica was inserted into the tip of the horizontal segments either vertically (A, B) or horizontally (C, D). The mica was introduced either into the cap (A, C) or the tip (B, D). The presence of the barrier reduced the 6-h curvature, but no differences were obtained whether the vertical mica was inside the cap or the tip. Similar data were found with a horizontal barrier placed in the cap only. But a significant decrease in the reaction was observed when the mica was inserted horizontally into the cap and the apex (D). These results confirm (16, 17) that the quantitiy of inhibitors produced by the lower part of the cap of a horizontal root is greater than that produced by the upper part and that the lateral transport of these inhibiting substances occurs inside the apex only.

As summarized in Fig. 4, elongation and georeaction of the horizontal root segments kept in light are controlled, at least in part, by a growth inhibitor system transported basipetally. Its production is found to be greater in the lower part of the cap (arrows 1 + 2) and it moves laterally inside the apex only (arrow 3). Consequently, the asymmetrical distribution of these inhibitors (arrows 4 + 5) induces the positive georeaction of the geostimulated apical root segments. A second flow – which may also act in the regulation of root growth and geotropism – corresponds to an acropetal transport of auxin (arrows 6 + 7) which accumulates in the root cap (5, 9, 17).

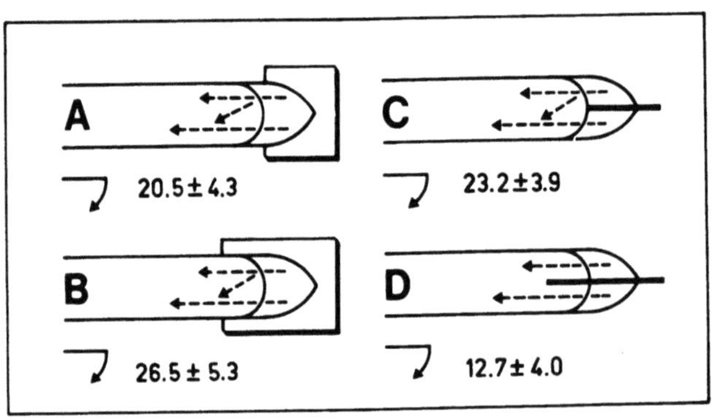

Fig. 3. Curvature in degrees (± standard error), after 6 h, of horizontal segments into which mica had been inserted vertically (A, B) or horizontally (C, D) inside the cap (A, C) or the tip (B, D)

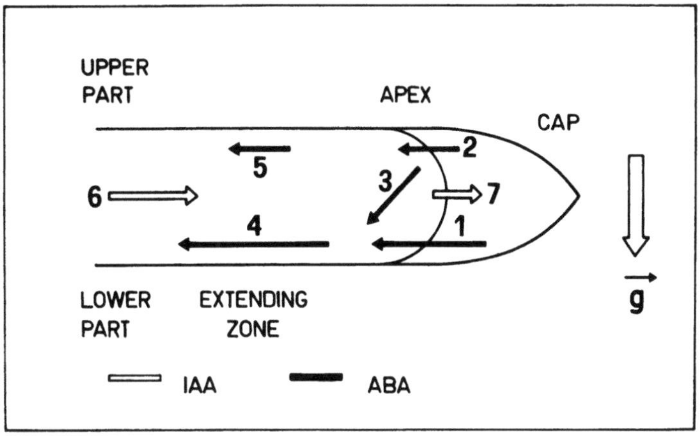

Fig. 4. Diagram showing the two hormone flows inside the root of maize horizontally maintained. For further details see text

Light Effects on Gravitropism

In the next experiments, apical root segments were prepared from two maize varieties, Anjou 210 and Kelvedon 33.

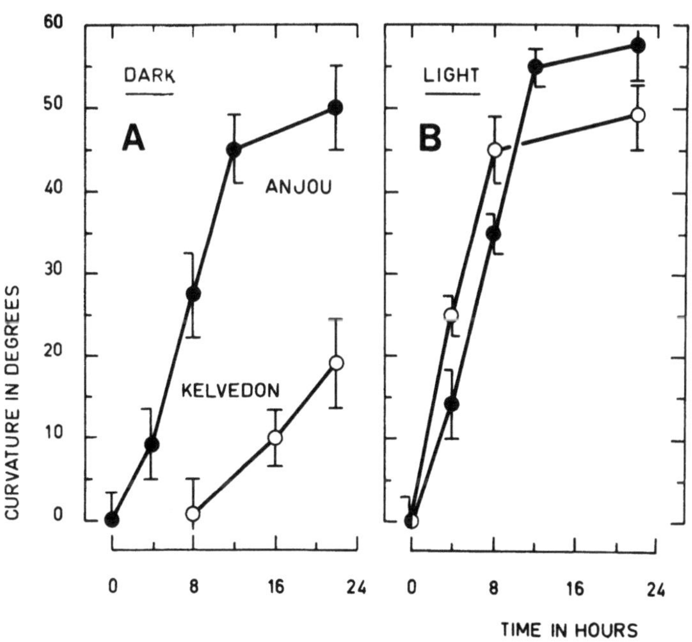

Fig. 5A and B. The georeaction (downward curvature in degrees) of horizontal root segments, prepared from two maize varieties (Anjou 210 and Kelvedon 33), with time. Segments kept in darkness (**A**) and in white light (**B**). The points and vertical lines represent the mean value ± standard error

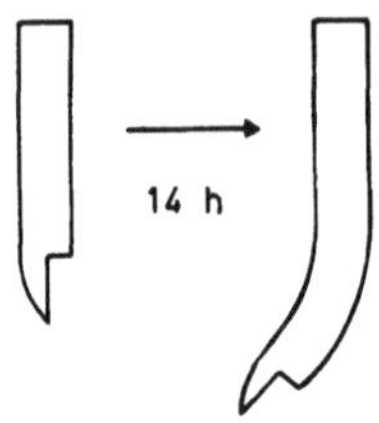

	DARK	LIGHT
ANJOU	22.7 ± 3.5	28.4 ± 5.6
KELVEDON	4.6 ± 3.9	24.5 ± 4.1

Fig. 6. The curvature (toward the remaining half-tip) of vertical half-decapitated segments of maize roots prepared from two varieties (cv. Anjou 210 and Kelvedon 33). Measurements after 14 ± 0.5 h

At first, georeaction of these segments was analyzed in the dark and in light. As can be seen in Fig. 5, in darkness, segments of the Anjou maize were strongly georeactive, while those of Kelvedon maize reacted slightly only after 8 h. However, in the light, both were very georeactive (14).

Then, vertical segments were half-decapitated (at 0.7 ± 0.2 mm). The curvature due to the asymmetrical production of the growth inhibitors from the cap was measured after 14 ± 0.5 h (Fig. 6). Both in dark and light, root segments from Anjou developed a significant curvature toward the side on which the remaining half-tip was located. It was also the case for the Kelvedon maize in light, while the segments of this variety gave only a very slight reaction in dark (14, 15).

Consequently, the georeaction of horizontal apical root segments was quite similar to the bending of vertical half-decapitated segments in relation to treatment with light or darkness. As mentioned already, the root cap contains a critical part of the geotropic mechanism, since it is the source of at least one inhibiting substance. In darkness, the segments of Kelvedon maize were not georeactive for the first few hours of the horizontal presentation, and the production (or activity) of their growth inhibitor was low.

In another assay, horizontal segments were decapitated (at 0.7 ± 0.2 mm) and their tips immediately replaced on the cut sections. As presented in Fig. 7, the georeaction (measured after 14 ± 0.5 h) of the Anjou maize (A) in darkness and in light and of the Kelvedon maize (B) in light was still significant. But the re-headed roots curved less than the intact roots and began to bend later on. Then the tips were exchanged. As can be seen, georeactivity was only related to the nature of the cap. In the dark, the curvature was directly associated with the presence of the tip from Anjou maize (D), while the segments with a Kelvedon tip (C) did not react. Under light, however, all the segments were significantly georeactive (14, 18).

The necessity for light as a prerequisite for normal geotropism seems to differ for roots prepared from several maize varieties (18). In Fig. 8, downward curvature with time is reported for four varieties. As can be seen, the reaction of two of them, cv. Anjou and Kiowa, was the same either in darkness or in light, while roots of cv. Orla and Kelvedon georeacted strongly in light only.

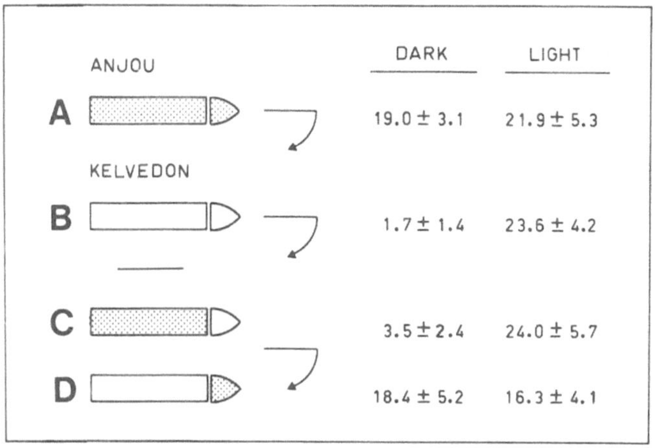

Fig. 7. The georeaction of horizontal segments decapitated with the tip exactly replaced on the apical cut section. Measurements after 14 ± 0.5 h in darkness and in white light. Downward bending in degrees ± standard error

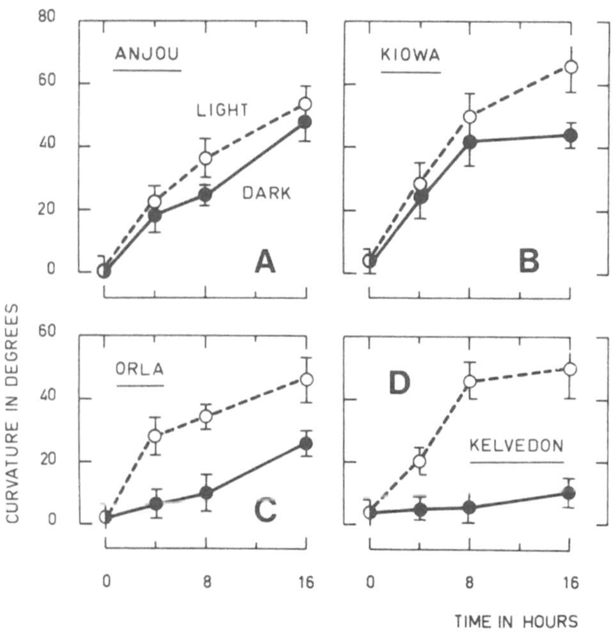

Fig. 8A–D. Downward curvature in degrees (± standard error) of horizontal segments (prepared from four maize varieties) with time, in darkness and in white light

Localized Light Action

To confirm the site of light perception in roots, optical fibers were used for the localized light treatment on maize (cv. INRA 258) primary roots, and their elongation was measured every 5 min (Fig. 9). When the cap of dark-grown roots was illuminated (A),

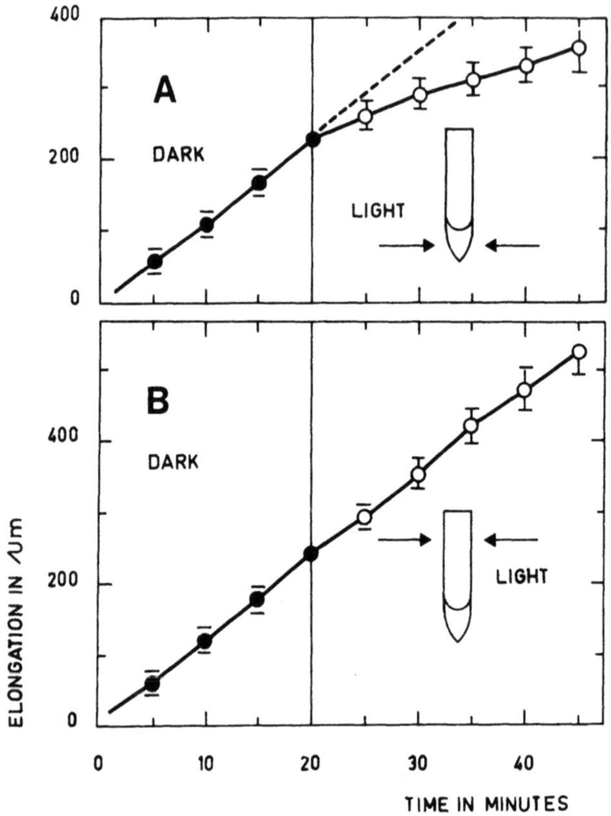

Fig. 9A and B. Elongation (measured in µm, every 5 min) of roots first kept in the dark and then exposed to microbeams of white light given on the root cap, between 0 and 0.5 mm from the tip (**A**) and on the growing zone, between 2.5 and 3.5 mm (**B**). The points and vertical lines represent the mean value ± standard error

a strong and very rapid reduction in growth rate was observed, the lower growth rate being reached after only a few minutes. When the microbeam irradiation was given only to the growing part of the root (B), light had no significant effect, and the roots continued to grow at the same rate as before illumination. Thus it is confirmed directly that the cap is the perception site of light that causes a decrease in the rate of root elongation (20).

Effects of the Light Pretreatment

Data on the magnitude and duration of the light-induced geosensitivity in root segments (prepared from the maize Orla 264 variety) will now be discussed.

At first (Fig. 10), segments were kept 150 min in vertical position and then placed horizontally in darkness for 6 h. While the segments were in the vertical position, they were illuminated for x min after having been in the dark for 150-x min. In complete darkness (x = 0), the segments exhibit a georeaction; however, the response increases greatly with increasing duration of the light pretreatment until it becomes constant at

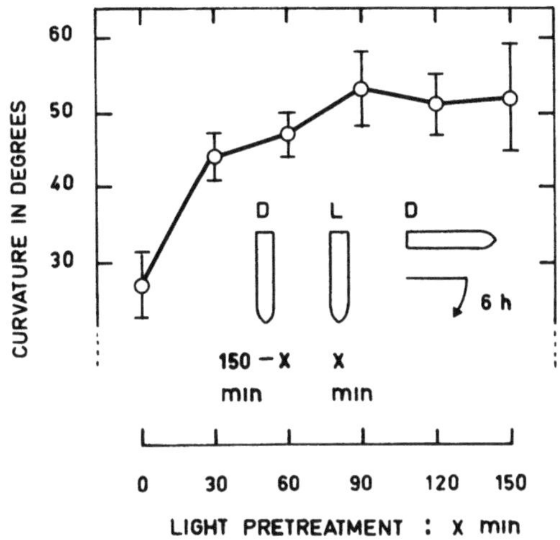

Fig. 10. Curvature (degrees ± standard error) of segments in relation to the duration (x min) of a light pretreatment. Segments kept 150 min vertically, first in darkness (*D*) for (150-x) min and then in light (*L*) for x min; then placed horizontally for 6 h in the dark

90 min pre-illumination. Thus the growth inhibitors produced in the root cap on exposure to light can be used by the root segments to bend in response to gravity in darkness (19).

The disappearance of the light effect on the subsequent georeaction in darkness will be analysed next (Fig. 11). Segments were kept vertically for 5 h and 45 min; then they were placed horizontally for 6 h in darkness and their curvature was measured. While in the vertical position, the segments received 45 min of light; this treatment was given at different times during this period so that the roots were in darkness for 5-y h prior to the light exposure and for y h after the latter. The results showed that the georeaction did not change during a relatively short dark period following the root illumination, but decreased when this dark period was extended. This decline in the geore-

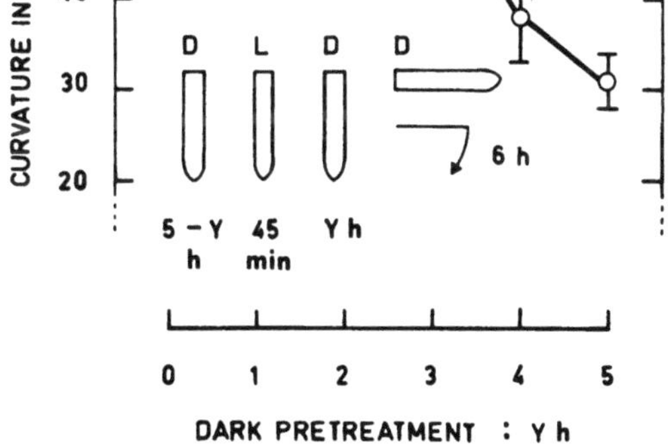

Fig. 11. Curvature (degrees ± standard error) of segments in relation to the duration (y h) of a dark (*D*) pretreatment following a light treatment of 45 min. Segments kept vertically for 5 h and 45 min, first in darkness (*D*) for 5-y h, then in light (*L*) for 45 min and in the dark (*D*) again for y h. Segments were placed in horizontal position for 6 h

action increased with increasing duration of the dark treatment, and after 5 h of darkness dropped to the level in controls, i.e., in roots which had not been exposed to light at all.

Consequently, the light-induced geotropic responsiveness can be completely lost during a dark period between the light treatment and the geostimulation of the root. This loss may be explained by assuming either that the inhibiting substances formed in the cap in light are of only limited stability in the dark or that they are transported beyond the elongation zone of the roots where they cannot exert an influence on growth and curvature.

Chemical Nature of the Growth Inhibitors

It has been shown that abscisic acid (ABA) occurs in the whole root of maize. Kundu and Audus (2, 3) detected the presence of a cap inhibitor whose chromatographic properties were similar to those of ABA. This hormone appeared to be produced in the cap in response to light (14, 15, 27). ABA, which inhibits root growth (7, 10), seems to have the same effects on root elongation as the root cap inhibitors (13, 14, 15, 29).

By using GC/MS analysis, it was found that ABA is present in the cap and in the apex of maize (cv. Kelvedon 33) exposed to light (22). On the other hand, the total amount of ABA in maize (cv. Golden Cross Bantam 70), measured by gas-liquid chromatography, increased after irradiation by a factor of ca. 1.8 (23).

Data will first be reported about the ABA effect on the curvature of decapitated segments and then about the endogenous ABA tested by the GC/MS technique.

In a first set of assays (Figs. 12 and 13), apical root segments – immediately after decapitating (at 0.7 ± 0.1 mm) – were fixed with their basal cut sections covered with a moist filter-paper buffered at pH 6.1. Their apical cut ends were half-covered (lower part) by a half-agar block; the cylindrical agar block (diameter: 2.5 mm; thickness: 1 mm) with or without ABA. Buffered (pH 6.1) agar at 1.8% was used. Curvature of the root segments was recorded after 10 ± 0.5 h. To reduce the standard error of the mean, a control assay (no ABA) was run for each series of experiments related to one tested concentration of ABA.

As an example, one set of assays (segments prepared from the Anjou variety) is presented in Fig. 12. As can be seen, the curvature obtained increased with increasing concentration of ABA (from 10^{-8} M to 10^{-4} M). For each concentration used, no significant differences could be noted between the bending of segments maintained in darkness and in light. Because the control assays varied 6.4 to 12.5 degrees, the data are also expressed in per cent of the control with their calculated standard deviation. Only these values should be used for comparing the ABA effect on bending (13, 15).

The other maize varieties were similarly assayed (Fig. 8), and the data (only relative values are reported) are summarized in Fig. 13. As can be seen, the georeaction of the decapitated root segments prepared from the four varieties was induced by an asymmetrical application of ABA supplied on the apical cut ends. This reaction, as already observed (Fig. 12), was found to be always enhanced by an increasing (from 10^{-8} to 10^{-4} M) ABA concentration. The ability of ABA – when moving basipetally – to inhibit elongation of the extending lower part of the root is clearly demonstrated

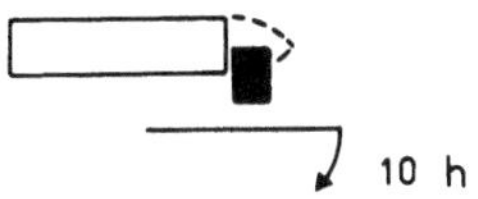

	DARK	LIGHT
$10^{-\infty}$ M : C	10.4 ± 2.3	7.2 ± 3.2
10^{-8} M : T	16.7 ± 5.9	13.5 ± 3.1
% = 100 (T − C) / C	60.6 ± 11.7	17.5 ± 15.9
$10^{-\infty}$ M	8.5 ± 6.0	6.4 ± 2.1
10^{-6} M	19.9 ± 4.2	14.4 ± 3.2
%	134.1 ± 28.6	125.0 ± 15.9
$10^{-\infty}$ M	9.8 ± 4.5	12.5 ± 6.2
10^{-4} M	27.6 ± 5.2	33.3 ± 5.9
%	181.6 ± 24.3	166.4 ± 23.9

Fig. 12. Curvature (in degrees and in % ± standard error) after 10 ± 0.5 h of decapitated segments (horizontally kept) prepared from maize (cv. Anjou 210) roots with their apical cut ends half-covered, on the lower part, by a buffered agar block containing ABA (at several concentrations in M)

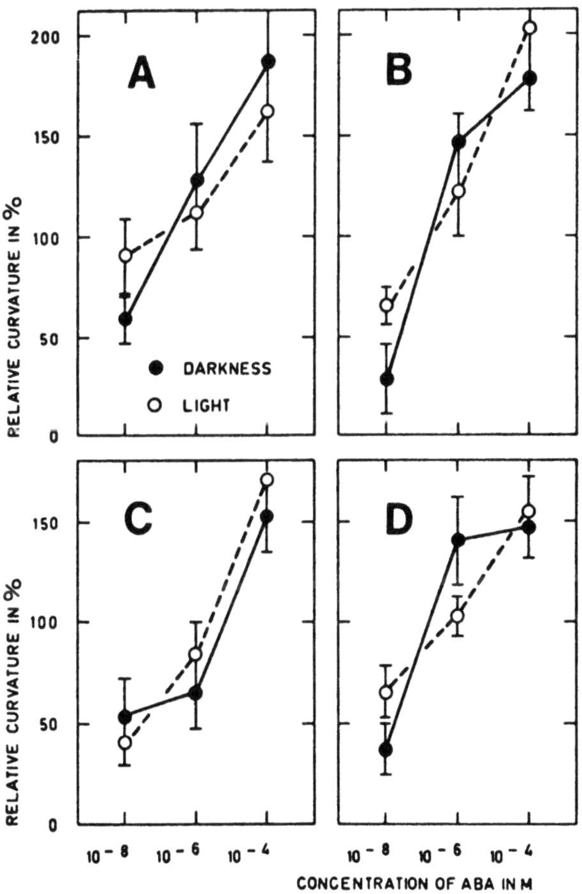

Fig. 13A–D. Curvature (relative values in % ± standard error) after 10 ± 0.5 h of decapitated maize (four varieties used: **A** Anjou 210; **B** Kiowa; **C** Orla 264; **D** Kelvedon 33) root segments with their apical cut sections half-covered, on the lower part, by a buffered agar block (see Fig. 12) containing ABA at three concentrations in M

here: ABA, applied asymmetrically to the lower part of the apical cut end, induced a positive georeaction. This obviously indicates (as already discussed above) that ABA, in the cap, could be one of the inhibitors involved in the growth response of roots to gravity.

On the other hand, no significant difference could be observed between the georeaction of the segments from the Anjou (A) and the Kiowa (B) varieties, in the dark and in light. This is not surprising because these varieties georeacted similarly in darkness and in light (Fig. 8). Equal curvatures due to ABA were obtained both in dark and light for segments prepared from Orla (C) and Kelvedon (D) which, as reported in Fig. 8, developed only positive bending (at least for the first few hours when in a horizontal position (18).

In the last series of experiments, ABA content was analyzed in the segments by using the GC/MS technique (22). First, the ABA level was studied in the tip of Kelvedon maize roots, kept in light before analyses. As shown in Table 1, ABA was present in the cap, the apex, and the elongating part of roots. In terms of μg of ABA per kg of fresh weight, the maximum concentration of this growth inhibitor was found in the root apex.

Table 1. ABA content (GC/MS determinations) of different regions of maize[a] roots

mm from the tip	ng per 100 fragments	μg per kg fr. wt.
0.0–0.5	0.95 ± 0.20	36.1 ± 7.6
0.5–1.0	2.46 ± 0.68	66.5 ± 18.4
1.0–4.0	13.71 ± 0.36	33.3 ± 0.9

[a] cv. Kelvedon
Roots kept in light

When roots of maize (cv. LG 11) kept in a horizontal position in the dark and in light for 2 h were analyzed, significant differences (Table 2) were obtained in the ABA content of the 5-mm tips of these roots. It is clear that the light treatment increased the ABA level expressed as ng per 100 fragments or as μg per mg of dry weight.

Table 2. ABA content (GC/MS determinations) of the 5-mm tips of maize[a] roots

	ng per 100 fragments	μg per mg of dry wt.
Dark	22.8 ± 2.2	0.568 ± 0.003
Light	31.1 ± 4.0	0.694 ± 0.034

[a] cv. LG 11
Roots kept horizontal for 2 h

References

1. Burström, H.: Physiol. Plant. *13*, 597–615 (1960)
2. Kundu, K.K., Audus, L.J.: J. Exp. Bot. *25*, 479–489 (1974)
3. Kundu, K.K., Audus, L.J.: Planta *117*, 183–186 (1974)
4. Lake, J.V., Slack, G.: Nature (London) *191*, 300–302 (1961)
5. Martin, H.V., Elliott, M.C., Wangermann, E., Pilet, P.E.: Planta *141*, 179–181 (1978)
6. Masuda, Y.: Physiol. Plant. *15*, 780–790 (1962)
7. Milborrow, B.V.: Annu. Rev. Plant Physiol. *25*, 259–307 (1974)
8. Ohno, Y., Fujiwara, A.: Plant Cell Physiol. *8*, 141–150 (1967)
9. Pernet, J.J., Pilet, P.E.: Planta *128*, 183–184 (1976)
10. Pilet, P.E.: J. Exp. Bot. *21*, 446–451 (1970)
11. Pilet, P.E.: Planta *111*, 275–278 (1973)
13. Pilet, P.E.: Planta *122*, 299–302 (1975)
14. Pilet, P.E.: Physiol. Plant. *33*, 94–97 (1975)
15. Pilet, P.E.: Planta *130*, 245–249 (1976)
16. Pilet, P.E.: Planta *131*, 91–93 (1976)
17. Pilet, P.E.: In: Plant Growth Regulation. Pilet, P.E. (ed.). pp. 115–128. Berlin-Heidelberg-New York: Springer 1977
18. Pilet, P.E.: Z. Pflanzenphysiol. *89*, 411–426 (1978)
19. Pilet, P.E.: Planta *145*, 403–403 (1979)
20. Pilet, P.E., Ney, D.: Planta *144*, 109–110 (1978)
21. Pilet, P.E., Went, F.: Am. J. Bot. *43*, 190–205 (1956)
22. Rivier, L., Milon, H., Pilet, P.E.: Planta *134*, 23–27 (1977)
23. Suzuki, T., Kondo, N., Fujii, T.: Planta *145*, 323–329 (1979)
24. Torrey, J.G.: Plant Physiol. *27*, 591–602 (1952)
25. Wilkins, H., Burden, R.S., Wain, R.L.: Ann. Appl. Biol. *78*, 339–340 (1974)
26. Wilkins, H., Larque-Saavedra, A., Wain, R.L.: Ann. Appl. Biol. *78*, 169–177 (1974)
27. Wilkins, H., Wain, R.L.: Planta *121*, 1–8 (1974)
28. Wilkins, H., Wain, R.L.: Planta *123*, 217–222 (1975)
29. Wilkins, H., Wain, R.L.: Planta *126*, 19–23 (1975)
30. Wilkins, M.B.: Curr. Adv. Plant Sci. *13*, 317–328 (1975)
31. Wilkins, M.B.: Sci. Prog. (Oxford) *63*, 187–217 (1976)
32. Wilkins, M.B.: In: Integration of Activity in Higher Plant. Jennings, D.H. (ed.). pp. 275–335. Cambridge: University Press 1977
33. Wilkins, M.B.: In: Plant Growth Regulation. Pilet, P.E. (ed.). pp. 199–207. Berlin-Heidelberg-New York: Springer 1977

Action Potentials and Rapid Plant Movements

T. SIBAOKA [1]

Rapid bending in the main pulvinus of *Mimosa pudica* (1–3), shutting movement in the trap-lobes of *Dionaea muscipula* (4–6), and visible curvatures in the filaments of stamens (6, 7) and in the bilobate stigmas of pistils (8) of some plants seem to be mediated by specific bioelectric events interposed between the stimulus and the loss of turgor in the motor cells. In this paper I will describe some recent findings in *Mimosa* and in two insectivores, *Dionaea* and *Aldrovanda vesiculosa*, and intend to gain some understanding about the bioelectric events occurring prior to the rapid plant movements.

Main Pulvinus in Mimosa pudica

In my previous study (1), when an action potential propagated basipetally through the petiole arrived at the joint between the slender part of the petiole and the main pulvinus, another type of action potential in the pulvinus (pulvinar action potential) was elicited after a latent period of 0.2–0.4 s. This result seems to indicate that excitable cells in the pulvinus may be stimulated at the joint by the action potential coming down through the petiole. The pulvinar action potential which is also elicited about 0.05 s after a slight touch is applied to the lower surface of the pulvinus (1) has a more sharply rising phase than that in the petiolar action potential and a long-lasting plateau (1, 3). Within a period of about 0.3 s after eliciting the pulvinar action potential at the joint,it is propagated throughout the pulvinus.

Simultaneous recordings of the action potential picked up at the middle of the pulvinus and the movement showed that the former arises 0.05–0.08 s before the latter occurs (3). This fact shows that the pulvinar action potential has not reached the basal end of the main pulvinus when the movement starts. An abrupt 5%–10% decrease in the electric impedance measured between upper and lower surfaces of the middle of the pulvinus occurs 0.05–0.10 s after the pulvinar action potential is elicited (9). This decrease seems to be due to a leakage of water containing ions from the motor cells to the intercellular space, and the start of the decrease seems to coincide with the beginning of movement.

A leaf together with a minute piece of stem attached to its base was excised and both lateral surfaces of its main pulvinus were peeled off with a razor to expose cortical parenchyma. The prepared leaf was placed on a lucite assembly so that its petiole

1 Biological Institute, Faculty of Science, Tohoku University, Sendai, 980 Japan

was immobilized with soft wax, keeping it horizontal. The pulvinus, with the attached stem piece, was immersed in a small pool of dilute saline solution. After a few hours for recovery, the pulvinus moved almost normally upon stimulation. The movement was recorded with a mechano-electric transducer which was connected to the piece of stem. To measure the membrane potentials at rest and during activity, a microelectrode was inserted into a cell at the peeled surface. Action potentials were picked up from cells in a defined location about 0.1 cm from the apical end of the pulvinus where the tip of the microelectrode could be held firmly in a cell with minimal deformation caused by movement.

The cortical cells in both the upper and lower halves of the pulvinus had an identical resting potential, about – 150 mV. Upon stimulation they generated an identical action potential of about 100 mV in amplitude and consisting of a fast-rising spike followed by a long-lasting plateau (Fig. 1). These results agree well with reports by

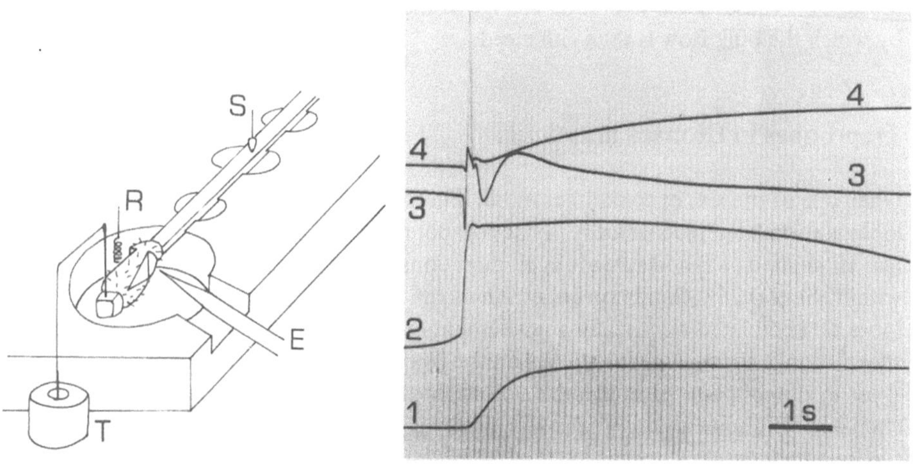

Fig. 1. Main pulvinus in *Mimosa pudica.* Recordings of movement *(trace 1),* action potential from an excitable cell *(trace 2),* electronically differentiated trace 4 *(trace 3),* and increase in [Cl^-] in cortical tissue of the lower half *(trace 4).* Traces 3 and 4 are distorted by local electric current due to action potential during a few tenth second at the start of the increase. Therefore, accurate determination of the starting time of [Cl^-] increase cannot be made. In this record, movement corresponds about 70° bending, potential changes from – 140 mV at rest to – 42 mV at spike, and [Cl^-] increases about 50 mM. Experimental arrangement: *E* microelectrode for potential recording or Cl-sensitive electrode, showing one of them; *S* cathode of electric stimulation or ice water application; *T* mechano-electric transducer, a differential transformer

previous workers (10), who secured the pulvinus with a fine needle to prevent movement, and show clearly that the cortical cells in both the upper and lower halves of the pulvinus are excitable. The action potentials recorded from the excitable cells in the position specified above rose 0.07–0.11 (mean, 0.09) s prior to the start of the pulvinar movement.

When a Cl^--sensitive electrode, an electrolytically polished fine silver wire, was introduced into the cortical tissue, it was observed that Cl^- concentration in the tissue began to increase almost coincidentally within the movement, and after about 1 s the rate of increase reached a maximum (Fig. 1). The increase in [Cl^-] was observed only in the lower half of the pulvinus while the action potential in the excitable cells was detected in both halves. Previous workers (11) have reported that following movement remarkable increases occur in both K^+ and Cl^- content of the fluid which was perfused through a small vessel in which the peeled pulvinus of an isolated *Mimosa* leaf was immersed. Membrane resistance of an excitable cell in the lower half of the pulvinus declined to about half the original value after the spike of the action potential. The start of the decrease in resistance and the increase in [Cl^-] seemed to coincide.

The type of experiment described above shows that excitable cells in the lower half of the main pulvinus act as real motor cells which may eject electrolyte-containing fluid, probably vacuolar sap, into the intercellular space immediately prior to the pulvinar movement. Fluid ejection from the motor cells, which must result in a loss of turgor and then in rapid movement, seems to imply a rapid and large increase in bulk flow across the cell membrane. It seems likely that the action potential elicited in the motor cells causes some structural change in the membrane as a secondary response, by which the bulk flow is then enhanced.

Trap-Lobes in Dionaea muscipula

When one of the six, or sometimes more, sensory hairs on the inner surface of the trap-lobes is stimulated mechanically, an action potential is generated at the base of the hair stimulated (12) and spreads in all directions over the whole surface of both lobes within about 0.25 s. Similar propagation of the action potential is observed when the lobes are stimulated electrically anywhere on their inner or outer surface. Electrical stimulation is convenient for studies on the propagation of action potentials, especially those in closed lobes. Although the action potential recorded intracellularly from an excitable cell shows a simple form (5), patterns of action potentials picked up from the lobe surfaces are somewhat complicated, especially those from the inner surface.

A rapid shutting of the trap-lobes usually requires two stimuli, disturbing either the same sensory hair twice or two different hairs, or applying two electrical stimuli, at intervals less than about 20 s. This fact indicates that a development of tension in the trap for the movement takes place after the second action potential spreads over the whole area of both lobes. Questions as to how the first action potential affects the closure of the trap and how the trap memorizes the response due to the first action potential are still unsolved.

It has been clearly demonstrated, however, that the spread of the first action potential facilitates the second one. Two electrical stimuli were applied at various time intervals to the margin of one lobe and the velocities of the two propagated action potentials were compared. The experiment can be carried out only with closed traps, the action potentials of which are recorded on the outer surfaces. The absolute refractory period for perception of the stimulus was usually less than 1 s. After the lapse of this period, propagation velocities of the second action potentials were remarkably higher than those of the first. When the second action potential occurred 3–5 s after the first one, the velocity of the former showed a maximum increase, 3–4 times that of the latter. As the time interval between the two stimuli was lengthened, the velocity of the second action potential gradually decreased and returned to the same value as the first action potential at an interval of more than 60 s.

To record separately the isometric tension developments during the shutting movement in both the lobes, two strain-gauge transducers were each connected with fine wires to both margins of a trap, which was detached at the base of its petiole and appropriately fixed on a lucite plate. By applying an electrical stimulus to the middle of the left marginal region of the trap an action potential was propagated from the point of stimulation through the left to the right lobe. When the second stimulus was applied < 20 s later, simultaneous records of propagated action potential and tension development in both lobes were obtained. Tensions in both lobes developed almost synchronously 0.32–0.57 (mean, 0.45) s after the beginning of the second action potential as recorded at the middle of the midrib.

As shown in Fig. 2, tensions develop gradually at the start and development in the left lobe begins 0.02 s earlier than in the right one. Since the time interval between the stimulus and the start of the action potential is 0.06 s, and since the distance between stimulating and recording electrodes is 1.5 cm, the velocity of the propagated action potential shown in this figure was 25 cm/s which is 2.3 times higher than that of the first one. This velocity and the time interval between the start of the tension development in each lobe seem to indicate that the motor zones in them are at least about 0.5 cm apart.

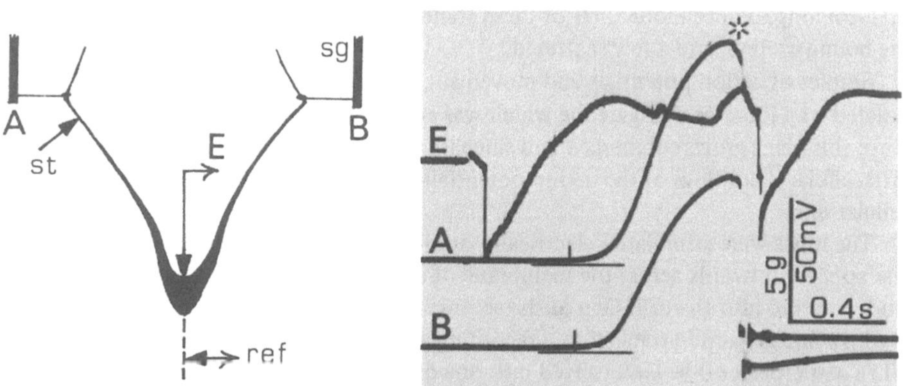

Fig. 2. Trap-lobes in *Dionaea muscipula*. Transmitted action potential *(E)* and tension developments in both lobes *(A, B)*. Tension in A is developed 0.02 s before that in B. Transducer A slips off accidentally from the lobe margin at the time indicated by an *asterisk*. Immediately after the slipping, another action potential is elicited probably by bending of sensory hair due to abrupt shutting of the trap. Experimental arrangement: *sg* strain gauges A and B; *st* electrical stimulation; *E* recording electrode at the middle of midrib; *ref* reference electrode on the petiole. An isolated leaf is used

No change in electrical potential was observed at the moment of tension development (Fig. 2). This fact indicates that the closure of the trap-lobes occurs as a result of certain mechanical changes in the same cells which elicited the action potential immediately prior to the change. When the left lobe is stimulated, action potentials are elicited first in the excitable cells in the motor zone of this lobe and then in those of the right lobe. After a certain lag time (about 0.5 s as described above, which is markedly longer than that in *Mimosa* or *Aldrovanda*), mechanical changes occur first in the same cells of the left lobe, then in those of the right lobe. The fast propagation of the second action potential ensures an almost synchronous movement in the two lobes.

Trap-Lobes in Aldrovanda vesiculosa

Aldrovanda is a small aquatic insectivore and consists of a slender stem which is covered with many whorls of leaves. Each whorl has usually eight leaves. The petioles are thin and wedge-shaped, and terminate in a leaf-blade and several bristles. The leaf-blade is in the form of a trap and consists of a pair of lobes which border each other at the midrib, about 0.4 cm in length. No vascular bundle is seen in the lobe, unlike that in *Dionaea.* Each lobe is nearly semicircular and convex outwards, and consists of two portions, central and marginal (13). The central portion is composed of three cell layers: slender-celled inner and outer epidermis, enclosing a single middle layer of large cells. The long axes of the cells in this portion run almost perpendicular to the midrib. The marginal portion is membranous and consists of only the two epidermal layers juxtaposed. The boundary between these two portions is visible as an arciform line. The inner surface of the central portion possesses about 20 sensory hairs (0.8–0.15 cm long) on each lobe: half of them stand along the midrib, about eight along the boundary line, and a few scattered.

Studies of action potentials and movement in *Aldrovanda* were carried out with isolated trap-lobes or a single lobe which was cut off near and parallel to the midrib. Since this plant grows submerged and since its transparent trap has a simple structure, intracellular recordings of the action potentials are more easily obtained than extracellular ones.

The lobes were stimulated electrically or mechanically. An electric pulse current was applied outwards across the membrane of a cell in the lobe, through a microelectrode inserted into the cell. One of the sensory hairs was bent by laterally displacing it with a fine glass-rod attached to a recording glavanometer driven by an electric pulse. When a microelectrode was inserted into one of the four thin-walled joint cells located toward the middle of the sensory hair, a graded depolarization of its membrane resulted from bending the hair at the joint. This seems to be the receptor potential in the joint cell. The magnitude of the potential depended on the degree of bending. After depolarization reached a certain level, and after a latent period, an action potential was noted in a cell of the lobe located at the base of the hair. The greater the bending, the shorter the latent period. The site where the action potential is first elicited by the receptor potential has not yet been determined.

By inserting microelectrodes into cells of the lobes, it was shown that all cells in the central and marginal portions of the epidermis and in the middle layer were excitable. These cells had an identical resting potential, – 130 mV. Upon stimulation they elicited action potentials, with identical pattern and amplitude (105 mV) (Fig. 3). The action potential spread in all directions over both the central and marginal portions of the trap-lobes.

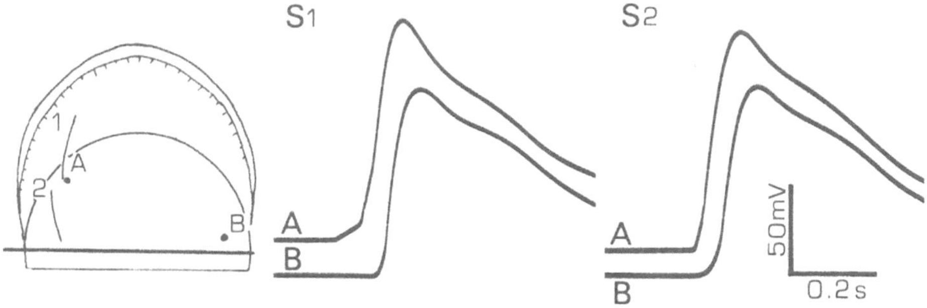

Fig. 3. Action potentials in the trap-lobes in *Aldrovanda vesiculosa.* Two microelectrodes are inserted one by one into cells located at A and B shown in the diagram. S_1 sensory hair 1 in the diagram is stimulated. Note the gradual depolarization seen before action potential at A. S_2 sensory hair 2 is stimulated. A single lobe is used

Fig. 3 shows two simultaneous recordings of the action potentials from two cells located at points A and B in the same lobe, as shown in the diagram. Point A was located close to the base of sensory hair-1, which was stimulated when the left record was taken. In this record a gradual depolarization was seen immediately before the rising phase of the action potential from the cell at point A. This is probably an electrotonic potential change due to the receptor potential in the stimulated hair. No gradual depolarization was seen in the right record which was obtained by stimulation of hair-2, situated at some distance from point A. The distance between A and B was 0.26 cm, and the propagation velocity estimated from the left record was 7.5 cm/s.

Propagation of the action potentials in the trap-lobes was faster in the direction perpendicular to the midrib than in the direction parallel to it. In a single trap the velocity measured in the former direction was 12 cm/s as compared with 4 cm/s in the latter direction. It follows that higher velocity is found in the direction of the long axis of each of the excitable cells mentioned above. Therefore, the time required for the transmission of an action potential between two points seems to depend on the number of cells between them. If a sensory hair standing near the middle of the midrib is stimulated, the action potential spreads throughout the trap-lobes within a period of 0.05 s.

One lobe of an isolated trap was glued with cyanoacrylate resin onto the bottom of a small lucite box. A mechano-electric transducer was placed at a point on the marginal portion of the other lobe, which was free to bend (Fig. 4). A microelectrode was inserted into a cell located at the middle of the motor zone, which will be described below, in the free lobe. As shown in Fig. 4, when a sensory hair on the fixed lobe was stimulated, bending of the unfixed lobe began 0.04 s after eliciting the action potential. From recordings in ten traps, the average time interval between the beginning of the action potential picked up from the motor zone and the start of bending was 0.046 ± 0.004 s. Unlike the response in *Dionaea,* bending of the lobes was always caused by the first action potential.

From sketches and microphotographs reported by Ashida (13), the closing movement is induced by the bending of the lobes in the zones parallel to the midrib and be-

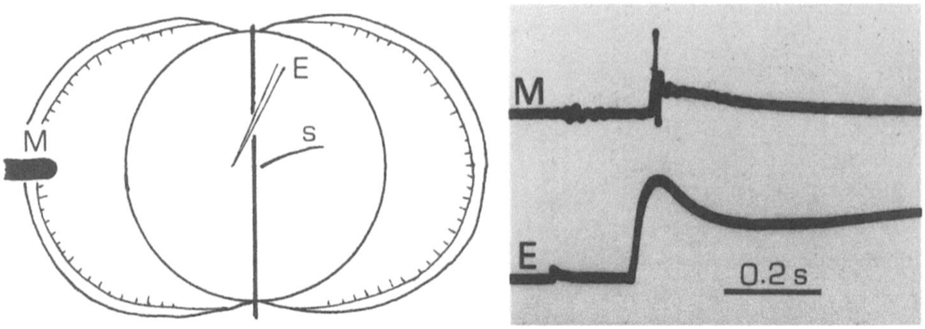

Fig. 4. Trap-lobes in *Aldrovanda vesiculosa.* Recordings of action potential *(E)* from a cell in the motor zone and start of movement *(M)* in the left lobe. Experimental arrangement: *E* microelectrode; *s* sensory hair stimulated; *M* mechano-electric transducer, a fine glass probe mounted on a piezoelectric crystal. An isolated trap is used

tween about 0.025 and 0.055 cm distant from it on both sides. In his case, the distance between the center lines of the motor zones in the two lobes can be estimated to be about 0.075 cm. If a sensory hair standing at the middle of the boundary between central and marginal portions is stimulated and if an action potential is propagated from here in a direction perpendicular to the midrib with a velocity of 10 cm/s, a significant time interval (about 0.008 s) would seem to exist between the start of bending in the two lobes. On the other hand, if a sensory hair standing along the midrib is stimulated both lobes must start to bend almost simultaneously.

Concluding Remarks

Our studies have demonstrated dual functions for the action potentials which are interposed between the stimulus and rapid movement in the three plants discussed above. The first one is responsible for the transmission of a signal from the stimulated site to the motor organ or zone. The second is involved in triggering mechanical responses in the motor cells. In *Mimosa,* the action potentials, generated in the petiole and in the pulvinus respectively, perform the two functions. In contrast, in *Dionaea* and *Aldrovanda,* a single action potential elicited in the trap-lobes actually serves both functions.

Rapid transmission of the action potential of the first type between the two motor zones ensures almost simultaneous bending of the paired lobes in *Dionaea* and *Aldrovanda.* Rapid spreading of the action potential of the second type is required for almost synchronous development of tension all along the motor cells in an organ or zone.

Although in rapid plant movements the question of what does occur as the first mechanical change in the motor cells is still far from solved, it can be said at present that the action potential in the membrane of the motor cells is the first response to the stimulus, which leads eventually to a mechanical change in these cells as a secondary response apparently mediated via a mechanism involving ion exchange through the membrane.

Acknowledgments. Research and preparation of the manuscript made possible by grants from the Ministry of Education (948253, 054034, 154217) and from the Yamada Science Foundation. I am indebted to M. Samejima and T. Iijima for their permission to quote unpublished illustrations (Figs. 1, 3, and 4) and for valuable cooperation.

References

1. Sibaoka, T.: Sci. Rep. Tohoku Univ. Biol. *19,* 133–139 (1951)
2. Dutt, B.K., Guhathakurta, A.: Trans. Bose Res. Inst. *25,* 181–198 (1962)
3. Oda, K., Abe, T.: Bot. Mag. *85,* 135–145 (1972)
4. DiPalma, J.R., Mohl, R., Best, W., Jr.: Science *133,* 878–879 (1961)
5. Sibaoka, T.: Symp. Soc. Exp. Biol. *20,* 49–74 (1966)
6. Sibaoka, T.: Proc. 39th Annu. Meet. Bot. Soc. Jpn. 189 (1974)
7. Bünning, E.: Planta *22,* 251–268 (1934)
8. Sinyukhin, A.M., Britikov, E.A.: Nature (Lond.) *215,* 1278–1280 (1967)
9. Sibaoka, T.: Proc. 40th Annu. Meet. Bot. Soc. Jpn. 189 (1975)
10. Abe, T., Oda, K.: Plant Cell Physiol. *17,* 1343–1346 (1976)
11. Oda, K., Abe, T., Tabe, K.: Proc. 41st Annu. Meet. Bot. Soc. Jpn. 210 (1976)
12. Jacobson, S.L.: J. Gen. Physiol. *49,* 117–129 (1965)
13. Ashida, J.: Mem. Coll. Kyoto Imp. Univ. Ser. B *9,* 141–244 (1934)

The Role of Action Potentials in the Control of Capture Movements of Drosera and Dionaea

S.E. WILLIAMS [1] and B.G. PICKARD [2]

Introduction: The Roles of Darwin, Burdon-Sanderson and Dionaea in the Discovery of Plant Action Potentials

The first demonstration of action potentials in plants was of those that control the rapid movements of the trap of the carnivorous plant *Dionaea*. The experiments in which physician John Burdon-Sanderson made this discovery were stimulated by conversations with Charles Darwin (1). As a result of these discussions, Darwin sent Burdon-Sanderson on September 8, 1873 "two plants with five goodish leaves" (2). Burdon-Sanderson, using a galvanometer, recorded action events as currents from the surface of stimulated traps and demonstrated for the first time that these signals are not unique to the animal kingdom (3). In 1876 (4) he used a capillary electrometer to measure the events as action potentials. Only in posthumously published letters is Darwin's role in this work recorded, but it must be included as one of Darwin's significant contributions to the understanding of plant movements.

General Comparison of the Trapping Mechanisms of Drosera and Dionaea

Although *Drosera* and *Dionaea* are in the same family, if examined superficially the trapping mechanisms of these two carnivorous plants appear dramatically different. In *Dionaea* the leaves are bi-lobed and the lobes snap together in response to mechanical disturbance of a sensitive hair (Fig. 1). If prey is caught between the lobes as they close, secretion of digestive fluid is stimulated. In comparison, the capture movements of the unlobed, tentacle-covered leaves of *Drosera* are slower, and capture movements are restricted to the tentacles (Fig. 1); and although the effective natural stimulus for such movements is now known to be mechanical, this was far from obvious to early workers [see Darwin's account (5) of lengthy experimentation on the problem]. The movements of the tentacles are slow in comparison to those of prey animals, but this does not matter because prey is restrained by the sticky mucilage secreted by the tentacles. The tentacles merely force the prey toward the center of the leaf and bring the digestive apparatus on the globular heads of the tentacles in contact with it. Digestive enzymes and mucilage occur on the heads at all times, but their secretion is enhanced after prey capture (5).

1 Biology Department, Lebanon Valley College, Annville, Pennsylvania 17003, USA
2 Biology Department, Washington University, St. Louis, Missouri 63130, USA

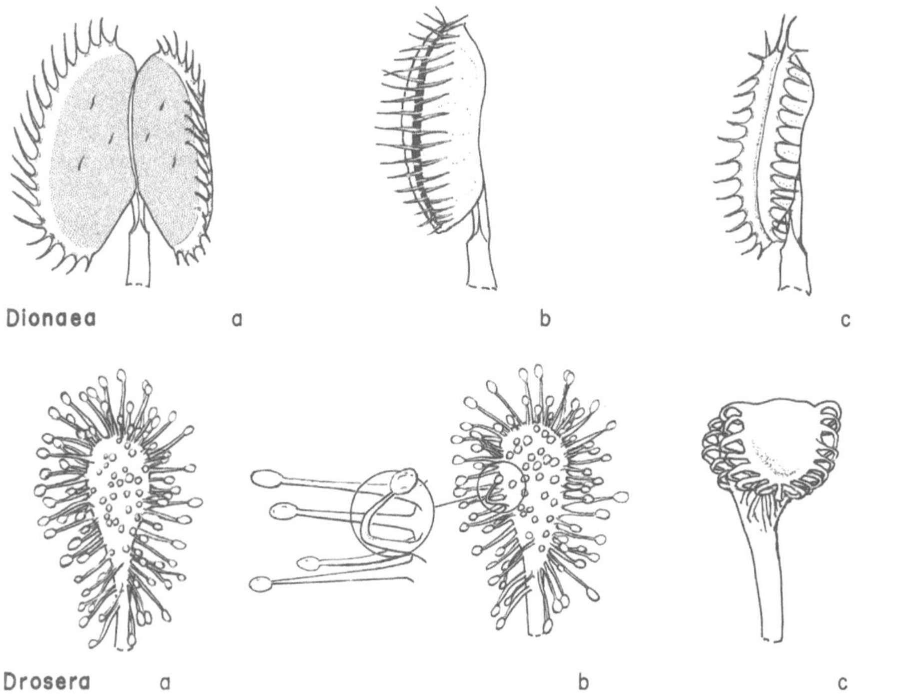

Fig. 1. Movements of *Dionaea* and *Drosera. Dionaea,* **a** Trap open; note three sensory hairs on each lobe. If a hair is touched, an action potential spreads across both lobes. **b** Trap closed by capture movement in response to passage of two action potentials in rapid succession. **c** Trap narrowed by slow post-capture movement which is to be distinguished from the faster capture movement discussed in this paper. (Postcapture movement is stimulated both mechanically and chemically, and the respective mediational mechanisms are thought to be electrical and hormonal.) *Drosera,* exemplified by *D. intermedia,* **a** Unstimulated leaf. **b** Leaf with a single tentacle bent as a result of a mechanical stimulus delivered to its head; the action potential which mediates this capture movement cannot propagate beyond the base of the tentacle, hence the confinement of the response. Normally, a prey animal would stimulate numerous tentacles. **c** Postcapture leaf with pad doubled over and all outer tentacles bent. As for *Dionaea,* chemical stimulation and hormonal mediation are thought to be involved in the slow postcapture movements, which are to be distinguished from the faster mechanically stimulated and electrically mediated capture movement discussed in this paper

The superficial differences between *Drosera* and *Dionaea* have long led those working on the physiology of either of the two organisms to restrict their attention to one genus or the other, ignoring the close taxonomic relationship and numerous subtle similarities between the two. However, the recent discovery that action potentials control capture movements in both plants focuses new interest on the similarities. This paper will elaborate on some of these similarities by comparing details of capture mechanisms in the two organisms.

Mechanoreception

Carnivory is initiated by stimulation of mechanoreceptors which are located in the trigger hairs of *Dionaea* and in the tentacles of *Drosera.* In both plants, the receptor cells respond by producing graded nonpropagating changes in transmembrane potential called receptor potentials which induce action potentials when their rises are sufficiently rapid or their amplitudes are sufficiently great.

The anatomical details of the trigger hair and the tentacle bear a close resemblance to each other (7), and it seems probable that both have evolved from a glandular structure in a common ancestor. In *Dionaea* the receptor potentials have been shown by intracellular recording to occur in modified epidermal cells located at a notch in the base of each trigger hair (8). It is clear that these cells are situated so as to be maximally stretched and compressed during flexure of the hair. In *Drosera* there are apparently homologous epidermal cells positioned at the base of the expanded, globular tip of the tentacle where they also receive a great deal of mechanical stress during tentacle stimulation, but no intracellular measurements have been made in the receptive region of the tentacle so it is not yet possible to be sure in which cells receptor potentials originate (7, 9).

The receptor potential in *Dionaea* can reach its maximum and initiate a propagating action potential within a second [Fig. 2, 3; (8)]. The receptor potential decays

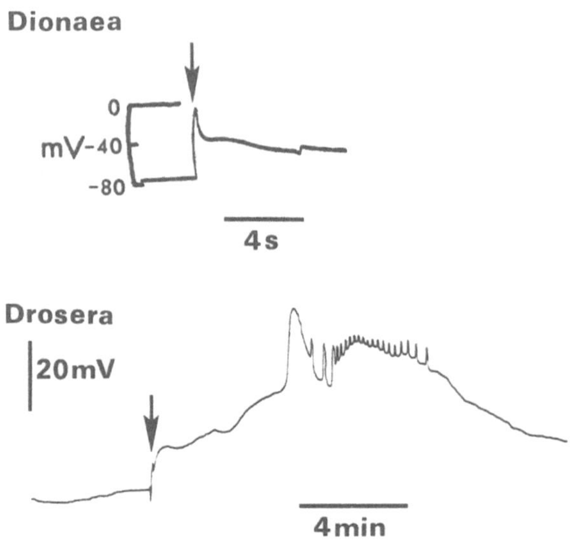

Fig. 2. Receptor potentials recorded from the mechanoreceptors of *Dionaea muscipula* [after Benolken and Jacobson (8)] and *Drosera intermedia* [after Williams and Pickard (9)]. Mechanical stimulation was initiated in both structures at the time indicated the *arrows.* The receptor potential of *Dionaea* rises very quickly, induces a single action potential, and then drops below threshold in a few seconds, while that of *Drosera* rises very slowly and remains above threshold long enough to induce a series of action potentials. The *Dionaea* recording is measured intracellularly from a sensory cell, while for *Drosera* the potential changes are measured extracellularly from the mucilage surrounding the head of the tentacle

rapidly, so that by the time cells have recovered from the refractory stage of the first action potential it is too low to initiate a second action potential. In *Drosera* the rise of the receptor potential is two orders of magnitude slower and far more variable (Fig. 2), and it may be minutes before stimulation elicits action potentials. However, once the receptor potential has attained sufficient rate of rise or sufficient magnitude to induce action potentials, its longer duration results in the production of a series of action potentials (Fig. 2). Thus, the critical differences between the mechanoreceptors of the two carnivores would appear to be quantitative rather than qualitative.

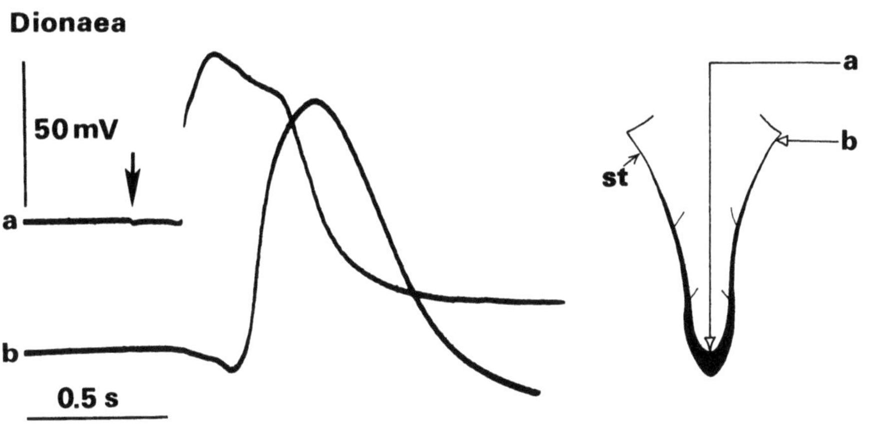

Fig. 3. Action potentials induced in *Dionaea muscipula* by a pulse of electrical current delivered to the trap lobe at point *st*. The surface potential of the trap lobe was recorded from points *a* and *b* relative to the petiole. The 0.2 s lag between the recorded rise of the action potential at point *a* and the rise at point *b* indicate a rate of 50 mm^{-1} s across the 10 mm trap lobe. After Sibaoka (10)

Action Potentials

For both *Dionaea* and *Drosera*, action potentials are the immediate cause of rapid leaf movements (4, 9). These signals are similar in their 100 mV intracellularly recorded amplitudes (10, 11) but they differ by an order of magnitude in their rates of propagation and duration. These values are, respectively, 50–100 mm/s versus 3–5 mm/s and 1–2 s versus 3–30 s at room temperature (Figs. 3, 4). The most important difference between the action potentials of the two plants is the path along which they propagate.

In the *Dionaea* trap, an action potential originating in any one of the six sensitive hairs on the inner surface of the leaf will radiate over the entire leaf blade (4, 10). The laminar pathway has not been elucidated intracellularly, but action potentials can be recorded easily with surface electrodes (4, 10). The cells in the trigger hair within which action potentials are normally initiated are modified epidermal cells (8). Anatomical evidence suggests that propagation continues in the epidermis at least until it leaves the sensitive hair (7). Ultimately, the action potentials reach the motor cells all over the leaf: these are the epidermal cells and the ones primarily responsible for the initial bending are those of the abaxial epidermis (see Sect. on Capture Movements). Thus, somehow the signals propagate across the chlorenchyma as well as through the

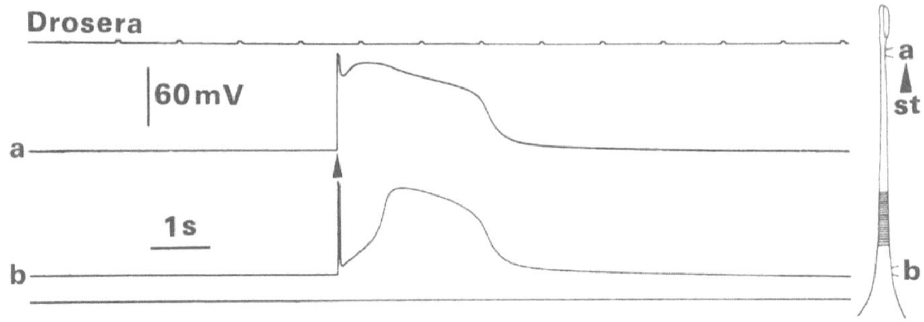

Fig. 4. Action potentials induced in *Drosera rotundifolia* by a pulse of electrical current delivered to the tentacle at point *a*. The surface potential of the tentacle stalk was recorded at points *a* and *b* relative to the water at the base of the plant. The 0.6 s lag before the rise of the action potential in the lower part of the tentacle indicates that the action potential propagated at about 3.3 mm s^{-1} down the 2 mm of stalk between the electrodes. (Two–three s action potentials are typical of *D. rotundifolia; D. intermedia* has action potentials of longer durations.) After Williams and Spanswick (11)

epidermis. Interestingly, Burdon-Sanderson and Page reported that the signals propagate more rapidly across the lower (abaxial) than across the upper surface (4), but this result has not been checked by recent workers.

In the *Drosera* tentacle, action potentials are initiated by a receptor potential just below the swollen head of the tentacle, and propagate only to its base (9). Thus, action potentials of each tentacle or "sensory hair" of *Drosera* influence motor behavior independently of those of its neighbors (9), in contrast to the situation in *Dionaea* where an action potential from one sensory hair invades the same motor area of the leaf influenced by action potentials from the others (4, 10).

The propagative pathway down the stalk of a tentacle potentially includes cells in both the epidermal layer and the inner stalk cell layer, since both are excitable (11). Cells in both layers are axially connected by numerous plasmodesmata and evince electrotonic continuity (11, 13). However, only the epidermis and not the inner cell layer includes the cells at the base of the head in which the generator potential is presumed to originate, and lateral connectivity between the two stalk layers is relatively weak. The work of Hooker (14, 15) (see Sect. on Capture Movements) indicates that it is primarily the abaxial epidermal cells of the stalk that are responsible for most of its rapid bending. Therefore, the epidermis would seem to be the critical part of the conductive pathway regardless of the extent of participation of the inner cells (11). In support of this view, it is notable that *Drosera pygmaea* has normal tentacle function even though subepidermal cells are lacking along much of the stalk (7).

Whether in *Drosera* the exclusion of conduction from the leaf lamina is determined by lack of electrotonic connectivity, lack of excitable cell membranes, or both has not been studied. Certainly, chlorenchymal cells of some nonmotile leaves appear to be excitable (16). Even for the conductive lobes of *Dionaea* there is information neither on the distribution of plasmodesmata and of excitable cells nor on their relations to the conductive pathway through the leaf. Indeed, a role for intercellular transmitter substances is not excluded in the *Dionaea* leaf, though it seems unlikely.

Capture Movements

In both *Drosera* and *Dionaea,* reasonably strong mechanical stimulation leads to nastic movements which for healthy plants at normal summer temperatures are typically quite rapid. These initial capture movements must be distinguished from the relatively slower movements which follow them if prey is successfully trapped; the later movements serve to expedite digestion and absorption, and their control is more complex than the control of capture movements (6, 7, 9).

Though *Drosera* and *Dionaea* do not lend themselves equally well to all kinds of kinetic experimentation, to the extent that parallel data are available, many aspects of the response patterns of the tentacle and the trap lobe are closely comparable except for speed. In both *Drosera* and *Dionaea* the response to the first action potential varies from undetectable to barely perceptible [Figs. 5, 6; (4, 9)]. However, the first action

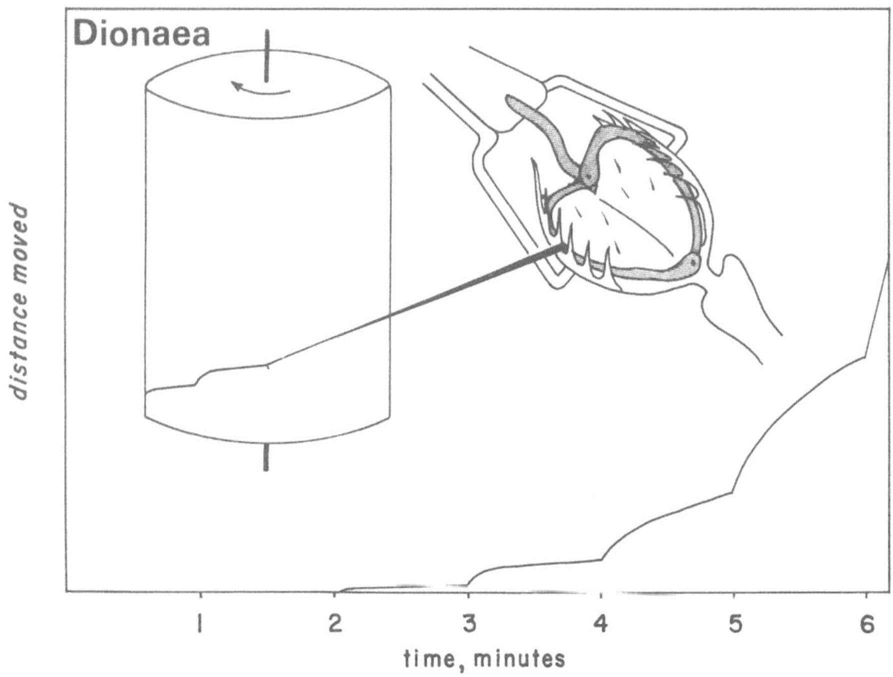

Fig. 5. Response of a *Dionaea* trap to action potentials at 1 min intervals. Stimuli were delivered with a camel's hair brush to a sensory hair and movement of the trap was recorded as illustrated in the inset. As shown in Figs. 2 and 3, each stimulus results in the spread of a single action potential. A movement occurs about 0.5 s after each stimulus except the first; each stimulus facilitates the response to the stimulus that follows it. After Burdon-Sanderson and Page (4)

potential enhances the ability to respond to a second, should the second follow without excessive lag; and indeed, the influence of any action potential in a series depends on the pattern of recently preceding stimulation (4, 9, 17). Typical responses of *Dionaea* leaves and *Drosera* tentacles to a sequence of evenly spaced action potentials are shown in Figs. 5 and 6.

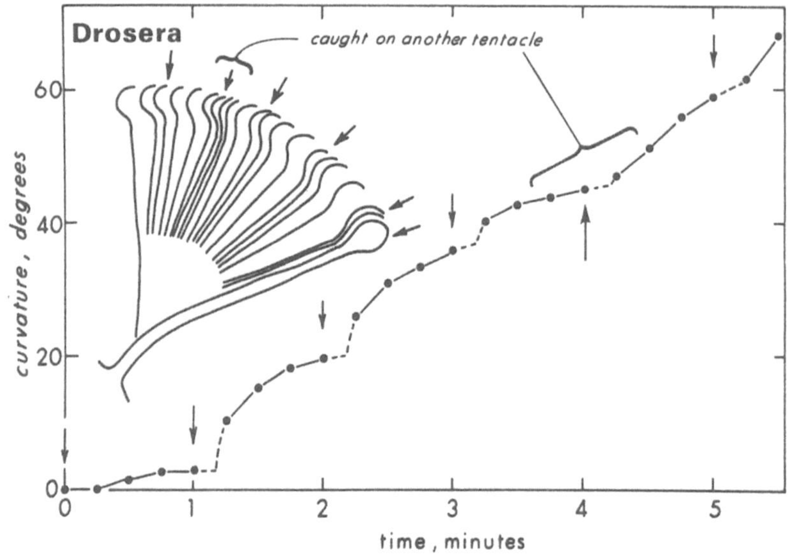

Fig. 6. Response of a *Drosera intermedia* tentacle to action potentials electrically triggered at 1-min intervals. A movement begins about 10 s after each stimulus and is well underway by 15 s, although recording equipment could not be activated at intervals briefer than 15 s. Facilitation is less pronounced than in *Dionaea* (cf. Fig. 5) but is nevertheless apparent. From Williams and Pickard (9)

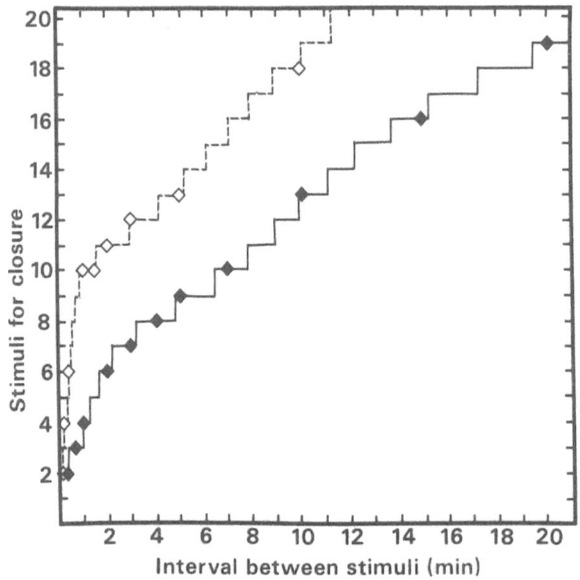

Fig. 7. Number of stimuli required to close *Dionaea* traps vs. interval between stimuli. Stimulation was by touching a trigger hair. ▪ are data of Brown (17) rounded to the nearest integer, and ▫ are rounded data from a repetition of Brown's experiment by John Kolater (unpublished). Both sets of data have been fit with functions derived by James B. Cooper, William M. Fleishman and Stephen E. Williams (unpublished) from a quantitative treatment of Ashida's hypothesis (18) for trap closure. Kolater's traps, which were probably larger than Brown's, required for closure a larger number of stimuli at any selected interval

Figure 7 shows for *Dionaea* how response increases as a function of decreasing lag between action potentials. Commonly, prey animals stimulate trigger hairs at such a high frequency that the second action potential closes the trap, giving the appearance of an all-or-none response to the second signal. It is primarily this feature of the response which, coupled with the greater rapidity, has given rise to the misconception that trap movements are qualitatively different from those of tentacles.

Different mechanisms for the movements of the two plants are proposed in the literature. However, the differential changes of the surfaces of the trap and tentacle have features in common. The most basic data to be explained are that during a typical capture movement the abaxial side of the moving organ expands about 10% in length while the adaxial side remains essentially unchanged. This differential expansion is documented in Table 1 with data from Brown (17) and Hooker (14); also illustrated is the compensating autotropic expansion of the adaxial (convex) side during trap opening or tentacle unbending (which occurs if and only if no prey is caught). The overall expansion following recovery is indicated in Table 2; evidently, the capture activity of a trap or tentacle results in a permanent net increase in length.

Table 1. Percent increase in length of abaxial and adaxial sides of *Dionaea* trap lobes [from data of Brown (17)] and of *Drosera* tentacles [from data of Hooker (14)] during closure or bending and recovery versus the increase of unstimulated controls

	Dionaea		*Drosera*	
	Abaxial	Adaxial	Abaxial	Adaxial
Closing or bending	8.4 (5) [a]	− 1.0 (2)	11.2 (5)	0.4 (5)
Recovery	0.8 (3)	9.4 (2)	− 0.9 (5)	8.1 (5)
Growth of control during 1/2 day	0.5 (3)	—	0.0 (1)	0.0 (1)

[a] The number of replications is given in parenthesis

Table 2. Percent increase in length of *Dionaea* trap lobes [data from Brown (17)] and *Drosera* [data from Hooker (14)] after recovery from each of three successive closings or bendings, respectively

	Dionaea	*Drosera*
% After 1st recovery	13.0 (1) [a]	11.0 (4)
% After 2nd recovery	10.6 (1)	6.6 (4)
% After 3rd recovery	3.3 (1)	6.9 (3)

[a] The number of replicates is given in parenthesis

Table 3 presents a second basic set of data, available for *Drosera* only: during tentacle bending, the osmotic concentration measured by inducing incipient plasmolysis with glucose or KNO_3 drops perhaps 10%–25% on the abaxial side but is little changed on the adaxial side. (Only the lower part of each tentacle stalk is considered because initially bending is more vigorous there than in the upper region.) Hooker (15), who obtained the measurements, interpreted them to mean that the cells on the abaxial surface grow because of increased plasticity of their cell walls. He pointed out that the drop in concentration of solutes was maximally 25% for the most rapidly bending tentacles and thus corresponds within experimental error to the calculated increase in cell volume of about 27%. He noted that lower osmotic drops might result from competing compensatory uptake of solute during slower bending.

Table 3. Osmotic potential (in bar) at incipient plasmolysis of epidermal cells of the basal half of the tentacle of *Drosera* before and during bending [data from Hooker (15)]

	Straight	Bending
Abaxial side	– 8 to – 9	– 6 to – 8
Adaxial side	– 8 to – 9	– 8 to – 9

Not included in Table 3 are Hooker's data on bent, unbending, and unbent tentacles; they are of somewhat more complex interpretation but may be summarized as indicating that when a tentacle finishes bending, symmetry of osmotic potential is ultimately restored, presumably by solute pumping, and that such pumping tends to keep pace with the slow expansion which occurs on the convex side during autotropic unbending.

It is unfortunate that no appropriate measurements of solute potential have been made for *Dionaea,* for these might permit distinction between a plasticity mechanism comparable with that which seems valid for *Drosera* and the commonly accepted explanation that capture movements result from differential changes of turgor. Two variants of the turgor hypothesis will be considered.

Brown, who made the measurements of *Dionaea* extension presented in Tables 1 and 2, proposed that the abaxial side of the leaf takes up solute and water rapidly and expands in consequence. Ultimately, the distended cell walls must stabilize in their new configuration. (Autotropic unbending was attributed by Brown not to repetition of this process in the adaxial tissue but to ordinary growth.) The model lacks appeal because the sudden and massive uptake of solute it requires is improbable.

The second explanation based on turgor shifts was presented by Ashida (18) not only for *Dionaea* but also for its extremely close trap-bearing relative *Aldrovanda.* He imagined that the cells of the abaxial epidermis are turgid but that they remain poised with walls of high plasticity because their size is maintained by compression from turgid adaxial cells. Stimulation might cause sudden loss of adaxial turgor and hence disappearance of constraint, resulting in abrupt abaxial extension and trap closure. This model when applied to *Dionaea* seems inconsistent with the observation in Table 1 that there is not much shrinkage of the inner epidermis. Since it might be argued that the values are uncertain because they are derived from only 2–5 traps, these critical measurements should be repeated.

Probably the main reason for the popularity of mechanisms involving turgor shifts rather than plasticization is that in the past it seemed counter-intuitive that plasticization could occur rapidly enough to permit detectable growth in less than 1 s in *Dionaea* and 0.1 s in *Aldrovanda*, whereas the movements of *Drosera* were thought to be slow and hence compatible with growth. However, recently it has been shown that even the "slow" tentacles of *Drosera* typically respond within 10–15 s (Fig. 6); previous measurements were of the lag from the start of stimulation rather than from passage of an action potential (cf. Fig. 6 with Fig. 2). Thus, all three species have relatively rapid movements. Furthermore, in the light of modern understanding of growth [e.g., (19)] such speed seems less implausible. If, in ordinary tissue, growth results from many short localized pulses of plasticization intermingled with many similar pulses of rigidification, it may be that in tentacle or trap capture movements these processes are separated in time. Perhaps the function of the action potentials is to release a fair quantity of a plasticizing factor. Application of acid causes the stalks to bend (15), so it is not impossible that the factor is simply the hydrogen ion. The reason for differential release or differential response on the abaxial side must of course be sought in the asymmetry of the tissue rather than in the pathway of the action potentials; this would be true for any model that can be formulated.

As an important step in sorting out these three models for *Dionaea* and confirming the plasticization model for *Drosera*, these early and limited data of Tables 1, 2, and especially 3 must be extended. No matter whether they speak for plasticization or turgor shift, more extensive data should provide a firmer basis for understanding how the action potentials effect movement.

Once the participation of plasticization or solute flux can be established for each plant, the abruptly triggered expansion process should prove rewarding for application of modern chemical techniques. A good start for a comprehensive study has been made by Lea (20), who has found in freshly stimulated traps of *Dionaea* an increase in a phospholipid which may be lysophosphatidic acid. Table 4 shows that the level of

Table 4. The effect of touching trigger hairs on the appearance of a water soluble phospholipid, assesed by a bioassay [data from Lea (20)]

Time after touching of hairs	Ratio of bioactivity to control
15 s	2.80 ± 0.33
600 s	2.16 ± 0.13
3 h	1.60 ± 0.14
24 h	1.00 ± 0.07

the compound rises within 15 s to 2.8 times that in unstimulated traps and gradually returns to unity as the trap recovers from closure. In animal cells, phosphatidic acid causes an increase in permeability of plasma membranes: is it released during excitation of the Droseraceae either increasing permeability to solutes or directly or indirectly leading to bond breakage which would result in plasticization?

One more chemical observation which might influence future work must be mentioned: Jaffe (21) has reported for *Dionaea* that ATP levels drop from "950 ± 50 μM per midrib" to "650 ± 50 μM per midrib" following trap closure. Although most of the bending movement is known to occur in the lobes (22), some may occur in the midrib also. This observation on ATP should be checked and extended in order to see if indeed ATP participates massively in the rapid movement of *Dionaea,* and of *Drosera* as well.

In conclusion, it is interesting that Darwin is celebrated more for his work which stimulated study of "classical", hormonally controlled growth in the coleptile than for his work on carnivorous plants. Yet, his pioneering interest in carnivorous plants initiated studies on atypical, electrically controlled expansion which may in the end help us to sort out what is fundamental to any kind of growth.

References

1. Burdon-Sanderson, G.: Sir John Burdon-Sanderson, a Memoir . . . with a Selection from his Papers and Addresses. Oxford, 1911
2. Darwin, F.: More Letters of Charles Darwin (2 Vols.). London, 1903
3. Burdon-Sanderson, J.S.: Proc. R. Soc. London *21,* 495–496 (1873)
4. Burdon-Sanderson, J.S., Page, J.F.M.: Proc. R. Soc. London *25,* 411–434 (1876)
5. Darwin, C.: Insectivorous Plants. London, 1875
6. Lichtner, F.T., Williams, F.E.: Am. J. Bot. *64,* 881–886 (1977)
7. Williams S.E.: Am. Philos. Soc. *120,* 187–204 (1976)
8. Benolken, R.M., Jacobsen, S.L.: J. Gen. Physiol. *56,* 64–82 (1970)
9. Williams, S.E., Pickard, B.G.: Planta *103,* 193–221 (1972)
10. Sibaoka, T.: Symp. Soc. Exp. Biol. *20,* 49–73 (1966)
11. Williams, S.E., Spanswick, R.M.: J. Comp. Physiol. A *108,* 211–223 (1976)
12. Williams, S.E., Pickard, B.G.: Planta *103,* 222–240 (1972)
13. Williams, S.E., Pickard, B.G.: Planta *116,* 1–16 (1974)
14. Hooker, H.D., Jr.: Bull. Torrey Bot. Club *43,* 1–27 (1916)
15. Hooker, H.D., Jr.: Bull. Torrey Bot. Club *44,* 389–403 (1917)
16. Cheeseman, J.M., Pickard, B.G.: Can. J. Bot. *55,* 497–510 (1977)
17. Brown, W.H.: Am. J. Bot. *3,* 68–90 (1916)
18. Ashida, J.: Mem. Coll. Sci. Kyoto Imp. Univ. Ser. B *9,* 141–244 (1934)
19. Penny, P., Penny, D.: In: Phytohormones and Related Compounds: A Comprehensive Treatise. Letham, D.S., Goodwin, P.B., Higgins, T.J.V. (eds.). Vol. II, pp. 537–597. Amsterdam: Elsevier/North Holland 1978
20. Lea, H.W.: Planta *129,* 39–41 (1976)
21. Jaffe, M.J.: Plant Physiol. *51,* 17–18 (1973)
22. Stuhlman, O., Jr.: Bull. Torrey Bot. Club. *75,* 22–44 (1948)

On the Mechanism of Contact Coiling of Tendrils

M.J. JAFFE [1,2]

"... the tendrils: their irritability is beautiful, as beautiful in all its modification as anything in orchids." – C. Darwin
[Letter to Asa Gray, August 4, 1863 (7)]

Although Charles Darwin is best known for his discovery of biological evolution, he was the eclectic natural scientist par excellence, and in the tradition of the classical naturalists such as Aristotle and Pliny the Elder, considered all the world of nature within the domain of his interests. Among these, his observations of plants, recorded in several books [e.g., (10–13)], papers and letters (14), apparently gave him special joy, because he writes in his autobiography: "it has always pleased me to exalt plants in the scale of organized beings"; (8).

One of the broadest themes of these studies dealt with the movements of plants, and as was so often the case in his work, his thinking led to a unifying concept of plant movements against which adequate evidence has not been presented to this day. This theory, presented in *The Power of Movement in Plants* (13) asserts that:

> "It has now been shown that the following important classes of movement all arise from modified circumnutation, which is omnipresent whilst growth lasts, and after growth has ceased, whenever pulvini are present. These classes of movement consist of those due to epinasty and hyponasty, – those proper to climbing plants, commonly called revolving nutation, – the nyctotropic or sleep movements of leaves and cotyledons, – and the two immense classes of movement excited by light and gravitation." (13)

As evidence for this statement he showed numerous observations wherein the translational rotation continued, decreased, and gradually blended into the types of movements described in the quote above, when the appropriate environmental cue was given.

Darwin obviously enjoyed studying the variety and behavior of plant tendrils, and studied both their circumnutation and contact coiling (12, 13). The contact coiling of excised pea tendrils, as an example, is shown in Fig. 1. However, he did not believe that the thigmonastic or thigmotropic coiling of tendrils followed his general rule, and said that contact coiling was not a modification of circumnutation:

> "Lastly, the curling of the extremity of a tendril when touched seems to be independent of its revolving or circumnutating movement. This is best shown by the part which is the most sensitive to contact, circumnutating much less than the lower parts, or apparently not at all." (13)

However, although he published copious recordings of the type mentioned above for other types of movements, little evidence was presented for the conclusion related to tendrils, other than the observation that circumnuation seemed to be greatest in the basal portion of the organ, whereas contact coiling appeared to begin in the apical region.

---

1 Laboratory of Sensory Physiology, Department of Botany, Ohio University, Athens, Ohio 45701, USA
2 Present Address: Biology Department, Wake Forest University, Winston-Salem, North Carolina 27109, USA

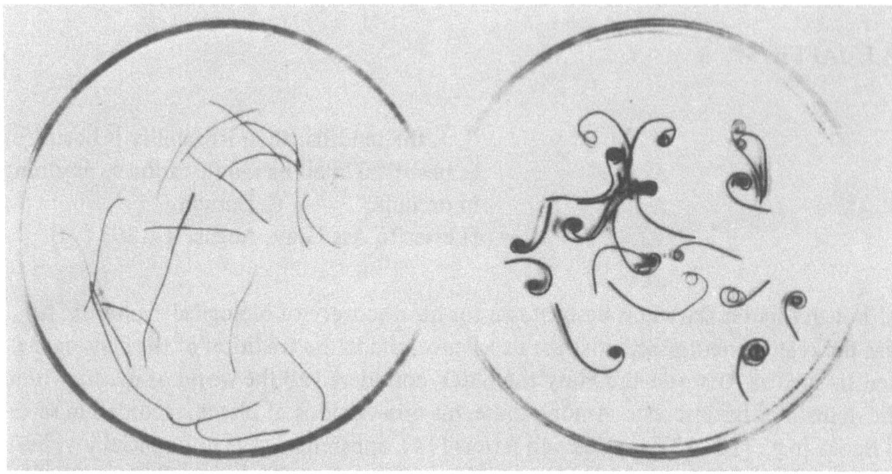

Fig. 1. The effect of rubbing the ventral side of excised pea tendrils *(right)* as compared to non-stimulated controls *(left)* after 1 h of incubation in phosphate buffer containing 0.01% Tween-20

Because of the fundamental and unifying nature of Darwin's theory, the first part of this paper will deal with efforts in the writer's laboratory to address the question of circumnutation and contact coiling.

Of the many questions concerning tendril movements raised by Charles Darwin, three will be discussed in this paper: The relationship between circumnutation and contact coiling, the nature of the sensory function in contact coiling, and aspects of the motor function. As we shall see, the first two of these are central to an understanding of both the ecology and the physiology of the tendril as a grasping and guying organ.

Relationship of Circumnutation and Coiling

The pea tendril is one of the most rapidly circumnutating plant organs known, sweeping through space so quickly that its movement can often be seen with the unaided eye (12, 26). One of the characteristics of this movement, noted by Darwin (12), is that the long axis of the circumnutating elipse is always at right angles to the direction of the sun. He called this phenomenon "heliotropism", and wisely so, for it was later shown that the response is to heat and not to light (17).

As the tendril travels through its circumnutatory elipses, it always moves with the concave surface of its terminal hook facing forward. Since it is this ventral surface that causes coiling when mechanically stimulated (18, 19, 26), it ist in the best position to respond to a support when it encounters one. Characteristically, when a circumnutating tendril touches a potential support, the movement of circumnutation causes the tendril to rub against it. Then, whether the support remains or is removed, the tendril stops circumnutating (Fig. 2, top and bottom) and begins coiling (Fig. 2, middle) (21). The cessation of circumnutation assures that the tendril will not pass beyond the support before it can coil around it. Furthermore, it seems that the potential for move-

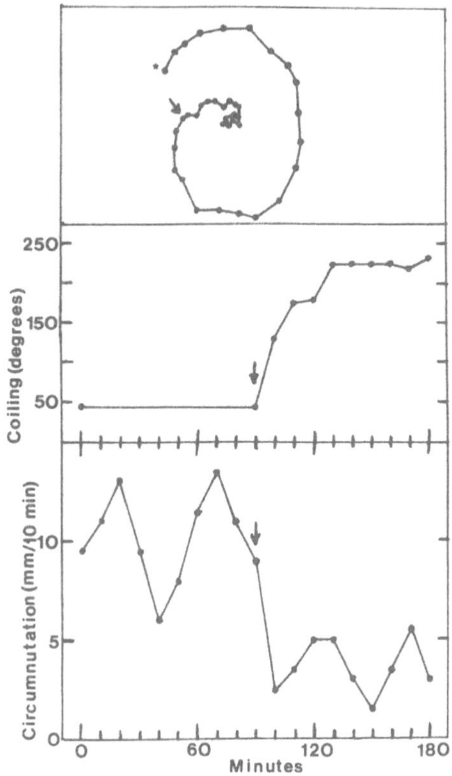

Fig. 2. The effect of a mechanical stimulus (rub) on circumnutation and coiling. *Top* The data were abstracted from a time lapse motion picture looking down upon the plant. The interval between points is 5 min. *Middle* Average contact coiling of two typical tendrils that happened to be in phase. *Bottom* Average circumnutation rate of the same two tendrils shown in the middle graph. In the top graph the *star* marks the beginning of the observations, and in all the graphs, the *arrow* indicates when the tendril was stimulated mechanically. Adapted from Jaffe (21)

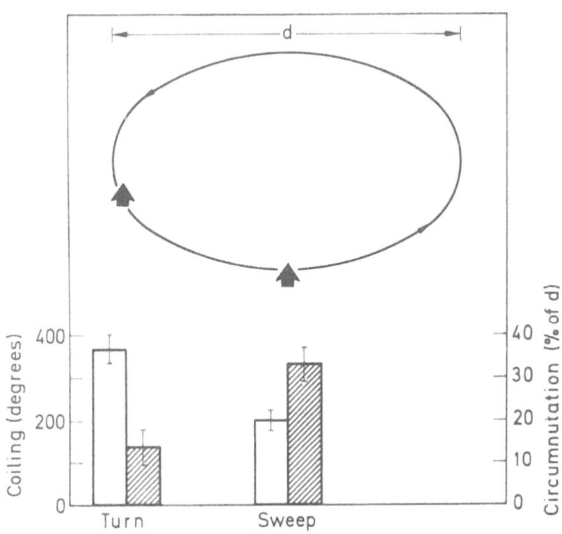

Fig. 3. The effect of the rate of circumnutation *(dark columns)* on the subsequent rate of coiling following mechanical stimulation *(unfilled columns)*. Adapted from Jaffe (21)

ment is limited by some unknown factor, possibly ATP (27), since when the tendril is circumnutating the fastest (i.e., the long axis of the elipse) it coils the slowest after mechanical stimulation, whereas it coils much faster if it encounters a support during the slow turn at either end of the elipse (Fig. 3) (21). If the support which stimulated the tendril is withdrawn at this time, the tendril will continue to coil for about a half hour and then uncoil in one or two hours (26). As soon as it does, circumnutation begins again (21).

Thus, the behavior of the pea tendril is admirably suited to its function: Rapid circumnutation during elongation of the plant providing optimum opportunity to encounter a support; rapid coiling around the support; and if the support proves to be illusory, uncoiling within an hour or two, followed by a resumption of circumnutation.

Because of the relationships described above in pea tendrils, it seems that its behavior is not at all inconsistent with Darwin's unifying theory of plant movements. For, as Fig. 2 shows, mechanical stimulation does indeed cause a rapid cessation of circumnutation, at the same time that contact coiling begins (21). Furthermore, in the case of this organ, circumnutation does, in fact, occur in the apical region (Fig. 4) (5).

The pea plant growing in the natural environment in which it evolved would undoubtedly be associated with other plants. If we picture it growing in the midst of these plants, and reaching a height where its stem will no longer support it, the ecological significance of the behavior of its tendrils becomes apparent. The path of the moving tendril will tend to become oriented towards places where foliage shields the tendril from the sun's heat; that is, towards a potential support. If wind blows the tendril about so that its dorsal as well as ventral surface is mechanically stimulated, the tendril will not coil (18, 19, 26). This minimizes its chances of starting to coil when the wind might blow it away from its potential support. Finally, since the ability of the tendril to encounter a support must be related to the availability of a support, plants growing in brighter light (i.e., not shielded by foliage which represent potential supports) tend to have more branches, and hence a greater chance for contact coiling (Table 1) (26).

Table 1. The effect of ambient light intensity during growth of the pea plant, on the relative occurrence of different branching patterns of the youngest mature tendril. Adapted from Jaffe and Galston (26)

Fraction of tendrils	Incident irridiation (kiloergs × cm^{-2} × s^{-1})		
	96	108	121
With no branches	0.78	0.69	0.62
With 2 branches	0.06	0.11	0.11
With 3 branches	0.17	0.23	0.37

Fig. 4. The circumnutation (seen from the side) of a pea tendril culminating in contact coiling around the support. *From left to right* the times are 15:40, 16:20, 16:40 and 17:10. Adapted from Cravens (5)

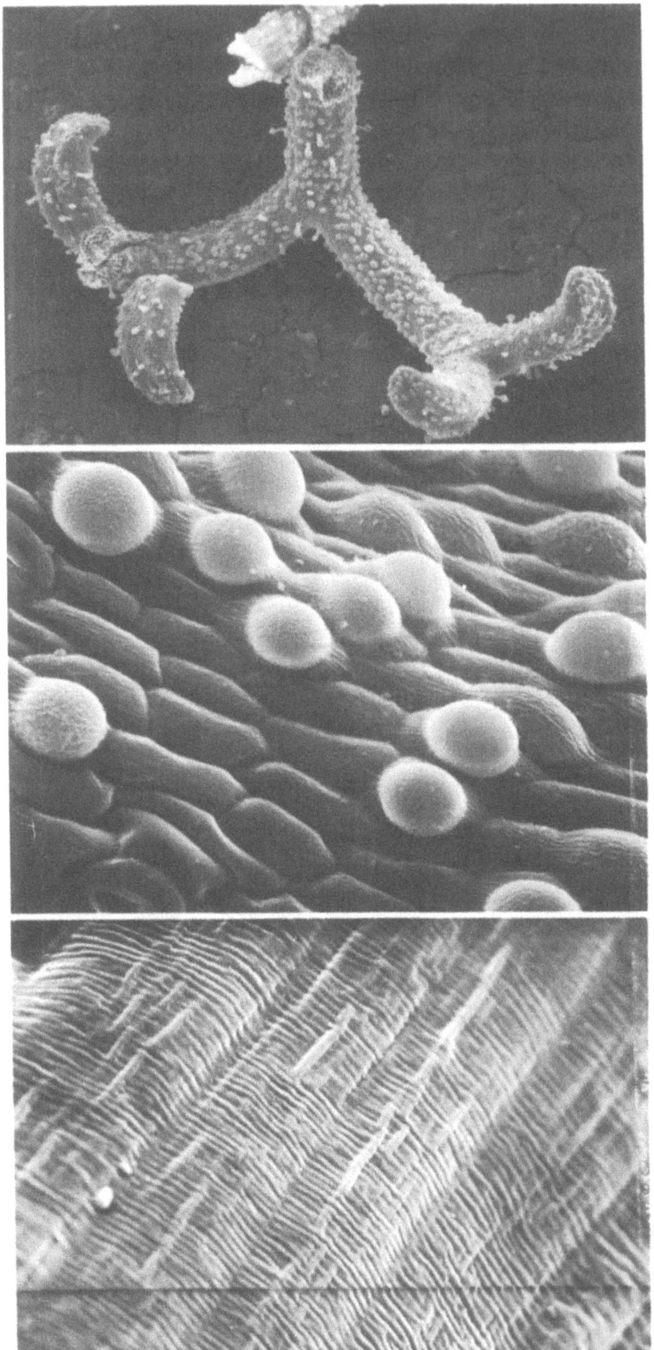

Fig. 5. Scanning electron micrographs of tendrils of *Eccromocarpus scaber* showing the whole tendril (*top* 50 ×) and the so-called "tactile papillae (*middle* 800 ×), and epidermal cells of *Pisum sativum* (*bottom* 2000 ×)

The Sensory Function

The sensory act may be described as the absorption by a receptor of the kinetic energy associated with the mechanical stimulus. It is an interesting fact that tendrils of different species will respond best to the type of stimulus that is found on the support which they normally encounter in nature (9). Although the receptor is unknown, it is believed by a number of workers to reside in specialized (Fig. 5, top) or unspecialized (Fig. 5, bottom) epidermal cells (24,33,37). Although the intracellular receptor is as yet unknown, we shall see that its activity may be associated with membranes of the endomembrane system.

Although exceeded in rate by closure of the Venus flytrap or sensitive *Mimosa* leaves, the contact çoiling of tendrils is still a rapid plant movement (29). In fact, it is so rapid that until recently it has been impossible to separate experimentally the sensory from the motor functions. Building upon the observation by Jaffe and Galston (27) that tendrils tend to coil less after the nightly dark period, Jaffe (23) found that if tendrils are dark adapted for 2 or 3 days, and kept in the dark, they will not coil following mechanical stimulation. However, as soon as they are illuminated, they begin to coil (Fig. 6). This Light Activation Effect (LAE) will occur even if the illumination is given up to 90 min after mechanical stimulation, although the rate of coiling begins to fall off after 60 min (Fig. 6) (25). Thus, the tendril can somehow store the sensory information of its having been touched, but cannot use this information until it is illuminated. By plotting the ability of the tendril to coil against the time interval between touching and illumination, a "memory curve" has been constructed (Fig. 7) (25). Theoretically, this curve, with its initial rise and eventual decay should be paralleled by similar biochemical changes which are part of the sensory function.

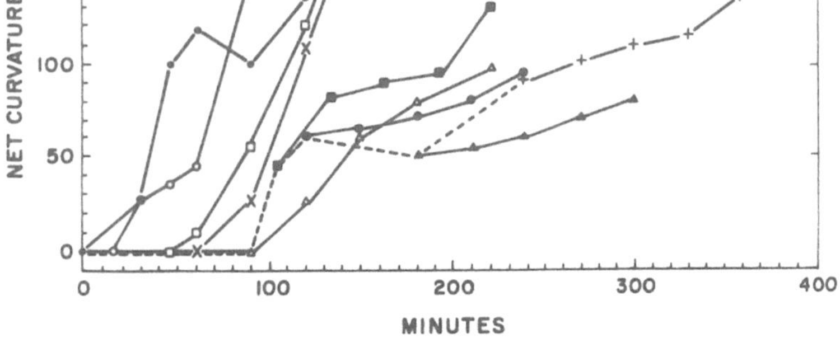

Fig. 6. The Light Activation Effect (LAE) in 3-day dark-adapted tendrils. At zero minutes, the tendrils were excised, rubbed and placed in petri dishes (see Fig. 1). They were then held in the dark for various lengths of time, and then illuminated with white light (5,000 lux) and allowed to coil. Adapted from Jaffe and Shotwell (25)

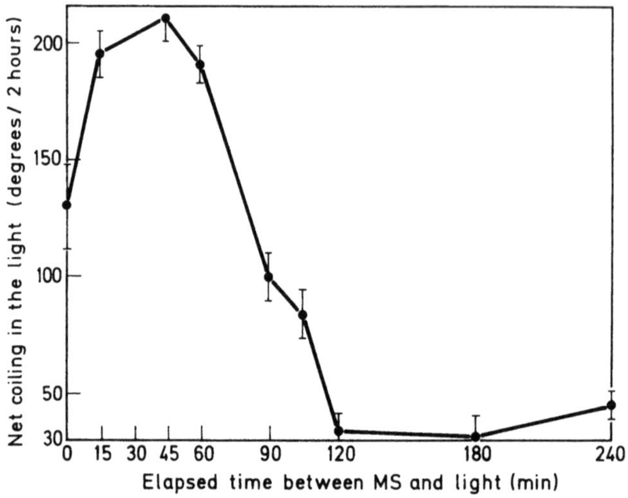

Fig. 7. The time course of retention of the "memory" of mechanical stimulation. The data points, calculated from the curves in Fig. 6, indicate the amount of coiling in 2 h in white light, following different lengths of time in the dark after mechanical stimulation

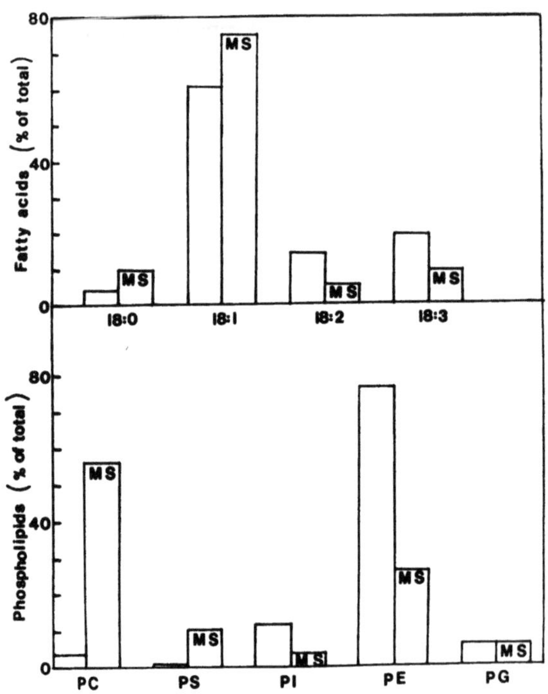

Fig. 8. The effects of rubbing *(MS)* on the 18 carbon fatty acids *(top)* and the phospholipids *(bottom)* of microsomal membrane preparations made from zero time control tendrils *(left bars)* or tendrils rubbed and held in the dark for 45 min before homogenization *(right bars)*. The fatty acids are: *18:0* stearic; *18:1* oleic; *18:2* linoleic; and *18:3* linolenic acid. The phospholipds are: *PC* Phosphatidyl-choline; *PS* P-serine; *PI* P-inositol; *PE* P-ethanolamine; and *PG* P-glycerol

Umrath (38) has published evidence that a propagated action potential can be observed in some tendrils following mechanical stimulation, and Jaffe and Galston have shown that efflux of a tracer from the base of a mechanically stimulated tendril occurs within the first 2 min (28). Thus, it seemed worthwhile to look for changes in cellular membranes due to mechanical stimulation. According to the "memory curve", such changes should be apparent within the first hour after touching. Accordingly, control or mechanically stimulated tendrils were homogenized, filtered, centrifuged at 6,000 *g* and then at 100,000 *g*, from which a microsomal membrane pellet, previously shown to contain the membranes of the endomembrane system, were extracted for their lipid components. The results of this experiment are shown in Fig. 8. Within 45 min after touching, the free fatty acids stearic and oleic have increased, whereas linoleic and linolenic have decreased. Even more strikingly, the phosphatidyl-choline (PC) content of the membranes has increased, and the phosphatidyl-ethanolamine (PE) titer has decreased. Thus far, we have tested the ability of the same fatty acids to substitute for mechanical stimulation, and find that the unsaturated 18 carbon fatty acids, but not the saturated stearic acid, can all mimic the effect of touch (Fig. 9). However, oleic

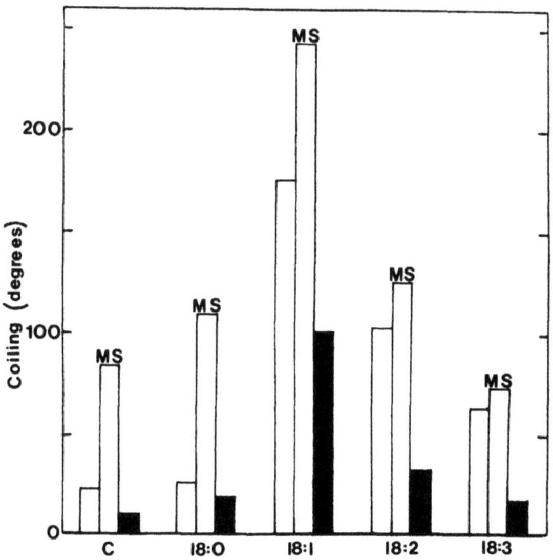

Fig. 9. The effects on coiling of incubating tendrils in 30 mM fatty acids, with or without mechanical stimulation by rubbing *(MS)*. The *left two unfilled columns* were irradiated with 20 min of blue light and then incubated for 40 min in the dark. The *right dark columns* represent tendrils that were not irradiated, but were incubated for 60 min in the dark

acid can also substitute for light. We have not yet tested the ability of exogenous PC or PE to mimic the effect of touch, but it is clear that changes in the composition of the endomembranes do occur during the period of the "memory curve" associated with the sensory function. It is as yet too early to draw any conclusions from these data, and work is currently in progress which hopefully will help in understanding the biochemical events responsible for thigmosensory perception.

The Motor Function

As we have seen, thigmosensing can be separated from the motor function in dark adapted tendrils by witholding illumination. Thus, mechanical stimulation coupled to absorption of radiation causes coiling. In order to understand the nature of this coupling, the part of the system which absorbs light must be known. Therefore, we have done some experiments designed to elucidate this mechanism. The first step in solving any photobiological problem is to ascertain the identity of the light absorbing pigment or pigments. Accordingly, a visible light action spectrum has been constructed (36) which reveals the LAE to appear to be typical of the many blue light effects encountered in plants (Fig. 10) (4). In addition, when the tendrils are irradiated with a very

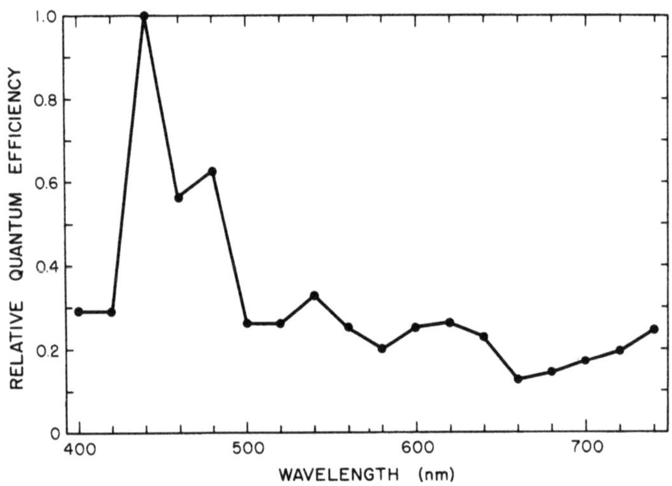

Fig. 10. The visible action spectrum of the LAE. The amount of light necessary to induce 30° of coiling was calculated for each wavelength from fluence response equations. Adapted from Shotwell and Jaffe (36)

bright flash of blue light, the LAE is inhibited, and the time course of recovery of the LAE (Fig. 11) is characteristic of those blue light effects which have been shown to be due to absorption of light by a membrane-associated flavoprotein (3, 15). Such systems have been shown to involve the reduction of a b cytochrome by the photooxidized flavorprotein (2). Thus, it should be possible to substitute reducing agents for light in such a system. When this was tried with the LAE, reducing agents but not oxidizing agents were indeed capable of mimicking the effect of blue light (Table 2).

Thus it seems that we can begin to sort out some of the early events involved in contact coiling and construct a first model of the phenomenon. According to this model, shown in Fig. 12, two parallel sets of reactions must occur: absorption of the energy of the mechanical stimulus and absorption of the energy of the blue light. Briggs and his co-workers (30) have shown that the blue light photoreceptor-cytochrome-b system is associated with the plasma membrane, and we have seen that within 45 min of being touched, dramatic changes occur in the lipid components of the

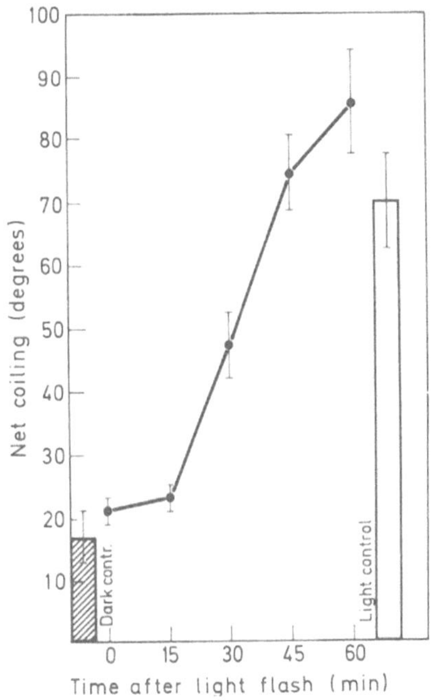

Fig. 11. The time course of recovery of LAE sensitivity to blue light following irradiation with an inhibiting single high intensity flash of blue-white light. Adapted from Jaffe and Shotwell (25)

Table 2. The effects of oxidizing or reducing agents as substitutes for blue light in the LAE of pea tendrils. Dark-adapted tendrils were stimulated and then either irradiated with blue light or placed in solutions of oxidizing or reducing agents for 1 h

Additive	Net coiling (degrees)
None (MS [a] + dark)	24 ± 3
None (MS + light)	71 ± 4
Oxidizing agents	
0.1 mM H_2O_2	11 ± 12
0.1 mM $MnCl_2$	28 ± 9
Reducing agents	
0.1 mM $FeSO_4$	64 ± 4
0.1 mM Tetrazolium blue	58 ± 5
0.06 mM Methylene blue	65 ± 8

[a] MS = mechanical stimulation

tendril's microsomal membranes. Thus it may be that both sensory systems are associated with membranes, although we have some preliminary evidence which suggests that both systems do not occur at the same site. According to the model, each sensory system produces a product ("X" due to touch and "Y" due to reduction of cytochrome b) and the coincidence of these two principles or components initiates coiling.

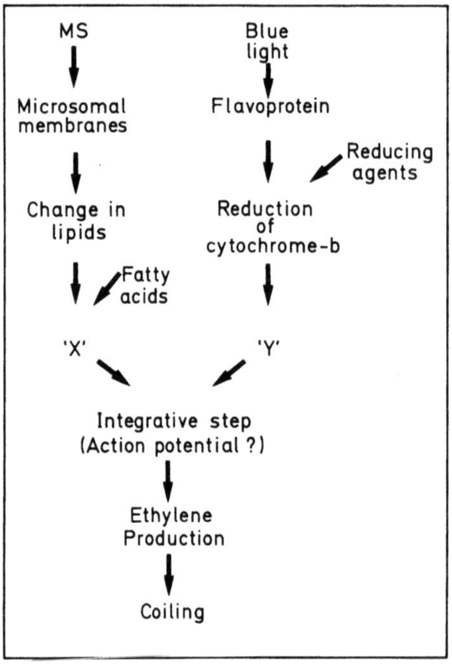

Fig. 12. Proposed model of the possible mechanism of contact coiling in pea tendrils. The *angled arrows* indicate exogenous additives that can substitute for the environmental cues

I shall now outline the evidence for what some of the subsequent steps in the motor process may be.

The final question for which we have evidence is: "In what way do the membrane changes induced by mechanical stimulation induce the motor cells to act?" It is possible that the answer to this question lies in the participation of ethylene in the mechanism. When the tendril is mechanically stimulated, it produces prodigious amounts of ethylene, and when exogenous ethylene (in the form of ethrel) is added to the ventral surface, coiling is induced (20). Furthermore, dorsal application of ethrel inhibits coiling (20), and the ventral surface produces more ethylene when stimulated by auxin than does the dorsal (1). Therefore, it may be that the membrane changes (possibly including an integrating action potential) due to mechanical stimulation may act to induce ethylene production in or around the motor cells, which in turn causes them to contract (26, 29) due to their loss of water (28) (Fig. 12).

The Guying Function of Tendrils

In this analysis of contact coiling, the process has been discussed up to the point wherein the ventral surface contracts to throw a first coil around the support. But no examination of tendrils can be complete without an appreciation of how they attach themselves firmly to the support and act as "guy wires" to hold the stem erect. This aspect of tendril function is very important and will be briefly described.

After the first 15 to 30 min of contact coiling, elongation of the tendril begins. This is very different from the basal elongation that occurs during maturation, for it happens apically, with greater elongation taking place on the dorsal (convex) surface than on the ventral (26). As one can imagine, therefore, the tendril literally grows round and round the support, until it has wrapped itself around many times (29). The question then is: what is the mechanism involved in this curving growth response? It is possible that the answer may lie in experiments concerning auxin that have been performed with several species. However, as we shall see, these studies by no means provide a clear or complete solution to the problem.

Reinhold (35), later confirmed by Jaffe (22), showed that the brief immersion of the tip of a tendril in a solution of auxin may rapidly initiate coiling (22). Although it seemed possible at first that coiling was therefore due to the basipetal transport of auxin, both Jaffe (22) and Junker (31) independently showed that there is no increase in such transport of $[^{14}C]$-IAA, and Junker further demonstrated no dorsi-ventral asymmetrical redistribution of auxin (32). However, Jaffe (22) was able to demonstrate that the auxin binding principle (16) which is a protein (6) associated with the endoplasmic reticulum (34) was primarily found in the apical, thigmosensitive and responding portion of the tendril, and that its activity disappeared after coiling. Because of the rapidity of the response to exogenous auxin, and the small number of auxin molecules necessary for the response, Jaffe suggested that the result was to locally release auxin to its receptor (22). Although it might be tempting to think that auxin is part of the early contractile phase, the lack of basipetal or dorsi-ventral auxin transport, together with the great differences in the nature of the time courses of auxin- versus tendril-induced coiling suggest that such is not the case. Whether or not auxin is involved in the later phase of growth curvature, still remains to be decided.

After the curvature due to growth around the support nears completion, the tendrils of some species (although not those of the pea plant) spontaneously form themselves into a helical spring along the part of the organ between the support and the tendril-bearing stem (29), so that they not only pull the stem closer to the support, but also become very elastic, and less liable to break when pulled.

The last phase of the tendril's guying mechanism is senescence and death. As the stem grows beyond the tendril, it senesces and becomes quite woody (29), so that its hold on the support can hardly be broken. This occurs at a region low enough down on the stem where elasticity is less of a requirement than holding strength. As Darwin describes it:

> "Tendrils soon after catching a support grow much stronger and thicker; and sometimes more durable to a wonderful degree; and this shows how much their internal tissues must be changed. Occasionally it is the part which is wound around a support which chiefly becomes thicker and stronger; I have seen, for instance, this part of a tendril of *Bignonia aequinoctialis* twice as thick and rigid as the free basal part." (12)

Later, in the same passage, he describes beautifully and concisely the rationale for tendril function, as follows:

> "The chief or sole reason for the belief that the curvature of a tendril when touched is due to rapid growth, seems to be that tendrils lose their sensitiveness and power of movement after they have grown to their full length; but this fact is intelligible, if we bear in mind that all the functions of a tendril are adapted to drag up the terminal growing shoot toward the light. Of what use would it be, if an old and full grown tendril, arising from the lower part of a shoot, were to retain its power of clasping a support? This would be of no use; and we have seen with tendrils so many instances of close adaptation and of the economy of means, that we may feel

assured that they would acquire irritibility and the power of clasping a support at the proper age – namely, youth – and would not uselessly retain such power beyond the proper age." (12)

To this writer, it is no wonder that Charles Darwin was so beguiled by these organs of such great activity and variety which enable some vining plants to grow a hundred feet high, although the tendrils themselves may be no greater in thickness than the lead of a pencil.

Acknowledgments. I am grateful to Meg Hummon and Amy Jaffe for devoted and expert technical assistance in the study of the light activation effect presented in this paper. Supported by grant PCM 77/24798 from the National Science Foundation (USA).

References

1. Bangerth, F.: Planta *117*, 329–338 (1974)
2. Brain, R.D., Freeberg, J.A., Weiss, C.V., Briggs, W.R.: Plant Physiol. *59*, 948–952 (1977)
3. Briggs, W.: Plant Physiol. *35*, 951–962 (1960)
4. Briggs, W.R.: In: Light and Plant Development. Smith, H. (ed.), pp. 7–8. London: Butterworths 1975, 516 pp.
5. Cravens, R.H.: Vines. 159 pp. Time-Life books, Alexandria, Va (1979)
6. Cross, J.W., Briggs, W.R.: Annu. Rep. Director Dep. Plant Biol. Carnegie Inst. pp. 353–356 (1978)
7. Darwin, C.: In: The Life and Letters of Charles Darwin – Including an Autobiographical Chapter. Darwin, F. (ed.), vol. II, 562 pp. New York: Appleton 1887
8. Darwin, C.: In: The Life and Letters of Charles Darwin – Including an Autobiographical Chapter. Darwin, F. (ed.), vol. I, 558 pp. New York: Appleton 1887
9. Darwin, C.: In: The Life and Letters of Charles Darwin – Including an Autobiographical Chapter. Darwin, F. (ed.), vol. II, 562 pp. New York: Appleton 1887
10. Darwin, C.: Insectivorous Plants. London: Appleton 1876
11. Darwin, C.: The Different Forms of Flowers on Plants of the Same Species. London: Appleton 1877
12. Darwin, C.: The Movements and Habits of Climbing Plants. 2nd ed, revised, 208 pp. New York: Appleton 1884
13. Darwin, C.: The Power of Movement in Plants. 592 pp. New York: Appleton 1897
14. Darwin, F.: The Life and Letters of Charles Darwin – Including an Autobiographical Chapter. Vol. II, 562 pp. New York: Appleton 1887
15. Diehn, B.: Biochim. Biophys. Acta *177*, 136–143 (1969)
16. Hertel, R., Thomson, K.-St., Russo, V.E.A.: Planta *107*, 325–340 (1972)
17. Jaffe, M.J.: Planta *92*, 146–151 (1970)
18. Jaffe, M.J.: Plant Cell Physiol. *11*, 47–53 (1970)
19. Jaffe, M.J.: Plant Physiol. *45*, 756–760 (1970)
20. Jaffe, M.J.: Plant Physiol. *46*, 631–633 (1970)
21. Jaffe, M.J.: Physiol. Plant. *26*, 73–80 (1972)
22. Jaffe, M.J.: Plant Sci. Lett. *5*, 217–225 (1975)
23. Jaffe, M.J.: Science *195*, 191–192 (1977)
24. Jaffe, M.J.: In: Encyclopedia of Plant Physiology (New Series). Berlin, Heidelberg New York: Springer 1980, in press
25. Jaffe, M.J., Shotwell, M.: In: Physiol. Plant. (1980, in press)
26. Jaffe, M.J., Galston, A.W.: Plant Physiol. *41*, 1014–1025 (1966)
27. Jaffe, M.J., Galston, A.W.: Plant Physiol. *41*, 1152–1158 (1966)
28. Jaffe, M.J., Galston, A.W.: Plant Physiol. *43*, 537–542 (1968)
29. Jaffe, M.J., Galston, A.W.: Annu. Rev. Plant Physiol. *19*, 417–434 (1968)

30. Jesaitis, A.J., Heners, P.R., Hertel, R., Briggs, W.: Plant Physiol. *59*, 941–947 (1977)
31. Junker, S.: Physiol. Plant. *37*, 258–262 (1976)
32. Junker, S.: Physiol. Plant. *41*, 51–54 (1977)
33. Junker, S., Reinhold, L.: J. Microsc. Biol. Cell. *23*, 175–180 (1975)
34. Normand, G., Hartmann, M.A., Schuber, F., Benveniste, P.: Physiol. Veg. *13*, 743–761 (1975)
35. Reinhold, L.: Science *158*, 971 (1967)
36. Shotwell, M., Jaffe, M.J.: Photochem. Photobiol. (1980, in press)
37. Tronchet, A.: Proces-verbaux et memoires de l'Academie des Sciences, Belle-Lettres et Arts de Besancon. Tome 175, 39 pp. 1962–63
38. Umrath, K.: Planta *23*, 47–50 (1934)

Movement by Bacteria: On the Mechanism of Sensory Transduction in Bacterial Chemotaxis

J. ADLER [1]

Introduction

The "plant" I am working with is *Escherichia coli* (Fig. 1). Ever since being introduced to bacteria by Professor K.V. Thimann's course on *The Behavior of the Lower Plants,* I have been convinced that I am studying a plant!

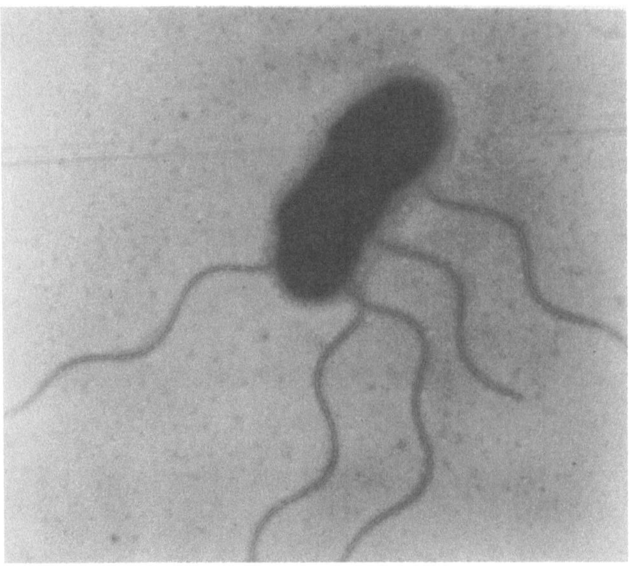

Fig. 1. Electron micrograph of the bacterium used in this study, *Escherichia coli.* By means of 6–10 flagella, the bacterium swims

About 100 years ago Charles Darwin published the book we are celebrating, *The Power of Movement in Plants.* In it he states in the concluding chapter, "Finally, it is impossible not to be struck with the resemblance between the foregoing movements of plants and many of the actions performed unconsciously by the lower animals. . . The habit of moving at certain periods is inherited both by plants and animals; and several

1 Departments of Biochemistry and Genetics, College of Agricultural and Life Sciences, University of Wisconsin-Madison, Madison, Wisconsin 53706, USA

points of similitude have been specified. But the most striking resemblance is the localisation of their sensitivities, and the transmission of an influence from the excited part to another which consequently moves. Yet plants do not of course possess nerves or a central nervous system; and we may infer that with animals such structures serve only for the more perfect transmission of impressions, and for the more complete intercommunication of the several parts."

Several aspects of this quotation are especially noteworthy. First, Darwin stressed the importance of inheritance in behavior. Second, he was among the earliest to indicate the universality in behavior: that there are universal principles which operate to explain behavior in the higher plants – presumably also in the lower plants – and in animals.

At the same time that Darwin's book appeared, the famous German plant physiologist, W. Pfeffer, showed that motile bacteria were attracted to a capillary containing nutrients (Fig. 2) (1). Also at the same time (1881) another well-known plant physiologist, Th.W. Engelmann, showed that bacteria are attracted by oxygen, for example, that produced by algae during photosynthesis (2). By shining a microspectrum on a *Spyrogyra* filament and observing the attraction of the bacteria as a function of wavelength, he was able in a famous experiment to determine the action spectra for photosynthesis in an alga.

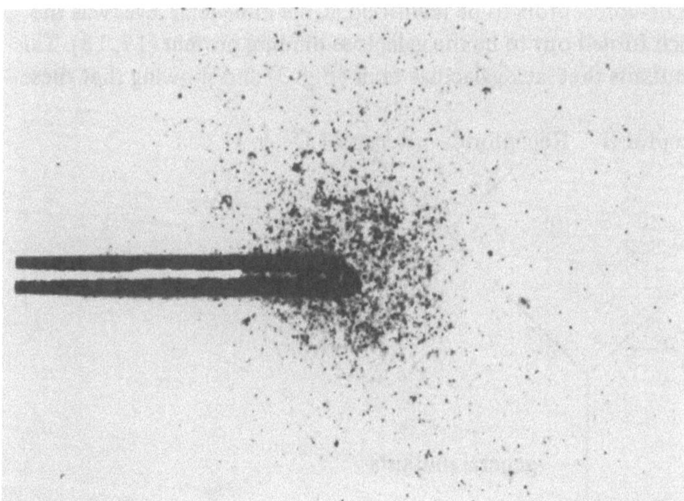

Fig. 2. Experiment of Pfeffer (1) showing attraction of motile bacteria to a capillary filled with an attractant. The bacteria are attracted to the mouth of the capillary, then later inside. The author has made this the basis for an objective, quantitative assay by counting the number of bacteria found inside the capillary after a period of time

Thus, chemotaxis, the movement of motile cells or organisms toward or away from a chemical source, was discovered in bacteria nearly a century ago by Pfeffer (1) and Engelmann (2). Bacterial chemotaxis was actively studied for about 50 years, but later there were very few reports until in the early 1960's, when I began to devise quantitative, objective assays and to apply modern concepts of biochemistry and genetics.

Bacterial chemotaxis can be dissected by means of the following questions: How do bacteria detect the chemicals that attract or repel them? How do individual bacteria move in a gradient of attractant or repellent? How do bacterial flagella produce motion? How is the sensory information communicated to the flagella to bring about motion in the appropriate direction? A number of reviews of both genetic and biochemical aspects of motility and chemotaxis in bacteria have appeared during the past three years (3–7).

E. coli is attracted by a large number of nutritious chemicals, such as various sugars and amino acids (8–11), and is repelled by a large number of harmful chemicals (12). For the closely related bacterium *Salmonella typhimurium* there is now a very similar list of attractants (13) and repellents (14).

Chemoreceptors

As do higher plants, bacteria have sensory receptors – chemoreceptors (9) that detect the attractants per se, not some product of their metabolism (such as the ATP level) as had been previously believed; metabolism of the attractants is not required for chemotaxis (9). Mutants unable to metabolize the attractants were used to demonstrate this result (9).

The first of these chemoreceptors to be identified at the molecular level was the one for galactose, which turned out to be the galactose binding protein (15, 16). This was shown by using mutants that lack galactose taxis (Fig. 3) and showing that these

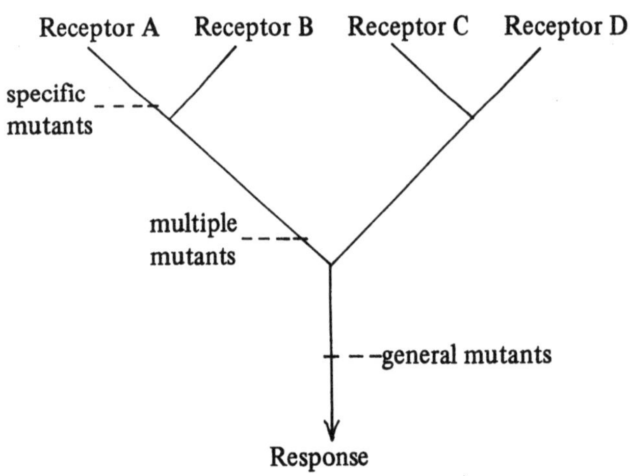

Fig. 3. Behavioral mutants. The author's laboratory discovered three types of motile but nonchemotactic mutants – specific, multiple, and general (see text). These were used to construct the pathway of information flow shown here

lack the binding protein. At the same time we found binding activities for maltose and ribose and indicated that they serve the respective chemoreceptors (15); subsequently the maltose- (17) and ribose- (18, 19) binding proteins were purified, and it was established by Hazelbauer (20) and Koshland's group (18) that they are indeed the chemoreceptors. Again, mutants lacking the binding proteins were a powerful tool. For other sugars we identified other chemoreceptors that are part of the phosphotransferase system by the use of mutants defective in that system (21).

All bacterial chemoreceptors known so far serve a dual function; they are required for transport and also for chemotaxis, but the transport function is not necessary for chemotaxis, as shown by use of mutants defective in transport (9). That is, chemicals can serve as attractants without entering the cell (9), since the chemoreceptors are located on the cell surface, in or at the cytoplasmic membrane.

Behavioral Mutants

As Darwin emphasized (see above), behavior is inherited. We made use of this by employing genetics extensively to dissect the mechanism of bacterial chemotaxis: we discovered behavioral mutants of bacteria (Fig. 3) (22). Some of these are *specifically* nonchemotactic because they lack specific chemoreceptors (15), some are *multiply* nonchemotactic because they lack a mechanism through which information from certain, but not all, chemoreceptors must flow (21a), while others are *generally* nonchemotactic because they are defective in a final common pathway through which information from all the chemoreceptors must flow (22). Due to recent work by Parkinson (6, 23) and most especially by Silverman and Simon (7, 24), it is now known that the generally nonchemotactic (*che*)mutants of *E. coli* fall into eight genes. The properties and use of behavioral mutants will be described later.

Central Role of Tumbling Frequency

Crucial studies from the laboratories of Berg (25) and Koshland (14, 26) established that the central feature of bacterial chemotaxis is the control of the frequency of tumbling. In the absence of a gradient of attractant or repellent, bacteria swim in a straight line ("smooth swimming" or "runs") for a second or two, then they tumble for a fraction of a second, then comes another run in a randomly chosen direction. In a spatial gradient of attractant (e.g., a higher concentration toward the right of a field than on the left), a bacterium that happens to be swimming toward higher concentration (toward the right) suppresses its tumbling and makes a very long run, while a bacterium that happens to be swimming in the "wrong" direction (toward the left) quickly tumbles. In this way, by means of a biased random walk, the bacterium sooner or later finds its way to the high concentration of attractant. For repellents exactly the opposite is true: an increasing concentration stimulates the tumbling frequency while a decreasing concentration suppresses tumbling, and so the bacterium eventually is able to escape from the source of the repellent (Fig. 4).

A remarkable discovery was made by Koshland's group (14, 26) and confirmed by Berg (27) (Fig. 5): the control of tumbling applies not only to spatial gradients but also to temporal gradients. Bacteria were mixed with attractant so rapidly that no

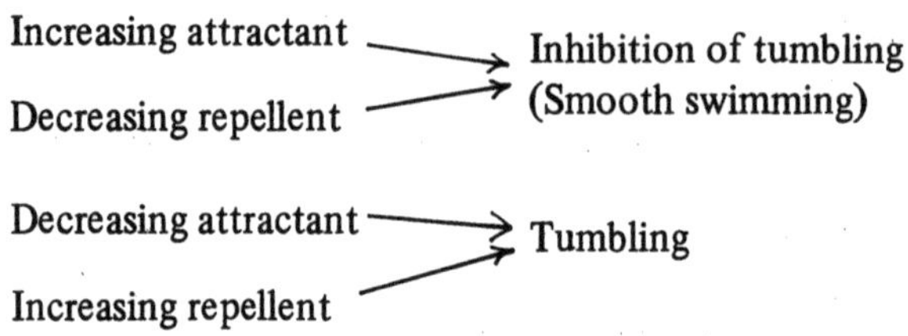

Fig. 4. Effect of change in concentration of stimulus on frequency of tumbling (14, 26, 27)

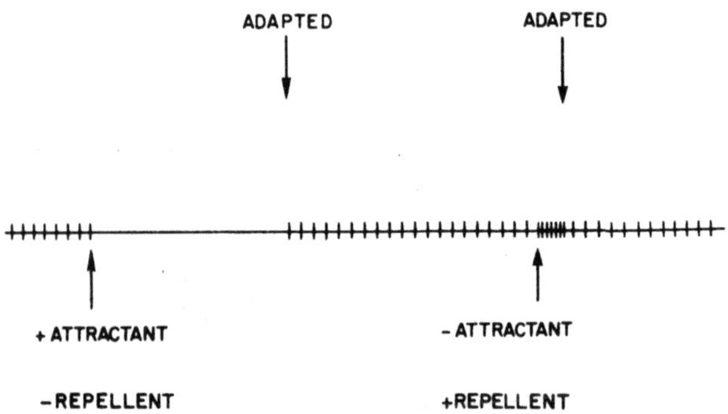

Fig. 5. Response to attractants and repellents, and the concept of adaptation. *Vertical lines* indicate tumbles, which occur at random, rather than evenly spaced, intervals (14, 26, 27)

spatial gradient survived; they were then observed under the microscope and found to be swimming without tumbling. Rapid dilution out of attractant caused them to tumble continuously, as did rapid mixing with repellent; rapid dilution out of repellent caused them to swim without tumbling. The same results were achieved when the increase or decrease in concentration was brought about by enzymatically generating or destroying the attractant (another way of creating a gradient in time instead of in space) (27). Thus, the tumbling frequency is controlled by temporal changes in concentration of attractant or repellent; even in spatial gradients, the bacterium most likely compares the concentration now to the concentration a short time ago, instead of the concentration at the two ends of the bacterium.

Some of the generally nonchemotactic *(che)* mutants never tumble, and others always tumble (4, 5, 22–24). These mutants are therefore very useful for learning about the mechanism that controls tumbling frequency.

Adaptation of Bacteria

The change in tumbling frequency brought about by a change in the concentration of attractant or repellent lasts only a certain length of time (14, 26, 28). Then the bacteria return to the tumbling frequency in effect before the concentration of stimulus was changed, the "unstimulated" tumbling frequency. The bacteria adapt to the stimulus: they stop responding even though the stimulus is still present, i.e., they recover (Fig. 5).

Central Role of Direction of Rotation of Flagella

Bacterial flagella work by rotating (29–31), a proposition advanced by Berg and Anderson on the basis of existing evidence (29). It was then proved by Silverman and Simon (30): they tethered cells to a glass slide by means of antibody to flagella (which of course combines with the flagella and just happens to stick to glass); now the flagella were no longer free to rotate and instead the cell rotated! Sometimes it would rotate counterclockwise, then clockwise, then counterclockwise, etc. (30). We proved that counterclockwise rotation leads to "runs" (absence of tumbling), while clockwise rotation results in tumbling (32). The reason for this has to do with the fact that bacterial flagella have rigid left-handed helices that act like propellers. One part of our proof consisted of showing that never-tumbling mutants rotate their flagella counterclockwise, while the flagella of always-tumbling mutants rotate clockwise.

Flagella of *E. coli* have been isolated in pure form (33, 34). They have a complex structure, including in the cell envelope a rod with four rings (33). The rod and rings serve to anchor the flagellum into the cell envelope and they may in addition be the motor that rotates the flagellum. Many mutants unable to synthesize flagella (*fla* mutants) are now known.

We showed that the energy for rotation of bacterial flagella is not ATP (unlike eukaryotic flagella), but rather the proton motive force (35). This finding has been confirmed and extended (36).

As summarized in Fig. 6, chemical gradients are detected by chemoreceptors. Information from the chemoreceptors is processed by a mechanism called "sensory transduction." This mechanism tells the flagella whether to rotate counterclockwise or clockwise: if counterclockwise, the cells swim in straight lines without tumbling; if clockwise, they tumble. After a while the cells return to their unstimulated tumbling frequency, i.e., they adapt.

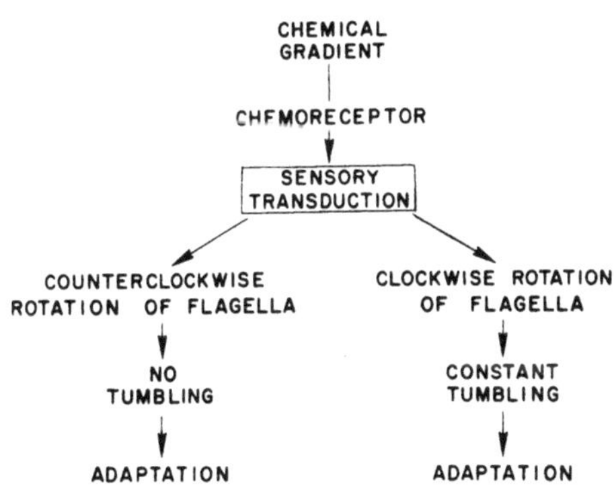

Fig. 6. Summary scheme of the mechanism of chemotaxis (3–5, 7)

Role of Methylation of Proteins in Chemotaxis

Some years ago I discovered that *L*-methionine is required for chemotaxis but not for motility (37). This was accomplished by use of a mutant that fails to synthesize *L*-methionine. Evidence was then presented that methionine appears to act via S-adenosylmethionine (35, 38, 39). Finally, we discovered the function of methionine in chemotaxis: it is used to methylate a protein in the cytoplasmic membrane of *E. coli*, the methyl-accepting chemotaxis protein (MCP) (40). Now it is known that there are really three MCPs of close molecular weight, MCP I, MCP II and MCP III (41, 42, 42a). [Each of them actually consists of several proteins apparently produced by processing a precursor protein (43).]. Certain attractants and repellents (Type I) employ receptors that use MCP I, other attractants and repellents (Type II) employ other receptors that use MCP II, while other attractants (Type III) use MCP III. Thus there are three complementary pathways of information flow in *E. coli* (41, 42) (Fig. 7).

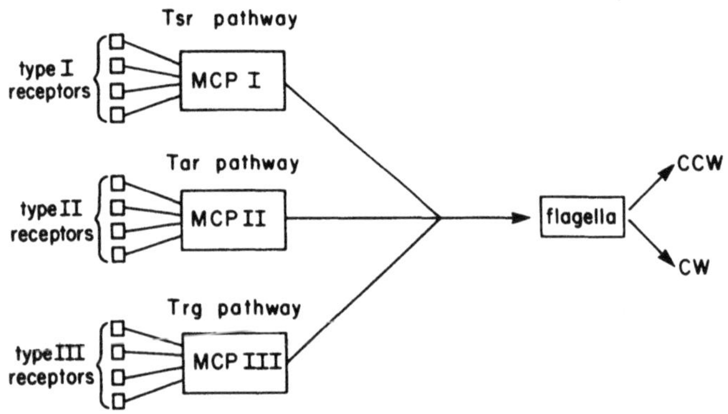

Fig. 7. Three complementary pathways of information flow. *MCP* methyl-accepting chemotaxis protein (41, 42, 42a)

Proof came from the use of the multiply nonchemotactic mutants (Fig. 3). *Tsr* mutants (*t*axis to *s*erine and certain *r*epellents) lack methylation of MCP I and respond abnormally to Type I stimuli; *tar* mutants (*t*axis to *a*spartate and certain *r*epellents) lack methylation of MCP II and fail to respond to Type II stimuli. *Trg* mutants (*t*axis to *r*ibose and *g*alactose) lack methylation of MCP III and fail to respond to Type III stimuli (see Fig. 3).

The moiety of MCP that becomes methylated is a glutamate residue to form the methyl ester (44, 45) (Fig. 8). Such a protein carboxyl methylation has been known in animals for some years (46, 47); on the basis of indirect but good evidence (46), in animals aspartate and glutamate residues are believed to be methylated to form the methyl esters.

Cell-free systems of *S. typhimurium* (48, 49) and *E. coli* (50) that carry out the protein-glutamate methylation and demethylation reactions (Fig. 8) have now been successfully established. Mutants that fail to carry out this methylation and demethylation have been identified (40–42, 48–50). These represent two genes of the generally nonchemotactic *(che)* mutants mentioned earlier.

Addition of an attractant to bacteria causes an *increase* in the methylation of MCP (one or the other MCP, depending on the attractant used), while dilution of the attractant causes a *decrease* in the methylation of MCP (51) (Fig. 9a). Repellents have the opposite effect (51) (Fig. 9b). Thus, when the chemoreceptors detect a change in

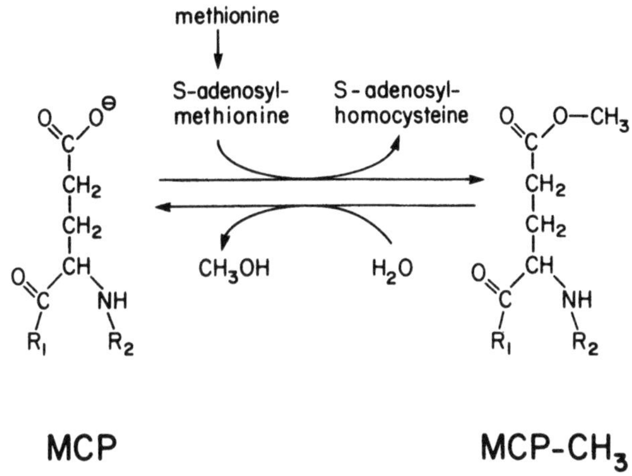

Fig. 8. Protein modification reactions discussed in this article

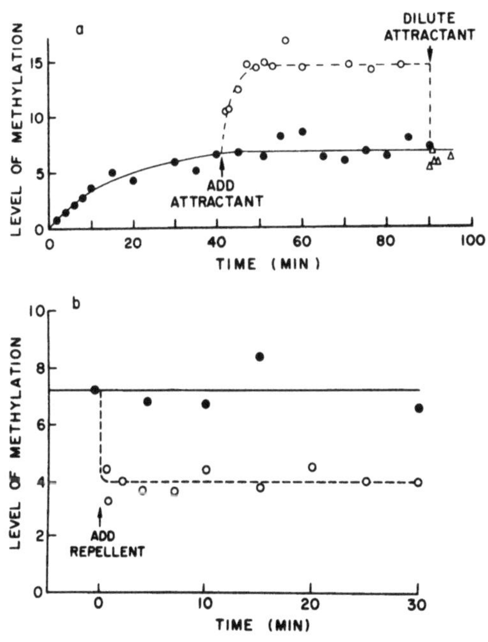

Fig. 9. Effect of attractants **(a)** and repellents **(b)** on the methylation of methyl-accepting chemotaxis protein (51, 53). Removal of the repellent causes remethylation (data not shown) (53)

concentration of attractant or repellent, they somehow tell the MCP to become more or less methylated. The change in methylation takes a certain time to occur, and then stops at a new plateau level of methylation (51) (Fig. 9). This new level remains constant until the bacteria encounter another change in concentration of attractants or repellents (51).

Excitation and Adaptation

Attractants or repellents interact with their specific chemoreceptor, and then this in turn interacts with its specific MCP (Fig. 10). As shown in Fig. 10, this interaction activates the MCP to send a signal to the flagella to tell them to rotate either counterclockwise or clockwise. We refer to this process as "excitation;" the nature of this signal is not yet known; it might be an ion flux [for example, Ca^{2+}, see (52)], or possibly there is only protein-protein interaction without mediation of any diffusible substance.

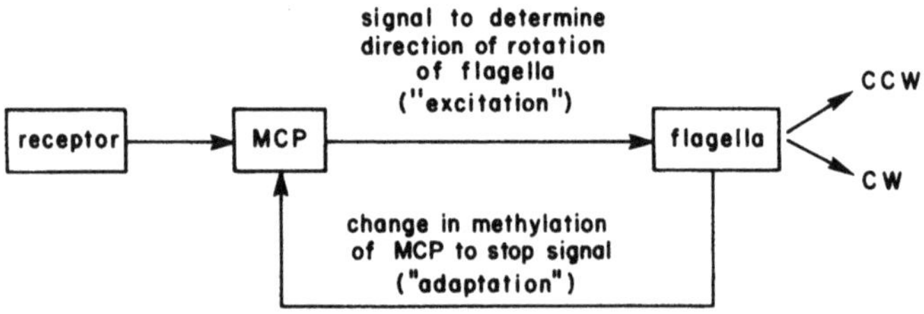

Fig. 10. Summary model indicating role of MCP and relating excitation and adaptation to response (53). See (53) for a detailed model

Then, possibly by a feedback mechanism, the signal sent by MCP to the flagella is shut off by a change in level of methylation of MCP (Fig. 10). This shut-off is "adaptation." If the signal requested counterclockwise rotation of the flagella, an *increase* in the level of methylation of the MCP is needed to stop this signal, while if the signal requested clockwise rotation a *decrease* in the level of methylation of the MCP is necessary to terminate the signal. Thus it is believed that the response time (or "adaptation time") is determined by the time it takes to reach a new plateau level of methylation; once this new level has been reached, adaptation has occurred (51).

Evidence for this comes from the study of mutants that do have MCP but are unable to methylate it. These mutants respond to attractants and repellents incessantly, i.e., they fail to adapt (4, 51). Thus the methylation of MCP is required for adaptation, but it is not required for the initiation of the response, "excitation" (51).

Although excitation does not require the methylation of MCP, it does require its presence. The evidence for this comes from the fact that mutants that lack MCP cannot even initiate a response to attractants or repellents (41, 42).

The role of methylation and demethylation of MCP in bacterial chemotaxis has been recently reviewed (53) and the extension of protein-carboxyl methylation to other organisms has been summarized there (53).

Summary of Bacterial Chemotaxis

The mechanism of bacterial chemotaxis is beginning to be understood. At the receptor end, we have considerable knowledge about the molecular properties of chemoreceptors. At the effector end, we know that flagella rotate and that the direction of rotation is determined by attractants and repellents, although we do not yet know the molecular features of the motor and the gear shift. Between the receptors and the effectors is a system for integrating the sensory information and transmitting a message to the flagella. This system, sensory transduction, somehow involves methylation of membrane proteins, but further details of how the mechanism works remain to be elucidated.

Concluding Remark

The inheritance of behavior and its underlying biochemical mechanisms are nowhere more amenable to genetic and biochemical investigation than in the bacteria. From the earliest studies of bacterial behavior to the present, people have hoped that this relatively simple system (Fig. 11) could tell us something about the mechanisms of behavior of plants, animals and man. Certainly, striking similarities exist between sensory reception in bacteria and in higher organisms. Fig. 11 encompasses well Darwin's statement (quoted above), perhaps the major conclusion of his book, that in higher plants there is transmission of an influence from the excited part to another part which consequently moves.

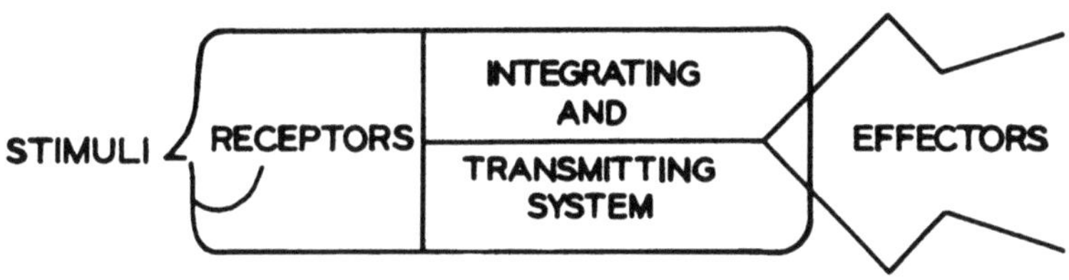

Fig. 11. The bacterial cell as a model for behavior. Like higher plants and animals, the bacterium has sensory receptors, effectors, and, connecting them, an integrating and transmitting system

Acknowledgments. The work reported here from my laboratory was carried out with a number of collaborators, to whom I am very highly indebted: Sylvia Zottu Schade, Margaret M. Dahl, John B. Armstrong, Melvin L. DePamphilis, Robert Mesibov, Gerald L. Hazelbauer, Wung-Wai Tso, Robert W. Reader, Edward N. Kort, Steven H. Larsen, J. Sandy Parkinson, George W. Ordal, Michael F. Goy, Wolfgang Epstein, Marc A. Muskavitch, David R. Repaske, Carl B. Ball, Martin S. Springer, Steven J. Kleene, Daniel J. Zagrodnik, Sevec Szmelcman, Myron L. Toews, Hisato Kondoh, and Ann C. Hobson (in chronological order). Supported by Public Health Service Grant AI08746 from the National Institute of Allergy and Infectious Diseases, National Science Foundation Grant PCM75-21007, and a grant from the Graduate School of the University of Wisconsin-Madison.

References

1. Pfeffer, W.: Untersuch. Bot. Inst. Tübingen *1*, 363–482 (1884)
2. Engelmann, T.W.: Pflügers Arch. Gesamte Physiol. *25*, 285–292 (1881)
3. Adler, J.: Annu. Rev. Biochem. *44*, 341–356 (1975)
4. Berg, H.C.: Annu. Rev. Biophys. Bioeng. *4*, 119–136 (1975)

5. Koshland, D.E., Jr.: In: Advances in Neurochemistry. Agranoff, B.W., Aprison, M.H. (eds.) Vol. 2, pp. 277–341. New York: Plenum Press 1977
6. Parkinson, J.S.: Annu. Rev. Genet. *11*, 397–414 (1977)
7. Silverman, M., Simon, M.: Annu. Rev. Microbiol. *31*, 397–419 (1977)
8. Adler, J.: Science *153*, 708–716 (1966)
9. Adler, J.: Science *166*, 1588–1597 (1969)
10. Adler, J., Hazelbauer, G.L., Dahl, M.M.: J. Bacteriol. *115*, 824–847 (1973)
11. Mesibov, R., Adler, J.: J. Bacteriol. *112*, 315–326 (1972)
12. Tso, W.-W., Adler, J.: J. Bacteriol. *118*, 560–576 (1974)
13. Melton, T., Hartman, P.E., Stratis, J.P., Lee, T.L., Davis, A.T.: J. Bacteriol. *133*, 708–716 (1978)
14. Tsang, N., Macnab, R , Koshland, D.E., Jr.: Science *181*, 60–63 (1973)
15. Hazelbauer, G.L., Adler, J.: Nature (Lond.) New Biol. *230* (12), 101–104 (1971)
16. Kalckar, H.M.: Science *174*, 557–565 (1971)
17. Kellerman, O., Szmelcman, S.: Eur. J. Biochem. *47*, 139–149 (1974)
18. Aksamit, R.R., Koshland, D.E., Jr.: Biochemistry *13*, 4473–4478 (1974)
19. Galloway, D.R., Furlong, C.E.: Arch. Biochem. Biophys. *184*, 496–504 (1977)
20. Hazelbauer, G.L.: J. Bacteriol. *122*, 206–214 (1975)
21. Adler, J., Epstein, W.: Proc. Natl. Acad. Sci. USA *71*, 2895–2899 (1974)
21a. Reader, R.W., Tso, W.-W., Springer, M.S., Goy, M.F., Adler, J.: J. Gen. Microbiol. *111*, 363–374 (1979)
22. Armstrong, J.B., Adler, J., Dahl, M.M.: J. Bacteriol. *93*, 390–398 (1967)
23. Parkinson, J.S.: J. Bacteriol. *126*, 758–770 (1976)
24. Silverman, M., Simon, M.: J. Bacteriol. *130*, 1317–1325 (1977)
25. Berg, H.C., Brown, D.A.: Nature (Lond.) *239*, 500–504 (1972)
26. Macnab, R.M., Koshland, D.E., Jr.: Proc. Natl. Acad. Sci. USA *69*, 2509–2512 (1972)
27. Brown, D.A., Berg, H.C.: Proc. Natl. Acad. Sci. USA *71*, 1388–1392 (1974)
28. Berg, H.C., Tedesco, P.M.: Proc. Natl. Acad. Sci. USA *72*, 3235–3239 (1975)
29. Berg, H.C., Anderson, R.A.: Nature (Lond.) *245*, 380–382 (1973)
30. Silverman, M., Simon, M.: Nature (Lond.) *249*, 73–74 (1974)
31. Berg, H.C.: Nature (Lond.) *249*, 77–79 (1974)
32. Larsen, S.H., Reader, R.W., Kort, E.N., Tso, W.-W., Adler, J.: Nature (Lond.) *249*, 74–77 (1974)
33. DePamphilis, M.L., Adler, J.: J. Bacteriol. *105*, 396–407 (1971)
34. Dimmitt, K., Simon, M.: J. Bacteriol. *105*, 369–375 (1971)
35. Larsen, S.H., Adler, J., Gargus, J.J., Hogg, R.W.: Proc. Natl. Acad. Sci. USA *71*, 1239–1243 (1974)
36. Manson, M.D., Tedesco, P., Berg, H.C., Harold, F.M., van der Drift, C.: Proc. Natl. Acad. Sci USA *74*, 3060–3064 (1977)
37. Adler, J., Dahl, M.M.: J. Gen. Microbiol. *46*, 161–173 (1967)
38. Armstrong, J.B.: Can. J. Microbiol. *18*, 1695–1701 (1972)
39. Aswad, D., Koshland, D.E., Jr.: J. Bacteriol. *118*, 640–645 (1974)
40. Kort, E.N., Goy, M.F., Larsen, S.H., Adler, J.: Proc. Natl. Acad. Sci. USA *72*, 3939–3943 (1975)
41. Springer, M.S., Goy, M.F., Adler, J.: Proc. Natl. Acad. Sci. USA *74*, 3312–3316 (1977)
42. Silverman, M., Simon, M.: Proc. Natl. Acad. Sci. USA *74*, 3317–3321 (1977)
42a. Kondoh, H., Ball, C.B., Adler, J.: Proc. Natl. Acad. Sci. USA *76*, 260–264 (1979)
43. Silverman, M., Matsumura, P., Hilmen, M., Simon, M.: J. Bacteriol. *130*, 877–887 (1977)
44. Van Der Werf, P., Koshland, D.E., Jr.: J. Biol. Chem. *252*, 2793–2795 (1977)
45. Kleene, S.J., Toews, M.L., Adler, J.: J. Biol. Chem. *252*, 3214–3218 (1977)
46. Kim, S., Paik, W.K.: J. Biol. Chem. *245*, 1806–1813 (1970)
47. O'Dea, R.F., Viveros, O.H., Axelrod, J., Aswanikumar, S., Schiffman, E., Corcoran, B.A.: Nature (Lond.) *272*, 462–464 (1978)
48. Springer, W.R., Koshland, D.E.: Proc. Natl. Acad. Sci. USA *74*, 533–537 (1977)
49. Stock, J.B., Koshland, D.E., Jr.: Proc. Natl. Acad. Sci. USA *75*, 3659–3663 (1978)

50. Kleene, S.J., Hobson, A.C., Adler, J.: Proc. Natl. Acad. Sci. USA *76*, 6309–6313 (1979)
51. Goy, M.F., Springer, M.S., Adler, J.: Proc. Natl. Acad. Sci. USA *74*, 4964–4968 (1977)
52. Ordal, G.W.: Nature (Lond.) *270*, 66–67 (1977)
53. Springer, M.S., Goy, M.F., Adler, J.: Nature (Lond.) *280*, 279–284 (1979)

Participants

ABU-TABIKH, A.A.T.
Urbana, Illinois 61801, USA

ADDICOTT, F.T.
Davis, California 95616, USA

ADLER, J.
Madison, Wisconsin 53706, USA

ALDRIDGE, E.G.
Kansas City, Kansas 66102, USA

ALONI, B.
Bet Dagan, Israel

ALONI, O.
Tel Aviv 61390, Israel

ALONI, R.
Tel Aviv 61390, Israel

AMENDT, B.A.
Iowa City, Iowa 52240, USA

AMRHEIN, N.
4630 Bochum, FRG

ANDERSON, J.D.
Beltsville, Maryland 20705, USA

ANDERSON, P.C.
St. Paul, Minnesota 55108, USA

ARMSTRONG, D.J.
Corvallis, Oregon 97331, USA

ARTECA, R.N.
Pullman, Washington 99163, USA

ATHERTON, J.G.
Loughborough, United Kingdom

AUDERSET, G.L.
1211 Geneva 4, Switzerland

AUSICH, R.L.
Bloomington, Indiana 47401, USA

AWAD, A.R.E.
Davis, California 95616, USA

BAKER, F.A.
St. Paul, Minnesota 55108, USA

BALL, E.A.
Irvine, California 92717, USA

BANDURSKI, R.S.
East Lansing, Michigan 48823, USA

BARENDSE, E.M.
Toernooiveld, Nijmegen, The Netherlands

BARENDSE, G.W.M.
Toernooiveld, Nijmegen, The Netherlends

BASSI, P.K.
Edmonton, Alberta, Canada T6G 2E3

BEARDER, J.R.
Bristol, United Kingdom

BECKER, W.M.
Madison, Wisconsin 53706, USA

BEDNAR, T.W.
Western Springs, Illinois 60558, USA

BELKE, C.
Brandon, Manitoba, Canada

BERNIER, G.
B4000 Liege, Belgium

BEYER, E.M.
Wilmington, Delaware 19898, USA

BHALLA, P.R.
Princeton, New Jersey 08520, USA

BIGGS, R.H.
Gainesville, Florida 32611, USA

BIRO, R.L.
Athens, Ohio 45701, USA

BLAYDES, D.F.
Morgantown, West Virginia 26506, USA

BOCK, R.M.
Madison, Wisconsin 53706, USA

BOGERS, R.J.
Leiden, The Netherlands

BOPP, M.
6900 Heidelberg, FRG

BOTTGER, M.
2000 Hamburg, FRG

BOWN, A.W.
St. Catharines, Canada

BOYER, G.L.
Madison, Wisconsin 53706, USA

BRADFORD, K.J.
Davis, California 95616, USA

BRAY, E.A.
St. Paul, Minnesota 55108,
USA

BRENNER, M.L.
St. Paul, Minnesota 55108,
USA

BRIDGE, K.
Jacksonville, Florida 32216,
USA

BRIDGES, I.G.
Cheshire, United Kingdom

BRUINSMA-HORREE, J.
Wageningen, Netherlands

BRUINSMA, J.
Wageningen, Netherlands

BRUMMELL, D.A.
Brighton BN1 9QG,
United Kingdom

BRUN, W.A.
St. Paul, Minnesota 55108,
USA

BRYANT, St.D.
Memphis, Tennessee 38138,
USA

BUCKHOUT, T.J.
Lafayette, Indiana 47094,
USA

BURRIS, R.H.
Madison, Wisconsin 53706,
USA

BUSHNELL, W.R.
St. Paul, Minnesota 55108,
USA

BUTA, J.G.
Beltsville, Maryland 20705,
USA

CAMPBELL, J.A.
Madison, Wisconsin 53706,
USA

CARR, D.J.
Canberra City, Australia

CHAILAKHYAN, M.K.
Moscow 127106, USSR

CHANDRA, G.R.
Beltsville, Maryland 20705,
USA

CHEN, C.-M.
Kenosha, Wisconsin 53141,
USA

CHERRY, J.H.
Lafayette, Indiana 47907,
USA

CHINNICI, M.F.
Painesville, Ohio 44077, USA

CHOINSKI, J.S.
Corvallis, Oregon 97331, USA

CIHA, A.J.
Pullman, Washington 99164,
USA

CLELAND, C.F.
Rockville, Maryland 20852,
USA

CLELAND, R.F.
Seattle, Washington 98195,
USA

COLLET, G.
1260 Changius/Nyou,
Switzerland

COLLET, G.F.
1260 Changius/Nyou,
Switzerland

COOLBAUGH, R.C.
Monmouth, Oregon 97361,
USA

CORCORAN, M.R.
Northridge, California 91344,
USA

CROSBY, K.E.
Blacksburg, Virginia 24061,
USA

CROSS, J.W.
Groton, Connecticut 06340,
USA

CROZIER, A.
Corvallis, Oregon 97331, USA

CURTIS, E.M.
West Lafayette,
Indiana 47907, USA

CURTIS, R.W.
West Lafayette,
Indiana 47907, USA

DAHL, G.A.
Madison, Wisconsin 53706,
USA

DARROW, L.S.
Iowa City, Iowa 52240, USA

DARVILL, A.G.
Boulder, Colorado 80309,
USA

DAVENPORT, T.L.
Homestead, Florida 33031,
USA

DAVIES, P.J.
Ithaca, New York 14853,
USA

DAVIES, W.J.
Bailrigg, Lancaster,
United Kingdom

DAVIS, D.G.
Fargo, North Dakota 58105,
USA

DELEUZE, G.G.
Cumana, Venezuela

DENNIS, F.G., Jr.
East Lansing,
Michigan 48824, USA

DIGBY, J.
Heslington, York YO1 5DD,
United Kingdom

DILLEY, D.R.
East Lansing,
Michigan 48824, USA

DIMALLA, G.G.
Davis, California 95616,
USA

DIVAN, C.L.
Madison, Wisconsin 53706, USA

DODDS, J.H.
Columbus, Ohio 43210, USA

DORFFLING, K.
2000 Hamburg, FRG

DOWLING, T.E.
Riverside, California 92507, USA

DUBOUCHET, J.
25042 Besancon, France

DUMBROFF, E.B.
Waterloo, Ontario, Canada

DUNCAN, J.D.
Los Angeles, California 90024, USA

DURLEY, R.C.
Saskatoon, Sask., Canada

DUSKY, J.A.
Fargo, North Dakota 58105, USA

DUTE, R.R.
Urbana, Illinois 61801, USA

EBERT, E.
Basel, Switzerland

EINSET, J.W.
Riverside, California 92521, USA

EL-BELTAGY, A.S.
Shubra El-Khima, Egypt

EL-OTMANI, M.A.
St. Paul Minnesota 55108, USA

ELLIOTT, M.C.
Leicester, England

EMERY, C.
Los Angeles, California 90027, USA

ENTSCH, B.
Canberra City, Australia

EVENSEN, K.B.
Columbia, Missouri 65211, USA

EVERETT, N.P.
Leicester, England

FELDMAN, L.J.
Berkeley, California 94704, USA

FIRN, R.D.
Heslington, York YO1 5DD, United Kingdom

FLEISCHMAN, W.
Annville, Pennsylvania 17003, USA

FOSKET, D.E.
Irvine, California 92664, USA

FOX, J.E.
Lawrence, Kansas 66045, USA

FREAR, D.S.
Fargo, North Dakota 58105, USA

FREGEAU, J.A.
Ottawa, Ontario, Canada

GALSTON, A.W.
New Haven, Connecticut 06520, USA

GARCIA-MARTINEZ, J.L.
Davis, California 95616, USA

GARDNER, G.M.
Modesto, California 95352, USA

GASPAR, T.E.C.
4020 Liege, Belgium

GEPSTEIN. S.
Santa Cruz, California 95064, USA

GERLOFF, G.C.
Madison, Wisconsin 53706, USA

GIMMLER, H.
8700 Würzburg, FRG

GOH, C.J.
Singapore, 10, Singapore

GOLDSCHMIDT, E.
Rehovot, Israel

GOLDSMITH, M.H.M.
New Haven, Connecticut 06520, USA

GOREN, N.
Rehovot, Israel

GOREN, R.
Rehovot, Israel

GRAEBE, J.E.
3400 Göttingen, FRG

GRAEBE, U.
3400 Göttingen, FRG

GREENE, B.
Madison, Wisconsin 53706, USA

GREGERSON, E.L.
Santa Cruz, California 95064, USA

GREPPIN, H.
1211 Geneva 4, Switzerland

GREYSON, R.I.
London, Ontario, Canada

GROVE, M.D.
Peoria, Illinois 61604, USA

GUILFOYLE, T.J.
St. Paul, Minnesota 55108, USA

GUO, Q.
Fukien, China

HACKETT, W.
Davis, California 95616, USA

HALEVY, A.H.
Rehovot, Israel

HALL, M.A.
Aberystwyth SY23 3DA
Dyfed, United Kingdom

HAMIEDA, A.B.
Tripoli, Libya

HANGARTER, R.P.
East Lansing, Michigan 48824, USA

HARRISON, M.A.
Ann Arbor, Michigan 48105, USA

HASHIZUME, T.
Tokyo 183, Japan

HECHT, S.M.
Charlottesville, Virginia 22901, USA

HEDDEN, P.
Los Angeles,
California 90024, USA

HEIN, M.B.
St. Paul, Minnesota 55108, USA

HELGESON, J.P.
Madison, Wisconsin 53706, USA

HESLOP-HARRISON, J.
Aberystwyth,
United Kingdom

HESLOP-HARRISON, P.
Aberystwyth,
United Kingdom

HESLOP-HARRISON, Y.
Aberystwyth,
United Kingdom

HEUPEL, A.M.
Los Angeles,
California 90024, USA

HEUPEL, R.C.
Los Angeles
California 90024, USA

HOFFMANN, O.L.
Merriam, Kansas 66216, USA

HONDA, S.I.
Dayton, Ohio 45435, USA

HOOVER, E.E.
St. Paul, Minnesota 55108, USA

HORTON, R.F.
Guelph, Ontario, Canada

HUBER, D.J.
Ames, Iowa 50010, USA

HUMBLE, G.D.
St. Louis, Missouri 63166, USA

HURTER, J.
8820 Wadenswil, Switzerland

IMASEKI, H.
Nagoyga 464, Japan

ISOGAI, Y.
Tokyo 158, Japan

IWAMURA, H.
Kyoto 606, Japan

JAFFE, M.
Athens, Ohio 45701, USA

JACKSON, M.B.
Oxon, OX12 9JT,
United Kingdom

JACOBS, W.P.
Princeton, New Jersey 08540, USA

JOHNSON, M.A.
Appleton, Wisconsin 54912, USA

JOHRI, M.M.
Columbia, Missouri 65211, USA

JONES, A.M.
Urbana, Illinois 61801, USA

JONES, J.F.
East Lansing, Michigan 48824, USA

JONES, M.G.
East Lansing, Michigan 48824, USA

JONES, R.L.
Berkeley, California 94708, USA

JOUANNEAU, J.P.
Luminy-13288 Marseille, France

JULIN-TEGELMAN, A.
10691 Stockholm, Sweden

JUNG, J.
6703 Limburgerhof, FRG

KAMIEN, E.N.
Lowell, Massachusetts 01854, USA

KAMISAKA, S.
Osaka 558, Japan

KANG, B.G.
Seoul 120, Korea

KANNANGARA, T.
Saskatoon, S7N OWO,
Canada

KAO, C.H.
Taipei, Taiwan

KARSSEN, C.M.
Wageningen, Netherlands

KATEKAR, G.F.
Canberra City, Australia

KATSUMI, A.
Tokyo, Japan

KATSUMI, M.
Tokoy, Japan

KAUFMAN, P.B.
Ann Arbor, Michigan 48109, USA

KAWADA, K.
Lake Alfred, Florida 33850, USA

KEARNS, A.W.
Heslington, York YO1 5DD, United Kingdom

KEEGSTRA, K.
Madison, Wisconsin 53706, USA

KENDE, H.
East Lansing, Michigan 48824, USA

KESSLER, R.W.
Iowa City, Iowa 52240, USA

KHAN, A.A.
Geneva, New York 14456, USA

KLAMBT, D.
5300 Bonn, FRG

KOEHLER, D.E.
College Station, Texas 77843, USA

KOSHIMIZU, K.
Kyoto 606, Japan

KOSSUTH, S.V.
Olustee, Florida 32072, USA

KOZLOWSKI, T.T.
Madison, Wisconsin 53706, USA

KRAMER, W.
44 Bitterfeld, DDR

KRIKORIAN, A.D.
Stony Brook, New York 11794, USA

KRUL, W.R.
Narragansett, Rhode Island 02882, USA

KULAEVA, O.N.
Moscow 127106, USSR

KUO, C.G.
Shanhua 741, Taiwan

LALOUE, C.
91190 Gif sur Yvette, France

LALOUE, M.A.
91190 Gif sur Yvette, France

LAMOTTE, C.E.
Ames, Iowa 50011, USA

LANG, A.
East Lansing, Michigan 48824, USA

LANG, L.
East Lansing, Michigan 48824, USA

LARQUE-SAAVEDRA, A.
Chapingo, Mexico

LAUDE, H.M.
Davis, California 95616, USA

LAUDE, J.
Davis, California 95616, USA

LAVEE, A.
Davis, California 95616, USA

LAVEE, S.
Davis, California 95616, USA

LAWRENCE, D.K.
Bracknell, Berks, RG12 GEY, England

LEATHAM, G.F.
Madison, Wisconsin 53706, USA

LEE, S.G.
Seoul, Korea

LEE, T.T.
London, Ontario, Canada

LEGUAY, J.-J.
75011 Paris, France

LENNART, E.
90187 Umea, Sweden

LEONARD, N.J.
Urbana, Illinois 61801, USA

LEOPOLD, A.C.
Ithaca, New York 14850, USA

LESHEM, Y.
Ramat Gan, Israel

LIEBERMAN, M.
Beltsville, Maryland 20705, USA

LIN, P.P.C.
Lexington, Kentucky 40546, USA

LINSMAIER-BEDNAR, E.M.
Western Springs, Illinois 60558, USA

LITTLE, C.H.A.
Fredericton, New Brunswick, Canada

LOONEY, N.E.
Summerland, B.C., Canada

LOROS, J.
Santa Cruz, California 95064, USA

LURSSEN, K.
5090 Leverkusen, FRG

MacMILLAN, J.
Bristol BS8 1TS, United Kingdom

MAHADEVAN, S.
Bangalore-560012, India

MANDAVA, B.
Beltsville, Maryland 20705, USA

MARAVOLO, N.C.
Appleton, Wisconsin 54912, USA

MARTIN, H.V.
1005 Lausanne, Switzerland

MARVEL, J.T.
St. Louis, Missouri 63166, USA

MASUDA, Y.
Osaka 558, Japan

McCHESNEY, J.D.
University, Mississippi 38677, USA

MEDOUAR, M.
St. Paul, Minnesota 55108, USA

METRAUX, J.-P.
Santa Cruz, California 95064, USA

METZGER, J.D.
East Lansing, Michigan 48824, USA

MEUDT, W.J.
Beltsville, Maryland 20705, USA

MIASSOD, R.L.
East Lansing, Michigan 48824, USA

MIGINIAC, E.
91190 Gif sur Yvette, France

MILBORROW, B.V.
Kensington, N.S.W., Australia

MILLER, C.O.
Bloomington, Indiana 47405, USA

MILLINGTON, W.F.
Milwaukee, Wisconsin 53233, USA

MINOCHA, S.C.
Durham, New Hampshire 03824, USA

MOK, D.W.S.
Corvallis, Oregon 97331, USA

MOK, M.C.
Corvallis, Oregon 97331, USA

MOORE, T.C.
Corvallis Oregon 97330, USA

MORGAN, P.W.
College Station, Texas 77843, USA

MORRÉ, D.J.
West Lafayette, Indiana 47907, USA

MORRIS, D.A.
Southampton, England

MUIR, R.M.
Iowa City, Iowa 52242, USA

MUIR, R.
Iowa City, Iowa 52242, USA

MUKHERJEE, B.B.
Calcutta-700009, India

MULKEY, T.J.
Fayetteville, Arkansas 72701, USA

MULLINS, M.G.
Sydney, NSW 2006, Australia

MURASHIGE, T.
Riverside, California 92521, USA

MUROFUSHI, N.
Tokyo 113, Japan

MUTHUKRISHNAN, S.
Bethesda, Maryland 20205, USA

NARAYAN, G.S.
Calcutta-700009, India

NETHERY, A.A.
Middleport, New York 14105, USA

NEVINS, D.J.
Ames, Iowa 50011, USA

NEWCOMB, E.H.
Madison, Wisconsin 53706, USA

NICKELL, L.G.
Chicago, Illinois 60611, USA

NISHITANI, K.
Osaka 558, Japan

NITSCH, C.M.
91190 Gif sur Yvette, France

NOGGLE, G.R.
Bethesda, Maryland 20014, USA

NOMA, M.M.
Calgary, Alberta T2N 1N4, Canada

NOODEN, L.D.
Ann Arbor, Michigan 48109, USA

OHLROGGE, J.B.
Davis, California 95616, USA

PALEG, B.
Adelaide, S.A., Australia

PALEG, L.G.
Adelaide, S.A., Australia

PALMER, C.E.
Winnipeg, Manitoba R3T 2N2, Canada

PALMER, J.H.
Kensington 2033, Australia

PALTA, A.
Tuxedo Park, New York 10987, USA

PARDOS, J.
Madrid, Spain

PARKER, C.W. (Bill)
Canberra City, ACT 2601, Australia

PARUPS, E.V.
Ottawa, K1A 0C6, Canada

PARUPS, M.
Ottawa, K1A 0C6, Canada

PASTERNAK, G.C.
Ithaca, New York 14853, USA

PAULET, P.
Cedex, France

PEAUD-LENOEL, C.
Luminy-13288 Marseille,
France

PEMBERTON, H.B.
St. Paul, Minnesota 55108,
USA

PERNET, J.J.
1005 Lausanne, Switzerland

PHINNEY, B.O.
Los Angeles,
California 90024, USA

PIERCE, M.L.
East Lansing, Michigan 48824,
USA

PIERIK, R.L.M.
Wageningen, Netherlands

PIETRAFACE, W.J.
Morgantown,
West Virginia 26506, USA

PILET, P.E.
1005 Lausanne, Switzerland

POOVAIAH, B.W.
Pullman, Washington 99164,
USA

POOVAIAH, S.
Pullman, Washington 99164,
USA

POPE, D.G.
Irving, Texas 75062, USA

POTTS, J.R.
Mountain View,
California 94042, USA

POWELL, B.B.
Ithaca, New York 14853, USA

POWELL, L.E.
Ithaca, New York 14853, USA

PRESLEY, H.
Chicago, Illinois 60680, USA

PRIESTLEY, D.A.
Ithaca, New York 14853, USA

PURSE, J.G.
Sittingbourne, Kent ME9 8AG,
United Kingdom

QUARRIE, S.A.
Cambridge CB2 2LQ,
United Kingdom

RADEMACHER, E.
3400 Göttingen, FRG

RADEMACHER, W.
3400 Göttingen, FRG

RAILTON, I.
Grahamstown, South Africa

RAILTON, N.
Grahamstown, South Africa

RAJAGOPAL, R.
Copenhagen, Denmark

RAPPAPORT, L.
Davis, California 95616,
USA

RASMUSSEN, G.K.
Orlando, Florida 32803, USA

RAUSCH, T.
6000 Frankfurt/M., FRG

RAVIV, M.
Bet-Dagan, Israel

REGIER, D.A.
Corvallis, Oregon 97331, USA

REID, D.M.
Calgary, Alberta T2N 1N4,
Canada

REUVENI,O.
Bet-Dagan, Israel

RIKIN, A.
Rehovot, Israel

RIVIER, L.
1005 Lausanne, Switzerland

ROGOYSKI, M.K.
Ithaca, New York 14853, USA

ROSS, C.W.
Fort Collins, Colorado 80523,
USA

ROUSSELL, D.L.
West Lafayette,
Indiana 47906, USA

RUBERY, P.H.
Cambridge, United Kingdom

RUDDAT, M.
Chicago, Illinois 60607, USA

SAKS, Y.
Ramat-Aviv, Israel

SAKURAI, A.
Saitama-Ken 351, USA

SAKURAI, N.J.
Osaka 558, Japan

SALTVEIT, M.E., Jr.
Raleigh, North Carolina 27650,
USA

SANCHEZ, E.
Ciudad Universitaria, Mexico

SANGER, M.P.
East Lansing, Michigan 48824,
USA

SANKHLA, N.
Jodhpur-342001, India

SASSE, J.M.
Parkville, Vie 3052, Australia

SAUNDERS, P.F.
Penglais, Aberystwyth Dyfed,
United Kingdom

SCHMID, A.
1700 Fribourg, Switzerland

SCHMITZ, R.Y.
Madison, Wisconsin 53706,
USA

SCOTT, T.K.
Chapel Hill,
North Carolina 27514, USA

SEKIYA, J.
East Lansing, Michigan 48824, USA

SEMBDNER, G.
402 Halle/S., DDR

SEQUEIRA, L.
Madison, Wisconsin 53706, USA

SETTER, T.L.
St. Paul, Minnesota 55108, USA

SEYEDIN, N.
Ames, Iowa 50010, USA

SHIMABUKURO, R.H.
Fargo, North Dakota 58105, USA

SHINOZAKE, M.
Kyoto City 606, Japan

SHUDO, K.
Tokyo, Japan

SIBAOKA, T.
Sendai 980, Japan

SIERRA, M.G.
University, Mississippi 38677, USA

SIMMONDS, J.A.
Ottawa, Ontario, Canada

SING, V.O.
Omaha, Nebraska 68137, USA

SJUT, V.
7000 Stuttgart 70, FRG

SKOOG, F.
Madison, Wisconsin 53706, USA

SLAPNICK, S.
Madison, Wisconsin 53706, USA

SMITH, A.R.
Warwick CV35 9EF, England

SPENCER, H.
Edmonton, Alberta T6G 2N2C, Canada

SPENCER, M.E.S.
Edmonton, Alberta T6G 2N2C, Canada

SPIESS, L.D.
Evanston, Illinois 60201, USA

SPONSEL, V.M.
Bristol B58 1TS, England

SPONSEL, W.E.
Bristol B58 1TS, England

STANGE, L.M.C.
3500 Kassel, FRG

STEARNS, E.M.
Shakopee, Minnesota 55379, USA

STEFFENS, G.L.
Beltsville, Maryland 20705, USA

STEWART, I.
Lake Alfred, Florida 33850, USA

STODDART, J.L.
Aberystwyth, Dyfed SY23 3EB, United Kingdom

STOWE, B.K.
New Haven, Connecticut 06520, USA

STOWE, B.B.
New Haven
Connecticut 06520, USA

STRUCKMEYER, B.E.
Madison, Wisconsin 53706, USA

SUSSMAN, M.R.
New Haven, Connecticut 06520, USA

SUTTLE, J.C.
East Lansing, Michigan 48824, USA

TAIZ, L.
Santa Cruz, California 95064, USA

TAKAHASHI, K.
Tokyo 113, Japan

TAKAHASHI, N.N.T.
Tokyo 113, Japan

TAKASHI, O.
Toyama 939-03, Japan

TALLER, B.
Corvallis, Oregon 97331, USA

TAMAS, I.A.
Ithaca, New York 14850, USA

TAMAS, M.J.
Ithaca, New York 14850, USA

TANG, Y.
Shanghai, China

TAUTVYDAS, K.J.
St. Paul, Minnesota 55101, USA

TELEWSKI, F.W.
Athens, Ohio 45701, USA

TERRY, M.E.
Albany, California 94706, USA

THIMANN, K.V.
Santa Cruz, California 95064, USA

THOMAS, A.
Leeds LS2 9JT, United Kingdom

THOMAS, T.H.
Warwick CV35 9E, England

THOMPSON, M.J.
Beltsville, Maryland 20705, USA

TILLBERG, E.
Stockholm, Sweden

TSAO, T.H.
Peking, China

TSUI, C.
Peking, China

TUCKER, D.J.
West Sussex BN163PU,
England

VALDOVINOS, J.G.
Bronx, New York 10468,
USA

VALIO, I.F.M.
Sao Paulo, Brazil

Van BRAGT, J.J.
Wageningen, Netherlands

Van OVERBEEK, J.
Bryan, Texas 77801, USA

Van VOLKENBURGH, E.
Seattle, Washington 98195,
USA

VANDERHOEF, L.N.
Urbana, Illinois 61801, USA

VANSTADEN, J.
Davis, California 95616, USA

VARNER, J.
St. Louis, Missouri 63130,
USA

VEALE, J.A.
Palmerston North,
New Zealand

VENDRIG, J.C.
Leuven, Belgium

VENIS, M.A.
Sittingbourne, Kent ME9 8AG,
United Kingdom

VERMA, D.C.
Appleton, Wisconsin 54912,
USA

VISSCHER, S.N.
Bozeman, Montana 59717,
USA

VREMAN, H.J.
Palo Alto, California 94302,
USA

VREUGDENHIL, D.
2311 VJ Leiden, Netherlands

WALALI, L.D.W.
St. Paul, Minnesota 55108,
USA

WANG, T.L.
Leeds LS2 9JT,
United Kingdom

WAREING, P.F.
Dyfed, United Kingdom

WEILER, E.W.
4630 Bochum 10, FRG

WEST, C.V.
Los Angeles, California 90024,
USA

WEST, C.A.
Los Angeles, California 90024,
USA

WIGHTMAN, F.
Ottawa K1S 586, Ontario,
Canada

WILLIAMS, S.E.
Annville, Pennsylvania 17003,
USA

WITHAM, F.H.
University Park,
Pennsylvania 16802, USA

WOLTER, K.E.
Madison, Wisconsin 53705,
USA

WOOLHOUSE, H.
Leeds, England

WURTELLE, E.S.
Los Angeles, California 90024,
USA

YAMAMOTO, M.
Osaka 558, Japan

YAMAMOTO, R.
Osaka 558, Japan

YANG, S.F.
Davis, California 95616, USA

YEE, V.F.
Greenfield, Indiana 46140,
USA

YOKOTA, T.
Tokyo, Japan

YOPP, J.H.
Carbondale, Illinois 62901,
USA

ZAERR, J.B.
Corvallis, Oregon 97331,
USA

ZEEVAART, J.A.
East Lansing, Michigan 48824,
USA

ZIESLIN, N.
Rehovot, Israel

ZURFLUH, L.L.
St. Paul, Minnesota 55108,
USA

ZWAR, J.A.
Canberra City, ACT2601,
Australia

Subject Index

Planta

An International Journal of Plant Biology

ISSN 0032-0935 Title No. 425

Planta publishes original articles in structural and functional botany, covering all aspects of plant physiology from biochemistry and ultrastructure to studies with cells, tissues, organs, whole plants, and populations (crop physiology and physiological ecology). Papers in genetics and papers from applied fields such as phytopathology will be accepted if contributing to the understanding of specifically botanical problems.

For subscription information and/or sample copy please write to:
Springer-Verlag,
Journal Promotion,
P.O. Box 10 52 80,
D-6900 Heidelberg 1, FRG

Springer-Verlag
Berlin
Heidelberg
New York

MIX
Papier aus verantwortungsvollen Quellen
Paper from responsible sources
FSC® C105338

If you have any concerns about our products,
you can contact us on
ProductSafety@springernature.com

In case Publisher is established outside the EU,
the EU authorized representative is:
Springer Nature Customer Service Center GmbH
Europaplatz 3, 69115 Heidelberg, Germany

Printed by Libri Plureos GmbH
in Hamburg, Germany